Spon's External Works and Landscape Price Book

2020

Spon's External Works and Landscape Price Book

Edited by

A≡COM

in association with
LandPro Ltd
Landscape Surveyors

2020

Thirty-ninth edition

CRC Press
Taylor & Francis Group

First edition 1978
Thirty-ninth edition published 2020
by CRC Press
2 Park Square, Milton Park, Abingdon, Oxon, OX14 4RN

and by CRC Press
Taylor & Francis, 6000 Broken Sound Parkway, NW, Suite 300, Boca Raton, FL 33487

CRC Press is an imprint of the Taylor & Francis Group, an informa business

British Library Cataloguing in Publication Data
A catalogue record for this book is available from the British Library

ISBN13: 978-0-367-27106-0
Ebook: 978-0-429-29479-2
ISSN: 0267-4181

Typeset in Arial by Taylor & Francis Books

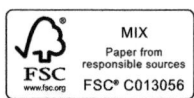
MIX
Paper from
responsible sources
FSC
www.fsc.org   FSC® C013056

Printed and bound in Great Britain by
TJ International Ltd, Padstow, Cornwall

Contents

Preface to the Thirty-Ninth Edition

The External Works and Landscape sector

There is a general but normal disparity across the country. London and home county rates continue to be high with acute labour shortages driving tender rates. Availability of skilled site management in the sector further exacerbates price risers in the south east. Established contractors in the south east have full order books driven by house-building. Our enquiries report that this is not reflected northwards both in wage rates and tender pricing. Established contractors are still buoyant with work from the designer sector working on high value domestic work.

Recent Construction Activity Trends

Construction all work output recorded no discernible yearly change in 2018 Q4, according to the Office for National Statistics' (ONS) latest data. This extends a period through 2018 of variable year-on-year change rates, where no definitive trend was evident. A conclusion to draw from the most recent data point is that little change in the yearly rate implies no growth or expansion of industry output. Equally, the nominal output value also indicates still respectable levels of overall workload.

All major construction sectors recorded declines in the Q4 new orders data — housing (−7%), infrastructure (−38%) and commercial (−3%). Combined, these sectors account for well over 80% of all new work output, so the recorded falls will be felt eventually by the industry. Only the private industrial and public sectors offered positive yearly change in the 2018 Q4 data. However, these sectors account for less than 14% of all new work output, so these increases in work are not expected to move the national construction output dial materially

Key influences

Some manufacturers and suppliers have advised us to keep our database prices for their products the same in many cases for this publication. Major suppliers do show a 5% increase in rates which is expected and these are often negotiable at time of purchase. However many suppliers of rates to this edition instructed us to keep their rates the same as last year. This edition has made no allowances for Brexit driven influences that may arise during the next year.

Labour rates

This year we have increased the price of the base cost of labour by £2.75 per hour (Approximately £4730.00 per annum) for the South East. A labour rate of approximately £24.00 per hour would be appropriate for prices north of the London contracting area.

This is due to demand mostly in the housing sector for contractors capable of management and delivery on South East projects which have absorbed much of contractor's capacity.

The standard rate for this years' calculations for Major works is £28.00 per hour, an increase of 10.8%

It should be noted as usual that there is an element of overhead within the labour rate used and readers should refer to this section in the 'Computation of Labour Rates' section following this preface.

Preliminaries

We have devised some preliminary cost models in the Approximate Estimates section of this edition and we will continue to update and monitor this over future editions.

Prices included within this edition do not include for VAT, professional fees etc. which must be added if appropriate.

Supplier prices

Similarly to last year, there have been many suppliers who have only reflected minimal increases or even no increase in their prices especially for material supplies. The larger suppliers have increased their list costs by a standard 5%–7.5% but we perceive that there is room to negotiate on material cost rates.

Plant rates

Plant rates have been stable within this edition with no major impacts on last year's prices.

Prices for suppliers and services

We acknowledge the support afforded to us by the suppliers of products and services who issue us with the base information used. Their contact details are published in the acknowledgements section at the front of this book. We advise that readers wishing to evaluate the cost of a product or service should approach these suppliers directly should they wish to confirm prices prior to submission of tenders or quotations.

Whilst every effort is made to ensure the accuracy of the information given in this publication, neither the editors nor the publishers in any way accept liability for loss of any kind resulting from the use made by any person of such information.

We remind readers that we would be grateful to hear from them during the course of the year with suggestions or comments on any aspect of the contents of this year's book and suggestions for improvements. Our contact details are shown below.

AECOM Ltd	SAM HASSALL
Aldgate Tower	LANDPRO LTD
2 Leman Street	Landscape Surveyors
London	14 Upper Bourne Lane
E1 8FA	Farnham
	Surrey
	GU10 4RQ
Tel: 020 7645 2000	Tel: 01252 795030
e-mail: spons.europe@aecom.com	e-mail: info@landpro.co.uk

Acknowledgements

This list has been compiled from the latest information available but as firms frequently change their names, addresses and telephone numbers due to reorganization, users are advised to check this information before placing orders.

ABG Ltd
Tel: 01484 852096
Website: www.abg-geosynthetics.com
E-mail: enquiries@abgltd.com
Reinforced Turf

ACO Technologies Plc
Tel: 01462 816666
Website: www.aco.co.uk
E-mail: technologies@aco.co.uk
Linear drainage

Addagrip Terraco Ltd
Tel: 01825 761333
Website: www.addagrip.co.uk
E-mail: sales@addagrip.co.uk
Resin bound and bonded surfacing

Adezz BV
Netherlands
Tel: 0031 413 745 423
Website: www.adezz.com
E-mail: info@adezz.uk
Corten edgings

Agripower Ltd
Tel: 01494 866776
Website: www.agripower.co.uk
E-mail: info@agripower.co.uk
Land drainage and sports grounds

Albion Stone Plc
Tel: 01737 771772
Website: www.albionstone.com
E-mail: enquiries@albionstone.com
Portland stone

ALS Contract Services
Tel: 01952 259281
Website: www.alscontracts.co.uk
E-mail: contracts@amenity.co.uk
Horticultural products supplier

Anderton Concrete Products Ltd
Tel: 01606 535300
Website: www.andertonconcrete.co.uk
E-mail: civils@andertonconcrete.co.uk
Concrete retaining wall systems

Architectural Heritage Ltd
Tel: 01386 584414
Website: www.architectural-heritage.co.uk
E-mail: office@architectural-heritage.co.uk
Ornamental stone buildings and water features

Axtell Concrete
Tel: 01483 498 192
Website: www.mixtrucks.com
E-mail: info@mixtrucks.com
Concrete

Bauder Ltd
Tel: 01473 257671
Website: www.bauder.co.uk
Email: info@bauder.co.uk
Green roof systems

Bituchem Group
Tel: 01594 826768
Website: www.bituchem.com
E-mail: tech@bituchem.com
Coloured hard landscaping materials

Breedon Special Aggregates
(A division of Breedon Aggregates England Ltd)
Tel: 01332 694001
Website: www.breedon-special-aggregates.co.uk
E-mail: specialaggs@breedongroup.com
Specialist gravels

British Sugar Topsoil
Tel: 0870 240 2314
Website: www.bstopsoil.co.uk
E-mail: topsoil@britishsugar.com
Topsoil

Browns Plant Hire
Tel 01563 549001
Website: www.brownplanthire.com
E-mail: admin@brownplanthire.com
Machinery hire

Broxap Street Furniture
Tel: 01782 564411
Website: www.broxap.com
E-mail: web.sales@broxap.com
Street furniture

Buzon UK
Tel:020 8614 0874
Website: www.buzonuk.com
E-mail: sales@buzonuk.com
Pedestals

Capital Garden Products Ltd
Tel: 01580 201092
Website: www.capital-garden.com
E-mail: sales@capital-garden.com
Plant containers

CED Ltd
Tel: 01708 867237
Website: www.cedstone.co.uk
E-mail: sales@cedstone.co.uk
Natural stone

Cedar Nursery
Tel: 01932 862473
Website: www.landscaping.co.uk
E-mail: sales@landscaping.co.uk
Aggregate containment grids

DLF Seeds Ltd
Tel: 01386 793135
Website: www.dlf.co.uk
E-mail: amenity@dlf.co.uk
Grass and wildflower seed

Earth Anchors Ltd
Tel: 020 8684 9601
Website: www.earth-anchors.com
E-mail: info@earth-anchors.com
Anchors for site furniture

Easigrass
Tel: 020 8843 4180
Website: www.easigrass.com
E-mail: reception@easigrass.com
Artificial grass

Easimix
Tel: 01536 674064
Website: www.easimix.co.uk
E-mail: enquires@easimix.co.uk
Concrete

Elite Balustrade Systems
Tel: 01254 825594
Website: www.elitebalustrade.com
Email: neil@elitebalustrade.com
Glass balustrades

Elliott Workspace Ltd
Tel: 01733 298700
Website: www.elliottuk.com
Email: hirediv@elliottuk.com
Site offices

English Woodlands
Tel: 01435 862992
Website: www.ewburrownursery.co.uk
E-mail: info@ewburrownursery.co.uk
Plant protection

En Tout Cas Tennis Court Ltd
Tel: 01832 274199
Website: www.tenniscourtsuk.co.uk
E-mail: sales@etc.courts.com
Tennis Courts

Everedge
Tel: 01630 417120
Website: www.everedge.co.uk
E-mail: enquiries@everedge.co.uk
Metal edgings

Exterior Solutions Ltd
Tel: 01494 711800
Website: www.exterior.supplies.co.uk
E-mail: via website
Decking

Fairwater Ltd
Tel: 01903 892228
Website: www.fairwater.co.uk
E-mail: info@fairwater.co.uk
Water garden specialists

Flagpole Express Ltd
Tel: 01159 44 22 55
Website: www.flagpoleexpress.co.uk
Email: sales@flagpoleexpress.co.uk
Flagpoles

FP McCann Ltd
Tel: 01530 240013
Website: www.fpmccann.co.uk
E-mail: sales@fpmccann.co.uk
Precast concrete pipes

Furnitubes Street Furniture
Tel: 020 8378 3200
Website: www.furnitubes.com
E-mail: sales@furnitubes.com
Street furniture

Gatic Slotdrain
Tel: 01304 203545
Website: www.gatic.com
E-mail: info@gatic.com
Linear Drainage

George Lines Ltd
Tel: 01753 685354
Website: www.georgelines.co.uk
E-mail: sales@georgelines.co.uk
Building materials

Germinal GB
Tel: 01522 868714
Website: www.germinalamenity.com
E-mail: expert@germinalamenity.com
Grass and wildflower seed

Grass Concrete Ltd
Tel: 01924 379443
Website: www.grasscrete.com
E-mail: info@grasscrete.com
Grass block paving

Greenblue Urban Ltd
Tel: 0800 018 7797
Website: www.greenblue.com
E-mail: enquiries@greenblueurban.com
Root directors

Greenfix Soil Stabilization and Erosion Control Ltd
Tel: 01386 881493
Website: www.greenfix.co.uk
E-mail: info@greenfix.co.uk
Seeded erosion control mats

Greentech Ltd
Tel: 01423 332100
Website: www.green-tech.co.uk
E-mail: sales@green-tech.co.uk
Landscaping suppliers

Griffin Nurseries
Tel: 01428 741655
Website: www.griffinnurseries.co.uk
E-mail: enquiries@griffinnurseries.co.uk
Hedging

Grundon Waste Management Ltd
Tel: 01494 834311
Website www.grundon.com
E-mail: info@grundon.com
Gravel

Guncast Swimming Pools Ltd
Tel: 01798 343725
Website: www.guncast.com
E-mail: info@guncast.com
Swimming pools

Haddonstone Ltd
Tel: 01604 770711
Website: www.haddonstone.com
E-mail: info@haddonstone.co.uk
Landscape ornaments and architectural cast stone

ICL (formerly Everris Limited)
Tel: 01473 237100
Website: www.icl-sf.com
E-mail: prof.sales@icl-group.com
Landscape chemicals

Hags -SMP Ltd
Tel: 01784 489100
Website: www.hags.co.uk
E-mail: sales@hags.co.uk
Playground equipment

Inturf
Tel: 01759 321000
Website: www.inturf.com
E-mail: info@inturf.co.uk
Turf

Hahn Plastics Ltd
Tel: 0161 850 1965
Website: www.hahnplastics.com
E-mail: info@hahnplastics.co.uk
Recycled products

J Toms Ltd
Tel: 01233 771923
Website: www.jtoms.co.uk
E-mail: jtoms@btopenworld.com
Plant protection

Headland Amenity Ltd
Tel: 01223 491090
Website: www.headlandamenity.com
E-mail: info@headlandamenity.com
Chemicals

Jacksons Fencing
Tel: 01233 750393
Website: www.jacksons-fencing.co.uk
E-mail: sales@jacksons-fencing.co.uk
Fencing

Hepworth Wavin Plc
Tel: 01709 856200
Website: www.wavin.com
E-mail: info@wavin.com
Drainage

James Coles & Sons (Nurseries) Ltd
Tel: 01162 412115
Website: www.colesnurseries.co.uk
E-mail: sales@colesnurseries.co.uk
Tree Nurseries

Hillier Nurseries
Tel: 01794 368733
Website: www.hillier.co.uk
E-mail: trees@hillier.co.uk
Trees

Johnsons Wellfield Quarries Ltd
Tel: 01484 652311
Website: www.johnsons-wellfield.co.uk
E-mail: sales@johnsons-wellfield.co.uk
Natural Yorkstone pavings

HSS Hire
Tel: 020 8936 7307
Website: www.hss.com
E-mail: hire@hss.com
Tool and plant hire

Kompan Ltd
Tel: 01908 201002
Website: www.kompan.co.uk
E-mail: kompan.uk@kompan.com
Play equipment

Hydroblox
Tel: 0203 189 1468
Website: www.hydrobloxinternational.co.uk
E-Mail: sales@hydrobloxinternational.co.uk
Recycled plastic drainage

Landmark Timber Products
Tel: 0808 129 3773
Website: www.landmarktimber.co.uk
E-mail: enquiries@madebylandmark.com
Finger Post signs

Landscape Plus
Tel: 01666 577577
Website: www.landscapeplus.com
E-mail: help@landscapeplus.com
Lighting

Lang and Fulton
Tel: 0131 441 1255
Website: www.langandfulton.co.uk
Email: sales@langandfulton.co.uk
Steel gratings and louvres

Lappset UK Ltd
Tel: 01536 412612
Website: www.lappset.com
E-mail: lappset@lappset.com
Play equipment

Leaky Pipe Systems Ltd
Tel: 01622 746495
Website: www.leakypipe.co.uk
E-mail: sales@leakypipe.co.uk
Irrigation systems

Leca UK
Tel: 08443 351770
Website: www.leca.co.uk
E-mail: enquiries@leca.co.uk
Light Aggregates

Lindum Turf
Tel: 01904 448675
Website: www.turf.co.uk
Email: lindum@turf.co.uk
Meadow and specialized turf

LiveTrakway
Tel: 08700 767676
Website: www.livetrakway.com
E-mail: marketing@livetrakway.com
Portable roads

London Stone
Tel: 01753 212950
Website: www.londonstone.co.uk
E-mail: info@londonstone.co.uk
Natural stone paving

Lorenz von Ehren
Tel: 01865 910261
Website: www.lve.de
E-mail: sales@lve.de
Topiary

Marshalls Plc
Tel: 01422 312000
Website: www.marshalls.co.uk
E-mail: info@marshalls.co.uk
Concrete products, hard landscape materials, linear drainage, street furniture and street lighting

Melcourt Industries Ltd
Tel: 01666 502711
Website: www.melcourt.co.uk
E-mail: mail@melcourt.co.uk
Mulch and compost

Meon UK
Tel: 023 9220 0606
Website: www.meonuk.com
E-mail: mail@meonuk.com
Resin surfaces and safety surfaces

Milton Precast
Tel: 01795 425191
Website: www.miltonprecast.co.uk
E-mail: sales@miltonprecast.com
Soakaway rings

Mobilane UK
Tel: 0203 741 8049
Website: www.mobilane.com
E-mail: sales@mobilane.co.uk
Green screens

Neptune Outdoor Furniture Ltd
Tel: 01962 777799
Website: www.neptunestreetfurniture.co.uk
E-mail: enquiries@neptunestreetfurniture.co.uk
Street furniture

Nomix Enviro
A Division of Frontier Agriculture Ltd
Tel: 0800 032 4352
Website: www.nomixenviro.co.uk
E-mail: sales@nomixenviro.co.uk
Herbicides

Oakover Nurseries Ltd
Tel: 01233 713016
Website: www.oakovernurseries.co.uk
E-mail: enquiries@oakovernurseries.co.uk
Native plant nurseries

Outdoor Design
Tel: 01903 716960
Website: www.outdoordesign.co.uk
E-mail: info@outdoordesign.co.uk
Metal Fabricator

PBA Solutions
Tel: 01202 816134
Website: www.pba-solutions.com
E-mail: info@pba-solutions.com
Japanese Knotweed consultants

Pennyhill Timber Ltd
Tel: 01483 486 739
Website: www.pennyhilltimber.co.uk
Email: david@pennyhilltimber.co.uk
Timber Products

Phi Group
Tel: 01242 707600
Website: www.phigroup.co.uk
E-mail: southern@phigroup.co.uk
Retaining wall specialists

Platipus Anchors Ltd
Tel: 01737 762300
Website: www.platipus-anchors.com
E-mail: info@platipus-anchors.com
Tree anchors

Pop up Power Supplies Ltd
Tel: 020 8551 8363
Website: www.popuppower.co.uk
E-mail: info@popuppower.co.uk
Public area power supplies

Readyhedge Ltd
Tel: 01386 750585
Website: www.readyhedgeltd.com
E-mail: info@readyhedgeltd.com
Pre-grown Hedging

Rigby Taylor Ltd
Tel: 0800 424 919
Website: www.rigbytaylor.com
E-mail: sales@rigbytaylor.com
Horticultural supply

RIW Ltd
Tel: 01344 397777
Website: www.riw.co.uk
E-mail: enquiries@riw.co.uk
Waterproofing products

Rolawn Ltd
Tel: 01904 757372
Website: www.rolawn.co.uk
E-mail: info@rolawn.co.uk
Industrial turf

Saint-Gobain PAM UK Ltd
Tel: 0115 930 5000
Website: www.saint-gobain-pam.co.uk
E-mail: sales.uk.pam@saint-gobain.com
Cast iron drainage

Scotscape Ltd
Tel: 02082 545000
Website: www.scotscape.net
E-mail: sales@scotscape.net
Living walls

Screwfix Ltd
Tel: 03330 112 112
Website: www.screwfix.com
E-mail: online@screwfix.com
Hardware

Shorts Group
Tel: 01344 620 316
Website: www.shorts-group.co.uk
Email: info@shorts-group.co.uk
Plant Hire

Southern Conveyors
Tel: 0800 278 5389
Website: www.coveya.co.uk
E-mail: support@coveya.co.uk
Conveyors

Steelway Brickhouse
Tel: 01215 214500
Website: www.steelway.co.uk
E-mail: sales@steelwaybrickhouse.co.uk
Access covers

Steelway Fensecure Ltd
Tel: 01902 490919
Website: www.steelway.co.uk
E-mail: sales@fensecure.co.uk
Fencing

SteinTec
Tel: 01708 860049
Website: www.steintec.co.uk
E-mail: webenquiries@tuffbau.com
Specialized mortars

SureSet UK Limited
Tel: 0800 612 6501
Website: www.sureset.co.uk
E-mail: website@sureset.co.uk
Solid steel products

Tensar International
Tel: 01254 262431
Website: www.tensar.co.uk
E-mail: info@tensar.co.uk
Erosion control, soil stabilization

Terranova Lifting Ltd
Tel: 0118 931 2345
Website: www.terranovagroup.co.uk
E-mail: cranes@terranovagroup.co.uk
Crane hire

The Garden Trellis Company
Tel: 01255 688361
Website: www.gardentrellis.co.uk
Email: info@gardentrellis.co.uk
Garden joinery experts

TK Solutions AG Ltd
Tel: 020 3853 1285
Email: tksukltdnew@gmail.com
Skype Name: tomer.kniznik
Irrigation supplies

TLC Electrical Supplies
Tel: 01293 565630
Website: www.tlc-direct.co.uk
E-mail: sales@tlc-direct.co.uk
Electrical goods

Townscape Products Ltd
Tel: 01623 513355
Website: www.townscape-products.co.uk
E-mail: sales@townscape-products.co.uk
Hard landscaping

Wade International Ltd
Tel: 01787 475151
Website: www.wadedrainage.co.uk
E-mail: sales@wadeint.co.uk
Stainless steel drainage

Wavin Plastics Limited
Tel: 0844 856 9400
Website: www.wavin.com
E-mail: info@wavin.com
Drainage products

Weavo Fencing Products Ltd
Tel: 01823 480 571
Website: www.weavo.co.uk
Email: sales@weavo.com
Fencing

Wicksteed Leisure Ltd
Tel: 01536 517028
Website: www.wicksteed.co.uk
E-mail: sales@wicksteed.co.uk
Play equipment

Willowbank Services
Tel: 01823 690113
Waterside bank stabilization

Woodscape Ltd
Tel: 01254 685185
Website: www.woodscape.co.uk
E-mail: sales@woodscape.co.uk
Street furniture

Wybone Ltd
Tel: 01226 352333
Website: www.wybone.co.uk
E-mail: hello@wybone.co.uk
Street furniture

Yeoman Aggregates Ltd
Tel: 020 8896 6820
Website: www.aggregate.com
E-mail: via website
Aggregates

Zinco Green Roof Systems Ltd
Tel: 01223 853843
Website: www.zinco-greenroof.com
E-mail: info@zinco-greenroof.com
Green roof systems

Item – Overhead and Profit Included	Unit	Range £
HOW TO USE THIS BOOK		
How this book is presented		
Prices in this book generally include overhead and profit. Market prices where shown exclude overhead and profit. The price information sections of this book are split into two subsections as follows: Approximate Estimates – This is presented in a 3 column format. This section combines prices from the Measured Works items to form a composite rate for the work described. Measured Works – This is presented in an 8 column format and is a detailed analysis of single tasks commonly employed throughout the external works and landscape sector.		
Column formats		
Cost and Marked up values shown in this edition		
Materials	15%	–
Labour	15%	–
Plant	15%	–
Subcontractors (generally)	10%	–
Note some high value subcontractors such as swimming pools and sportsfields and some street furniture and playground items which have a high unit cost may be marked up at lower rates (5%)		
Description of the column formats used in this book. Approximate Estimates Prices shown include overhead and profit since the 2018 edition at the above rates: Only the 3 column format is used; users of the book cannot see the components which make up the prices in the book format. An electronic version is supplied free with this book. Users of the electronic system can see more detail in the electronic format. Measured Works Where an 8 column analysis is shown these are tasks carried out directly by the external works or landscape contractor and are shown at cost. Where prices are shown in the 8 column section without analysis, prices have been supplied by specialist subcontractors and are assumed to have the specialist subcontractor's selling price of the item included.		
Explanation of the columns used in the Measured Works analysis Prime Cost Commonly known as the 'PC'. Prime Cost is the actual price of the material item being addressed such as paving, shrubs, bollards or turves, as sold by the supplier. Prime Cost is given 'per square metre', 'per 100 bags' or 'each' according to the way the supplier sells his product. In researching the material prices for the book we requested that the suppliers price for full loads of their product delivered to a site close to the M25 in London. Spon's rates do not include VAT. Some companies may be able to obtain greater discounts on the list prices than those shown in the book. Prime Cost prices for those products and plants which have a wide cost range will be found under the heading of Market Prices in the main sections of this book, so that the user may select the product most closely related to their specification.		

Item – Overhead and Profit Included	Unit	Range £

Column formats – cont

Explanation of the columns used in the Measured Works analysis – cont
Labour Hours
    This is the total amount of man hours required to carry out the quantity and unit of the measured works item which is being analysed.
Labour £ – Shown inclusive of Overhead & Profit – Generally 15%
    This is the value of the labour used. Please refer to the labour rates used in this edition which are detailed preceding the Approximate estimates section
Plant – Shown inclusive of Overhead & Profit – Generally 15%
    This covers the cost of machinery used to work out the cost of the item addressed. Plant can range from heavy excavators to light electrical breakers. All plant is assumed as hired in except for concrete mixers and normal site tools which a contractor would be expected to carry in his or her company vehicle (hand drills, grinders etc.). Plant includes for fuel and down time. The rates used are shown in the section of this book 'Computation of Mechanical Plant Costs' which follows this section.
    Delivery and removal to site costs are not included in any rates.
Materials – Shown inclusive of Overhead & Profit – Generally 15%
    The PC material plus the additional materials required to fix the PC material. Every job needs materials for its completion besides the product bought from the supplier. Paving needs sand for bedding, expansion joint strips and cement pointing; fencing needs concrete for post setting and nails or bolts; tree planting needs manure or fertilizer in the pit, tree stakes, guards and ties. If these items were to be priced out separately, *Spon's External Works and Landscape Price Book* (and the Bill of Quantities) would be impossibly unwieldy, so they are put together under the heading of Materials.
Subcontract rates or subcontractors used within an item build-up – Shown inclusive of Overhead & Profit – Generally 10%
Where only a figure is shown in the columns
    This is a subcontract rate which has been received from a specialist subcontractor.
    This rate would generally include the subcontractor's overhead and profit but not the main contractor's oncosts.
Delivery Prices
    These are not generally included. Exceptions would be aggregates or loose materials which are delivered by the load. Our specification to suppliers of information for this book is that all materials are delivered in full loads to a site close to the M25 in North London.
Unit
    The unit being used to measure the item.
Total Price £
    The total price 'inclusive of Overhead & Profit
Where a figure is shown in the Material column only
    This is generally the market price of a single material only.

Item – Overhead and Profit Included	Unit	Range £
Worked examples using data in this edition		
The following operations describe the trench excavation, pipe laying and backfilling operations as contained in section NRM Section 34 of this edition.		
Example 1		
Pipe laying 100 m		
Remove topsoil 150 × 300 mm wide	100 m	43.50
Excavate for drain 150 × 450 mm; excavated material to spoil heaps on site by machine	100 m	1475.00
Lay flexible plastic pipe 100 mm dia. wide	100 m	200.00
Backfilling with gravel rejects, blinding with sand and topping with 150 mm topsoil	100 m	460.00
TOTAL FOR ALL OF THE ABOVE COMBINED	100 m	2175.00
This total above item is a composite of all of the above 'Measured Works' items combined and is typical of an entry in the Approximate Estimates section of this book.		
Example 2		
The following individual items are shown as components of a slab paving specification for one m². These individual items are all extracted from the paving section of this book and include a profit margin of 15%.		
The PC of 900 × 600 × 50 mm precast concrete flags is shown in the PC (Prime cost) column as:	m^2	10.87
Costs of bedding and pointing materials are added to give a total Materials cost of:	m^2	6.61
Further costs for labour and mechanical plant give a resultant price of:	m^2	30.50
If a quotation of PC £9.00/m^2 was received from a supplier of the flags the resultant price would be calculated as:		
Original price of the flags:	m^2	10.87
Gives a reduced cost of	m^2	−0.52
Less the original cost PC £9.45 plus the revised cost £9.00 to give	m^2	29.98
Example 3		
The following samples are extracted from the tree planting section of this edition.		
The prime cost of an Acer platanoides 10–12 cm bare root tree is given as:	nr	28.00
To which the cost of handling and planting only to treepits excavated separately is added:	nr	18.70
Giving an overall cost of:	nr	47.00
The following items are also shown as component prices in the Measured Works section		
labour and mechanical plant for mechanical excavation of a 600 × 600 × 600 mm pit	nr	5.50
a single tree stake	nr	8.30
importing Topgrow compost in 75 litre bags	nr	4.10
backfilling with imported topsoil	nr	7.50
transporting the excavated material to spoil heaps on site 25 m distant	nr	0.70
and disposing of the excavated material	nr	9.00

Item – Overhead and Profit Included	Unit	Range £
Worked examples using data in this edition – cont		
The following samples are extracted from the tree planting section of this edition. – cont		
The following price which is a composite of all the above might be shown in the Approximate Estimates section:		
Tree planting; supply and plant Acer platanoides 8–10 cm in 600 × 600 mm tree pit excavated mechanically; allow for a single tree stake and Topgrow compost and backfilling with imported topsoil; all excavated material removed from site	nr	82.00
Example 4 – Composite labour rates		
The worked example below shows the build-up for Keyblock paviours as used elsewhere in this book. Readers will note that there appears to be an anomaly between the labour time and the labour rate. This however is due to the useage of a composite labour rate. In this case a 3 man team which is 3 × the hourly rate shown in the labour rates section. 2020 = £28.00 / hour		
Precast concrete paving blocks; Keyblok Marshalls Plc; on prepared base (not included); on 50 mm compacted sharp sand bed; blocks laid in 7 mm loose sand and vibrated; joints filled with sharp sand and vibrated; level and to falls only		
Herringbone bond		
200 × 100 × 60 mm; natural grey	m²	44.50
The labour value is shown as £27.06 but the labour hours are 0.28 hrs. The labour rate is (£28.00 × 0.28 hrs) × 3 man team × 15% profit. = £27.06. Please ignore minor rounding calculations. Readers should extrapolate in these instances.		
ADJUSTMENT AND VARIATION OF THE RATES IN THIS BOOK		
It will be appreciated that a variation in any one item in any group will affect the final Measured Works price. Any cost variation must be weighed against the total cost of the contract. A small variation in Prime Cost where the items are ordered in thousands may have more effect on the total cost than a large variation on a few items. A change in design that necessitates the use of earth moving equipment which must be brought to the site for that one job will cause a dramatic rise in the contract cost. Similarly, a small saving on multiple items will provide a useful reserve to cover unforeseen extras.		
Using the tree planting, 'Example 3', above, if the tree size was to be increased from a 10–12 cm tree to a 12–14 cm tree of the same variety, this might necessitate an increase in tree pit size, more compost, a larger tree stake and more excavated material being disposed of off site.		
The resultant variation would again be referenced from the tree planting section as follows:		
The revised PC (prime cost) of an Acer platanoides 12–14 cm bare root tree is given as:	nr	48.00
to which the cost of handling and planting only to treepits excavated separately is added:	nr	27.00
Giving an overall cost of:	nr	75.00
Add to this the following variations as described:		
Tree pit increased in size from 600 × 600 × 600 mm to 900 × 900 × 600 mm	nr	12.40
Additional compost	nr	13.60
Increased staking requirements to two stakes	nr	11.50

Item – Overhead and Profit Included	Unit	Range £
Increased volume of imported topsoil	nr	16.70
Increased disposal volume	nr	21.50
Giving a resultant price for the increased tree size as	nr	150.00
The variation therefore from the original specification for the 10–12 cm tree to the revised 12–14 cm tree is calculated as an additional sum per tree of	nr	50.00
Labour Rates Used in this Edition		
General landscape construction; The labour rates shown below are shown at cost and are deemed to include company overhead as per the analysis of labour rates in the front of this book. Profit margins are added to these rates as per the rates shown below to produce the final labour rate		
Specialist labour rate for brickwork	hr	36.00
General labour rate for all other works	hr	28.00
Landscape maintenance		
General labour rate for all works	hr	23.00
Cost and Marked up values shown in this edition		
Materials	15%	–
Labour	15%	–
Plant	15%	–
Subcontractors (generally)	10%	–
Computation of Mechanical Plant Costs		
Hired Plant; Prices exclusive of Overhead & Profit		
All equipment below is based on a productivity rate of 32 hours per week and include costs per hour for an operator and fuel		
Dumper 3 tonne	hr	34.00
Dumper 5 tonne Thwaites self drive	hr	38.00
Dumper 6 tonne Thwaites self drive	hr	37.50
Excavator 360 Tracked 21 tonne operated	hr	64.00
Excavator 360 Tracked 8 tonne	hr	47.00
Excavator Tracked 5 tonne	hr	42.00
Fork lift telehandler (24 hours/week)	hr	37.50
Mini Excavator 1.5 tonne	hr	4.50
Mini Excavator JCB 803 Rubber tracks	hr	9.20
Mini Excavator JCB 803 Steel tracks	hr	7.80
Skip loader 1 tonne	hr	3.30
Fuel charge Red diesel	l	0.70
HSS Ltd		
Access tower; alloy; 5.2 m	day	35.00
Cultivator; 110 kg	hr	4.60
Diamond blade consumable; 450 mm; concrete	mm	13.00
Heavy-duty breaker; 110 V; 2200 W	hr	9.00
Heavy-duty breaker; petrol; 5 hrs/day; single tool	hr	8.35
Mesh fence; temporary security fencing; 2.85 × 2.0 m high	1 m/wk	1.15
Petrol masonry saw bench; 350 mm	hr	3.70

Item – Overhead and Profit Included	Unit	Range £
Computation of Mechanical Plant Costs – cont		
Hired Plant – cont		
HSS Ltd – cont		
Petrol poker vibrator + 50 mm head	hr	4.00
Post hole borer; 1 man; weekly rate	hr	6.40
Vibrating plate compactor	hr	4.20
Vibrating roller; 136 kg; 10.1 kN; 4 hrs/day	hr	4.35
Vibration damped breaker (light); 1600 W; 110 V	hr	1.50

How this Book is Compiled

INTRODUCTION

First-time users of *Spon's External Works and Landscape Price Book* and others who may not be familiar with the way in which prices are compiled may find it helpful to read this section before starting to calculate the costs of landscape works.

Please also refer to the 'How to Use this Book' notes on page xvii

The cost of an item of landscape construction or planting is made up of many components:

- the cost of the product
- the labour and additional materials needed to carry out the job
- the cost of running the contractor's business

These are described more fully below.

IMPORTANT NOTES ON THE PROFIT ELEMENT OF RATES IN THIS BOOK

The rates shown in the Approximate Estimates and Measured Works sections of this book now include a profit element. Prices are shown at marked up rates. This is the assumed installed price including the costs of company overhead and profit required to perform this task or project. For reference please see the tables on pages 6–9.

Analysed Rates Versus Subcontractor Rates

As a general rule if a rate is shown as an analysed rate in the Measured Works section, i.e. it has figures shown in the columns other than the 'Unit' and 'Total Rate' column, it can be assumed that this has the subcontractors profit and overhead element as well as the landscape subcontractors profit on top of this. This calculation is shown as a direct labour/material supply/plant task being performed by a contractor.

On the other hand, if a rate is shown as a Total Rate only, this would normally be a subcontractor's rate and would contain the profit and overhead for the subcontractor and the landscape contractors profit element for management of the subcontractor.

The foregoing applies for the most part to the Approximate Estimates section, however in some items there may be an element of subcontractor rates within direct works build-ups.

As an example of this, to excavate, lay a base and place a macadam surface, a general landscape contractor would normally perform the earthworks and base installation. The macadam surfacing would be performed by a specialist subcontractor to the landscape contractor.

The Approximate Estimate for this item uses rates from the Measured Works section to combine the excavation, disposal, base material supply and installation with all associated plant. There is profit included on these elements. The cost of the surfacing, however, as supplied to us in the course of our annual price enquiry from the macadam subcontractor, does include the subcontractor's profit element plus a portion for the landscape contractor of 10%.

The landscape contractor would add this to his section of the work but would normally apply a mark-up at a lower percentage to the subcontract element of the composite task. Users of this book should therefore allow for the profit element at the prevailing rates. Please see the worked examples below and the notes on overheads for further clarification.

INTRODUCTORY NOTES ON PRICING CALCULATIONS USED IN THIS BOOK

There are two pricing sections to this book:

Approximate Estimating Rates – Average contract value £100,000.00–£300,000.00

Measured Works – Average contract value £100,000.00–£300,000.00

Typical Project Profile Used in this Book

Contract value	£100,000.00–£300,000.00
Labour rate (see page 3)	£28.00 per hour
Labour rate for maintenance contracts	£23.00 per hour
Number of site staff	35
Project area	6000 m^2
Project location	Outer London
Project components	50% hard landscape 50% soft landscape and planting
Access to works areas	Very good
Contract	Main contract
Delivery of materials	Full loads

As explained in more detail later the prices are generally based on wage rates and material costs current at June 2019. They do not allow for preliminary items, which are dealt with in the Preliminaries section of this book, or for any Value Added Tax (VAT) which may be payable.

Adjustments should be made to standard rates for time, location, local conditions, site constraints and any other factors likely to affect the costs of a specific scheme.

Term contracts for general maintenance of large areas should be executed at rates somewhat lower than those given in this section.

There is now a facility available to readers that enables a comparison to be made between the level of prices given and those for projects carried out in regions other than outer London; this is dealt with on pagexx

Units of Measurement

The units of measurement have been varied to suit the type of work and care should be taken when using any prices to ascertain the basis of measurement adopted.

The prices per unit of area for executing various mechanical operations are for work in the following areas and under the following conditions:

Prices per m^2 relate to areas not exceeding $100\,m^2$

(any plan configuration)

Prices per $100\,m^2$ relate to areas exceeding $100\,m^2$ but not exceeding ¼ ha

(generally clear areas but with some subdivision)

Prices per ha relate to areas over ¼ ha

(clear areas suitable for the use of tractors and tractor-operated equipment)

The prices per unit area for executing various operations by hand generally vary in direct proportion to the change in unit area.

Approximate Estimating

These are combined Measured Work prices which give an approximate cost for a complete section of landscape work. For example, the construction of a car park comprises excavation, levelling, road-base, surfacing and marking parking bays. Each of these jobs is priced separately in the Measured Works section, but a comprehensive price is given in the Approximate Estimates section, which is intended to provide a quick guide to the cost of the job. It will be seen that the more items that go to make up an approximate estimate, the more possibilities there are for variations in the PC prices and the user should ensure that any PC price included in the estimate corresponds to his specification.

The figures given in *Spon's External Works and Landscape Price Book* are intended for general guidance, and if a significant variation appears in any one of the cost groups, the Price for Measured Work should be recalculated.

Measured Works

Prime Cost: Commonly known as the 'PC'. Prime Cost is the actual price of the material item being addressed such as paving, shrubs, bollards or turves, as sold by the supplier. Prime Cost is given 'per square metre', 'per 100 bags' or 'each' according to the way the supplier sells his product. In researching the material prices for the book we requested that the suppliers price for full loads of their product delivered to a site close to the M25 in London. Spon's rates do not include VAT. Some companies may be able to obtain greater discounts on the list prices than those shown in the book. Prime Cost prices for those products and plants which have a wide cost range will be found under the heading of Market Prices in the main sections of this book, so that the user may select the product most closely related to his specification.

Materials: The PC material plus the additional materials required to fix the PC material. Every job needs materials for its completion besides the product bought from the supplier. Paving needs sand for bedding, expansion joint strips and cement pointing; fencing needs concrete for post setting and nails or bolts; tree planting needs manure or fertilizer in the pit, tree stakes, guards and ties. If these items were to be priced out separately, *Spon's External Works and Landscape Price Book* (and the Bill of Quantities) would be impossibly unwieldy, so they are put together under the heading of Materials.

Labour: This figure covers the cost of planting shrubs or trees, laying paving, erecting fencing etc. and is calculated on the wage rate (skilled or unskilled) and the time needed for the job. Extras such as highly skilled craft work, difficult access, intermittent working and the need for labourers to back up the craftsman all add to the cost. Large regular areas of planting or paving are cheaper to install than smaller intricate areas, since less labour time is wasted moving from one area to another.

Plant, consumable stores and services: This rather impressive heading covers all the work required to carry out the job which cannot be attributed exactly to any one item. It covers the use of machinery ranging from JCBs to shovels and compactors, fuel, static plant, water supply (which is metered on a construction site), electricity and rubbish disposal. The cost of transport to site is deemed to be included elsewhere and should be allowed for elsewhere or as a preliminary item. Hired plant is calculated on an average of 36 hours working time per week.

Subcontract rates: Where there is no analysis against an item, this is deemed to be a rate supplied by a subcontractor. In most cases these are specialist items where most external works contractors would not have the expertise or the equipment to carry out the task described. It should be assumed that subcontractor rates include for the subcontractor's profit. An example of this may be found for the tennis court rates in Section 35.

Labour Rates Used in this Edition

The rates for labour used in this edition have been based on surveys carried out on a cross section of external works contractors. These rates include for company overheads such as employee administration, transport insurance, and oncosts such as National Insurance. Please see the calculation of these rates shown on page xxiv. The rates do now include for profit.

Overheads: An allowance for this is included in the labour rates which are described above. The general overheads of the contract such as insurance, site huts, security, temporary roads and the statutory health and welfare of the labour force are not directly assignable to each item, so they are distributed as a percentage on each, or as a separate preliminary cost item. The contractor's and subcontractor's profits are now included in this group of costs. Site overheads, which will vary from contract to contract according to the difficulties of the site, labour shortages, inclement weather or involvement with other contractors, have not been taken into account in the build-up of these rates, while overhead (or profit) may have to take into account losses on other jobs and the cost to the contractor of unsuccessful tendering.

In many instances a composite team rate is used in the calculations.

ADJUSTMENT AND VARIATION OF THE RATES IN THIS BOOK

It will be appreciated that a variation in any one item in any group will affect the final Measured Works price. Any cost variation must be weighed against the total cost of the contract. A small variation in Prime Cost where the items are ordered in thousands may have more effect on the total cost than a large variation on a few items. A change in design that necessitates the use of earth moving equipment which must be brought to the site for that one job will cause a dramatic rise in the contract cost. Similarly, a small saving on multiple items will provide a useful reserve to cover unforeseen extras.

Worked examples
These are shown on pages xix and xx.

HOW THIS BOOK IS UPDATED EACH YEAR

The basis for this book is a database of Material, Labour, Plant and Subcontractor resources each with its own area of the database.

Material, Plant and Subcontractor Resources

Each year the suppliers of each material, plant or subcontract item are approached and asked to update their prices to those that will prevail in September of that year.

These resource prices are individually updated in the database. Each resource is then linked to one or many tasks. The tasks in the task library section of the database are automatically updated by changes in the resource library. A quantity of the resource is calculated against the task. The calculation is generally performed once and the links remain in the database.

On occasions where new information or method or technology are discovered or suggested, these calculations would be revisited. A further source of information is simple time and production observations made during the course of the last year.

Labour Resource Update

Most tasks, except those shown as subcontractor rates (see above), employ an element of labour. The Data Department at Aecom conducts ongoing research into the costs of labour in various parts of the country.

Tasks or entire sections are then re-examined and recalculated. Comments on the rates published in this book are welcomed and may be submitted to the contact addresses shown in the Preface.

It is one thing to imagine a better world.
It's another to deliver it.

Creating iconic structures. Planning entire communities. Connecting people with transport systems. Rescuing historic buildings. Our team of circa 900 cost and project managers collaborate to turn imagination into reality, providing innovative solutions to complex challenges across the globe.

Imagine it. Delivered.

PART 1

General

This part of the book contains the following sections:

Estimator's Pocket Book, 2nd edition

Duncan Cartlidge

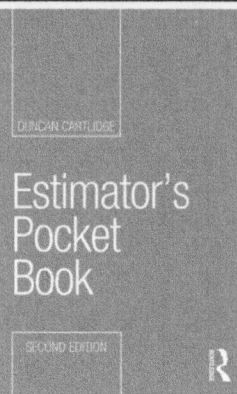

The *Estimator's Pocket Book, Second Edition* is a concise and practical reference covering the main pricing approaches, as well as useful information such as how to process sub-contractor quotations, tender settlement and adjudication. It is fully up to date with NRM2 throughout, features a look ahead to NRM3 and describes the implications of BIM for estimators.
It includes instructions on how to handle:

*   the NRM order of cost estimate;
*   unit-rate pricing for different trades;
*   pro-rata pricing and dayworks;
*   builders' quantities;
*   approximate quantities.

Worked examples show how each of these techniques should be carried out in clear, easy-to-follow steps. This is the indispensable estimating reference for all quantity surveyors, cost managers, project managers and anybody else with estimating responsibilities. Particular attention is given to NRM2, but the overall focus is on the core estimating skills needed in practice. Updates to this edition include a greater reference to BIM, an update on the current state of the construction industry as well as up-to-date wage rates, legislative changes and guidance notes.

Routledge Pocket Books

February 2019: 198 × 129 mm: 292pp
Pb: 978-1-138-36670-1 : £24.99

To Order: Tel: +44 (0) 1235 400524 Fax: +44 (0) 1235 400525
or Post: Taylor and Francis Customer Services,
Bookpoint Ltd, Unit T1, 200 Milton Park, Abingdon, Oxon, OX14 4TA UK
Email: book.orders@tandf.co.uk

For a complete listing of all our titles visit:
www.tandf.co.uk

Labour Rates Used in this Edition

Based on surveys carried out on a cross section of external works contractors, the following rates for labour have been used in this edition.

These rates include for company overheads such as employee administration, transport insurance, and oncosts such as National Insurance.

The rates used include for overhead & profit. Please see the calculations on page 7.

The rates used may include a subcontractor's profit where subcontractor's rates are used to build up the costs of an item in the book.

Please use the Regional Variations table published on page 27 to adjust the costs of the Item being selected from this book for the geographical area of interest. The factor should be applied to the whole project cost and not just the labour rate. The rates in this book are based on a project on the London outer periphery

The rates for all labour used in this edition are as follows:

Labour Type	Unit	Cost incl Overhead	Mark-up %	Rate incl. Overhead & Profit
General contracting	/man-hour	£28.00	15%	£32.20
Brickwork 1+1 man gang	/team-hour	£64.00	15%	£73.60
Brickwork 1+2 man gang	/team-hour	£92.00	15%	£105.80
Maintenance contracting	/man-hour	£23.00	15%	£26.45

Other resource categories

Materials generally	15%
Plant	15%
Subcontractors generally	10%

Quality Auditing in Construction Projects

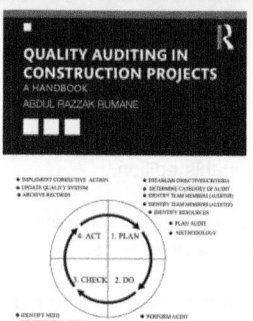

Abdul Razzak Rumane

This book provides construction professionals, designers, contractors and quality auditors involved in construction projects with the auditing skills and processes required to improve construction quality and make their projects more competitive and economical.

The processes within the book focus on auditing compliance to ISO, corporate quality management systems, project specific quality management systems, contract management, regulatory authorities' requirements, safety, and environmental considerations. The book is divided into seven chapters and each chapter is divided into numbered sections covering auditing-related topics that have importance or relevance for understanding quality auditing concepts for construction projects.

No other book covers construction quality auditing in such detail and with this level of practical application. It is an essential guide for construction and quality professionals, but also for students and academics interested in learning about quality auditing in construction projects.

June 2019: 234 × 156 mm: 600pp
Hb: 978-0-8153-8531-8 : £100.00

To Order: Tel: +44 (0) 1235 400524 Fax: +44 (0) 1235 400525
or Post: Taylor and Francis Customer Services,
Bookpoint Ltd, Unit T1, 200 Milton Park, Abingdon, Oxon, OX14 4TA UK
Email: book.orders@tandf.co.uk

For a complete listing of all our titles visit:
www.tandf.co.uk

Taylor & Francis
Taylor & Francis Group

Computation of Labour Rates Used in this Edition

Different organizations will have varying views on rates and costs, which will in any event be affected by the type of job, availability of labour and the extent to which mechanical plant can be used. However, this information should assist the reader to:

(1)   compare the prices to those used in his own organization
(2)   calculate the effect of changes in wage rates or prices of materials
(3)   calculate prices for work similar to the examples given

From June 2017 – assumed basic weekly rates of pay for craft and general operatives are £465.27 and £349.83 respectively; to these rates have been added allowances for the items below in accordance with the recommended procedure of the Chartered Institute of Building in its Code of Estimating Practice. The resultant hourly rates are £15.82 and £11.79 for craft operatives and general operatives respectively.

The items for which allowances have been made are:

- Lost time
- Construction Industry Training Board Levy
- Holidays with pay
- Accidental injury, retirement and death benefits scheme
- Sick pay
- National Insurance
- Severance pay and sundry costs
- Employer's liability and third-party insurance

The tables that follow illustrate how the above hourly rates have been calculated. Productive time has been based on a total of 1720 hours worked per year for daywork calculations above.

How the Labour Rate has been calculated

A survey of typical landscape/external works companies indicates that they are currently paying above average wages for multi-skilled operatives regardless of specialist or supervisory capability. In our current overhaul of labour constants and costs we have departed from our previous policy of labour rates based on national awards and used instead a gross rate per hour for all rate calculations. Estimators can readily adjust the rate if they feel it is inappropriate for their work.

The productive labour of any organization must return their salary plus the cost of the administration which supports the labour force.

The labour rate used in this edition is calculated as follows:

Team size

A three man team with annual salaries as shown and allowances for National Insurance, uniforms, site tools and a company vehicle, returns a basic labour rate of £21.26 per hour to which company overhead is added as per the following tables.

Basic labour rate calculation for Spon's 2020 – See table on page 10

Standard Working Hours per year (2019/2020)
        = 1794

Less allowance for sick leave and down time
for inclement weather
        = −74

Actual Working Hours per year (2019/2020)
        =1720 (1719.9) hrs

Major Works							
Working Hours per year 2019/2020		**1720** *Allows sick and down time of 39 hours*					
	Number of	**Labour team**	**NI**	**Nett Cost**	**Clothing**	**Site Tools**	**TOTAL 3 Man Team**
Foreman	1	35000.00	4830.00	39830.00	200.00	150.00	
Craftsman	1	28000.00	3864.00	31864.00	200.00	150.00	
Labourer	1	20000.00	2760.00	22760.00	200.00	150.00	
	3.00	83000.00	11454.00	94454.00	600.00	450.00	**£ 95,504.00**
Overtime Hours based on working							
nominal hours	40.00	30.52	3.91	1377.21			
/annum	40.00	24.42	3.13	1101.77			
at 1.5 times normal	40.00	17.44	2.23	786.98			
standard rate							
Total staff in Team	3.00			97719.95	600.00	450.00	98769.95
Vehicle Costs Inclusive of Fuel Insurances etc.	**Working Days**	**£ /Day**					
	240.00	45.50					£10,920.00
						TOTAL	**£109,689.95**
Nett Average Cost per man Hour		£21.26					

The basic average labour rate excluding overhead costs (below) is

$$\frac{£\ 109{,}689.95}{\text{Total staff in team (3) x Working Hours per year (1720)}} = £21.26 \text{ per hour}$$

Add to the above the overhead costs as per the table on the next page

Overhead costs

COST CENTRE	Number of	Cost	Total
MD: Salary only: excludes profits + NI + Vehicle	1	80000.00	80000.00
Senior contracts managers + NI + Vehicle (50 K basic)	1	58000.00	58000.00
Other contracts managers + NI + Vehicle (£35 K basic)	1	40000.00	40000.00
Business administrator/secretary	1	25000.00	25000.00
Bookkeeper	1	25000.00	25000.00
Rental	12	1200.00	14400.00
Insurances	1	5000.00	5000.00
Telephone and Mobiles	12	250.00	3000.00
Office Equipment	1	1000.00	1000.00
Stationery	12	50.00	600.00
Advertising	12	300.00	3600.00
Other vehicles not allocated to contract teams	3	9000.00	27000.00
Other consultants	1	6000.00	6000.00
Accountancy	1	2500.00	2500.00
Lights, heating, water	12	200.00	2400.00
Other expenses	1	8000.00	8000.00
TOTAL OFFICE OVERHEAD			301500.00

Add to the above basic rate the company overhead costs for a small company as per the table below. These costs are absorbed by the number of working men multiplied by the number of working hours supplied in the table below. This then generates an hourly overhead rate which is added to the net cost rate above.

Illustrative overhead for company employing 12–50 landscape operatives

The final labour rate used is then generated as follows:

$$\text{Net labour rate} = \frac{\text{Total Office Overhead (£301,500.00)}}{\text{Working hours per year (1720)} \times \text{Labour resources employed}}$$

TOTAL NR OF SITE STAFF	Admin Cost per hour	Total Rate per Man Hour
12	14.61	35.87
15	11.69	32.94
18	9.74	31.00
20	8.76	30.02
25	7.01	28.27
35	5.01	26.27
40	4.38	25.64
45	3.90	25.15
50	3.51	24.76

Hourly labour rates used for *Spon's External Works and Landscape Price Book 2020*

This year's **'at-cost'** rates are calculated on the rounded value of a 25-person working team rounded to £28.00

Labour Type	Unit	Cost incl Overhead	Mark-up %	Rate incl. Overhead & Profit
General contracting	/man-hour	£28.00	15%	£32.20
Brickwork 1+1 man gang	/team-hour	£64.00	15%	£73.60
Brickwork 1+2 man gang	/team-hour	£92.00	15%	£105.80
Maintenance contracting	/man-hour	£23.00	15%	£26.45

Other resource categories

Materials generally		15%
Plant		15%
Subcontractors generally		10%

Some high cost material or subcontract items may be at lower rates of 5%

Smaller organizations

Smaller organizations should adjust the rates shown in this book to suit a labour rate which is generally slightly higher due to the balance of company overhead and number of employees. The costs for smaller organizations are no longer shown in this edition but users may adjust the rates using the resultant labour rates shown below.

The calculation of this labour rate is as follows:

- A typical smaller works labour calculation is based on the above but the administration cost of the organization is £115,640.00 per annum
- The owner of the business does not carry out site works and is assumed to be paid a salary of £58,000 including National insurance. The balance of his income is taken in dividend and profit share figure constitutes part of the £115,640.00 shown
- There are between 6 and 15 site operatives

Illustrative overhead for small company employing 6–15 landscape operatives

COST CENTRE	Number of	Cost	Total
MD: Salary only: excludes profits + NI + Vehicle	1	58000.00	58000.00
Secretary	1	21000.00	21000.00
Bookkeeper	1	10000.00	10000.00
Rental	12	325.00	3900.00
Insurances	1	4000.00	4000.00
Telephone and Mobiles	12	200.00	2400.00
Office Equipment	1	1000.00	1000.00
Stationery	12	20.00	240.00
Advertising	12	200.00	2400.00
Other vehicles not allocated to contract teams	1	4000.00	4000.00
Other consultants	1	1000.00	1000.00
Accountancy	1	1500.00	1500.00
Lights, heating, water	12	100.00	1200.00
Other expenses	1	8000.00	8000.00
TOTAL OFFICE OVERHEAD			118640.00

The final labour rate used is then generated as follows:

$$\text{Net labour rate} = \frac{\text{Total Office Overhead (£118640.00)}}{\text{Working hours per year (1720)} \times \text{Labour resources employed}}$$

Total nr of site staff	Admin Cost per hour	Total Rate per Man Hour
3	22.99	44.25
6	11.50	32.75
9	7.66	28.92
12	5.75	27.01
15	4.60	25.86

Hourly labour rates for a typical small company working in London for 2020 could be calculated on the rounded value of a 9-man working team (£27.50).

General and Maintenance Contracting (Smaller organizations)	29.00

Calculation of Annual Hours Worked See full table on page 6

Normal hours worked		
Normal hours worked per week		39 hours
Normal hours worked per year	39 hrs × 52 wks	2028 hours
Less holidays		
Annual holidays	39 hrs × 4.2 wks	−163.8 hours
Public holidays	8 days @7.8 hours	−62.4 hours
Less sick and downtime	39 hrs × 1 wk	−74 hours
Actual hours worked per annum		1720 hours

Notes on the New Rules of Measurement (NRM) Used in this Edition

New Rules of Measurement (NRM)

The NRM suite covers the life cycle of cost management and means that, at any point in a building's life, quantity surveyors will have a set of rules for measuring and capturing cost data. In addition, the BCIS Standard Form of Cost Analysis (SFCA) 4th edition has been updated so that it is aligned with the NRM suite.

The three volumes of the NRM are as follows.

- **NRM1** – Order of cost estimating and cost planning for capital building works.
- **NRM2** – Detailed measurement for building works.
- **NRM3** – Order of cost estimating and cost planning for building maintenance works.

NRM1

In May 2009 the New Rules of Measurement (NRM1) Order of cost estimating and elemental cost planning were introduced by the RICS. The NRM have been written to provide a standard set of measurement rules that are understandable by all those involved in a construction project, including the employer; thereby aiding communication between the project/design team and the employer. In addition, the NRM should assist the quantity surveyor/cost manager in providing effective and accurate cost advice to the employer and the rest of the project team.

NRM2

Developed to provide a detailed set of measurement rules to support the procurement of construction works. These rules are a direct replacement for the *Standard Method of Measurement for Building Works (SMM7)*.

The NRM suite is published by the RICS and readers are strongly advised to read the documents.

The Client Role in Successful Construction Projects

Jason Challender *et al.*

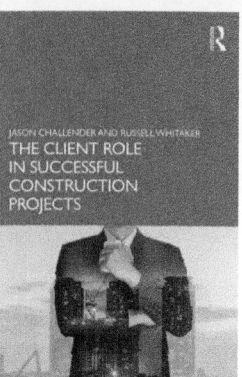

The Client Role in Successful Construction Projects is a practical guide for clients on how to initiate, procure and manage construction projects and developments. This book is written from the perspective of the client initiating a construction project as part of a business venture and differs from most available construction literature which can externalise the client as a risk to be managed by the design team. The book provides a practical framework for new and novice clients undertaking construction, giving them a voice and enabling them to:

- Understand the challenges that they and the project are likely to face.
- Communicate and interact effectively with key stakeholders and professionals within the industry.
- Understand in straightforward terms where they can have a positive impact on the project.
- Put in place a client-side due diligence process.
- Reduce their institutional risk and the risk of project failure.
- Discover how their standard models are able to co-exist and even transfer to a common client-side procedure for managing a construction project.

Written by clients, for clients, this book is highly recommended not only for clients, but for construction industry professionals who want to develop their own skills and enhance their working relationship with their clients. A supporting website for the book will be available, which will give practical examples of the points illustrated in the book and practical advice from specialists in the field.

May 2019: 234 × 156 mm: 318pp
Pb: 978-1-138-05821-7 : £29.99

To Order: Tel: +44 (0) 1235 400524 Fax: +44 (0) 1235 400525
or Post: Taylor and Francis Customer Services,
Bookpoint Ltd, Unit T1, 200 Milton Park, Abingdon, Oxon, OX14 4TA UK
Email: book.orders@tandf.co.uk

For a complete listing of all our titles visit:
www.tandf.co.uk

Computation of the Cost of Materials

Percentages of default waste are placed against material resources within the supplier database. These range from 2.5% (for bricks) to 20% (for topsoil). An allowance for the cost of unloading, stacking etc. should be added to the cost of materials.

The following are typical hours of labour for unloading and stacking some of the more common building materials.

Material	Unit	Labourer hours
Cement	tonne	0.67
Lime	tonne	0.67
Common bricks	1000	1.70
Light engineering bricks	1000	2.00
Heavy engineering bricks	1000	2.40

New Aspects of Quantity Surveying Practice, 4th edition

Duncan Cartlidge

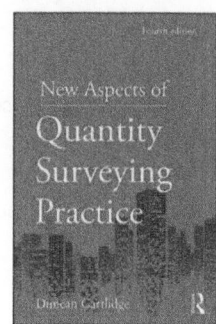

In this fourth edition of New Aspects of Quantity Surveying Practice, renowned quantity surveying author Duncan Cartlidge reviews the history of the quantity surveyor, examines and reflects on the state of current practice with a concentration on new and innovative practice, and attempts to predict the future direction of quantity surveying practice in the UK and worldwide.

The book champions the adaptability and flexibility of the quantity surveyor, whilst covering the hot topics which have emerged since the previous edition's publication, including:

- the RICS 'Futures' publication;
- Building Information Modelling (BIM);
- mergers and acquisitions;
- a more informed and critical evaluation of the NRM;
- greater discussion of ethics to reflect on the renewed industry interest;
- and a new chapter on Dispute Resolution.

As these issues create waves throughout the industry whilst it continues its global growth in emerging markets, such reflections on QS practice are now more important than ever. The book is essential reading for all Quantity Surveying students, teachers and professionals. It is particularly suited to undergraduate professional skills courses and non-cognate postgraduate students looking for an up to date understanding of the industry and the role.

December 2017: 234 x 156 mm: 282 pp
Pb: 978-1-138-67376-2: £48.99

To Order: Tel: +44 (0) 1235 400524 Fax: +44 (0) 1235 400525
or Post: Taylor and Francis Customer Services,
Bookpoint Ltd, Unit T1, 200 Milton Park, Abingdon, Oxon, OX14 4TA UK
Email: book.orders@tandf.co.uk

For a complete listing of all our titles visit:
www.tandf.co.uk

Taylor & Francis
Taylor & Francis Group

The Landfill Tax

The Tax

The Landfill Tax came into operation on 1 October 1996. It is levied on operators of licensed landfill sites in England, Wales and Northern Ireland at the following rates with effect from 1 April 2019

- Inactive or inert wastes   £2.90 per tonne   Included are soil, stones, brick, plain and reinforced
  [Lower Rate]   concrete, plaster and glass – lower rate

- All other taxable wastes   £91.35 per tonne   Included are timber, paint and other organic wastes
  [Standard Rate]   generally found in demolition work and builders skips –
                                                           standard rate

The standard and lower rates of Landfill Tax were due to increase in line with the RPI, rounded to the nearest 5 pence, from 1 April 2016. However, legislation introduced in the Finance Bill 2016 provided for new rates of Landfill Tax until 31 March 2019 as follows:

- Lower Rate   From 1 April 2017
                     £2.70/tonne
- Lower Rate   From 1 April 2018
                     £2.80/tonne
- Lower Rate   From 1 April 2019
                     £2.90/tonne
- Standard Rate   From 1 April 2017
                     £86.10/tonne
- Standard Rate   From 1 April 2018
                     £88.95/tonne
- Standard Rate   From 1 April 2019
                     £91.35/tonne

There has been no change in the loss on ignition threshold for fines which is a major eligibility factor for lower rate Landfill Tax since its introduction in April 2015. The threshold remains at 10%.

The Landfill Tax (Qualifying Material) Order 2011 came into force on 1 April 2011 and set out the qualifying criteria that are eligible for the lower rate of Landfill Tax. A waste will be lower rated for Landfill Tax only if it is listed as a qualifying material in the Landfill Tax (Qualifying Material) Order 2011, and is disposed of at an authorized waste site.

The principal qualifying materials are:

- Naturally occurring rocks and subsoils
- Ceramic or concrete materials
- Processed or prepared minerals
- Furnace slags
- Ash arising from wood or waste combustion and from the burning of coal and petroleum coke (including when burnt with biomass)
- Low activity inorganic compounds
- Calcium sulphate
- Calcium hydroxide and brine

From 1 April 2018, Landfill Tax is due on disposals of material at unauthorized sites in England and Northern Ireland. This also applies to disposals made prior to 1 April 2018 which are still on the site on 1 April 2018. Currently disposals on a site operated as a 'relevant regulated facility' which (as defined under the Environmental Permitting (England and Wales) Regulations 2016) includes sites which are 'a waste operation, such as a transfer station or treatment facility' are not considered to be taxable disposals. The Landfill Tax (Disposals of Material) Order 2018 provides more clarity on what constitutes a 'taxable disposal'. In particular, it clarifies which elements of a landfill cell are not taxable under the requirements of Part 3 of the Finance Act 1996, listing them as:

1.   material that forms a layer which performs the function of drainage at the base of a landfill cell;
2.   material that forms the impermeable layer that delimits the landfill cell;
3.   a pipe, pump or associated infrastructure inserted into a landfill cell for the purposes of the extraction or control of surplus liquid or gas from or within that cell;
4.   material used for restoration of a landfill cell that only contains inert material; or
5.   material placed in an information area within the meaning given by regulation 16 A(1) of the Landfill Tax Regulations 1996.

Calculating the Weight of Waste

There are two options:

- If licensed sites have a weighbridge tax will be levied on the actual weight of waste.
- If licensed sites do not have a weighbridge tax will be levied on the permitted weight of the lorry based on an alternative method of calculation based on volume to weight factors for various categories of waste.

Water

- If water greater than or equal to 25% of the material by weight has been added to the material to facilitate disposal, for the extraction of minerals or in the course of an industrial process, a landfill operator may apply to discount the water content of the material before tax is levied.

Effect on Prices

The tax is paid by landfill site operators only. Tipping charges reflect this additional cost.

As an example, *Spon's A & B* rates for mechanical disposal will be affected as follows:

• Inactive waste	*Spon's A & B 2020* net rate	£29.40 per m³
	Tax, 1.9 m³/tonne (unbulked) @ £2.90/tonne	£5.51 per m³
	Spon's rate including tax	£34.91 per m³
• Active waste	Active waste will normally be disposed of by skip and will probably be mixed with inactive waste. The tax levied will depend on the weight of materials in the skip which can vary significantly.	

Exemptions

The following disposals are exempt from Landfill Tax subject to meeting certain conditions:

- dredgings which arise from the maintenance of inland waterways and harbours
- naturally occurring materials arising from mining or quarrying operations
- pet cemeteries
- material from the reclamation of contaminated land (see below)
- inert waste used to restore landfill sites and to fill working and old quarries where a planning condition or obligation is in existence
- waste from visiting NATO forces

The exemption for waste from contaminated land has been phased out completely from 1 April 2012 and no new applications for Landfill Tax exemption are now accepted.

Devolution of Landfill Tax to Scotland from 1 April 2015

The Scotland Act 2012 provided for Landfill Tax to be devolved to Scotland. From 1 April 2015, operators of landfill sites in Scotland are no longer liable to pay UK Landfill Tax for waste disposed at their Scottish sites. Instead, they will be liable to register and account for Scottish Landfill Tax (SLfT).

The Scottish Government announced rates of Landfill Tax from 1 April 2019 to be as follows:

- Lower Rate – £2.90/tonne
- Standard Rate – £91.35/tonne

Both rates mirror those applicable in England and Northern Ireland.

Devolution of Landfill Tax to Wales from April 2018

The Wales Act 2014 provides for Landfill Tax to be devolved to Wales. This took effect from 1 April 2018. Operators of landfill sites in Wales are liable to register and account for Landfill Disposals Tax (LDT); information is available on the Welsh Revenue Authority website, but in summary there are 3 rates of LDT:

- a lower rate of £2.90/tonne
- a standard rate of £91.35/tonne
- an unauthorized disposals rate of £137.00/tonne

For further information contact the HMRC VAT and Excise Helpline, telephone 0300 200 3700.

The Devolution of Landfill Tax (Wales) (Consequential Transitional and Savings Provisions) Order 2018 provides for the closure of the Landfill Communities Fund (LCF) in Wales and the arrangements for a two year transitional period during which environmental bodies may continue to spend funds on projects in Wales.

Introduced in 1996 the LCF is a mechanism whereby landfill operators can receive a credit against their landfill tax liability by making contributions to environmental bodies involved in the scheme to improve the environment in the vicinity of landfill sites. On or after 1 April 2018 landfill tax credits provided under the existing landfill tax legislation will only be available to operators registered for and paying landfill tax in England and Northern Ireland, landfill tax having been devolved to Scotland in 2015 (see above). A transitional regime provides that landfill operators who have previously made contributions to environmental bodies carrying out work in Wales will remain eligible for tax credit provided that any unspent funds held by those environmental bodies and derived from contributions under the scheme are spent within 2 years of the date upon which devolution of landfill tax took place.

Contractual Procedures in the Construction Industry, 7th edition

Allan Ashworth and Srinath Perera

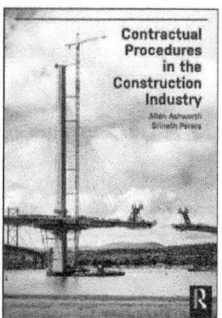

Contractual Procedures in the Construction Industry 7th edition aims to provide students with a comprehensive understanding of the subject, and reinforces the changes that are taking place within the construction industry. The book looks at contract law within the context of construction contracts, it examines the different procurement routes that have evolved over time and the particular aspects relating to design and construction, lean methods of construction and the advantages and disadvantages of PFI/PPP and its variants. It covers the development of partnering, supply chain management, design and build and the way that the clients and professions have adapted to change in the procurement of buildings and engineering projects.

Key features of the new edition include:

- A revised chapter covering the concept of value for money in line with the greater emphasis on added value throughout the industry today.
- A new chapter covering developments in information technology applications (building information modelling, blockchains, data analytics, smart contracts and others) and construction procurement.
- Deeper coverage of the strategies that need to be considered in respect of contract selection.
- Improved discussion of sustainability and the increasing importance of resilience in the built environment.
- Concise descriptions of some the more important construction case laws.

March 2018: 246 x 174 mm: 458 pp
Pb: 978-1-138-69393-7: £48.99

To Order: Tel: +44 (0) 1235 400524 Fax: +44 (0) 1235 400525
or Post: Taylor and Francis Customer Services,
Bookpoint Ltd, Unit T1, 200 Milton Park, Abingdon, Oxon, OX14 4TA UK
Email: book.orders@tandf.co.uk

For a complete listing of all our titles visit:
www.tandf.co.uk

The Aggregates Levy

The Aggregates Levy came into operation on 1 April 2002 in the UK, except for Northern Ireland where it was phased in over five years from 2003.

It was introduced to ensure that the external costs associated with the exploitation of aggregates are reflected in the price of aggregate, and to encourage the use of recycled aggregate. There continues to be strong evidence that the levy is achieving its environmental objectives, with sales of primary aggregate down and production of recycled aggregate up. The Government expects that the rates of the levy will at least keep pace with inflation over time, although it accepts that the levy is still bedding in.

The rate of the levy will continue to be £2.00 per tonne from 1 April 2019 and is levied on anyone considered to be responsible for commercially exploiting 'virgin' aggregates in the UK and should naturally be passed by price increase to the ultimate user.

All materials falling within the definition of 'Aggregates' are subject to the levy unless specifically exempted.

It does not apply to clay, soil, vegetable or other organic matter.

The intention is that it will:

- Encourage the use of alternative materials that would otherwise be disposed of to landfill sites
- Promote development of new recycling processes, such as using waste tyres and glass
- Promote greater efficiency in the use of virgin aggregates
- Reduce noise and vibration, dust and other emissions to air, visual intrusion, loss of amenity and damage to wildlife habitats

Definitions

'Aggregates' means any rock, gravel or sand which is extracted or dredged in the UK for aggregates use. It includes whatever substances are for the time being incorporated in it or naturally occur mixed with it.

'Exploitation' is defined as involving any one or a combination of any of the following:

- Being removed from its original site, a connected site which is registered under the same name as the originating site or a site where it had been intended to apply an exempt process to it, but this process was not applied
- Becoming subject to a contract or other agreement to supply to any person
- Being used for construction purposes
- Being mixed with any material or substance other than water, except in permitted circumstances

The definition of 'aggregate being used for construction purposes' is when it is:

- Used as material or support in the construction or improvement of any structure
- Mixed with anything as part of a process of producing mortar, concrete, tarmacadam, coated roadstone or any similar construction material

Incidence

It is a tax on primary aggregates production – i.e. 'virgin' aggregates won from a source and used in a location within the UK territorial boundaries (land or sea). The tax is not levied on aggregates which are exported or on aggregates imported from outside the UK territorial boundaries.

It is levied at the point of sale.

Exemption from tax

An 'aggregate' is exempt from the levy if it is:

- Material which has previously been used for construction purposes
- Aggregate that has already been subject to a charge to the Aggregates Levy
- Aggregate which was previously removed from its originating site before the start date of the levy
- Aggregate which is moved between sites under the same Aggregates Levy Registration
- Aggregate which is removed to a registered site to have an exempt process applied to it
- Aggregate which is removed to any premises where china clay or ball clay will be extracted from the aggregate
- Aggregate which is being returned to the land from which it was won provided that it is not mixed with any material other than water
- Aggregate won from a farm land or forest where used on that farm or forest
- Rock which has not been subjected to an industrial crushing process
- Aggregate won by being removed from the ground on the site of any building or proposed building in the course of excavations carried out in connection with the modification or erection of the building and exclusively for the purpose of laying foundations or of laying any pipe or cable
- Aggregate won by being removed from the bed of any river, canal or watercourse or channel in or approach to any port or harbour (natural or artificial), in the course of carrying out any dredging exclusively for the purpose of creating, restoring, improving or maintaining that body of water
- Aggregate won by being removed from the ground along the line of any highway or proposed highway in the course of excavations for improving, maintaining or constructing the highway otherwise than purely to extract the aggregate
- Drill cuttings from petroleum operations on land and on the seabed
- Aggregate resulting from works carried out in exercise of powers under the New Road and Street Works Act 1991, the Roads (Northern Ireland) Order 1993 or the Street Works (Northern Ireland) Order 1995
- Aggregate removed for the purpose of cutting of rock to produce dimension stone, or the production of lime or cement from limestone
- Aggregate arising as a waste material during the processing of the following industrial minerals:
  - anhydrite
  - ball clay
  - barites
  - calcite
  - china clay
  - clay, coal, lignite and slate
  - feldspar
  - flint
  - fluorspar
  - fuller's earth
  - gems and semi-precious stones
  - gypsum
  - any metal or the ore of any metal
  - muscovite
  - perlite
  - potash
  - pumice
  - rock phosphates
  - sodium chloride
  - talc
  - vermiculite
  - Spoil from the separation of the above industrial minerals from other rock after extraction
  - Material that is mainly but not wholly the spoil, waste or other by-product of any industrial combustion process or the smelting or refining of metal.

Anything that consists 'wholly or mainly' of the following is exempt from the levy (note that 'wholly' is defined as 100% but 'mainly' as more than 50%, thus exempting any contained aggregates amounting to less than 50% of the original volumes:

- clay, soil, vegetable or other organic matter
- drill cuttings from oil exploration in UK waters
- material arising from utility works, if carried out under the New Roads and Street Works Act 1991

However, when ground that is more than half clay is mixed with any substance (for example, cement or lime) for the purpose of creating a firm base for construction, the clay becomes liable to Aggregates Levy because it has been mixed with another substance for the purpose of construction.

Anything that consists completely of the following substances is exempt from the levy:

- Spoil, waste or other by-products from any industrial combustion process or the smelting or refining of metal – for example, industrial slag, pulverized fuel ash and used foundry sand. If the material consists completely of these substances at the time it is produced it is exempt from the levy, regardless of any subsequent mixing
- Aggregate necessarily arising from the footprint of any building for the purpose of laying its foundations, pipes or cables. It must be lawfully extracted within the terms of any planning consent
- Aggregate necessarily arising from navigation dredging
- Aggregate necessarily arising from the ground in the course of excavations to improve, maintain or construct a highway or a proposed highway
- Aggregate necessarily arising from the ground in the course of excavations to improve, maintain or construct a railway, monorail or tramway

Relief from the levy either in the form of credit or repayment is obtainable where:

- it is subsequently exported from the UK in the form of aggregate
- it is used in an exempt process
- it is used in a prescribed industrial or agricultural process
- it is waste aggregate disposed of by dumping or otherwise, e.g. sent to landfill or returned to the originating site

An exemption for aggregate obtained as a by-product of railway, tramway and monorail improvement, maintenance and construction was introduced in 2007.

Discounts

Water which is added to the aggregate after the aggregate has been won (washing, dust dampening etc.) may be discounted from the tax calculations. There are two accepted options by which the added water content can be calculated.

The first is to use HMRC's standard added water percentage discounts listed below:

- washed sand 7%
- washed gravel 3.5%
- washed rock/aggregate 4%

Alternatively a more exact percentage can be agreed for dust dampening of aggregates.

Whichever option is adopted, it must be agreed in writing in advance with HMRC.

Impact

The British Aggregates Association suggested that the additional cost imposed by quarries is more likely to be in the order of £3.40 per tonne on mainstream products, applying an above average rate on these in order that by-products and low grade waste products can be held at competitive rates, as well as making some allowance for administration and increased finance charges.

With many gravel aggregates costing in the region of £20.00 per tonne, there is a significant impact on construction costs.

Avoidance

An alternative to using new aggregates in filling operations is to crush and screen rubble which may become available during the process of demolition and site clearance as well as removal of obstacles during the excavation processes.

Example: Assuming that the material would be suitable for fill material under buildings or roads, a simple cost comparison would be as follows (note that for the purpose of the exercise, the material is taken to be 1.80 tonnes per m³ and the total quantity involved less than 1000 m³):

Disposing of site material:	£/m³	£/tonne
Cost of removing materials from site	30.92	17.18
Importing fill material:		
Cost of 'new' aggregates delivered to site	34.31	19.06
Addition for Aggregates Tax	3.60	2.00
Total cost of disposing waste aggregate and importing fill materials	68.83	38.24
Crushing site materials:	£/m³	£/tonne
Transportation of material from excavations or demolition to stockpiles	0.89	0.50
Transportation of material from temporary stockpiles to the crushing plant	2.39	1.33
Establishing plant and equipment on site; removing on completion	2.39	1.33
Maintain and operate plant	10.75	5.97
Crushing hard materials on site	15.53	8.62
Screening material on site	2.39	1.33
Total cost of crushing site materials ready for reuse	34.34	19.08

From the above it can be seen that potentially there is a great benefit in crushing site materials for filling rather than importing fill materials.

Setting the cost of crushing against the import price would produce a saving of £3.57 per m³. If the site materials were otherwise intended to be removed from the site, then the cost benefit increases by the saved disposal cost to £30.92 per m³.

Even if there is no call for any or all of the crushed material on site, it ought to be regarded as a useful asset and either sold on in crushed form or else sold with the prospects of crushing elsewhere.

Specimen Unit rates	Unit[3]	£
Establishing plant and equipment on site; removing on completion		
Crushing plant	trip	1,450.00
Screening plant	trip	750.00
Maintain and operate plant		
Crushing plant	week	9,000.00
Screening plant	week	2,200.00
Transportation of material from excavations or demolition places to temporary stockpiles	m^3	2.00
Transportation of material from temporary stockpiles to the crushing plant	m^3	2.00
Breaking up material on site using impact breakers		
mass concrete	m^3	50.00
reinforced concrete	m^3	71.00
brickwork	m^3	36.00
Crushing material on site		
mass concrete not exceeding 1000 m^3	m^3	16.00
mass concrete 1000–5000 m^3	m^3	15.00
mass concrete over 5000 m^3	m^3	14.00
reinforced concrete not exceeding 1000 m^3	m^3	19.00
reinforced concrete 1000–5000 m^3	m^3	17.00
reinforced concrete over 5000 m^3	m^3	16.00
brickwork not exceeding 1000 m^3	m^3	15.00
brickwork 1000–5000 m^3	m^3	14.00
brickwork over 5000 m^3	m^3	13.00
Screening material on site	m^3	3.00

More detailed information can be found on the HMRC website (www.hmrc.gov.uk) in Notice AGL1 Aggregates Levy published 1 April 2014 (updated 9 February 2017).

Metric Handbook:
Planning and Design Data,
6th edition

Pamela Buxton

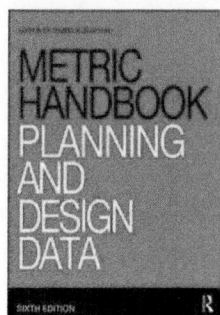

- Significantly updated in reference to the latest construction standards and new building types
- Sustainable design integrated into chapters throughout
- Over half of the entire book has now been updated since 2015
- Over 100,000 copies sold to successive generations of architects and designers
- This book belongs in every design office.

The Metric Handbook is the major handbook of planning and design data for architects and architecture students. Covering basic design data for all the major building types it is the ideal starting point for any project. For each building type, the book gives the basic design requirements and all the principal dimensional data, and succinct guidance on how to use the information and what regulations the designer needs to be aware of.

As well as buildings, the Metric Handbook deals with broader aspects of design such as materials, acoustics and lighting, and general design data on human dimensions and space requirements. The Metric Handbook is the unique reference for solving everyday planning problems.

March 2018: 297 x 210 mm: 890 pp
Pb: 978-1-138-71468-7: £41.99

To Order: Tel: +44 (0) 1235 400524 Fax: +44 (0) 1235 400525
or Post: Taylor and Francis Customer Services,
Bookpoint Ltd, Unit T1, 200 Milton Park, Abingdon, Oxon, OX14 4TA UK
Email: book.orders@tandf.co.uk

For a complete listing of all our titles visit:
www.tandf.co.uk

Cost Indices

The purpose of this section is to show changes in the cost of carrying out landscape work (hard surfacing and planting) since 2010. It is important to distinguish between costs and tender prices: the following table reflects the change in cost to contractors but does not necessarily reflect changes in tender prices. In addition to changes in labour and material costs, which are reflected in the indices given below, tender prices are also affected by factors such as the degree of competition at the time of tender and in the particular area where the work is to be carried out, the availability of labour and materials, and the general economic situation. This can mean that in a period when work is scarce, tender prices may fall despite the fact that costs are rising, and when there is plenty of work available, tender prices may increase at a faster rate than costs.

The Constructed Cost Index

A Constructed Cost Index has been developed based on NEDO Price Adjustment Formula Indices (PAFI) (Building Series 2). Cost indices for the various building trades are published monthly by BCIS and are available via a subscription service.

The indices comprise 49 Building Work indices plus seven 'Appendices' and other specialist indices. The Building Work indices are compiled by monitoring the cost of labour and materials for each category and applying a weighting to these to calculate a single index.

Although the PAFI indices are prepared for use with price-adjustment formulae for calculating reimbursement of increased costs during the course of a contract, they also present a time series of cost indices for the main components of landscaping projects. They can therefore be used as the basis of an index for landscaping costs.

The method used here is to construct a composite index by allocating weightings to the indices representing the usual work categories found in a landscaping contract, the weightings being established from an analysis of actual projects. These weightings totalled 100 in 2015 and the composite index is calculated by applying the appropriate weightings to the appropriate PAFI indices on a monthly basis, which is then compiled into a quarterly index and rebased to 2015 = 100.

Constructed Landscaping Cost Index REBASE 2015 = 100

Year	Quarter 1	Quarter 2	Quarter 3	Quarter 4	Annual Average	Annual change
2000	55.3	55.7	57.6	57.8	56.6	0.0%
2001	57.5	57.8	59.6	59.5	58.6	3.5%
2002	59.6	60.6	64.3	64.5	62.3	6.3%
2003	64.9	65.4	67.7	67.7	66.4	6.6%
2004	67.7	68.5	71.7	71.8	69.9	5.3%
2005	72.0	72.7	76.8	76.7	74.6	6.6%
2006	77.2	77.7	79.8	79.8	78.6	5.5%
2007	80.1	80.7	82.8	83.4	81.7	4.0%
2008	84.5	86.6	90.2	89.2	87.6	7.2%
2009	88.7	89.0	89.4	90.0	89.3	1.9%
2010	90.4	91.0	91.4	91.8	91.1	2.1%
2011	92.7	93.7	95.0	95.7	94.3	3.5%
2012	96.4	96.4	96.6	96.9	96.6	2.4%
2013	98.2	98.0	98.2	98.2	98.2	1.7%
2014	98.6	98.8	100.0	99.5	99.3	1.1%
2015	99.0	99.8	100.8	100.4	100.0	0.8%
2016	100.1	100.9	102.4	103.4	101.7	1.7%
2017	104.1	104.2	106.0	106.5	105.2	3.5%
2018	107	108	110	111	109	3.5%
2019	111	112 (P)	114 (F)	115	113 (F)	3.8%

(P) – Provisional; (F) – Forecast thereafter

This index is updated every quarter in *Spon's Price Book Update*. The updating service is available, free of charge, to all purchasers of Spon's Price Books (complete the reply-paid card enclosed).

Regional Variations

Prices in *Spon's External Works and Landscape Price Book* are based upon conditions prevailing for a competitive tender in the Outer London area. For the benefit of readers, this edition includes regional variation adjustment factors which can be used for an assessment of price levels in other regions.

Special further adjustment may be necessary when considering city centre or very isolated locations.

The regional variations shown in the following table are based on our forecast of price differentials in each of the economic planning regions at the time the book was published. Historically the adjustments vary little year-on-year; however there can be wider variations within any region, particularly between urban and rural locations. The following table shows suggested adjustments required to cost plans or estimates prepared using the *Spon's External Works and Landscape Price Book*.

Please note: these adjustments need to be applied to the total cost plan value, not to individual material or project resources or items.

Region	Adjustment Factor
Outer London (Spon's 2020)	1.00
Inner London	1.06
South East	0.96
South West	0.89
East Midlands	0.87
West Midlands	0.87
East of England	0.93
Yorkshire & Humberside	0.86
North East	0.84
North West	0.88
Scotland	0.87
Wales	0.86
Northern Ireland	0.75

The following example illustrates the adjustment of prices for regions other than Outer London, by use of regional variation adjustment factors.

A.	Value of items priced using *Spon's External Works and Landscape Price Book* for Outer London	£100,000
B.	Adjustment to value of A. to reflect South West region price level 100,000 × 0.89	£ 89,000

Business Principles for Landscape Contracting, 3rd edition

Steven Cohan

Business Principles for Landscape Contracting, fully revised and updated in its third edition, is an introduction to the application of business principles of financial management involved in setting up your own landscape contracting business and beginning your professional career. Appealing to students and professionals alike, it will build your knowledge of financial management tools and enable you to relate their applications to real-life business scenarios. Focusing on the importance of proactive financial management, the book serves as a primer for students in landscape architecture, contracting, and management courses and entrepreneurs within the landscape industry preparing to use business principles in practice.

Topics covered include:

- Financial management and accountability
- Budget development
- Profitable pricing and estimating
- Project management
- Creating a lean culture
- Personnel management and employee productivity
- Professional development
- Economic sustainability.

March 2018: 246 x 189 mm: 274 pp
Pb: 978-0-415-78820-5 : £33.99

To Order: Tel: +44 (0) 1235 400524 Fax: +44 (0) 1235 400525
or Post: Taylor and Francis Customer Services,
Bookpoint Ltd, Unit T1, 200 Milton Park, Abingdon, Oxon, OX14 4TA UK
Email: book.orders@tandf.co.uk

For a complete listing of all our titles visit:
www.tandf.co.uk

Taylor & Francis
Taylor & Francis Group

PART 2

Rates of Wages – Building Industry

Quantity Surveyor's Pocket Book, 3rd Edition

D. Cartlidge

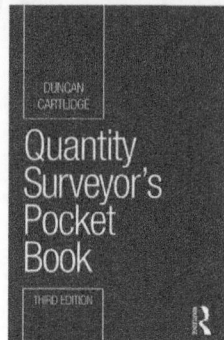

The third edition of the Quantity Surveyor's Pocket Book has been updated in line with NRM1, NRM2 and NRM3, and remains a must-have guide for students and qualified practitioners. Its focused coverage of the data, techniques and skills essential to the quantity surveying role makes it an invaluable companion for everything from initial cost advice to the final account stage.

Key features and updates included in this new edition:
- An up-to-date analysis of NRM1, 2 and 3;
- Measurement and estimating examples in NRM2 format;
- Changes in procurement practice;
- Changes in professional development, guidance notes and schemes of work;
- The increased use of NEC3 form of contract;
- The impact of BIM.

This text includes recommended formats for cost plans, developer's budgets, financial reports, financial statements and final accounts. This is the ideal concise reference for quantity surveyors, project and commercial managers, and students of any of the above.

March 2017; 186 × 123 mm, 466 pp
Pbk: 978-1-138-69836-9; £21.99

To Order: Tel: +44 (0) 1235 400524 Fax: +44 (0) 1235 400525
or Post: Taylor and Francis Customer Services,
Bookpoint Ltd, Unit T1, 200 Milton Park, Abingdon, Oxon, OX14 4TA UK
Email: book.orders@tandf.co.uk

For a complete listing of all our titles visit:
www.tandf.co.uk

Taylor & Francis
Taylor & Francis Group

BUILDING INDUSTRY WAGES

BUILDING CRAFT AND GENERAL OPERATIVES

N/I threshold/week £166

The following basic rates of pay are assumed effective from 24 June 2019

	Rate per 39-hour week (£)	Rate per hour (£)
Craft Rate	495.18	12.70
Skill Rate 1	470.34	12.06
Skill Rate 2	453.18	11.62
Skill Rate 3	423.93	10.87
Skill Rate 4	400.53	10.27
General operative	371.67	9.53

BUILDING CRAFT AND GENERAL OPERATIVES

Typical rate build-up Assumed Effective from 24 June 2019			Craft operatives		General operatives	
			£	£	£	£
Wages at standard basic rate						
Productive time	44.1	wks	495.18	21,837.57	371.67	16,390.65
Lost time allowance, say	0.9	wks	495.18	445.66	371.67	334.50
Overtime, say	0.0	wks	0.00	0.00	0.00	0.00
				22,283.24		16,725.15
Holiday Pay	6.0	wks	495.18	2,971.10	371.67	2,230.02
CITB Levy	0.50%	year		126.27		94.78
				25,380.60		19,049.95
Employer's contributions to:						
EasyBuild Stakeholder Pension (Death and accident cover is provided free)	52	wks	7.50	390.00	7.50	390.00
National Insurance (average weekly payment)	52	wks	44.45	2,311.31	27.65	1,437.68
				28,081.91		20,877.62
Severance pay and sundry costs	1.5%			421.23		313.16
				28,503.14		21,190.79
Employer's Liability and Third Party Insurance	2.0%			570.06		423.82
Total cost per annum				**29,073.20**		**21,614.60**
Total cost per hour				**16.90**		**12.57**
Total cost per week				**659.26**		**490.13**

NOTES:

1. Absence due to sickness has been assumed to be for periods not exceeding 3 days for which no payment is due.
2. EasyBuild Stakeholder Pension effective from 1 July 2002. Death and accident benefit cover is provided free of charge. The minimum employer contribution shall be £3.00 per week. Where the operative contributes between £3.00 and £13.50 per week the employer shall increase the minimum contribution to match that of the operative up to a maximum of £13.50 per week.
3. All N.I. Payments are at not-contracted out rates applicable from April 2019. National Insurance is paid for 52 weeks
4. Public holiday pay includes for 8 normal public holidays

NOTES: HOURS WORKED CALCULATIONS			
Actual hours worked			
Normal hours worked/week	39.0 hrs		
Normal weeks worked/year	52.0 weeks		**2028 hrs**
Less Holidays			
Annual holidays	−4.4 weeks	−171.6 hrs	
Public holidays	−1.6 weeks	−62.4 hrs	
		−234.0 hrs	**1794 hrs**
Less Sick			
1 week	−1.0 weeks	−39.0 hrs	
		−39.0 hrs	
		Balance	**1755.0 hrs**
Less time lost for inclement weather, say	−2%		−35.1 hrs
			1719.9 hrs
Actual hours worked per annum		**1719.9 hrs**	
Actual weeks worked per annum		**44.1 weeks**	

DAYWORK RATES

Example calculation of Price Cost of Labour in Daywork

Effective from 24 June 2019

		Craft operative		General operative	
		Rate	Annual cost	Rate	Annual cost
		£	£	£	£
Basic Wages	52 wks	495.18	25,749.52	371.67	19,326.84
CITB levy (0.5% of payroll)	0.5%		128.75		96.63
Pension & Welfare Benefit	52 wks	12.86	668.72	12.86	668.72
Employer's National Insurance contribution					
(National Insurance 13.8% above threshold of £166 per week)	46.2 wks		2,495.08		1,608.75
Annual labour cost			**29,042.07**		**21,700.95**
Hourly base rates as defined			**16.19**		**12.10**

This is the prime cost of employment per person, which the employer has to meet even if there is no work for the employee to do.

NOTES:

1. Calculated following Definition of Prime Cost of Daywork carried out under a Building Contract, published by the Royal Institution of Chartered Surveyors and the Construction Confederation.
2. Standard basic rates effective from June 2019.
3. Standard working hours per annum calculated as follows:

52 weeks @ 39 hours	=	2028
Less		
4.2 weeks holiday @ 39 hours	=	171.6
8 days public holidays @ 7.8 hours	=	62.4
		234
		1794.0

4. All labour costs incurred by the contractor in his capacity as an employer, other than those contained in the hourly base rate, are to be taken into account under Section 6.
5. The above example is for guidance only and does not form part of the Definition; all the basic costs are subject to re-examination according to the time when and in the area where the daywork is executed.
6. N.I. payments are at not-contracted out rates applicable from April 2019.
7. Basic rate and GMB number of weeks = 52 weeks – 4.4 weeks annual holiday – 1.6 weeks public holiday = 46 weeks.

Approximate Estimating Rates

Prices in this section are based upon the Prices for Measured Works, but allow for incidentals which would normally be measured separately in a Bill of Quantities. They do not include for Preliminaries which are priced elsewhere in this book.

Items shown as subcontract or specialist rates would normally include the specialist's overhead and profit. All other items which could fall within the scope of works of general landscape and external works contractors do now include profit in this edition.

Based on current commercial rates, profits of 10% to 15% have been added to these rates to indicate the likely 'with profit' values of the tasks below. The variation quoted above is dependent on the sector in which the works are taking place – domestic, public or commercial. Users should adjust these rates for their sector or location.

Labour Type	Unit	Cost incl Overhead	Mark-up %	Rate incl. Overhead & Profit
General contracting	/man-hour	£28.00	15%	£32.20
Brickwork 1+1 man gang	/team-hour	£64.00	15%	£73.60
Brickwork 1+2 man gang	/team-hour	£92.00	15%	£105.80
Maintenance contracting	/man-hour	£23.00	15%	£26.45

Other resource categories

Materials generally	15%
Plant	15%
Subcontractors generally	10%

Some high cost material or subcontract items may be at lower rates of 5%

Estimating and Tendering for Construction Work, 5th edition

Martin Brook

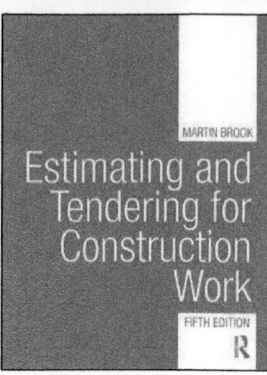

Estimators need to understand the consequences of entering into a contract, often defined by complex conditions and documents, as well as to appreciate the technical requirements of the project. Estimating and Tendering for Construction Work, 5th edition, explains the job of the estimator through every stage, from early cost studies to the creation of budgets for successful tenders.

This new edition reflects recent developments in the field and covers:

- new tendering and procurement methods
- the move from basic estimating to cost-planning and the greater emphasis placed on partnering and collaborative working
- the New Rules of Measurement (NRM1 and 2), and examines ways in which practicing estimators are implementing the guidance
- emerging technologies such as BIM (Building Information Modelling) and estimating systems which can interact with 3D design models

With the majority of projects procured using design-and-build contracts, this edition explains the contractor's role in setting costs, and design statements, to inform and control the development of a project's design.

Clearly-written and illustrated with examples, notes and technical documentation, this book is ideal for students on construction-related courses at HNC/HND and Degree levels. It is also an important source for associated professions and estimators at the outset of their careers.

July 2016; 246 × 189 mm, 334 pp
Pbk: 978-1-138-83806-2; £36.99

To Order: Tel: +44 (0) 1235 400524 Fax: +44 (0) 1235 400525
or Post: Taylor and Francis Customer Services,
Bookpoint Ltd, Unit T1, 200 Milton Park, Abingdon, Oxon, OX14 4TA UK
Email: book.orders@tandf.co.uk

For a complete listing of all our titles visit:
www.tandf.co.uk

Taylor & Francis
Taylor & Francis Group

NEW ITEMS FOR THIS EDITION

Item – Overhead and Profit Included	Unit	Range £
PRELIMINARY COST MODELS		
Preliminaries for a project value £100,000.00: Project length 6–8 weeks; Contract let as main contract; If the works are let as a subcontract some of the items listed below will be provided by the main contractor		
Management preliminaries		
Health and safety		
Health and safety file	nr	690.00
H&S Consultant visit – monthly	nr	250.00
Accommodations		
Office; welfare; drying room; storage and toilet including collection and delivery costs	nr	1075.00
Site meetings		
Attendance by Contracts manager; monthly	nr	370.00
Services		
Temporary water supply and electrical connections	nr	770.00
Other sundry site costs		
Signage	nr	330.00
Site setup costs and demobilization costs	nr	770.00
Drawing duplication and printing costs	nr	230.00
Plant preliminaries		
Hired in plant		
Deliveries and collections	nr	550.00
Safety fencing		
Heras fencing inclusive of delivery and collection; Site dimensions 30 m × 30 m	nr	1825.00
Fuel bowser	nr	480.00
Site skips	nr	810.00
Allowances for site plant standing time	nr	2175.00
Labour preliminaries		
Labour costs not task associated		
Unloading time	nr	1025.00
Site cleaning costs	nr	580.00
Non productive time	nr	1125.00
Health and safety inductions	nr	385.00
Material preliminaries		
Costs of materials that are not task associated		
Surface protection; boards or mats for longer term contracts	nr	450.00
General delivery allowances on materials not allowed elsewhere	nr	440.00
Performance bond; 10%; Based on Job value of £100,000.00	nr	830.00
Tree root zone protection	nr	1450.00
Samples	nr	385.00
Testing; Soil tests CBR and the like	nr	440.00
TOTAL FOR ALL THE ABOVE	nr	17500.00

NEW ITEMS FOR THIS EDITION

Item – Overhead and Profit Included	Unit	Range £
PRELIMINARY COST MODELS – cont		
Preliminaries for a project value £300,000.00: Project length 26–30 weeks; Contract let as main contract; If the works are let as a subcontract some of the items listed below will be provided by the main contractor		
Management preliminaries		
Health and safety		
Health and safety file	nr	690.00
H&S Consultant visit – monthly	nr	3000.00
Accommodations		
Office; welfare; Drying room; storage and toilet including collection and delivery costs	nr	3000.00
Site meetings		
Attendance by Contracts manager; monthly	nr	1275.00
Services		
Temporary water supply and electrical connections	nr	770.00
Internet connection	nr	660.00
Other sundry site costs		
Signage	nr	330.00
Site setup costs and demobilization costs	nr	770.00
Security guard; nights and weekends	nr	41000.00
Road cleaning machine hire fortnightly; during soft landscape portion of the works	nr	1325.00
Drawing duplication and printing costs	nr	320.00
Plant preliminaries		
Hired in plant		
Deliveries and collections	nr	2200.00
Heras fencing inclusive of delivery and collection; Site dimensions 30 m × 30 m	nr	8800.00
Fuel bowser	nr	2075.00
Site skips	nr	2700.00
Allowances for site plant standing time	nr	3650.00
Labour preliminaries		
Labour costs not task associated		
Unloading time	nr	2600.00
Site cleaning costs	nr	3850.00
Non productive time	nr	3850.00
Health and safety inductions	nr	770.00
Material preliminaries		
Costs of materials that are not task associated		
Surface protection; boards or mats for longer term contracts	nr	750.00
General delivery allowances on materials not allowed elsewhere	nr	1750.00
Performance bond; 10%; Based on Job value of £300,000.00	nr	3300.00
Tree root zone protection	nr	1450.00
Samples	nr	880.00
Testing; Soil tests CBR and the like	nr	1750.00
TOTAL FOR ALL THE ABOVE	nr	51000.00

NEW ITEMS FOR THIS EDITION

Item – Overhead and Profit Included	Unit	Range £
Preliminaries for a project value £300,000.00: Project length 26–30 weeks; Contract let as subcontract		
Management preliminaries		
Health and safety		
Health and safety file	nr	690.00
H&S Consultant visit – monthly	nr	3000.00
Site meetings		
Attendance by Contracts manager; monthly	nr	1275.00
Other sundry site costs		
Road cleaning machine hire fortnightly; during soft landscape portion of the works	nr	1325.00
Drawing duplication and printing costs	nr	320.00
Plant preliminaries		
Hired in plant		
Deliveries and collections	nr	2200.00
Heras fencing inclusive of delivery and collection; Site dimensions 30 m × 30 m	nr	8800.00
Fuel bowser	nr	2075.00
Site skips	nr	2700.00
Allowances for site plant standing time	nr	3650.00
Labour preliminaries		
Labour costs not task associated		
Unloading time	nr	2600.00
Site cleaning costs	nr	3850.00
Non productive time	nr	3850.00
Health and safety inductions	nr	770.00
Material preliminaries		
Costs of materials that are not task associated		
Surface protection; boards or mats for longer term contracts	nr	750.00
General delivery allowances on materials not allowed elsewhere	nr	1750.00
Performance bond; 10%; Based on Job value of £300,000.00	nr	3300.00
Tree root zone protection	nr	1450.00
Samples	nr	880.00
Testing; Soil tests CBR and the like	nr	1750.00
TOTAL FOR ALL THE ABOVE	nr	46000.00

NEW ITEMS FOR THIS EDITION

Item – Overhead and Profit Included	Unit	Range £
Blue roof		
Blue roof systems; ZinCo Green roof systems Ltd; (Stormwater management roof); For managment of stormwater under a green roof system; build-up of Blue roof system over roof surface with water proofing (not included) and root resistant protection		
Extensive Blue roof		
200 mm thick	m^2	54.00
300 mm thick	m^2	115.00
Intensive Blue roof		
200 mm thick	m^2	67.00
200 mm thick	m^2	130.00
Flow control unit and inspection chamber to the systems above		
RDS 28	m^2	400.00
Reconstituted stone Balustrade		
Bottle balustrades reconstituted stone; Haddonstone Ltd; to prepared brickwork walls or bases (not included)		
Top and bottom rails to support balustrades; laid to builders work of brick block or other stonework supports (not included); under coping; pier shafts; Piers 798 mm high		
Overall 1519 mm high; Balusters 457 mm high	m	680.00
Overall 1672 mm high; Balusters 610 mm high	m	750.00
Top and bottom rails to support balustrades; laid to builders work of brick block or other stonework supports (not included); under coping; pier shafts; Piers 898 mm high		
Overall 1619 mm high; Balusters 457 mm high	m	700.00
Overall 1712 mm high; Balusters 610 mm high	m	770.00
Composite decking		
Composite decking – Millboard; London Stone Ltd		
Decking type on Composite substructure; laid to blinded leveled ground; Average area 25–50 m^2		
Enhanced grain range	m^2	210.00
Weathered Oak range	m^2	225.00
Decking type on Composite substructure; average area 25–50 m^2; On support pedestals 28 mm high		
Enhanced grain range	m^2	220.00
Weathered Oak range	m^2	235.00
Decking type on Composite substructure; average area 25–50 m^2; On support pedestals 175–285 mm high		
Enhanced grain range	m^2	240.00
Weathered Oak range	m^2	255.00
Decking type on Composite substructure; average area 25–50 m^2; On support pedestals 285–400 mm high		
Enhanced grain range	m^2	260.00
Weathered Oak range	m^2	275.00

NEW ITEMS FOR THIS EDITION

Item – Overhead and Profit Included	Unit	Range £
RECYCLED PLASTIC RETAINING WALLS		
Hahn Recycled plastic retaining wall systems		
Excavate trench 300 mm wide to one third of the finished height of the retaining walls below; lay 100 mm hardcore; fix recycled retaining system; concrete; 1:3:6; remove excavated material from site; fix geofabric to rear of units; backfill with previously excavated material set aside in position; all works by machine		
Ogee C Section interlocking units; 100 mm dia.		
300 mm high	m	165.00
600 mm high	m	210.00
1.20 m high	m	320.00
Round palisades; 100 mm dia.; solid profile; retained height		
1.20 m high	m	280.00
1.60 m high (constructed from 2.40 m lengths)	m	455.00
FLAG PAVINGS		
Marshalls Flag pavings		
Supply and lay precast concrete flags; excavate ground and reduce levels; treat substrate with total herbicide; supply and lay granular fill Type 1 150 mm thick laid to falls and compacted		
Marshalls Urbex paving		
450 × 450 × 35 mm	m²	84.00
600 × 600 × 35 mm	m²	76.00
600 × 600 × 63 mm	m²	120.00
Marshalls Celestia paving		
450 × 450 × 50 mm	m²	120.00
600 × 600 × 50 mm	m²	110.00
600 × 200 × 50 mm	m²	130.00
Marshalls Modal paving		
450 × 450 × 50 mm	m²	97.00
600–900 × 600 × 50 mm	m²	97.00
Random lengths × 200/300/450/600/900/ × 50 mm	m²	110.00

Approximate Estimates

COST MODELS

Item – Overhead and Profit Included	Unit	Range £
HOUSING AND HOUSE BUILDER		
Cost Models for areas as described below. These models below exclude items such as demolitions, bulk groundworks, drainage unless specifically mentioned as included; Site or project preliminaries of 7.5%–12% may also be added to the rates; These items below are based on areas of 1000 m²		
Housing cost models meeting minimum planning requirements or to maximize sales potential at economic cost		
Soft landscaped space only space; planting: turf ratios as shown; planting sizes and densities as indicated; 1 tree of 18–20 cm girth at 1 tree per 100 m² overall external area; Mulch and watering after planting		
Landscaped area; Turf and trees only; imported topsoil to turf areas at 50 thick	100 m²	1675.00
Landscaped area 70: 30 turf to planted area with imported topsoil to turf areas at 50 thick	100 m²	2600.00
Landscaped area 50: 50 turf to planted area; imported topsoil to turf areas at 50 thick	100 m²	2900.00
Landscaped area; Planting and trees only; Planting of shrubs 100% at 400 mm centres	100 m²	3600.00
Landscaped area; Planting and trees only; Planting of shrubs and groundcovers 100% Shrubs: Groundcovers 70:30 at 400:250 mm centres	100 m²	4150.00
Landscaped area; Planting and trees only; Planting of shrubs and groundcovers 100% Shrubs: Groundcovers 30:70 at 400:250 mm centres	100 m²	4750.00
Mixed Hard and Soft landscaped Space with parking; with 1 tree of 18–20 cm every 100 m²; imported topsoil to turf areas at 50 thick		
Landscaped area where 50% of the external area is hard surface of macadam parking area; with surround of PC kerbs and 50% soft landscape of 70% turf and 30% planted area at 50 thick	100 m²	5600.00
Landscaped area where 50% of the external area is hard surface of block paved parking area; surround of PC kerbs 50% soft landscape of 70% turf and 30% planted area	100 m²	6800.00
External area where the area is 50% parking and 50% soft landscape. The soft landscape is 50% turf and 50% planting; planting at 400 mm ccs – 2–3 litre plants per m²; Trees at 1 nr 16–18 cm girth per 100 m² throughout the parking and landscape area		
macadam parking area	100 m²	5900.00
concrete block paved parking area	100 m²	7400.00
External area where the area is 50% parking and 50% soft landscape. The soft landscape comprises 100% planting; planting at 400 mm ccs – 2–3 litre plants per m²; Trees at 1 nr 16–18 cm girth per 100 m² throughout the parking and landscape area		
macadam parking area	100 m²	3700.00
concrete block paved parking area	100 m²	5000.00
External area where the area is 50% parking and 50% soft landscape. The soft landscape comprises 100% planting; planting at 400 mm ccs – 2–3 litre plants per m²; Trees at 1 nr 16–18 cm girth per 100 m² throughout the parking and landscape area; imported topsoil to planting areas 400 mm thick inclusive of removal of existing material and disposal off site		
macadam parking area	100 m²	5500.00
concrete block paved parking area	100 m²	6800.00

COST MODELS

Item – Overhead and Profit Included	Unit	Range £
Green infrastructure to new housing development; Comprising public greenspace with footpaths, play and amenity areas, mitigation of new internal spine roads; Landscaping to individual properties not included; Prices shown relate to the total site area Rural locations or not in major metropolitan areas; average property size 350 m²; Landscape impact to suit development for first time buyers; amenity areas catering for families with young children		
Rural location with 100 properties; public space, green buffers and green mitigation comprises 25% of the total area; Total site area 58000 m²; amenity space buffers and accesses 23000 m²; Green infrastructure space – 14600 m²; 800 m² of fenced equipped play area	100 m²	340.00
Economic housing – first time buyers; amenity catering for young children; Total site area 13500 m²; public space and accesses; 5100 m²; Green space including footways – 4,500 m²; fenced and gated green area with benches and litterbins with LEAP compliant play equipment	100 m²	530.00
Economic housing – first time buyers; amenity catering for young children; Total site area 13500 m²; public space and accesses; 5100 m²; Green space including footways – 4500 m²; fenced and gated green area with benches and litterbins with no play equipment but fenced, seeded play area	100 m²	330.00
Green infrastructure and access; fenced isolated site of 25000 m²; comprising 35 properties; prices shown include for creation of roadways 5.5 m wide with 500 mm roadway build-up inclusive of kerbs and soft verge 1 side; Allowance for development is 16250 m²; Roadways 2300 m²; footways 2 m wide 415 m (430 m²); All costs to sales related development areas excluded; services excluded No allowance for community shared space; Seeded footway 2.00 m wide one side only; Trees 14–16 at 1 tree per 15 lm		
greenfield site	1000 m²	12000.00
brownfield site	1000 m²	13000.00
Public open space internally comprising 5% of the site; with graded and levelled bow fenced and gated play area for informal ball games; grass cutting for 3 years at 22 times per year		
greenfield site	1000 m²	14000.00
brownfield site	1000 m²	12000.00
Green infrastructure for housing development; Existing greenfield site; Enhancement of existing woodland boundary; Structural enhancement of existing native species boundary to one side; New native tree and shrub boundary enhancement based on tree consultant report; Native species hedge regeneration and gapping; New accent tree planting along main vehicle access path; New occasional tree planting in public open spaces; creation of enhanced recreation open spaces with 5 year development size planting; Planting and turf/seeded/wildflower areas to open space and arrival areas; Unit frontages comprising ornamental hedges and ornamental planting; Wildflower open areas; seeded open space areas; Play areas NEAPs and LEAPS. Private internal gardens excluded. Plan area assumed as square		

COST MODELS

Item – Overhead and Profit Included	Unit	Range £
HOUSING AND HOUSE BUILDER – cont		
Arboriculture and clearances		
to one side of development 400 m	400 m	3150.00
to entire perimeter of development	1600 m	10500.00
Existing boundary tree works and planting enhancements		
Tree felling and tree removal; works as recommend by arboricultural consultant	400 m	4900.00
New native planting along boundary		
Structural whips and shrubs to reinforce existing wooded boundary at 1 whip/100 m^2 and 25% planting reinfocement of native shrubs; all planting in protective tubes; total area shown – only 25% planted	4000 m^2	11000.00
New hedge planting		
native hedge reinforcement within development along boundaries; fill gaps in existing hedges		
Turfing Seeding and Wildflower to public open space		
Open greenspace areas 95% of green infrastructure; cultivate and prepare for seeding and surfing to public open space areas in proportions of Seeded: Turf: Wildflower −70:10:20 (7.6 ha)	76000 m^2	465000.00
Ornamental Trees within development; allowances for varying sizes of structural or ornamental trees based on placement along drive routes; and within proximities of dwelling units		
Trees of 14–16 cm girth inclusive of timber stakes and protection with watering tubes and backfilling with composts and fertilizers	75 trees	19000.00
Trees of 16–18 cm girth inclusive of timber stakes and protection with watering tubes and backfilling with composts and fertilizers	75 trees	26000.00
Ornamental planting and hedges at unit frontages, parking areas development entrances, and amenity areas		
SCIENCE AND OFFICE PARKS		
Business or Science park models.		
Generally areas of more dense landscape incorporating relaxation areas and larger planting schemes; planting is generally more mature at implementation with larger trees; Planting schemes often incorporate larger swathes of groundcovers to give graded views over landscaped areas; planting density and sizes are generally of higher standard than housing		
Soft landscaped space only space; planting: turf ratios as shown; planting sizes and densities as indicated; 1 tree of 25–30 cm girth at 1 tree per 100 m^2 overall external area; Mulch and watering after planting		
Landscaped area; Turf and trees only; imported topsoil to turf areas at 50 thick	100 m^2	1975.00
Landscaped area 70:30 turf to planted area with imported topsoil to turf areas at 50 thick	100 m^2	2900.00
Landscaped area 50:50 turf to planted area; imported topsoil to turf areas at 50 thick	100 m^2	3200.00
Landscaped area; Planting and trees only; Planting of shrubs 100% at 400 mm centres	100 m^2	3900.00
Landscaped area; Planting and trees only; Planting of shrubs and groundcovers 100% Shrubs: Groundcovers 70:30 at 400:250 mm centres	100 m^2	4450.00
Landscaped area; Planting and trees only; Planting of shrubs and groundcovers 100% Shrubs: Groundcovers 30:70 at 400:250 mm centres	100 m^2	5100.00

COST MODELS

Item – Overhead and Profit Included	Unit	Range £
Mixed hard and soft landscaped space with parking; 15% of the external space are pathways or building surrounds paved with good quality textured concrete pavings. The remaining area of 85% of the external area is divided into 50:50 Soft landscape: Parking; 1 tree 25–30 cm per 100 m^2 of external area; Parking area treated as described		
External area where 50% of the remaining external area is parking area; with surround of PC kerbs and 50% soft landscape of 70% turf and 30% planted area; imported topsoil to turf areas at 50 thick;		
macadam parking area	100 m^2	8400.00
concrete block paved parking area	100 m^2	9400.00
The soft landscape is 50% turf and 50% planting; planting at 400 mm ccs – 2–3 litre plants per m^2;		
macadam parking area	100 m^2	8700.00
concrete block paved parking area	100 m^2	9900.00
The soft landscape comprises 100% planting; planting at 400 mm ccs – 2–3 litre plants per m^2;		
macadam parking area	100 m^2	8200.00
concrete block paved parking area	100 m^2	9400.00
External area where the area is 50% parking and 50% soft landscape. The soft landscape comprises 100% planting; planting at 400 mm ccs – 2–3 litre plants per m^2; imported topsoil to planting areas 400 mm thick inclusive of removal of existing material and disposal off site		
macadam parking area	100 m^2	9400.00
concrete block paved parking area	100 m^2	10500.00
CIVIC ENVIRONMENTS AND STREETSCAPE		
Civic and Public areas in the current climate encompass higher quality surfaces for public access. Planting is generally larger. There is a need for furniture and lighting		
Civic town centre area of paved open space; model area 500 m^2		
General open area; one 30–35 cm tree with root cell planting system every 400 m^2; Metal tree grid; One bench per 200 m^2 and one bin per 500 m^2; Granite kerb surround		
Conservation concrete slab paved PC £48.00 /m^2	100 m^2	18000.00
Granite paved	100 m^2	24500.00
Granite paved with stepped terraces. 3 steps of 10 m long each in an area of 500 m^2	100 m^2	29000.00
Diamond sawn Yorkstone paved	100 m^2	30000.00
Granite paved with stepped terraces. 3 steps of 10 m long each in an area of 500 m^2	100 m^2	34000.00
General open area; one 30–35 cm tree with root cell planting system every 200 m^2; Metal tree grid; One bench per 200 m^2 and one bin per 500 m^2; Granite kerb surround		
Granite paved	100 m^2	27000.00
Granite paved with stepped terraces. 3 steps of 10 m long each in an area of 500 m^2	100 m^2	31000.00
Diamond sawn Yorkstone paved	100 m^2	32500.00
Granite paved with stepped terraces. 3 steps of 10 m long each in an area of 500 m^2	100 m^2	36500.00

COST MODELS

Item – Overhead and Profit Included	Unit	Range £
CIVIC ENVIRONMENTS AND STREETSCAPE – cont		
Hard landscaped civic streetscape space of 1000 m² plan area with paved amenity areas comprising kerb surrounded paved area with street furniture. Raised beds filled with planting; trees of 30–35 cm at 100 m² centres		
Raised seating walls surrounding planted area with bollard lights at 100 m² centres; planting of low box hedges and shrubs comprising 10% of the total area		
Reconstituted stone slab paved	100 m²	19000.00
Granite paved	100 m²	28500.00
Yorkstone paved	100 m²	33000.00
Sunken area with long bullnosed steps with bollard and in-ground lighting; Planting area comprising 15% of the available area with box hedging and annual bedding plants and associated maintenance for 1 year		
Reconstituted stone slab paved	100 m²	26500.00
Yorkstone paved	100 m²	35000.00
Hard landscaped civic streetscape space of 1000 m² plan area with paved amenity areas comprising kerb surrounded paved area with street furniture. Raised water feature of 25 m²; lighting bollards; Trees of 30–35 cm at 10 m centres to the perimeter		
Sunken area with long bullnosed steps with bollard lighting; Planting area comprising 15% of the available area with box hedging and annual bedding plants and associated maintenance for 1 year		
Reconstituted stone slab paved	100 m²	28500.00
Yorkstone paved	100 m²	36500.00
PLAYGROUNDS		
New Lapset Play area 1200 m²; Sub to one of above; Play area comprising play equipment in landscape and fencepark; Picket fencing with 2 × swing gates, macadam pathway, and planting to 10% of the park area on the perimeter; safety surfaces as required by play equipment supplier; Gates (2x) single Tree 20–25 cm; bins and benches		
Climbing and play tower Q10945, ABC play tower Q10159, Tractor rocker 010515; Pig rocker O10516; Octopus seasaw; Q10259; Froggy spring equipment; Swing with cradle seats		
Total area 1200 m²	nr	77000.00
equivalent budget cost	m²	64.00
Urban play area Lappset Ltd; comprising fenced play equipment for children to 12 years; Cushioned fall play equipment as specified by equipment manufacturer. Bow fenced all sides with benches and litter bins; appropriate signage		
Traditional equipment swings slides and spring equipment; 6 activities; Safety surfacing below and surrounding equipment; 3 benches and litter bin; turfed where not surfaced		
Local parish traditional equipment	1200 m²	52000.00
City and urban medium sized play equipment; Modern design climbing, zipwire net climbing and medium sized theme multiplay units Lapset	1200 m²	90000.00
City and urban but with larger multiplay themed play equipment	1500 m²	140000.00
Mixed themed and traditional play area with fitness elements and adventure climbing equipment	3000 m²	165000.00

COST MODELS

Item – Overhead and Profit Included	Unit	Range £
LEAP Play areas for compliance with housing development; 35–100 units; Lappset Ltd Local Equipped Area for Play (LEAP) is a play area equipped for children of early school age (mainly 4–8 year olds). The activity zone should have a minimum area of 400 m², with grass playing space and at least five types of play equipment with appropriate safety surfacing. There should also be seating for accompanying adults. Contains at least 5 types of play equipment; Contains seating for parents and/or carers; Contains a litter bin signage fencing;		
minimum requirement	nr	24000.00
minimum requirement but with green seeded kickabout area	nr	24000.00
minimum requirement brownfield site	nr	28000.00
NEAP Play area for compliance for housing developers; Lappset Ltd A Neighbourhood Equipped Area for Play (NEAP) is a play area for 8–14 year olds which should include a grassed kickabout area, a hard surfaced area for ball games or wheeled activities, 8 types of play equipment appropriate to children in this age group and seating, including a youth shelter. This requires an activity zone of at least 1000 m²		
minimum requirement	nr	135000.00
minimum requirement brownfield site	nr	150000.00
DOMESTIC COST MODELS		
Suburban new or refurbished landscape comprising; Driveways and parking; Recreational pavings; Access pavings; Perimeter fencing; Trees; Planting beds and turf; Measured at per 1000 m² of total site area Total property size; 3300 m² of which building footprint comprises 280 m² (8.5%) Access and parking 330 m² (10%); Closeboard fencing to perimeter; Balance of landscape (2700 m² +/- 80%) area as follows		
Hard recreation and access areas/soft landscaped areas; standard size planting	1000 m²	110000.00
Hard recreation and access areas/soft landscaped areas; with enhanced planting and larger trees	1000 m²	135000.00
Hard recreation and access areas/soft landscaped areas; with enhanced planting and larger trees but with brick entrance walls, gates and piers	1000 m²	140000.00
Hard recreation and access areas/soft landscaped areas; with enhanced planting and larger trees but with brick entrance walls, gates and piers and with a lighting scheme	1000 m²	150000.00
Hard recreation and access areas/soft landscaped areas; with enhanced planting and larger trees and with brick entrance walls, gates and piers and with a lighting scheme; but with stone paved steps to the patio 5 steps high × 2.00 m wide	1000 m²	155000.00
Hard recreation and access areas/soft landscaped areas; with enhanced planting and larger trees and with brick entrance walls, gates and piers and with a lighting scheme and with stone paved steps to the patio 5 steps high × 2.00 m wide but with timber pergolas and water feature	1000 m²	160000.00

Approximate Estimates

COST MODELS

Item – Overhead and Profit Included	Unit	Range £
PARKING AREAS		
Parking area to office block; Surface of car park stated below inclusive of all excavations, bases and surface		
Macadam 625 m^2; 200 Type 1 base		
Open space with kerb surround 25 × 25 m	nr	44000.00
With trees 18–20 cm girth at 10.00 m centres	nr	48000.00
With trees 18–20 cm girth at 10.00 m centres and planting to 1.00 m boundary perimeter; (external to the parking area on 3 sides – 75 m^2)	nr	51000.00
With trees 18–20 cm at 10 m centres; planting to 1.00 m boundary and Marshalls Bega asymmetrical pole lighting 6.00 high	nr	62000.00
Concrete block surface of Marshalls Priora on 200 mm Type 3 base; 625 m^2		
Open space with kerb surround 25 × 25 m	nr	61000.00
With trees 18–20 cm girth at 10.00 m centres;	nr	65000.00
With trees 18–20 cm girth at 10.00 m centres and planting to 1.00 m boundary perimeter; (external to the parking area on 3 sides – 75 m^2)	nr	68000.00
With trees 18–20 cm at 10 m centres; planting to 1.00 m boundary and Marshalls Bega asymmetrical pole lighting 6.00 high	nr	79000.00
Clay brick; Wienerberger Dutch pavers laid flat; butt jointed on 200 mm Type 1 base 625 m^2		
Open space with kerb surround 25 × 25 m	nr	80000.00
With trees 18–20 cm girth at 10.00 m centres	nr	83000.00
With trees 18–20 cm girth at 10.00 m centres and planting to 1.00 m boundary perimeter; (external to the parking area on 3 sides – 75 m^2)	nr	87000.00
With trees 18–20 cm at 10 m centres; planting to 1.00 m boundary and Marshalls Bega asymmetrical pole lighting 6.00 high	nr	98000.00
Recycled Suds; SudsTech; on 200 mm Type 3 base 625 m^2		
Open space with kerb surround 25 × 25 m	nr	82000.00
With trees 18–20 cm girth at 10.00 m centres	nr	87000.00
With trees 18–20 cm girth at 10.00 m centres and planting to 1.00 m boundary perimeter; (external to the parking area on 3 sides – 75 m^2)	nr	90000.00
With trees 18–20 cm at 10 m centres; planting to 1.00 m boundary and Marshalls Bega asymmetrical pole lighting 6.00 high	nr	100000.00

PRELIMINARIES

Item – Overhead and Profit Included	Unit	Range £
PRELIMINARY COST MODELS		
Preliminaries for a project value £100,000.00: Project length 6–8 weeks; Contract let as main contract; If the works are let as a subcontract some of the items listed below will be provided by the main contractor		
Management preliminaries		
Health and safety		
Health and safety file	nr	690.00
H&S Consultant visit – monthly	nr	250.00
Accommodations		
Office; welfare; Drying room; storage and toilet including collection and delivery costs	nr	1075.00
Site meetings		
Attendance by Contracts manager; monthly	nr	370.00
Services		
Temporary water supply and electrical connections	nr	770.00
Other sundry site costs		
Signage	nr	330.00
Site setup costs and demobilization costs	nr	770.00
Drawing duplication and printing costs	nr	230.00
Plant preliminaries		
Hired in plant		
Deliveries and collections	nr	550.00
Safety fencing		
Heras fencing inclusive of delivery and collection; Site dimensions 30 m × 30 m	nr	1825.00
Fuel bowser	nr	480.00
Site skips	nr	810.00
Allowances for site plant standing time	nr	2175.00
Labour preliminaries		
Labour costs not task associated		
Unloading time	nr	1025.00
Site cleaning costs	nr	580.00
Non productive time	nr	1125.00
Health and safety inductions	nr	385.00
Material preliminaries		
Costs of materials that are not task associated		
Surface protection; boards or mats for longer term contracts	nr	450.00
General delivery allowances on materials not allowed elsewhere	nr	440.00
Performance bond; 10%; Based on Job value of £100,000.00	nr	830.00
Tree root zone protection	nr	1450.00
Samples	nr	385.00
Testing; Soil tests CBR and the like	nr	440.00
TOTAL FOR ALL THE ABOVE	nr	17500.00

Approximate Estimates

PRELIMINARIES

Item – Overhead and Profit Included	Unit	Range £
PRELIMINARY COST MODELS – cont		
Preliminaries for a project value £300,000.00: Project length 26–30 weeks; Contract let as main contract; If the works are let as a subcontract some of the items listed below will be provided by the main contractor		
Management preliminaries		
Health and safety		
Health and safety file	nr	690.00
H&S Consultant visit – monthly	nr	3000.00
Accommodations		
Office; welfare; Drying room; storage and toilet including collection and delivery costs	nr	3000.00
Site meetings		
Attendance by Contracts manager; monthly	nr	1275.00
Services		
Temporary water supply and electrical connections	nr	770.00
Internet connection	nr	660.00
Other sundry site costs		
Signage	nr	330.00
Site setup costs and demobilization costs	nr	770.00
Security guard; nights and weekends	nr	41000.00
Road cleaning machine hire fortnightly; during soft landscape portion of the works	nr	1325.00
Drawing duplication and printing costs	nr	320.00
Plant preliminaries		
Hired in plant		
Deliveries and collections	nr	2200.00
Heras fencing inclusive of delivery and collection; Site dimensions 30 m × 30 m	nr	8800.00
Fuel bowser	nr	2075.00
Site skips	nr	2700.00
Allowances for site plant standing time	nr	3650.00
Labour preliminaries		
Labour costs not task associated		
Unloading time	nr	2600.00
Site cleaning costs	nr	3850.00
Non productive time	nr	3850.00
Health and safety inductions	nr	770.00
Material preliminaries		
Costs of materials that are not task associated		
Surface protection; boards or mats for longer term contracts	nr	750.00
General delivery allowances on materials not allowed elsewhere	nr	1750.00
Performance bond; 10%; Based on Job value of £300,000.00	nr	3300.00
Tree root zone protection	nr	1450.00
Samples	nr	880.00
Testing; Soil tests CBR and the like	nr	1750.00
TOTAL FOR ALL THE ABOVE	nr	51000.00

PRELIMINARIES

Item – Overhead and Profit Included	Unit	Range £
Preliminaries for a project value £300,000.00: Project length 26–30 weeks; Contract let as subcontract;		
Management preliminaries		
Health and safety		
Health and safety file	nr	690.00
H&S Consultant visit – monthly	nr	3000.00
Site meetings		
Attendance by Contracts manager; monthly	nr	1275.00
Other sundry site costs		
Road cleaning machine hire fortnightly; during soft landscape portion of the works	nr	1325.00
Drawing duplication and printing costs	nr	320.00
Plant preliminaries		
Hired in plant		
Deliveries and collections	nr	2200.00
Heras fencing inclusive of delivery and collection; Site dimensions 30 m × 30 m	nr	8800.00
Fuel bowser	nr	2075.00
Site skips	nr	2700.00
Allowances for site plant standing time	nr	3650.00
Labour preliminaries		
Labour costs not task associated		
Unloading time	nr	2600.00
Site cleaning costs	nr	3850.00
Non productive time	nr	3850.00
Health and safety inductions	nr	770.00
Material preliminaries		
Costs of materials that are not task associated		
Surface protection; boards or mats for longer term contracts	nr	750.00
General delivery allowances on materials not allowed elsewhere	nr	1750.00
Performance bond; 10%; Based on Job value of £300,000.00	nr	3300.00
Tree root zone protection	nr	1450.00
Samples	nr	880.00
Testing; Soil tests CBR and the like	nr	1750.00
TOTAL FOR ALL THE ABOVE	nr	46000.00

Approximate Estimates

PRELIMINARIES

Item – Overhead and Profit Included	Unit	Range £
CONTRACT ADMINISTRATION AND MANAGEMENT		
Prepare tender bid for external works or landscape project; measured works contract; inclusive of bid preparation, provisional programme and method statements		
Measured works bid; measured bills provided by the employer; project value		
£30,000.00	nr	650.00
£50,000.00	nr	1075.00
£100,000.00	nr	2400.00
£200,000.00	nr	3750.00
£500,000.00	nr	4850.00
£1,000,000.00	nr	7000.00
Lump sum bid; scope document, drawings and specification provided by the employer; project value		
£30,000.00	nr	1400.00
£50,000.00	nr	1825.00
£100,000.00	nr	3600.00
£200,000.00	nr	6200.00
£500,000.00	nr	8500.00
£1,000,000.00	nr	13000.00
Prepare and maintain Health and Safety file; prepare risk and COSHH assessments, method statements, works programmes before the start of works on site; project value		
£35,000.00	nr	320.00
£75,000.00	nr	480.00
£100,000.00	nr	640.00
£200,000.00–£500,000.00	nr	1600.00
Contract management		
Site management by site surveyor carrying out supervision and administration duties only; inclusive of oncosts		
full time management	week	1200.00
managing one other contract	week	600.00
Parking		
Parking expenses where vehicles do not park on the site area; per vehicle		
metropolitan area; city centre	week	200.00
metropolitan area; outer areas	week	160.00
suburban restricted parking areas	week	40.00
Congestion charging		
London only	week	58.00

PRELIMINARIES

Item – Overhead and Profit Included	Unit	Range £
SITE SETUP AND SITE ESTABLISHMENT		
Site fencing; supply and erect temporary protective fencing and remove at completion of works		
Cleft chestnut paling		
1.20 m high 75 mm larch posts at 3 m centres	100 m	1975.00
Heras fencing		
2 week hire period	100 m	980.00
4 week hire period	100 m	1250.00
8 week hire period	100 m	1775.00
12 week hire period	100 m	2325.00
16 week hire period	100 m	2850.00
24 week hire period	100 m	3900.00
Site compounds		
Establish site administration area on reduced compacted Type 1 base; 150 mm thick; allow for initial setup of site office and welfare facilities for site manager and maximum 8 site operatives		
temporary base or hard standing; 20 × 10 m	200 m^2	4250.00
To hardstanding area above; erect office and welfare accommodation; allow for furniture, telephone connection, and power; restrict access using temporary safety fencing to the perimeter of the site administration area		
Establishment; delivery and collection costs only		
site office and chemical toilet; 20 m of site fencing	nr	1025.00
site office and chemical toilet; 40 m of site fencing	nr	1125.00
site office and chemical toilet; 60 m of site fencing	nr	1225.00
Weekly hire rates of site compound equipment		
site office 2.4 × 3.6 m; chemical toilet; 20 m of site fencing	week	100.00
site office 4.8 × 2.4 m; chemical toilet; 20 m of site fencing; armoured store 2.4 × 3.6 m	week	120.00
site office 2.4 × 3.6 m; chemical toilet; 40 m of site fencing	week	130.00
site office 2.4 × 3.6 m; chemical toilet; 60 m of site fencing	week	155.00
Removal of site compound		
Remove base; fill with topsoil and reinstate to turf on completion of site works	200 m^2	2075.00
Surveying and setting out		
Note: In all instances a quotation should be obtained from a surveyor. The following cost projections are based on the costs of a surveyor using EDM (electronic distance measuring equipment) where an average of 400 different points on site are recorded in a day. The surveyed points are drawn in a CAD application and the assumption used that there is one day of drawing up for each day of on-site surveying.		

PRELIMINARIES

Item – Overhead and Profit Included	Unit	Range £
SITE SETUP AND SITE ESTABLISHMENT – cont		
Surveying and setting out – cont		
Survey existing site using laser survey equipment and produce plans of existing site conditions		
site 1000 to 10000 m^2 with sparse detail and features; boundary and level information only; average five surveying stations	nr	750.00
site up to 1000 m^2 with dense detail such as buildings, paths, walls and existing vegetation; average five survey stations	nr	1500.00
site up to 4000 m^2 with dense detail such as buildings, paths, walls and existing vegetation; average 10 survey stations	nr	2250.00

PRELIMINARIES

Item – Overhead and Profit Included	Unit	Range £
PROJECT PRELIMINARY COST MODELS		
Preliminaries for a project value £100,000.00: Project length 6–8 weeks; Contract let as main contract; If the works are let as a subcontract some of the items listed below will be provided by the main contractor		
Management preliminaries		
Health and safety		
Health and safety file	nr	690.00
H&S Consultant visit – monthly	nr	225.00
Accommodations		
Office; welfare; Drying room; storage and toilet including collection and delivery costs	nr	930.00
Site meetings		
Attendance by Contracts manager; monthly	nr	320.00
Services		
Temporary water supply and electrical connections	nr	700.00
Internet connection	nr	–
Telephone	nr	–
Other sundry site costs		
Signage	nr	300.00
Site setup costs and demobilization costs	nr	670.00
Security guard	nr	–
Road cleaning machine hire fortnightly; during soft landscape portion of the works	nr	–
Drawing duplication and printing costs	nr	200.00
Plant preliminaries		
Hired in plant		
Deliveries and collections	nr	500.00
Safety fencing		
Heras fencing inclusive of delivery and collection; Site dimensions 30 m × 30 m	nr	1575.00
Fuel bowser	nr	420.00
Site skips	nr	700.00
Allowances for site plant standing time	nr	1900.00
Site logistics	nr	–
Temporary lighting	nr	–
Labour preliminaries		
Labour costs not task associated		
Unloading time	nr	900.00
Site cleaning costs	nr	500.00
Non productive time	nr	980.00
Health and safety inductions	nr	335.00
Material preliminaries		
Costs of materials that are not task associated		
Surface protection; boards or mats for longer term contracts	nr	390.00
General delivery allowances on materials not allowed elsewhere	nr	400.00
Performance bond; 10%; Based on Job value of £100,000.00	nr	750.00

PRELIMINARIES

Item – Overhead and Profit Included	Unit	Range £
PROJECT PRELIMINARY COST MODELS – cont		
Material preliminaries – cont		
Costs of materials that are not task associated – cont		
Tree root zone protection	nr	1250.00
Protection of existing building features; boards	nr	–
Samples	nr	350.00
Testing; Soil tests CBR and the like	nr	400.00
TOTAL FOR ALL THE ABOVE	nr	15000.00

PRELIMINARIES

Item – Overhead and Profit Included	Unit	Range £
Preliminaries for a project value £300,000.00: Project length 26–30 weeks; Contract let as main contract; If the works are let as a subcontract some of the items listed below will be provided by the main contractor		
Management preliminaries		
Health and safety		
Health and safety file	nr	690.00
H&S Consultant visit – monthly	nr	2700.00
Accommodations		
Office; welfare; Drying room; storage and toilet including collection and delivery costs	nr	2600.00
Site meetings		
Attendance by Contracts manager; monthly	nr	1125.00
Services		
Temporary water supply and electrical connections	nr	700.00
Internet connection	nr	600.00
Telephone	nr	–
Other sundry site costs		
Signage	nr	300.00
Site setup costs and demobilization costs	nr	670.00
Security guard; nights and weekends	nr	41000.00
Road cleaning machine hire fortnightly; during soft landscape portion of the works	nr	1200.00
Drawing duplication and printing costs	nr	280.00
Plant preliminaries		
Hired in plant		
Deliveries and collections	nr	2000.00
Heras fencing inclusive of delivery and collection; Site dimensions 30 m × 30 m	nr	7700.00
Fuel bowser	nr	1825.00
Site skips	nr	2350.00
Allowances for site plant standing time	nr	3200.00
Site logistics	nr	–
Temporary lighting	nr	–
Labour preliminaries		
Labour costs not task associated		
Unloading time	nr	2250.00
Site cleaning costs	nr	3350.00
Non productive time	nr	3350.00
Health and safety inductions	nr	670.00
Material preliminaries		
Costs of materials that are not task associated		
Surface protection; boards or mats for longer term contracts	nr	660.00
General delivery allowances on materials not allowed elsewhere	nr	1600.00
Performance bond; 10%; Based on Job value of £300,000.00	nr	3000.00
Tree root zone protection	nr	1250.00
Protection of existing building features; boards	nr	–
Samples	nr	800.00
Testing; Soil tests CBR and the like	nr	1600.00
TOTAL FOR ALL THE ABOVE	nr	45000.00

Approximate Estimates

PRELIMINARIES

Item – Overhead and Profit Included	Unit	Range £
PROJECT PRELIMINARY COST MODELS – cont		
Preliminaries for a project value £300,000.00: Project length 26–30 weeks; Contract let as subcontract		
Management preliminaries		
Health and safety		
Health and safety file	nr	690.00
H&S Consultant visit – monthly	nr	2700.00
Site meetings		
Attendance by Contracts manager; monthly	nr	1125.00
Services		
Internet connection	nr	600.00
Telephone	nr	–
Other sundry site costs		
Road cleaning machine hire fortnightly; during soft landscape portion of the works	nr	1200.00
Drawing duplication and printing costs	nr	280.00
Plant preliminaries		
Hired in plant		
Deliveries and collections	nr	2000.00
Heras fencing inclusive of delivery and collection; Site dimensions 30 m × 30 m	nr	7700.00
Fuel bowser	nr	1825.00
Site skips	nr	2350.00
Allowances for site plant standing time	nr	3200.00
Site logistics	nr	–
Temporary lighting	nr	–
Labour preliminaries		
Labour costs not task associated		
Unloading time	nr	2250.00
Site cleaning costs	nr	3350.00
Non productive time	nr	3350.00
Health and safety inductions	nr	670.00
Material preliminaries		
Costs of materials that are not task associated		
Surface protection; boards or mats for longer term contracts	nr	660.00
General delivery allowances on materials not allowed elsewhere	nr	1600.00
Performance bond; 10%; Based on Job value of £300,000.00	nr	3000.00
Tree root zone protection	nr	1250.00
Protection of existing building features; boards	nr	–
Samples	nr	800.00
Testing; Soil tests CBR and the like	nr	1600.00
TOTAL FOR ALL THE ABOVE	nr	41000.00

2.3 GREEN ROOFS

Item – Overhead and Profit Included	Unit	Range £
GREEN ROOF NOTES		
All items below do not include for time to deliver the materials to the point of installation		
EXTENSIVE PITCH ROOF		
Zinco Green Roof Systems Ltd; suitable for planting with Sedum plugs or overlaying with pre- grown cultivated Sedum Carpet. Waterproofing layer must contain a root barrier as a separate root barrier cannot be used on pitched roofs		
System saturated weight – 115 kg/m^2; system water storage capacity – 38 l/m^2; Slopes 10–25°		
System depth 130 mm	m^2	90.00
Pitched roofs with slopes from 25° up to 35°; System height – 120 mm; system saturated weight – 155 kg/m^2; water storage capacity – 64 l/m^2 (depending on slope)		
system depth 120 mm	m^2	89.00
EXTENSIVE FLAT ROOF		
Extensive Sedum type Green Roof System; Zinco Green Roof Systems Ltd. System for planting with Sedum plugs or overlaying with pre-grown cultivated Sedum Carpet. (ETA 13/0668)		
Flat roofs with no standing water up to 10° or for Inverted roofs and as an alternative to single layer systems. Drainage, water retention and protection mat with attached filter sheet. 20 mm thick		
system depth 120 mm	m^2	220.00
Extensive Rockery type Green Roof System; For planting with Sedum plugs or rockery type plants; flat roofs with no standing water up to 10°. (Not suitable for inverted roofs); Suitable system depth 90 mm. System saturated weight 110 kg/m^2; system water storage capacity: 36 l/m^2; (ETA 13/0668)		
Suitable system depth 90 mm. System saturated weight 110 kg/m^2; system water storage capacity: 36 l/m^2;		
system depth 90 mm; planted with sedum mat	m^2	230.00
system depth 90 mm; planted with wildflower	m^2	200.00
Zero° roof		
system depth 150 mm; planted with sedum mat	m^2	230.00
system depth 150 mm; planted with wildflower	m^2	200.00

2.3 GREEN ROOFS

Item – Overhead and Profit Included	Unit	Range £
INTENSIVE GREEN ROOF		
Flat roofs up to 10°; Intensive Simple and Wildflower Green Roof System; Zinco Green Roof Systems Ltd; (not suitable for Inverted roofs); for planting with small shrubs, blooming perennials and wildflowers; can be combined with patios and walkways if required; saturated weight – 145 kg/m^2–245 kg/m^2; water storage capacity – 62 l/m^2 to 80 l/m^2. (ETA 13/0668)		
Heather and lavender planting system		
System depth – 145 mm–200 mm	m^2	185.00
Small shrubs		
System depth – 500 mm	m^2	155.00
Turf		
System depth 300 mm	m^2	90.00
LAWN Green Roof System; Zinco Green Roof Systems Ltd. Lawns on roofs where build-up depth is restricted; Flat roofs – 0° up to 5°; Lawn surface treatment; System depth 150 mm system saturated weight min 165 kg/m^2. System water storage capacity; from 65 l/m^2		
Drip irrigated lawn system (excludes controller)		
system depth 150 mm	m^2	89.00
BLUE ROOFS		
Blue roof systems; ZinCo Green roof systems Ltd; (Stormwater management roof); For managment of stormwater under a green roof system; build-up of Blue roof system over roof surface with water proofing (not included) and root resistant protection		
Extensive Blue roof		
200 mm thick	m^2	54.00
300 mm thick	m^2	115.00
Intensive Blue roof		
200 mm thick	m^2	67.00
200 mm thick	m^2	130.00
Flow control unit and inspection chamber to the systems above		
RDS 28	m^2	400.00

8.1.1 SITE CLEARANCE

Item – Overhead and Profit Included	Unit	Range £
CLEARING VEGETATION		
Clear existing scrub vegetation including shrubs and hedges and stack on site		
By machine		
light scrub and grasses	100 m²	26.00
light scrub and grasses but remove vegetation to licensed tip	100 m²	74.00
undergrowth brambles and heavy weed growth	100 m²	61.00
small shrubs	100 m²	61.00
By hand		
light scrub and grasses to spoil heaps 25 m distance	100 m²	100.00
light scrub and grasses to spoil heap; subsequently disposed off site	100 m²	110.00
light scrub and grasses to tip	100 m²	60.00
Fell trees on site; grub up roots; all by machine; remove debris to licensed tip		
trees 600 mm girth	each	485.00
trees 1.5–3.00 m girth	each	1350.00
trees over 3.00 m girth	each	2075.00
Chip existing vegetation on site; remove to stockpile 100 m		
light scrub and bramble; average height 500 mm	100 m²	32.00
dense scrub and bramble; average height 1.00 m	100 m²	64.00
mixed scrub and shrubs average 1.50 m high	100 m²	110.00
dense overgrown shrubs and bramble; average 2.00 m high	100 m²	150.00
Turf stripping for preservation		
Cut and strip existing turf area by machine; turves 50 mm thick; turf to be lifted for preservation		
load to dumper or palette by hand; transport on site not exceeding 100 m travel to new location	100 m²	120.00
all by hand; including barrowing to stacking locally 10 m distance	100 m²	750.00
Turf stripping for disposal; spray turf with glyphosate and return to site after 3 weeks; strip dead grass; load to stockpile on site; rotavate to prepare area for new surface treatments to within 50 mm of final levels		
By knapsack sprayer; excavator and dumper; pedestrian rotavator		
5 tonne excavator and 6 tonne dumper; stock pile distance 25 m	100 m²	75.00
5 tonne excavator and 3 tonne dumper; stock pile distance 50 m	100 m²	77.00
5 tonne excavator and 3 tonne dumper; stock pile distance 100 m	100 m²	82.00
By tractor drawn equipment		
spraying with glyphosate; 2 passes	ha	570.00

8.1.1 SITE CLEARANCE

Item – Overhead and Profit Included	Unit	Range £
CLEARING FENCES, GATES AND LIGHT STRUCTURES		
Clear site; demolish existing closeboard fence; clear shrubs 50% in shrub beds and 50% free-standing; spray turf and return to clear ground of sprayed vegetation; chip vegetation on site and transport to spoil heaps 50 m; works by mechanical excavator		
Rectangular site with 70% turf and 30% shrub area		
1000 m²	nr	2475.00
2000 m²	nr	4550.00
4000 m²	nr	8500.00
Rectangular site with 60% turf grass/scrub area and 40% shrubs and overgrowth		
site area; 1000 m²	nr	2850.00
site area; 2000 m²	nr	5400.00
site area; 4000 m²	nr	10000.00
Demolish fencing and gates		
Demolish fencing; break out posts, remove fencing stack and load mechanically for removal to tip		
Chain link fencing; vegetation not present on angle iron stakes driven in	100 m	260.00
Chain link fencing; overgrown with vegetation on angle iron stakes driven in	100 m	930.00
DEMOLITIONS OF EXTERNAL STRUCTURES AND SURFACES		
Demolish existing free-standing walls; grub out foundations; remove arisings to tip		
Brick wall; 112 mm thick		
300 mm high	m	23.50
500 mm high	m	36.00
Brick wall; 225 mm thick		
300 mm high	m	30.00
500 mm high	m	39.00
1.00 m high	m	62.00
1.20 m high	m	66.00
1.50 m high	m	81.00
1.80 m high	m	91.00
Demolish existing structures		
Timber sheds on concrete base 150 mm thick inclusive of disposal and backfilling with excavated material; plan area of building		
10 m²	nr	380.00
15 m²	nr	540.00
Building of cavity wall construction; insulated with tiled roof; inclusive of foundations; allow for disconnection of electrical water and other services; grub out and remove 10 m of electrical cable, water and waste pipes		
10 m²	nr	2475.00
20 m²	nr	3300.00

8.1.1 SITE CLEARANCE

Item – Overhead and Profit Included	Unit	Range £
CLEARING SURFACES		
Demolish existing surfaces		
Break up plain concrete slab; remove to licensed tip		
150 mm thick	m²	17.80
200 mm thick	m²	24.00
300 mm thick	m²	35.50
Break up plain concrete slab and preserve arisings for hardcore; break down to maximum size of 200 × 200 mm and transport to location; maximum distance 50 m		
150 mm thick	m²	10.50
200 mm thick	m²	14.00
300 mm thick	m²	21.00
Break up reinforced concrete slab and remove to licensed tip		
150 mm thick	m²	24.50
200 mm thick	m²	32.50
300 mm thick	m²	49.00
Break out existing surface and associated 150 mm thick granular base load to 20 tonne muck away vehicle for removal off site		
macadam surface; 70 mm thick	m²	17.50
macadam surface; 150 mm thick	m²	24.00
block paving; 50 mm thick	m²	16.20
block paving; 80 mm thick	m²	17.80

8.1.2 PREPARATORY GROUNDWORKS

Item – Overhead and Profit Included	Unit	Range £
TREE ROOT PROTECTION		
Protect tree roots from compaction in 'Root Protection Zone (RPZ)		
Reduce levels by hand; load to machines external to the RPZ and remove off site;		
Lay cellular protection on non woven geofabric to depth;		
75 mm	m²	24.50
100 mm	m²	29.00
150 mm	m²	37.00
Extra over to remove excavated material off site	m³	37.00
EXCAVATING GENERALLY		
Prices for excavating and reducing to levels are for work in light or medium soils; multiplying factors for other soils are as follows		
Clay	1.5	–
Compact gravel	1.2	–
Soft chalk	2.0	–
Hard rock	3.0	–
Excavating topsoil for preservation		
Remove topsoil average depth 300 mm; deposit in spoil heaps not exceeding		
100 m travel to heap; turn heap once during storage		
by machine	m³	13.10
by hand	m³	195.00
Excavate to reduce levels; remove spoil to stockpile not exceeding 100 m travel; all by machine		
Average 0.25 m deep	m³	5.90
Average 0.30 m deep	m²	3.30
Average 0.30 m deep	m³	9.95
Average 1.0 m deep; using 21 tonne 360 tracked excavator	m³	5.70
Excavate to reduce levels; remove spoil to stockpile not exceeding 100 m travel; all by hand		
Average 0.10 m deep	m²	17.30
Average 0.20 m deep	m²	34.50
Average 0.30 m deep	m²	52.00
Average 100–300 mm deep	m³	170.00
Average 1.0 m deep	m³	195.00
Extra for carting spoil to licensed tip off site (20 tonnes)	m³	51.00
Extra for carting spoil to licensed tip off site (by skip; machine loaded)	m³	62.00
Extra for carting spoil to licensed tip off site (by skip; hand loaded)	m³	155.00
Spread excavated material to levels in layers not exceeding 150 mm; using scraper blade or bucket		
Average thickness 100 mm	m²	0.60
Average thickness 100 mm but with imported topsoil	m²	4.30
Average thickness 200 mm	m²	1.25
Average thickness 200 mm but with imported topsoil	m²	8.60
Average thickness 250 mm	m²	1.60
Average thickness 200–250 mm	m³	6.35
Average thickness 250 mm but with imported topsoil	m²	10.80
Average thickness 250 mm but with imported topsoil	m³	43.00
Extra for work to banks exceeding 30° slope	30%	–

8.1.2 PREPARATORY GROUNDWORKS

Item – Overhead and Profit Included	Unit	Range £
Rip subsoil using approved subsoiling machine to a depth of 600 mm below topsoil at 600 mm centres; all tree roots and debris over 150 × 150 mm to be removed; cultivate ground where shown on drawings to depths as shown		
100 mm	100 m²	17.90
200 mm	100 m²	18.30
300 mm	100 m²	18.80
400 mm	100 m²	23.00
Form landform from imported material; mounds to falls and grades to receive turf or seeding treatments; material dumped by 20 tonne vehicle to location; volume 100 m³ spread over area; thickness varies from 150–500 mm; prices to not include for surface preparations		
Imported standard grade topsoil		
by large excavator	m³	35.00
by 5 tonne excavator	m³	35.50
by 3 tonne excavator	m³	38.50
Form landform from excavated material; mounds to falls and grades to receive turf or seeding treatments; material moved by dumper to location maximum 50 m; volume 100 m³ spread over a 200 m² area; thickness varies from 150–500 mm; prices to not include for surface preparations		
Imported recycled topsoil		
by large excavator	m³	4.20
by 5 tonne excavator	m³	5.15
by 3 tonne excavator	m³	8.15
EXCAVATION OF TRENCHES		
Excavate trenches for foundations; trenches 225 mm deeper than specified foundation thickness; pour plain concrete foundations GEN 1 10 N/mm and to thickness as described; disposal of excavated material off site; dimensions of concrete foundations		
250 mm wide × 250 mm deep	m	16.20
300 mm wide × 300 mm deep	m	22.00
400 mm wide × 600 mm deep	m	52.00
600 mm wide × 400 mm deep	m	55.00
Excavate trenches for foundations; trenches 225 mm deeper than specified foundation thickness; pour plain concrete foundations GEN 1 10 N/mm and to thickness as described; disposal of excavated material off site; dimensions of concrete foundations		
250 mm wide × 250 mm deep	m³	320.00
300 mm wide × 300 mm deep	m³	240.00
600 mm wide × 400 mm deep	m³	230.00
400 mm wide × 600 mm deep	m³	195.00

8.1.2 PREPARATORY GROUNDWORKS

Item – Overhead and Profit Included	Unit	Range £
EXCAVATING GENERALLY – cont		
Excavate trenches for services; grade bottoms of excavations to required falls; remove 20% of excavated material off site		
Trenches 300 mm wide		
600 mm deep	m	11.80
750 mm deep	m	15.10
900 mm deep	m	18.10
1.00 m deep	m	19.90
Ha-ha		
Excavate ditch 1200 mm deep × 900 mm wide at bottom battered one side 45° slope; excavate for foundation to wall and place GEN 1 concrete foundation 150 × 500 mm; construct one and a half brick wall (brick PC £500.00/1000) battered 10° from vertical; laid in cement: lime: sand (1:1:6) mortar in English garden wall bond; precast concrete coping weathered and throated set 150 mm above ground on high side; rake bottom and sides of ditch and seed with low maintenance grass at 35 g/m²		
wall 600 mm high	m	415.00
wall 900 mm high	m	530.00
wall 1200 mm high	m	670.00
Excavate ditch 1200 mm deep × 900 mm wide at bottom battered one side 45° slope; place deer or cattle fence 1.20 m high at centre of excavated trench; rake bottom and sides of ditch and seed with low maintenance grass at 35 g/m²		
fence 1200 mm high	m	31.00
EXCAVATION OF BANKS		
Excavate existing slope to vertical to form flat area; grade bank behind to 45°; no allowance for removal or backfilling of excavated materials		
Existing slope angle 10°; excavation by 8 tonne machine; excavated material to location on site 50 m distant; linear measurement of flat area at the base of the slope		
2.00 m wide; height 350 mm	m	3.85
4.00 m wide; height 860 mm	m	15.40
6.00 m wide; height 1.28 m	m	35.00
8.00 m wide; height 1.71 m	m	62.00
10.00 m wide; height 2.14 m	m	97.00
Existing slope angle 20°; excavation by 8 tonne machine; excavated material to location on-site 50 m distant; linear measurement of flat area at the base of the slope		
2.00 m wide; height 1.14 m	m	10.40
4.00 m wide; height 2.29 m	m	41.00
6.00 m wide; height 3.43 m	m	93.00
8.00 m wide; height 4.58 m	m	165.00
10.00 m wide; height 5.72 m	m	255.00

8.1.2 PREPARATORY GROUNDWORKS

Item – Overhead and Profit Included	Unit	Range £
Excavate and grade banks to grade to receive stabilization treatments below; removal of excavated material not included		
By 21 tonne excavator	m^3	0.50
By 8 tonne excavator	m^3	1.60
By 5 tonne excavator	m^3	1.45
By 3 tonne excavator	m^3	4.45
Excavate sloped bank to vertical to receive retaining or stabilization treatment (priced separately); allow 1.00 m working space at top of bank; backfill working space and remove balance of arisings off site; measured as length along the top of the bank		
Bank height 1.00 m; excavation by 5 tonne excavator		
10°	m	125.00
30°	m	39.50
45°	m	22.50
60°	m	13.10
Bank height 1.00 m; excavation by 5 tonne excavator; arisings moved to stockpile on site		
10°	m	10.90
30°	m	4.70
45°	m	2.70
60°	m	1.55
Bank height 2.00 m; excavation by 5 tonne excavator		
10°	m	510.00
30°	m	155.00
45°	m	91.00
60°	m	52.00
Bank height 2.00 m; excavation by 5 tonne excavator but arisings moved to stockpile on site		
10°	m	61.00
30°	m	18.20
45°	m	10.70
60°	m	6.15
Bank height 3.00 m; excavation by 21 tonne excavator		
20°	m	550.00
30°	m	350.00
45°	m	195.00
60°	m	105.00
Excavate bank 3.00 m but arisings moved to stockpile on site		
20°	m	61.00
30°	m	38.50
45°	m	15.60
60°	m	8.45

8.1.2 PREPARATORY GROUNDWORKS

Item – Overhead and Profit Included	Unit	Range £
SOIL STABILIZATION		
Please refer to slope excavation above for preparation of banks		
REVETMENTS		
Gabion mattress revetments		
Construct revetment of Reno mattress gabions laid on firm level ground; tightly packed with broken stone or concrete and securely wired; all in accordance with manufacturer's instructions		
gabions 6 × 2 × 0.17 m; one course	m²	56.00
gabions 6 × 2 × 0.17 m; two courses	m²	95.00
Grass concrete; bank revetments; excluding bulk earthworks		
Bring bank to final grade by machine; lay regulating layer of Type 1; lay filter membrane; lay 100 mm drainage layer of broken stone or approved hardcore 28–10 mm size; blind with 25 mm sharp sand; lay grass concrete surface (price does not include edgings or toe beams); fill with approved topsoil and fertilizer at 35 g/m²; seed with dwarf rye based grass seed mix		
Grasscrete in situ reinforced concrete surfacing GC 1; 100 mm thick	m²	69.00
Grasscrete in situ reinforced concrete surfacing GC 2; 150 mm thick	m²	84.00
Grassblock 103 mm open matrix blocks; 406 × 406 × 103 mm	m²	80.00
Grade slope to even grade; Stabilize river bank edge with woven willow walling; lay bank stabilization mat pinned to graded embankment 5 m long (measured per m of river bank edge)		
Slope edge to river bank 5 m long seeded with Greenfix stabilization matting		
Brushwood faggott and pre-planted coir revetment edge	m	200.00
Brushwood faggott revetment edge	m	175.00
StableGRASS grass reinforcement: cellular slope retention; clear existing bank of light scrub, shrubs and grasses; spray with glyphosate; grade surface only to even grade by machine and final hand grade; lay soil retaining system		
Backfilled with excavated material		
38 mm deep	m²	30.50
Backfilled with imported topsoil and seeded		
100 mm deep	m²	2.60
SOIL REINFORCEMENT		
Geogrid soil reinforcement (Note: measured per metre run at the top of the bank)		
Backfill excavated bank; lay Tensar geogrid; cover with excavated material as work proceeds		
2.00 m high bank; geogrid at 1.00 m vertical lifts		
10° slope	m	990.00
20° slope	m	550.00
30° slope	m	400.00

8.1.2 PREPARATORY GROUNDWORKS

Item – Overhead and Profit Included	Unit	Range £
2.00 m high bank; geogrid at 0.50 m vertical lifts		
45° slope	m	110.00
60° slope	m	89.00
3.00 m high bank		
10° slope	m	990.00
20° slope	m	550.00
30° slope	m	400.00
45° slope	m	860.00
EROSION CONTROL		
Bank stabilization; erosion control mats		
Excavate fixing trench for erosion control mat 300 × 300 mm; backfill after placing mat selected below	m	9.60
Place mat to slope to be retained in anchor trench; allow extra over to the slope for the area of matting required to the anchor trench		
By machine; final grade only of excavated bank by excavator; lay erosion control solution; sow with low maintenance grass at 35 g/m²; spread imported topsoil 25 mm thick incorporating medium grade sedge peat at 3 kg/m² and fertilizer at 30 g/m²; water lightly		
seeded Covamat Standard	m²	4.30
lay Greenfix biodegradable pre-seeded erosion control mat; fix with 6 × 300 mm steel pegs at 1.0 m centres	m²	4.90
Tensar open textured erosion mat	m²	7.15
By hand; final grade only of excavated bank by hand; lay erosion control solution; sow with low maintenance grass at 35 g/m²; spread imported topsoil 25 mm thick incorporating medium grade sedge peat at 3 kg/m² and fertilizer at 30 g/m²; water lightly		
seeded Covamat Standard	m²	5.10
lay Greenfix biodegradable pre-seeded erosion control mat; fix with 6 × 300 mm steel pegs at 1.0 m centres	m²	5.70
Tensar open textured erosion mat	m²	8.00
Cellular retaining systems; Greenfix Ltd; final grade to excavated bank by machine and by hand; lay bank retaining system		
Geoweb; filling with angular aggregate	m	30.00
Haul road for site enablement		
Reduce existing soil to stockpile on site; Lay Tensar Biaxial geogrid on 200 mm compacted hardcore base; backfill with 200 mm Type 1 granular fill and compact	m²	24.00
Reduce existing soil to stockpile on site; Lay 200 mm compacted hardcore base blinded with 50 mm sharp sand; Lay Cellweb load support	m²	43.00

8.2.1 ROADS, PATHS AND PAVINGS

Item – Overhead and Profit Included	Unit	Range £
KERBS AND EDGINGS TO PAVED SURFACES		
Excavate for pathways or roadbed to receive surface 100 mm thick (priced separately); grade; 100 mm hardcore; kerbs PC 125 × 255 mm on both sides; including foundations haunched		
Excavation 350 mm deep; Type 1 base 150 mm thick		
1.50 m wide	m	140.00
2.00 m wide	m	165.00
3.00 m wide	m	200.00
4.00 m wide	m	235.00
5.00 m wide	m	270.00
6.00 m wide	m	310.00
7.00 m wide	m	345.00
Excavation 450 mm deep; Type 1 base 250 mm thick		
1.50 m wide	m	270.00
2.00 m wide	m	325.00
3.00 m wide	m	440.00
4.00 m wide	m	560.00
5.00 m wide	m	680.00
6.00 m wide	m	790.00
7.00 m wide	m	910.00
EDGINGS TO PATHS AND PAVINGS		
PRECAST CONCRETE KERBS		
Note: excavation is by machine unless otherwise mentioned		
Excavate trench and construct concrete foundation 300 mm wide × 300 mm deep; lay precast concrete kerb units bedded in semi-dry concrete; slump 35 mm maximum; haunching one side; disposal of arisings off site		
Kerbs laid straight; site mixed concrete bases		
125 mm high × 125 mm thick; bullnosed; type BN	m	41.00
125 × 255 mm; ref HB2; SP	m	45.00
150 × 305 mm; ref HB1	m	54.00
Kerbs laid straight; ready mixed concrete bases		
125 mm high × 125 mm thick; bullnosed; type BN	m	39.00
125 × 255 mm; ref HB2; SP	m	43.00
150 × 305 mm; ref HB1	m	52.00
Straight kerbs 125 mm × 255 mm laid to radius ready mixed concrete bases		
4.5 m radius	m	54.00
6.1 m radius	m	53.00
9.15 m radius	m	53.00
12.20 m radius	m	52.00
Dropper kerbs		
125 mm × 255–150 mm; left or right handed	m	49.00
Quadrants		
305 mm radius	nr	57.00
455 mm radius	nr	61.00

8.2.1 ROADS, PATHS AND PAVINGS

Item – Overhead and Profit Included	Unit	Range £
Excavate trench and construct concrete foundation 300 mm wide × 150 mm deep; lay precast concrete kerb units bedded in semi-dry concrete; slump 35 mm maximum; haunching one side; disposal of arisings off site		
Conservation kerbs laid straight		
225 mm wide × 150 mm high	m	80.00
155 mm wide × 255 mm high	m	79.00
Conservation kerbs laid to radius; 145 mm × 255 mm		
radius 3.25 m	m	100.00
radius 6.50 m	m	90.00
radius 9.80 m	m	89.00
solid quadrant; 305 × 305 mm	nr	94.00
STONE KERBS		
Excavate and construct concrete foundation 450 mm wide × 150 mm deep; lay kerb units bedded in semi-dry concrete; slump 35 mm maximum; haunching one side		
Granite kerbs		
straight; 125 × 250 mm	m	82.00
curved; 125 × 250 mm	m	96.00
Precast concrete edging on concrete foundation 100 × 150 mm deep and haunching one side 1:2:4 including all necessary excavation disposal and formwork		
Rectangular, chamfered or bullnosed; to one side of straight path		
50 × 150 mm	m	37.00
50 × 200 mm	m	46.50
50 × 250 mm	m	50.00
Rectangular, chamfered or bullnosed to both sides of straight paths		
50 × 150 mm	m	81.00
50 × 200 mm	m	93.00
50 × 250 mm	m	100.00
Brick or concrete block edge restraint; excavate for foundation; lay concrete 1:2:4 250 mm wide × 150 mm deep; on 50 mm thick sharp sand bed; lay blocks or bricks; Haunched one side		
Blocks 200 × 100 × 60 mm; butt jointed		
header course	m	44.00
stretcher course	m	39.00
Bricks Wienerberger Blue Baggeridge 200 × 100 × 62 mm; with flowable mortar joints		
header course; brick on flat	m	65.00
header course; brick on edge	m	66.00
stretcher course; flat or brick on edge	m	65.00
stretcher course; 2 rows brick on edge	m	50.00

8.2.1 ROADS, PATHS AND PAVINGS

Item – Overhead and Profit Included	Unit	Range £
KERBS AND EDGINGS TO PAVED SURFACES – cont		
STONE EDGINGS		
Sawn Yorkstone edgings; excavate for foundations; lay concrete 1:2:4 150 mm deep × 33.3% wider than the edging; on 35 mm thick mortar bed		
Yorkstone 50 mm thick		
100 mm wide × random lengths	m	61.00
100 × 100 mm	m	62.00
100 mm wide × 200 mm long	m	73.00
250 mm wide × random lengths	m	73.00
500 mm wide × random lengths	m	110.00
Granite edgings; excavate for footings; lay concrete 1:2:4 150 mm deep × 33.3% wider than the edging; on 35 mm thick mortar bed		
Granite 50 mm thick		
100 mm wide × 200 mm long	m	56.00
100 mm wide × random lengths	m	65.00
250 mm wide × random lengths	m	68.00
setts; 100 × 100 mm	m	82.00
300 mm wide × random lengths	m	75.00
TIMBER EDGINGS		
Softwood treated boards; 150 mm × 22 mm; with timber pegs at 1.00 m centres; Excavated and backfilled by machine		
Straight		
150 × 22 mm	m	9.90
150 × 50 mm	m	12.90
Curved; over 5 m radius		
150 × 22 mm	m	18.00
Curved; 4–5 m radius		
150 × 22 mm	m	21.00
Curved; 3–4 m radius		
150 × 22 mm	m	26.00
Curved; 1–3 m radius		
150 × 22 mm	m	34.00
Softwood treated boards; 150 mm × 22 mm; with timber pegs at 1.00 m centres; Excavated and backfilled by hand		
Straight		
150 × 22 mm	m	13.00
Curved; over 5 m radius		
150 × 22 mm	m	24.00
Curved; 4–5 m radius		
150 × 22 mm	m	24.00
Curved; 3–4 m radius		
150 × 22 mm	m	29.00
Curved; 1–3 m radius		
150 × 22 mm	m	37.00

8.2.1 ROADS, PATHS AND PAVINGS

Item – Overhead and Profit Included	Unit	Range £
METAL EDGINGS		
Everedge – Steel edgings		
Steel edgings for macadam		
50 mm	m	19.50
65 mm	m	25.50
75 mm	m	22.00
100 mm	m	24.50
Edgings for paths; L shaped Edgings with 3.2 thick		
100 mm; 4.75 mm exposed upper lip	m	14.50
50 mm with 3.18 exposed upper lip	m	19.50
75 mm; 5.2 mm exposed upper lip	m	22.00
Steel Edgings – Outdoor space design		
Galvainized steel; 6 mm thick		
100 mm	m	29.00
150 mm	m	40.00
For Corten Edgings 200–600 mm – Please see Section 8.4.3 – Retaining Walls		
BASES FOR PAVING		
Excavate ground and reduce levels to receive 38 mm thick slab and 25 mm mortar bed; dispose of excavated material off site		
Lay granular fill Type 1; limestone aggregate 150 mm thick laid to falls and compacted		
all by machine	m²	24.50
all by hand except disposal by grab	m²	78.00
Lay granular fill Type 1; recycled material; 150 mm thick laid to falls and compacted		
all by machine	m²	18.80
all by hand except disposal by grab	m²	73.00
Lay granular fill Type 1; limestone aggregate 200 mm thick laid to falls and compacted		
all by machine	m²	33.00
all by hand except disposal by grab	m²	99.00
Lay granular fill Type 1; recycled material; 200 mm thick laid to falls and compacted		
all by machine	m²	25.00
all by hand except disposal by grab	m²	93.00
Lay 1:2:4 concrete base 150 mm thick laid to falls		
all by machine	m²	41.00
all by hand except disposal by grab	m²	100.00
150 mm hardcore base with concrete base 150 mm deep		
concrete 1:2:4 site mixed	m²	61.00
concrete PAV 1 35 ready mixed	m²	51.00
concrete PAV 1 35 ready mixed; but for 50 mm thick slab	m²	53.00

8.2.1 ROADS, PATHS AND PAVINGS

Item – Overhead and Profit Included	Unit	Range £
BASES FOR PAVING – cont		
Excavate ground and reduce levels to receive 38 mm thick slab and 25 mm mortar bed – cont		
Hardcore base 150 mm deep; concrete reinforced with A142 mesh		
site mixed concrete 1:2:4; 150 mm deep	m²	75.00
site mixed concrete 1:2:4; 250 mm deep	m²	100.00
concrete PAV 1 35 N/mm² ready mixed; 150 mm deep	m²	60.00
concrete PAV 1 35 N/mm² ready mixed; 250 mm deep	m²	88.00
CHANNELS		
Excavate and construct concrete foundation 450 mm wide × 150 mm deep; lay precast concrete kerb units bedded in semi-dry concrete; slump 35 mm maximum; haunching one side; lay channel units bedded in 1:3 sand mortar; jointed in cement: sand mortar 1:3		
Kerbs; 125 mm high × 125 mm thick; bullnosed; type BN		
dished channel; 125 × 225 mm; Ref CS	m	77.00
square channel; 125 × 150 mm; Ref CS2	m	73.00
bullnosed channel; 305 × 150 mm; Ref CBN	m	91.00
Brick channel; class B engineering bricks		
Excavate and construct concrete foundation 600 mm wide × 200 mm deep; lay channel to depths and falls bricks to be laid as headers along the channel; bedded in 1:3 sand mortar bricks close jointed in cement: sand mortar 1:3		
3 courses wide	m	100.00
3 coursed granite setts; dished		
Excavate and construct concrete foundation 400 mm wide × 200 mm deep; lay channel to depths and falls; bedded in 1:3: sand mortar		
3 courses wide	m	105.00
CONCRETE ROADS		
Excavate weak points of excavation by hand and fill with well rammed Type 1 granular material		
100 mm thick	m²	16.80
200 mm thick	m²	21.50
Reinforced concrete roadbed; excavation and disposal off site; reinforcing mesh A142; expansion joints at 50 m centres; on 150 mm hardcore; sand blinded; kerbs 155 × 255 mm to both sides; haunched one side		
Concrete 150 mm thick		
4.00 m wide	m	280.00
5.00 m wide	m	335.00
6.00 m wide	m	390.00
7.00 m wide	m	450.00

8.2.1 ROADS, PATHS AND PAVINGS

Item – Overhead and Profit Included	Unit	Range £
Concrete 250 mm thick		
4.00 m wide	m	355.00
5.00 m wide	m	435.00
6.00 m wide	m	510.00
7.00 m wide	m	600.00
FOOTPATHS		
Note: For pathway surfaces; select appropriate finish from the Measured Works section		
Excavate 250 mm; disposal off site; grade to levels and falls; lay 150 mm Type 1 granular material; all disposal off site; Lay or fix edge as described		
Lay Permaloc Asphalt edge 100 mm metal edgings; path width		
1.50 m wide	m	73.00
3.00 m wide	m	105.00
Lay timber edging 150 × 38 mm; path width		
1.50 m wide	m	46.00
3.00 m wide	m	77.00
Lay precast edgings kerbs 50 × 150 mm on both sides; including foundations haunched one side in 11.5 N/mm² concrete with all necessary formwork; path width		
1.50 m wide	m	110.00
3.00 m wide	m	140.00
Lay concrete edging Keyblock header edging; path width		
1.50 m wide	m	71.00
3.00 m wide	m	100.00
Lay concrete edging Keyblock stretcher edging; path width		
1.50 m wide	m	75.00
3.00 m wide	m	130.00
Lay brick on flat header edging; brick £300.00/1000; path width		
1.50 m wide	m	120.00
3.00 m wide	m	150.00
Lay brick on flat header edging; brick £600.00/1000; path width		
1.50 m wide	m	120.00
3.00 m wide	m	155.00
Lay brick on edge header edging; brick £300.00/1000; path width		
1.50 m wide	m	140.00
3.00 m wide	m	170.00
Lay brick on edge header edging; brick £600.00/1000; path width		
1.50 m wide	m	150.00
3.00 m wide	m	180.00
Lay brick on flat or brick on edge stretcher edging; brick £300.00/1000; path width		
1.50 m wide	m	99.00
3.00 m wide	m	130.00
Lay brick on edge strecher edging; brick £600.00/1000; path width		
1.50 m wide	m	100.00
3.00 m wide	m	135.00

8.2.1 ROADS, PATHS AND PAVINGS

Item – Overhead and Profit Included	Unit	Range £
FOOTPATHS – cont		
Excavate 250 mm – cont		
Lay new granite setts 100 × 100 × 100 mm single row; path width		
1.50 m wide	m	175.00
3.00 m wide	m	210.00
Lay new granite setts 200 × 100 × 100 mm single row; path width		
1.50 m wide	m	140.00
3.00 m wide	m	170.00
BRICK PAVING		
WORKS BY MACHINE		
Excavate and lay base Type 1 150 mm thick remove arisings; all by machine;		
lay clay brick paving		
Wienerberger; butt jointed on 50 mm sharp sand bed		
Blue Baggeridge range; 200 × 50 × 62 mm laid flat	m²	130.00
Chamfered collection; 200 × 50 × 62 mm laid flat	m²	120.00
Dutch pavers; 200 × 48 × 85 mm laid on edge	m²	180.00
Dutch pavers; 200 × 64 × 85 mm laid on edge	m²	160.00
Wienerberger Dutch tumbled pavers 200 × 50 × 65 mm; laid on edge	m²	180.00
Wienerberger Dutch tumbled pavers 150 × 150 × 50 mm	m²	125.00
Wienerberger; flowable mortar jointed on 50 mm mortar bed		
Blue Baggeridge range; 200 × 50 × 62 mm laid flat	m²	150.00
Chamfered collection; 200 × 50 × 62 mm laid flat	m²	140.00
Dutch pavers; 200 × 64 × 85 mm laid on edge	m²	200.00
Dutch pavers; 200 × 64 × 85 mm laid on flat	m²	160.00
Excavate and lay base Type 1 250 mm thick all by machine; remove arisings;		
lay clay brick paving		
Wienerberger; flowable mortar jointed on 50 mm mortar bed		
Dutch pavers; 200 × 64 × 85 mm laid on edge	m²	210.00
Dutch pavers; 200 × 64 × 85 mm laid on flat	m²	170.00
Blue Baggeridge range; 200 × 50 × 62 mm laid flat	m²	165.00
Chamfered collection; 200 × 50 × 62 mm laid flat	m²	150.00
Excavate and lay 150 mm readymix concrete base reinforced with		
A393 mesh; all by machine; remove arisings; lay clay brick paving		
Wienerberger; flowable mortar jointed on 50 mm mortar bed		
Dutch pavers; 200 × 64 × 85 mm laid on edge	m²	220.00
Dutch pavers; 200 × 64 × 85 mm laid on flat	m²	180.00
Blue Baggeridge range; 200 × 50 × 62 mm laid flat	m²	170.00
Chamfered collection; 200 × 50 × 62 mm laid flat	m²	160.00

8.2.1 ROADS, PATHS AND PAVINGS

Item – Overhead and Profit Included	Unit	Range £
PATHS		
Brick paved paths; Wienerberger brick paviors with butt joints; prices are inclusive of excavation, 150 mm Type 1 and 50 mm sharp sand; jointing in kiln dried sand brushed in; inclusive of edge restraints of brick headers		
Path 1.50 mm wide; edging course of headers		
Dutch pavers; 200 × 64 × 85 mm laid on edge	m²	270.00
Blue Baggeridge range; 200 × 50 × 62 mm laid flat	m²	280.00
Wienerberger Dutch tumbled pavers 200 × 50 × 65 mm; laid on edge	m²	270.00
Chamfered collection; 200 × 50 × 62 mm laid flat	m²	235.00
WORKS BY HAND		
Excavate and lay base Type 1 150 mm thick by hand; arisings barrowed to spoil heap maximum distance 25 m and removal off site by grab; lay clay brick paving; edgings not included		
Wienerberger; butt jointed on 50 mm sharp sand bed		
Dutch pavers; 200 × 48 × 85 mm laid on edge	m²	240.00
Wienerberger Dutch tumbled pavers 200 × 50 × 65 mm; laid on edge	m²	230.00
Dutch pavers; 200 × 64 × 85 mm laid on edge	m²	210.00
Blue Baggeridge range; 200 × 50 × 62 mm laid flat	m²	180.00
Wienerberger Dutch tumbled pavers 150 × 150 × 50 mm	m²	180.00
Chamfered collection; 200 × 50 × 62 mm laid flat	m²	170.00
Excavate and lay 150 mm concrete base; 1:3:6: site mixed concrete reinforced with A393 mesh; remove arisings to stockpile and then off site by grab; lay clay brick paving		
Wienerberger; butt jointed on 50 mm sharp sand bed		
Dutch collection; 200 × 48 × 85 mm laid on edge	m²	240.00
Wienerberger Dutch tumbled pavers 200 × 50 × 65 mm; laid on edge	m²	235.00
Dutch pavers; 200 × 64 × 85 mm laid on edge	m²	215.00
Blue Baggeridge range; 200 × 50 × 62 mm laid flat	m²	185.00
Wienerberger Dutch tumbled pavers 150 × 150 × 50 mm	m²	180.00
Chamfered collection; 200 × 50 × 62 mm laid flat	m²	180.00

8.2.1 ROADS, PATHS AND PAVINGS

Item – Overhead and Profit Included	Unit	Range £
BLOCK PAVING		
CONCRETE SETT PAVING		
Excavate ground and bring to levels; supply and lay 150 mm granular fill Type 1 laid to falls and compacted; supply and lay setts bedded in 50 mm sand and vibrated; joints filled with dry sand and vibrated; all excluding edgings or kerbs measured separately		
Marshalls Mono; Tegula precast concrete setts		
random sizes 60 mm thick	m²	80.00
single size 60 mm thick	m²	76.00
random size 80 mm thick	m²	85.00
single size 80 mm thick	m²	105.00
Excavate 450 mm and prepare base of 100 hardcore and 250 mm Type 1; Work to falls, crossfalls and cambers not exceeding 15° mark out car parking bays 5.0 × 2.4 m with thermoplastic road paint; surfaces all mechanically laid; edgings not included		
Marshalls Keyblock concrete blocks 200 × 100 × 80 mm		
grey	m²	105.00
colours	m²	105.00
Marshalls Keyblock concrete blocks 200 × 100 × 60 mm		
grey	m²	100.00
colours	m²	105.00
Marshalls Tegula concrete blocks 200 × 100 × 80 mm		
random sizes	m²	115.00
single size	m²	110.00
BLOCK PAVING EDGINGS		
Edge restraint to block paving; excavate for groundbeam; lay concrete 1:2:4 200 mm wide × 150 mm deep; on 50 mm thick sharp sand bed; inclusive of haunching one side		
Blocks 200 × 100 × 60 mm; PC £11.76 /m²; butt jointed		
header course	m	24.50
stretcher course	m	19.00
Excavate ground; supply and lay granular fill Type 1 150 mm thick laid to falls and compacted; supply and lay block pavers; laid on 50 mm compacted sharp sand; vibrated; joints filled with loose sand excluding edgings or kerbs measured separately		
Concrete blocks		
200 × 100 × 60 mm	m²	71.00
200 × 100 × 80 mm	m²	76.00

8.2.1 ROADS, PATHS AND PAVINGS

Item – Overhead and Profit Included	Unit	Range £
Reduce levels; lay 150 mm granular material Type 1; lay vehicular block paving to 90° herringbone pattern; on 50 mm compacted sand bed; vibrated; jointed in sand and vibrated; excavate and lay precast concrete edging 50 × 150 mm; on concrete foundation 1:2:4		
Blocks 200 × 100 × 60 mm		
1.0 m wide clear width between edgings	m	145.00
1.5 m wide clear width between edgings	m	180.00
2.0 m wide clear width between edgings	m	220.00
Blocks 200 × 100 × 60 mm but blocks laid 45° herringbone pattern including cutting edging blocks		
1.0 m wide clear width between edgings	m	155.00
1.5 m wide clear width between edgings	m	190.00
2.0 m wide clear width between edgings	m	230.00
Blocks 200 × 100 × 60 mm but all excavation by hand disposal off site by grab		
1.0 m wide clear width between edgings	m	350.00
1.5 m wide clear width between edgings	m	400.00
2.0 m wide clear width between edgings	m	455.00
PERMEABLE BLOCK PAVING		
Excavate 450 mm and prepare base of 150 mm Type 3 open; Work to falls, crossfalls and cambers not exceeding 15° mark out car parking bays 5.0 × 2.4 m with thermoplastic road paint; surfaces all mechanically laid; edgings not included		
Tegula Priora		
160 × 120 × 80 mm	m²	71.00
240 × 160 × 80 mm; traditional herringbone	m²	76.00
Conservation Priora – Silver Grey		
200 × 400 × 65 mm	m²	105.00
400 × 400 × 65 mm	m²	93.00
PORCELAIN PAVING		
Porcelain paving; London Stone; including excavation and blinded reinforced concrete base 150 mm		
Porcelain pavings; exluding any cutting		
596 × 596 mm	m²	150.00
748 × 748 mm	m²	145.00
1194 × 596 mm	m²	140.00
MACADAM SURFACES		
Macadam roadway over 1000 m²		
Excavate 350 mm for pathways or roadbed; bring to grade; lay 100 mm well-rolled hardcore; lay 150 mm Type 1 granular material; lay precast edgings kerbs 50 × 150 mm on both sides; including foundations haunched one side in 11.5 N/mm² concrete with all necessary formwork; machine lay surface of 90 mm macadam 60 mm base course and 30 mm wearing course; all disposal off site		
1.50 m wide	m	175.00
2.00 m wide	m	185.00
3.00 m wide	m	270.00

8.2.1 ROADS, PATHS AND PAVINGS

Item – Overhead and Profit Included	Unit	Range £
MACADAM SURFACES – cont		
Macadam roadway over 1000 m² – cont		
Excavate 350 mm for pathways or roadbed – cont		
4.00 m wide	m	335.00
5.00 m wide	m	400.00
6.00 m wide	m	460.00
7.00 m wide	m	530.00
Macadam roadway 400–1000 m²		
Excavate 350 mm for pathways or roadbed; bring to grade; lay 100 mm well-rolled hardcore; lay 150 mm Type 1 granular material; lay precast edgings kerbs 50 × 150 mm on both sides; including foundations haunched one side in 11.5 N/mm² concrete with all necessary formwork; machine lay surface of 90 mm macadam 60 mm base course and 30 mm wearing course; all disposal off site		
1.50 m wide	m	200.00
2.00 m wide	m	240.00
3.00 m wide	m	320.00
4.00 m wide	m	400.00
5.00 m wide	m	480.00
6.00 m wide	m	560.00
7.00 m wide	m	640.00
MACADAM FOOTPATH		
Excavate footpath to reduce level; remove arisings to tip on site maximum distance 25 m; lay Type 1 granular fill 100 mm thick; lay base course of 28 mm size dense bitumen macadam 50 mm thick; wearing course of 10 mm size dense bitumen macadam 20 mm thick; timber edge 150 × 38 mm		
Areas over 1000 m²		
1.0 m wide	m	51.00
1.5 m wide	m	69.00
2.0 m wide	m	87.00
Areas 400–1000 m²		
1.0 m wide	m	66.00
1.5 m wide	m	92.00
2.0 m wide	m	140.00
CAR PARKS		
Excavate 450 mm and prepare base of 100 mm hardcore and 250 mm Type 1; Work to falls, crossfalls and cambers not exceeding 15° mark out car parking bays 5.0 × 2.4 m with thermoplastic road paint; surfaces all mechanically laid; edgings not included		
Macadam base course 60 mm thick and wearing course 40 mm thick		
per bay; 5.0 × 2.4 m	each	1050.00
Gangway	m²	87.00
Bay for disabled people; Macadam base course 60 mm thick and wearing course 40 mm thick		
per bay; 5.80 × 3.25 m ambulant	each	1650.00
per bay; 6.55 × 3.80 m wheelchair	each	2150.00

8.2.1 ROADS, PATHS AND PAVINGS

Item – Overhead and Profit Included	Unit	Range £
Excavate 280 mm and prepare base of 150 mm Type 1		
Surface treatments as shown		
Macadam 60:40	m²	62.00
Tegula setts 80 mm; random sizes	m²	92.00
Tegula setts 80 mm; single size	m²	88.00
Keyblock 80 mm; Colours	m²	82.00
Keyblock 80 mm; Grey	m²	81.00
IN SITU CONCRETE PAVINGS AND SURFACES		
CONCRETE PAVING		
Pedestrian areas		
Excavate to reduce levels; lay 100 mm Type 1; lay PAV 1 air entrained concrete; joints at maximum width of 6.0 m cut out and sealed with sealant; inclusive of all formwork and stripping		
100 mm thick	m²	75.00
Trafficked areas		
Excavate to reduce levels; lay 150 mm Type 1 lay PAV 2 40 N/mm² air entrained concrete; reinforced with steel mesh 200 × 200 mm square at 2.22 kg/m²; joints at max width of 6.0 m cut out and sealed with sealant		
150 mm thick	m²	98.00
Reinforced concrete paths; excavate path and dispose of excavated material off site; bring to grade; lay reinforcing mesh A142 lapped and joined; lay in situ reinforced concrete PAV1 35 N/mm²; with 15 impregnated fibreboard expansion joints sealant at 50 m centres; on 100 mm hardcore blinded sand; edgings 150 × 50 mm to both sides; including foundations haunched one side in 11.5 N/mm² concrete with all necessary formwork		
Concrete 150 mm thick		
1.00 m wide	m	88.00
2.00 m wide	m	140.00
3.00 m wide	m	190.00
WORKS BY HAND; Ready mixed concrete – mixed on site		
Reinforced concrete paths; excavate path by hand barrow to spoil heap 25 m and dispose by grab off site; bring to grade; lay reinforcing mesh A142 lapped and joined; lay in situ reinforced ready mixed concrete mixed on site (1:2:4); with 15 impregnated fibreboard expansion joints sealant at 50 m centres; on 100 mm hardcore blinded sand; edgings 150 × 50 mm to both sides; including foundations haunched one side in 11.5 N/mm² concrete with all necessary formwork		
Concrete 150 mm thick		
1.00 m wide	m	130.00
2.00 m wide	m	250.00
3.00 m wide	m	350.00

8.2.1 ROADS, PATHS AND PAVINGS

Item – Overhead and Profit Included	Unit	Range £
GRASS CONCRETE		
IN SITU GRASS CONCRETE TO CAR PARKS		
Grass Concrete Grasscrete in situ continuously reinforced cellular surfacing; fill with topsoil and peat (5:1) and fertilizer at 35 g/m²; seed with dwarf rye grass at 35 g/m²; Excludes marking		
GC1; 100 mm thick for cars and light traffic; per bay; 5.0 × 2.4 m	each	1175.00
GC1; 100 mm thick for cars and light traffic; gangway	m²	96.00
GC2; 150 mm thick for HGV traffic including dust carts		
per bay; 5.0 × 2.4 m	each	1375.00
gangway	m²	115.00
interlocking concrete blocks 200 × 100 × 60 mm; grey; but mark out bays		
per bay; 5.80 × 3.25 ambulant	each	840.00
per bay; 6.55 × 3.80 wheelchair	each	1100.00
gangway	m²	44.50
MODULAR GRASS CONCRETE PAVING		
Reduce levels; lay 150 mm Type 1 granular fill compacted; supply and lay precast grass concrete blocks on 20 sand and level by hand; fill blocks with 3 mm sifted topsoil and pre-seeding fertilizer at 50 g/m²; sow with perennial ryegrass/chewings fescue seed at 35 g/m²		
Grass concrete		
GB103 406 × 406 × 103 mm	m²	72.00
GB83 406 × 406 × 83 mm	m²	70.00
extra for geotextile fabric underlayer	m²	1.10
Firepaths		
Excavate to reduce levels; lay 300 mm well rammed hardcore; blinded with 100 mm type 1; supply and lay Marshalls Mono Grassguard 180 precast grass concrete blocks on 50 mm sand and level by hand; fill blocks with sifted topsoil and pre-seeding fertilizer at 50 g/m²		
firepath 3.8 m wide	m	470.00
firepath 4.4 m wide	m	540.00
firepath 5.0 m wide	m	620.00
turning areas	m²	120.00
RESIN BOUND AND RESIN BONDED SURFACES		
Resin bonded surfaces; excavate 230 mm; bring to grade; lay 150 mm Type 1 granular material; hand lay base course of 60 mm macadam and resin bonded wearing course; all disposal off site; edgings excluded		
Natratex resin bonded macadam wearing course		
general paved area	m²	84.00
pathway 1.50 m wide	m	135.00
pathway 3.00 m wide	m	270.00
Addagrip; resin bonded aggregate; 1–3 mm golden pea gravel with buff adhesive		
general paved area	m²	82.00
pathway 1.50 m wide	m	130.00
pathway 3.00 m wide	m	260.00

8.2.1 ROADS, PATHS AND PAVINGS

Item – Overhead and Profit Included	Unit	Range £
Addagrip; resin bonded aggregate; 6 mm chocolate gravel 18 mm depth		
general paved area	m^2	110.00
pathway 1.50 m wide	m	170.00
pathway 3.00 m wide	m	345.00
Addagrip; resin bonded aggregate; Cobalt blue recycled glass 15 mm depth		
general paved area	m^2	93.00
pathway 1.50 m wide	m	150.00
pathway 3.00 m wide	m	295.00
Resin bound surfaces; excavate 230 mm; bring to grade; lay 150 mm Type 1 granular material; hand lay base course of 60 mm macadam and resin bound wearing course; all disposal off site; edgings excluded		
Sureset; resin bound permeable surfacing; natural gravel 6 mm aggregate 18 mm thick		
paved area	m^2	100.00
pathway 1.50 m wide	m	160.00
pathway 3.00 m wide	m	320.00
Sureset; resin bound permeable surfacing; crushed rock 6 mm aggregate 18 mm thick		
paved area	m^2	140.00
pathway 1.50 m wide	m	160.00
pathway 3.00 m wide	m	520.00
Sureset; resin bound permeable surfacing; marble 6 mm aggregate 18 mm thick		
paved area	m^2	120.00
pathway 1.50 m wide	m	185.00
pathway 3.00 m wide	m	370.00
GRAVEL PATHWAYS		
Reduce levels and remove spoil to dump on site; lay 150 mm hardcore well rolled; lay geofabric; fix timber edge 150 × 38 mm to both sides of straight paths		
Lay Cedec gravel 50 mm thick; watered and rolled		
1.0 m wide	m	45.00
1.5 m wide	m	60.00
2.0 m wide	m	82.00
Lay Breedon gravel 50 mm thick; watered and rolled		
1.0 m wide	m	34.50
1.5 m wide	m	45.00
2.0 m wide	m	55.00
Reduce levels and remove spoil to dump on site; lay 150 mm Type 1 well rolled; lay geofabric; fix timber edge 150 × 38 mm to both sides of straight paths		
Lay Cedec gravel 50 mm thick; watered and rolled		
1.0 m wide	m	50.00
1.5 m wide	m	67.00
2.0 m wide	m	92.00

8.2.1 ROADS, PATHS AND PAVINGS

Item – Overhead and Profit Included	Unit	Range £
GRAVEL PATHWAYS – cont		
Reduce levels and remove spoil to dump on site – cont		
Lay Breedon gravel 50 mm thick; watered and rolled		
1.0 m wide	m	39.00
1.5 m wide	m	52.00
2.0 m wide	m	65.00
Lay 20 mm shingle; 30 mm thick		
1.0 m wide	m	34.50
1.5 m wide	m	44.00
2.0 m wide	m	61.00
CELLULAR CONFINED SURFACES		
Excavate ground and reduce levels to receive cellular confined paving; dispose of excavated material off site		
Lay granular fill Type 1; limestone aggregate 150 mm thick laid to falls and compacted; lay confined cellular surface		
Geoweb 100 thick	m²	54.00
StablePATH honeycomb cellular gravel grid; 25 mm deep for foot traffic and bicycles filled with shingle	m²	45.00
StablePATH honeycomb cellular gravel grid; 25 mm deep for foot traffic and bicycles filled with decorative aggregate	m²	50.00
StablePATH honeycomb cellular gravel grid 40 mm deep for vehicular traffic; filled with shingle	m²	48.00
StablePATH honeycomb cellular gravel grid 40 mm deep for vehicular traffic; filled with decorative aggregate	m²	52.00
FLAG PAVINGS		
SLAB/BRICK PATHS		
Stepping stone path inclusive of hand excavation and mechanical disposal, 100 mm Type 1 and sand blinding; slabs laid 100 mm apart to existing turf		
600 mm wide; 600 × 600 × 50 mm slabs		
natural finish	m	81.00
coloured	m	84.00
900 mm wide; 600 × 900 × 50 mm slabs		
natural finish	m	89.00
coloured	m	95.00
Pathway inclusive of mechanical excavation and disposal; 100 mm Type 1 and sand blinding; slabs close butted		
Straight butted path 900 mm wide; 600 × 900 × 50 mm slabs		
natural finish	m	36.00
coloured	m	40.00
Straight butted path 1200 mm wide; double row; 600 × 900 × 50 mm slabs; laid stretcher bond		
natural finish	m	62.00
coloured	m	71.00

8.2.1 ROADS, PATHS AND PAVINGS

Item – Overhead and Profit Included	Unit	Range £
Straight butted path 1200 mm wide; one row of 600 × 600 × 50 mm; one row 600 × 900 × 50 mm slabs		
natural finish	m	61.00
coloured	m	69.00
Straight butted path 1500 mm wide; slabs of 600 × 900 × 50 mm and 600 × 600 × 50 mm; laid to bond		
natural finish	m	75.00
coloured	m	85.00
Straight butted path 1800 mm wide; two rows of 600 × 900 × 50 mm slabs; laid bonded		
natural finish	m	83.00
coloured	m	105.00
Straight butted path 1800 mm wide; three rows of 600 × 900 × 50 mm slabs; laid stretcher bond		
natural finish	m	91.00
coloured	m	115.00
MARSHALLS FLAG PAVINGS		
Supply and lay precast concrete flags; excavate ground and reduce levels; treat substrate with total herbicide; supply and lay granular fill Type 1 150 mm thick laid to falls and compacted		
Marshalls Urbex paving		
450 × 450 × 35 mm	m²	84.00
600 × 600 × 35 mm	m²	76.00
600 × 600 × 63 mm	m²	120.00
Marshalls Celestia paving		
450 × 450 × 50 mm	m²	120.00
600 × 600 × 50 mm	m²	110.00
600 × 200 × 50 mm	m²	130.00
Marshalls Modal paving		
450 × 450 × 50 mm	m²	97.00
600–900 × 600 × 50 mm	m²	97.00
Random lengths × 200/300/450/600/900 × 50 mm	m²	110.00
FLAG PAVING TO PEDESTRIAN AREAS		
Prices are inclusive of all mechanical excavation and disposal.		
Supply and lay precast concrete flags; excavate ground and reduce levels; treat substrate with total herbicide; supply and lay granular fill Type 1 150 mm thick laid to falls and compacted		
Standard precast concrete flags bedded and jointed in lime: sand mortar (1:3)		
450 × 450 × 50 mm; chamfered	m²	67.00
600 × 300 × 50 mm	m²	71.00
600 × 450 × 50 mm	m²	61.00
600 × 600 × 50 mm	m²	58.00
750 × 600 × 50 mm	m²	57.00
900 × 600 × 50 mm	m²	55.00

8.2.1 ROADS, PATHS AND PAVINGS

Item – Overhead and Profit Included	Unit	Range £
FLAG PAVING TO PEDESTRIAN AREAS – cont		
Supply and lay precast concrete flags – cont		
Coloured flags bedded and jointed in lime: sand mortar (1:3)		
600 × 600 × 50 mm	m²	65.00
400 × 400 × 65 mm	m²	67.00
900 × 600 × 50 mm	m²	63.00
Marshalls Saxon; textured concrete flags; reconstituted Yorkstone in colours; butt jointed bedded in lime: sand mortar (1:3)		
300 × 300 × 35 mm	m²	130.00
450 × 450 × 50 mm	m²	97.00
600 × 300 × 35 mm	m²	100.00
600 × 600 × 50 mm	m²	85.00
Tactile flags; Marshalls blister tactile pavings; red or buff; for blind pedestrian guidance laid to designed pattern		
450 × 450 mm	m²	94.00
400 × 400 mm	m²	86.00
PEDESTRIAN DETERRENT PAVING		
Excavate ground and bring to levels; treat substrate with total herbicide; supply and lay granular fill Type 1 150 mm thick laid to falls and compacted; supply and lay precast deterrent paving units bedded in lime: sand mortar (1:3) and jointed in lime: sand mortar (1:3)		
Marshalls Mono		
Contour deterent paving; 600 × 600 × 75 mm	m²	95.00
BARK PAVINGS		
Excavate to reduce levels; remove all topsoil to dump on site; treat area with herbicide; lay 150 mm clean hardcore; blind with sand; lay 0.7 mm geotextile filter fabric water flow 50 l/m²/sec; supply and fix treated softwood edging boards 50 × 150 mm to bark area on hardcore base extended 150 mm beyond the bark area; boards fixed with galvanized nails to treated softwood posts 750 × 50 × 75 mm driven into firm ground at 1.0 m centres; edging boards to finish 25 mm above finished bark surface; tops of posts to be flush with edging boards and once weathered		
Supply and lay 100 mm Melcourt conifer walk chips 10–40 mm size		
1.00 m wide	m	40.00
2.00 m wide	m	61.00
3.00 m wide	m	81.00
4.00 m wide	m	98.00
Extra over for wood fibre	m²	−1.30
Extra over for hardwood chips	m²	−1.56

8.2.1 ROADS, PATHS AND PAVINGS

Item – Overhead and Profit Included	Unit	Range £
COBBLE PAVINGS		
BEACH COBBLE PAVING		
Excavate ground and bring to levels and fill with compacted Type 1 fill 100 mm thick; lay GEN 1 concrete base 100 mm thick; supply and lay cobbles individually laid by hand bedded in cement: sand mortar (1:3) 25 mm thick minimum; dry grout with 1:3 cement: sand grout; brush off surplus grout and water in; sponge off cobbles as work proceeds; all excluding formwork edgings or kerbs measured separately		
Scottish beach cobbles 50–75 mm	m²	200.00
Scottish beach cobbles 100–200 mm	m²	155.00
Kidney flint cobbles 75–100 mm	m²	170.00
NATURAL STONE PAVING		
GRANITE SETT PAVING – PEDESTRIAN		
Excavate and grade; 100 mm hardcore; Type 1 50 mm thick; 100 mm concrete 1:2:4; supply and lay granite setts 100 × 100 × 100 mm bedded in cement: sand mortar (1:3) 25 mm thick minimum; jointed in mortar 1:3		
Setts laid to bonded pattern and jointed; bedded in cement: sand mortar 1:3 25 mm thick		
new setts 100 × 100 × 100 mm	m²	210.00
second-hand cleaned 100 × 100 × 100 mm	m²	200.00
Setts laid to bonded pattern and jointed; bedded in SteinTec mortar 25 mm thick		
new setts 100 × 100 × 100 mm	m²	210.00
second-hand cleaned 100 × 100 × 100 mm	m²	200.00
Setts laid in curved pattern		
new setts 100 × 100 × 100 mm	m²	220.00
second-hand cleaned 100 × 100 × 100 mm	m²	210.00
GRANITE SETT PAVING – TRAFFICKED AREAS		
Excavate to reduce levels; lay 100 mm hardcore; compact; Type 150 mm thick; lay 150 mm site mixed concrete 1:2:4 reinforced with steel fabric; granite setts 100 × 100 × 100 mm bedded in cement: sand mortar (1:3) 25 mm thick minimum; jointed in mortar 1:3; all excluding edgings or kerbs measured separately		
Site mixed concrete		
new setts 100 × 100 × 100 mm	m²	220.00
second-hand cleaned 100 × 100 × 100 mm	m²	210.00
Ready mixed concrete		
new setts 100 × 100 × 100 mm	m²	220.00
second-hand cleaned 100 × 100 × 100 mm	m²	205.00

8.2.1 ROADS, PATHS AND PAVINGS

Item – Overhead and Profit Included	Unit	Range £
NATURAL STONE PAVING – cont		
NATURAL STONE SLAB PAVING		
WORKS BY MACHINE		
Excavate ground by machine and reduce levels; to receive 65 mm thick slab and 35 mm mortar bed; dispose of excavated material off site; lay granular fill Type 1 150 mm thick laid to falls and compacted; lay to random rectangular pattern on 35 mm mortar bed		
New riven slabs		
laid random rectangular on 35 mm 1:3 mortar bed	m²	225.00
laid random rectangular on 30 mm thick Steintec mortar bed	m²	275.00
New riven slabs; but to 150 mm plain concrete base		
laid random rectangular on 35 mm 1:3 mortar bed	m²	235.00
laid random rectangular on 30 mm thick Steintec mortar bed	m²	285.00
New riven slabs; disposal by grab		
laid random rectangular	m²	230.00
Reclaimed Cathedral grade riven slabs		
laid random rectangular	m²	280.00
Reclaimed Cathedral grade riven slabs; but to 150 mm plain concrete base		
laid random rectangular	m²	290.00
Reclaimed Cathedral grade riven slabs; disposal by grab		
laid random rectangular	m²	285.00
New slabs sawn 6 sides		
laid random rectangular	m²	235.00
three sizes; laid to coursed pattern	m²	215.00
New slabs sawn 6 sides; but to 150 mm plain concrete base		
laid random rectangular	m²	240.00
three sizes; laid to coursed pattern	m²	220.00
New slabs sawn 6 sides; disposal by grab		
laid random rectangular	m²	235.00
3 sizes, laid to coursed pattern	m²	220.00
Portland stone paving – Albion stone; inclusive of excavation and 150 mm reinforced blinded concrete base		
Portland stone pavings 40 mm thick		
coursed 3 sizes	m²	310.00
600 × 600 mm	m²	260.00
Portland stone pavings 50 mm thick		
coursed 3 sizes	m²	325.00
600 × 600 mm	m²	270.00
1200 × 750 mm	m²	360.00
WORKS BY HAND		
Excavate ground by hand and reduce levels, to receive 65 mm thick slab and 35 mm mortar bed; barrow all materials and arisings 25 m; dispose of excavated material off site by grab; treat substrate with total herbicide; lay granular fill Type 1 150 thick laid to falls and compacted; lay to random rectangular pattern on 35 mm mortar bed		

8.2.1 ROADS, PATHS AND PAVINGS

Item – Overhead and Profit Included	Unit	Range £
New riven slabs		
laid random rectangular	m²	280.00
New riven slabs laid random rectangular; but to 150 mm plain concrete base		
laid random rectangular	m²	300.00
New riven slabs; but disposal to skip		
laid random rectangular	m²	305.00
Reclaimed Cathedral grade riven slabs		
laid random rectangular	m²	340.00
Reclaimed Cathedral grade riven slabs; but to 150 mm plain concrete base		
laid random rectangular	m²	355.00
Reclaimed Cathedral grade riven slabs; but disposal to skip		
laid random rectangular	m²	360.00
New slabs sawn 6 sides		
laid random rectangular	m²	290.00
three sizes; sawn 6 sides laid to coursed pattern	m²	270.00
New slabs sawn 6 sides; but to 150 mm plain concrete base		
laid random rectangular	m²	305.00
three sizes; sawn 6 sides laid to coursed pattern	m²	285.00
DECKING – TIMBER AND COMPOSITE		
HARDWOOD DECKS		
Timber decking; Exterior Decking Limited; timber decking to roof gardens; cranage or movement of materials to roof top not included; coated one coat with Seasonite		
Joist size 47 × 47 mm		
Ipe; Exterpark; 21 mm invisibly fixed	m²	120.00
Ipe; pre-drilled and countersunk; screw fixed	m²	190.00
Rustic Teak; 21 mm invisibly fixed	m²	205.00
Timber decking to commercial applications; double beams at 300 mm centres; coated one coat with Seasonite; Exterior decking Limited		
Joist size 47 × 150 mm		
Ipe; Exterpark; 21 mm invisibly fixed	m²	125.00
Ipe; pre-drilled and countersunk; screw fixed	m²	190.00
Rustic Teak; 21 mm invisibly fixed	m²	210.00
SOFTWOOD DECKS		
Timber decking; support structure of timber joists for decking laid at 400 mm centres on blinded base (measured separately); Exterior decking Limited		
Yellow balau 125 mm × 21 mm		
joists 47 × 47 mm	m²	115.00
joists 47 × 100 mm	m²	120.00
joists 47 × 150 mm	m²	120.00
Red cedar decking boards 90 mm wide × 25 mm thick		
joists 47 × 47 mm	m²	150.00
joists 47 × 100 mm	m²	150.00
joists 47 × 150 mm	m²	155.00

8.2.1 ROADS, PATHS AND PAVINGS

Item – Overhead and Profit Included	Unit	Range £
DECKING – TIMBER AND COMPOSITE – cont		
Composite decking – Millboard; London Stone Ltd		
Decking type on Composite substructure; laid to blinded levelled ground; Average area 25–50 m²		
Enhanced grain range	m²	210.00
Weathered Oak range	m²	225.00
Decking type on Composite substructure; average area 25–50 m²; On support pedestals 28 mm high		
Enhanced grain range	m²	220.00
Weathered Oak range	m²	235.00
Decking type on Composite substructure; average area 25–50 m²; On support pedestals 175–285 mm high		
Enhanced grain range	m²	240.00
Weathered Oak range	m²	255.00
Decking type on Composite substructure; average area 25–50 m²; On support pedestals 285–400 mm high		
Enhanced grain range	m²	260.00
Weathered Oak range	m²	275.00
Handrails		
handrails fixed to posts 100 × 100 × 1370 mm high		
square balusters at 100 mm centres	m	110.00
square balusters at 300 mm centres	m	82.00
turned balusters at 100 mm centres	m	140.00
turned balusters at 300 mm centres	m	93.00
STEPS		
Excavate for new steps; grade to correct levels; remove excavated material from site and lay concrete base; Construct formwork; supply and lay site mixed concrete; Fix treads and risers to steps		
Stone clad steps; 1.20 m wide; tread depth 225 mm; Granite cut off site		
1 riser	nr	390.00
2 risers	nr	650.00
3 risers	nr	930.00
4 risers	nr	1275.00
6 risers	nr	1775.00
Stone clad steps; 1.20 m wide; tread depth 225 mm; Riven Yorkstone cut on site		
1 riser	nr	385.00
2 risers	nr	640.00
3 risers	nr	900.00
4 risers	nr	1225.00
6 risers	nr	1750.00
Stone clad steps; 2.40 m wide; tread depth 225 mm; Granite cut off site		
1 riser	nr	740.00
2 risers	nr	1250.00
3 risers	nr	1800.00
4 risers	nr	2475.00
6 risers	nr	3550.00

8.2.1 ROADS, PATHS AND PAVINGS

Item – Overhead and Profit Included	Unit	Range £
Steps; Portland stone; Albion stone; Cast bases of in situ concrete including excavation; Construct steps; risers of Portland stone 30 mm thick fixed to concrete step face		
tread depth 350 mm; 50 mm thick		
1st riser	m	530.00
subsequent risers	m	490.00
tread depth 500 mm; 50 mm thick		
1st riser	m	620.00
subsequent risers	m	600.00
Excavate for new steps; grade to correct levels; remove excavated material from site and lay concrete base; Construct formwork; supply and lay site mixed concrete; Fix treads and risers to steps		
Stone clad steps; tread depth 250 mm; Granite cut off site; Marshalls Bega LED Lights at 2.00 m centres		
1st riser	m	590.00
subsequent risers	m	570.00
Stone clad steps; tread depth 250 mm; Granite cut off site; Marshalls Bega LED Lights at 1.00 m centres		
1st riser	m	780.00
subsequent risers	m	760.00
Stone clad steps; tread depth 400 mm; Granite cut off site; Marshalls Bega LED Lights at 2.00 m centres		
1st riser	m	680.00
subsequent risers	m	680.00
Stone clad steps; tread depth 450 mm; Granite cut off site; Marshalls Bega LED Lights at 1.00 m centres		
1st riser	m	870.00
subsequent risers	m	850.00
Stone clad steps; tread depth 600 mm; Granite cut off site; Marshalls Bega LED Lights at 1.00 m centres		
1st riser	m	1000.00
subsequent risers	m	980.00

8.2.1 ROADS, PATHS AND PAVINGS

Item – Overhead and Profit Included	Unit	Range £
PLAY AREA FOR BALL GAMES		
Excavate for playground; bring to grade; lay 225 mm consolidated hardcore; blind with 100 mm type 1; lay macadam base of 40 mm size dense bitumen macadam to 75 mm thick; lay wearing course 30 mm thick		
Over 1000 m²	100 m²	7900.00
400–1000 m²	100 m²	8800.00
Excavate for playground; bring to grade; lay 100 mm consolidated hardcore; blind with 100 mm type 1; lay macadam base of dense bitumen macadam to 50 mm thick; lay wearing course 20 mm thick		
Over 1000 m²	100 m²	5200.00
400–1000 m²	100 m²	7300.00
Excavate for playground; bring to grade; lay 100 mm consolidated hardcore; blind with 100 mm type 1; lay macadam base of dense bitumen macadam to 50 mm thick; lay wearing course of Addagrip resin coated aggregate 3 mm dia. × 6 mm thick		
Over 1000 m²	100 m²	7600.00
400–1000 m²	100 m²	8900.00
SAFETY SURFACED PLAY AREAS		
Excavate ground and reduce levels to receive safety surface; dispose of excavated material off site; treat substrate with total herbicide; lay granular fill Type 1 150 mm thick laid to falls and compacted; lay macadam base 40 mm thick; supply and lay Rubaflex wet pour safety system to thicknesses as specified		
All by machine except macadam by hand		
black		
15 mm thick	m²	83.00
35 mm thick	m²	90.00
60 mm thick	m²	100.00
coloured		
15 mm thick	m²	91.00
35 mm thick	m²	80.00
60 mm thick	m²	110.00
All by hand except disposal by grab		
black		
15 mm thick	m²	120.00
35 mm thick	m²	130.00
60 mm thick	m²	145.00
coloured		
15 mm thick	m²	130.00
35 mm thick	m²	140.00
60 mm thick	m²	155.00
Excavate for playground; bring to grade; lay with 150 mm type 1; lay macadam base of dense bitumen macadam to 50 mm thick; lay safety surface		
Wetpour coloured; 0.50 m critical fall height	m	94.00
Wetpour black; 0.50 m critical fall height	m	84.00
Wetpour coloured; 1.50 m critical fall height	m	110.00
Wetpour black; 1.50 m critical fall height	m	98.00

8.2.1 ROADS, PATHS AND PAVINGS

Item – Overhead and Profit Included	Unit	Range £
Excavate for playground; bring to grade; lay with 150 mm type 1; lay safety surface		
Meon UK		
Black; 10 m² – 50 m²	m	99.00
Black; over 50 m²	m	94.00
Colours; 10 m² – 50 m²	m	130.00
Colours; over 50 m²	m	110.00
BARK PLAY AREA		
Excavate playground area to 450 mm depth; lay 150 mm broken stone or clean hardcore; lay filter membrane; lay bark surface		
Melcourt Industries		
Playbark 10/50; 300 mm thick	m²	60.00
Playbark 8/25; 300 mm thick	m²	67.00

8.2.2 SPECIAL SURFACINGS AND PAVINGS

Item – Overhead and Profit Included	Unit	Range £
SPORTS FIELD CONSTRUCTION		
Strip existing vegetation; grade to levels using laser guided equipment; cultivate ground; apply pre-seeding fertilizer at 900 kg/ha; apply pre-seeding selective weedkiller; seed in two operations with sports pitch type grass seed at 350 kg/ha; harrow and roll lightly; including initial cut; size; mark out and paint lines; erect goal posts		
New ground construction of existing field; greenfield or brownfield; existing lightweight or loam soils		
association football; senior 114 × 72 m	each	31500.00
association football; junior 106 × 58 m	each	24000.00
New ground construction of existing field; greenfield or brownfield; existing clay soil; replacement of rootzone 150 mm		
association football; senior 114 × 72 m	each	120000.00
association football; junior 106 × 58 m	each	91000.00
Plain sports pitches; site clearance, grading and drainage not included		
Agripower Ltd; cultivate ground and grade to levels; apply pre-seeding fertilizer at 900 kg/ha; apply pre-seeding selective weedkiller; seed in two operations with sports pitch type grass seed at 350 kg/ha; harrow and roll lightly; including initial cut; size		
Existing level and graded ground previously used for playing surface		
association football; senior 114 × 72 m	each	4600.00
association football; junior 106 × 58 m	each	3600.00
rugby union pitch; 156 × 81 m	each	6000.00
rugby league pitch; 134 × 60 m	each	6000.00
hockey pitch; 95 × 60 m	each	4600.00
Agripower Ltd; cricket square; excavate to depth of 100 mm; lay imported marl or clay loam; bring to accurate levels; apply pre-seeding fertilizer at 50 g/m; apply selective weedkiller; seed with cricket square type g grass seed at 50 g/m²; rake in and roll lightly; erect and remove temporary protective chestnut fencing; allow for initial cut and watering three times		
22.8 × 22.8 m	each	19000.00
JOGGING TRACK		
Excavate track 250 mm deep; lay filter membrane; lay 100 mm depth gravel waste or similar; lay 100 mm compacted gravel	100 m²	12000.00
Extra for treated softwood edging 50 × 150 mm on 50 × 50 mm posts		
both sides	m	13.80

8.3.1 SOFT LANDSCAPING, SEEDING AND TURFING

Item – Overhead and Profit Included	Unit	Range £
SPORTS GROUND FOR AMENITY SPORT		
To existing field; Spray to remove existing turf. Scarify top 25 mm; Grade and level using ripper power harrow cultivator and stone burier combination inclusive of grass seed and fertilizers		
Light to medium soils		
Existing area to 1:100 existing grade	6000 m²	3600.00
Clay soils		
Existing area to 1:100 existing grade	6000 m²	4550.00
SURFACE PREPARATION FOR SEEDING AND TURFING		
BY MACHINE		
Mechanized preparation; Stone burier and cultivator; Treat area with systemic herbicide in advance of operations (not included); rip subsoil using subsoiling machine to a depth of 500 mm at 1.20 m centres in light to medium soils; rotavate to 200 mm deep in two passes		
Larger areas		
light to medium soils	100 m²	28.00
heavier soils	100 m²	51.00
Rip area by tractor and subsoiler; soil preparation by pedestrian operated rotavator; clearance and raking by hand		
Light to medium soils	100 m²	105.00
Heavier soils	100 m²	115.00
Spread and lightly consolidate topsoil brought from spoil heap not exceeding 100 m; in layers not exceeding 150 mm; grade to specified levels; remove stones over 25 mm; all by machine		
100 mm thick	100 m²	120.00
Extra to the above for imported topsoil		
Topsoil to BS3882 PC £24.72 m³ calculation allows for 20% settlement		
100 mm thick	100 m²	340.00
Excavate existing topsoil and remove off site; fill with new good quality topsoil to BS3882; add 50 mm mushroom compost (delivered in 65 m³ loads) and cultivate in; grade to level removing stones		
For turfing		
200 mm deep	m²	22.00
300 mm deep	m²	30.00
SURFACE PREPARATION BY HAND		
To area previously cleared and sprayed; cultivate existing topsoil to receive new ornamental planting; rotavate using pedestrian rotavator; dig over by hand 1 spit deep; rake and grade to level to receive new planting		
For areas to be seeded or turfed	m²	1.35

8.3.1 SOFT LANDSCAPING, SEEDING AND TURFING

Item – Overhead and Profit Included	Unit	Range £
SURFACE PREPARATION FOR SEEDING AND TURFING – cont		
SURFACE PREPARATION BY HAND – cont		
Spread only and lightly consolidate topsoil brought from spoil heap in layers not exceeding 150 mm; grade to specified levels; remove stones over 25 mm; all by hand		
100 mm thick	100 m²	640.00
100 mm thick; but inclusive of loading to location by barrow maximum distance 25 m; finished topsoil depth	100 m²	1425.00
100 mm thick but loading to location by barrow maximum distance 100 m; finished topsoil depth	100 m²	1025.00
Extra to above for imported topsoil PC £24.72 m³; Calculation allows for 20% settlement		
100 mm thick	100 m²	340.00
SEEDING		
Cultivate topsoil 200 mm a fine tilth by tractor; remove stones over 25 mm by mechanical stone rake and bring to final tilth by harrow; apply pre-seeding fertilizer at 50 g/m² and work into top 50 mm during final cultivation; seed with certified grass seed in two operations; roll seedbed lightly after sowing		
General amenity grass areas; 35 g/m²; BSH A3		
general grass seeded areas	100 m²	80.00
lawn areas for public or domestic general use	100 m²	99.00
high grade lawn areas	100 m²	125.00
General amenity grass at 35 g/m²; DLF Seeds Ltd Promaster 10		
general turfed areas	100 m²	110.00
lawn areas for public or domestic general use	100 m²	125.00
high grade lawn areas	100 m²	150.00
Shaded areas; BSH A6 at 50 g/m²	100 m²	95.00
Motorway and road verges; DLF Seeds Ltd Promaster 120 at 25–35 g/m²	100 m²	93.00
Cultivate topsoil 200 mm a fine tilth by tractor; remove stones over 25 mm by mechanical stone rake and bring to final tilth by harrow; Add 50 mm of imported topsoil over surface of area to be seeded; apply pre-seeding fertilizer at 50 g/m² and work into top 50 mm during final cultivation; seed with certified grass seed in two operations; roll seedbed lightly after sowing		
General amenity grass areas; 35 g/m²; BSH A3		
general turfed areas	100 m²	280.00
lawn areas for public or domestic general use	100 m²	300.00
high grade lawn areas	100 m²	325.00

8.3.1 SOFT LANDSCAPING, SEEDING AND TURFING

Item – Overhead and Profit Included	Unit	Range £
Cultivate topsoil 200 mm a fine tilth by tractor; remove stones over 25 mm by mechanical stone rake and bring to final tilth by harrow; Add 50 mm of topsoil from site spoil heaps maximum distance 100 m over surface of area to be seeded; apply pre-seeding fertilizer at 50 g/m² and work into top 50 mm during final cultivation; seed with certified grass seed in two operations; roll seedbed lightly after sowing		
General amenity grass areas; 35 g/m²; BSH A3		
general turfed areas	100 m²	140.00
lawn areas for public or domestic general use	100 m²	160.00
high grade lawn areas	100 m²	185.00
Bring top 200 mm of topsoil to a fine tilth using pedestrian operated rotavator; remove stones over 25 mm; apply pre-seeding fertilizer at 50 g/m² and work into top 50 mm during final hand cultivation; seed with certified grass seed in two operations; rake and roll seedbed lightly after sowing		
General amenity grass areas; 35 g/m²; BSH A3		
general grass seeded areas	100 m²	180.00
lawn areas for public or domestic general use	100 m²	200.00
high grade lawn areas	100 m²	220.00
Extra to above for using imported topsoil spread by machine		
100 mm minimum depth	100 m²	390.00
150 mm minimum depth	100 m²	590.00
extra for slopes over 30°	50%	–
Extra to above for using imported topsoil spread by hand; maximum distance for transporting soil 100 m		
100 mm minimum depth	m²	990.00
150 mm minimum depth	m²	1475.00
extra for slopes over 30°	50%	–
Extra for mechanically screening top 25 mm of topsoil through 6 mm screen and spreading on seedbed; debris carted to dump on site not exceeding 100 m	m³	8.40
Bring top 200 mm of topsoil to a fine tilth; remove stones over 50 mm; apply pre-seeding fertilizer at 50 g/m² and work into top 50 mm of topsoil during final cultivation; seed with certified grass seed in two operations; harrow and roll seedbed lightly after sowing		
Areas inclusive of fine levelling by specialist machinery		
outfield grass at 350 kg/ha; DLF Seeds Ltd Promaster 40	ha	11000.00
outfield grass at 350 kg/ha; DLF Seeds Ltd Promaster 70	ha	10000.00
sportsfield grass at 300 kg/ha; DLF Seeds Ltd J Pitch	ha	10000.00
Seeded areas prepared by chain harrow		
low maintenance grass at 350 kg/ha	ha	11500.00
verge mixture grass at 150 kg/ha	ha	10000.00
Extra for wild flora mixture at 30 kg/ha		
BSH WSF 75 kg/ha	ha	4300.00
extra for slopes over 30°	50%	–
Cut existing turf to 1.0 × 1.0 × 0.5 m turves; roll up and move to stack not exceeding 100 m		
by pedestrian operated machine; roll up and stack by hand	100 m²	120.00
all works by hand	100 m²	350.00
Extra for boxing and cutting turves	100 m²	6.20

8.3.1 SOFT LANDSCAPING, SEEDING AND TURFING

Item – Overhead and Profit Included	Unit	Range £
TURFING		
Cultivate topsoil in two passes; remove stones over 25 mm and bring to final tilth; apply pre-seeding fertilizer at 50 g/m² and work into top 50 mm during final cultivation; roll turf bed lightly		
Using tractor drawn implements and mechanical stone rake	100 m²	67.00
Cultivation by pedestrian rotavator; all other operations by hand		
light or medium soils	m²	1.60
heavy wet or clay soils	m²	1.80
Cultivate topsoil in two passes; remove stones over 25 mm and bring to final tilth; apply pre-seeding fertilizer at 50 g/m² and work into top 50 mm during final cultivation; lay imported turves maximum distance from turf stack 25 m		
using tractor drawn implements and mechanical stone rake	m²	8.90
cultivation by pedestrian rotavator; all other operations by hand	m²	8.95
Cultivate topsoil in two passes; remove stones over 25 mm and bring to final tilth; apply pre-seeding fertilizer at 50 g/m² and work into top 50 mm during final cultivation; roll turf bed supply and lay imported turf; Rolawn Medallion		
using tractor drawn implements and mechanical stone rake	m²	8.20
cultivation by pedestrian rotavator; all other operations by hand	m²	8.15
Turfed areas; watering on two occasions and carrying out initial cut		
by ride on triple mower	100 m²	1.90
by pedestrian mower	100 m²	8.10
by pedestrian mower; box cutting	100 m²	9.05
Extra for using imported topsoil spread and graded by machine		
25 mm minimum depth	m²	1.10
75 mm minimum depth	m²	3.00
100 mm minimum depth	m²	4.00
150 mm minimum depth	m²	5.95
Extra for using imported topsoil spread and graded by hand; distance of barrow run 25 m		
25 mm minimum depth	m²	2.60
75 mm minimum depth	m²	6.50
100 mm minimum depth	m²	8.25
150 mm minimum depth	m²	12.40
Extra for work on slopes over 30° including pegging with 200 galvanized wire pins	m²	2.40
INTURF BIG ROLL		
Supply, deliver in one consignment, fully prepare the area and install in Big Roll format Inturf 553, a turfgrass comprising dwarf perennial ryegrass, smooth stalked meadowgrass and fescues; installation by tracked machine		
Preparation by tractor drawn rotavator	m²	8.05
Cultivation by pedestrian rotavator; all other operations by hand	m²	8.20
Erosion control		
On ground previously cultivated, bring area to level, treat with herbicide; lay 20 mm thick open texture erosion control mat with 100 mm laps, fixed with 8 × 400 mm steel pegs at 1.0 m centres; sow with low maintenance grass suitable for erosion control on slopes at 35 g/m²; spread imported topsoil 25 mm thick and fertilizer at 35 g/m²; water lightly using sprinklers on two occasions	m²	9.15
hand-watering by hose pipe instead of by sprinkler maximum distance from mains supply 50 m	m²	9.30

8.3.1 SOFT LANDSCAPING, SEEDING AND TURFING

Item – Overhead and Profit Included	Unit	Range £
ARTIFICIAL TURF		
Excavate ground and reduce levels to receive 20 mm thick artificial grass surface on a compacted 150 thick Type 1 base; dispose of excavated material off site		
Lay artificial turf; by machine		
Trulawn Value; 16 mm	m²	58.00
Trulawn Play; 20 mm; entry level; lawn and play areas	m²	51.00
Trulawn Continental; 20 mm; realistic appearance; lawns and patios	m²	52.00
Trulawn Optimum; 26 mm; softest pile; lush green; lawns and patios	m²	55.00
Trulawn Luxury; 30 mm; deep pile; realistic lawn	m²	61.00
All by hand except disposal by grab		
Trulawn Value; 16 mm	m²	100.00
Trulawn Play; 20 mm; entry level; lawn and play areas	m²	97.00
Trulawn Continental; 20 mm; realistic appearance; lawns and patios	m²	97.00
Trulawn Optimum; 26 mm; softest pile; lush green; lawns and patios	m²	100.00
Trulawn Luxury; 30 mm; deep pile; realistic lawn	m²	110.00
MAINTENANCE OF TURFED/SEEDED AREAS		
MAINTENANCE EXECUTED AS PART OF A LANDSCAPE CONSTRUCTION CONTRACT		
Grass cutting		
Grass cutting; fine turf; using pedestrian guided machinery; arisings boxed and disposed of off site		
per occasion	100 m²	8.15
per annum 26 cuts	100 m²	210.00
per annum 18 cuts	100 m²	150.00
Grass cutting; standard turf; using self propelled three gang machinery		
per occasion	100 m²	0.60
per annum 26 cuts	100 m²	16.10
per annum 18 cuts	100 m²	11.20
Grass cutting for one year; recreation areas, parks, amenity grass areas; using tractor drawn machinery		
per occasion	ha	41.00
per annum 26 cuts	ha	1075.00
per annum 18 cuts	ha	740.00
Maintain grass area inclusive of grass cutting; fertilization twice per annum; leaf and litter clearance		
using pedestrian guided machinery; arisings boxed and disposed of off site		
26 cuts; public areas	100 m²	260.00
26 cuts; private areas	100 m²	230.00
18 cuts; public areas	100 m²	190.00
18 cuts; private areas	100 m²	170.00

8.3.1 SOFT LANDSCAPING, SEEDING AND TURFING

Item – Overhead and Profit Included	Unit	Range £
MAINTENANCE OF TURFED/SEEDED AREAS – cont		
Maintain grass area inclusive of grass cutting; fertilization twice per annum; leaf and litter clearance		
Using ride-on three cylinder machinery; arisings left on site		
26 cuts; public areas	100 m²	57.00
26 cuts; private areas	100 m²	30.00
18 cuts; public areas	100 m²	54.00
18 cuts; private areas	100 m²	27.00
Aeration of turfed areas		
Aerate ground with spiked aerator; apply spring/summer fertilizer once; apply autumn/winter fertilizer once; cut grass, 16 cuts; sweep up leaves twice		
as part of a landscape contract, defects liability	ha	2600.00
as part of a long term maintenance contract	ha	1275.00

8.3.2 EXTERNAL PLANTING

Item – Overhead and Profit Included	Unit	Range £
SURFACE PREPARATIONS FOR PLANTING		
SURFACE PREPARATION BY MACHINE		
Rip subsoil using subsoiling machine to a depth of 250 mm below topsoil at 1.20 m centres in light to medium soils; rotavate to 200 mm deep in two passes; cultivate with chain harrow; roll lightly; clear stones over 50 mm		
By tractor	100 m²	32.50
By tractor but carrying out operations in clay or compacted gravel	100 m²	36.00
Ripping by tractor rotavation by pedestrian operated rotavator; clearance and raking by hand; herbicide application by knapsack sprayer	100 m²	74.00
Ripping by tractor; rotavation by pedestrian operated rotavator; clearance and raking by hand; herbicide application by knapsack sprayer but carrying out operations in clay or compacted gravel	100 m²	78.00
Spread and lightly consolidate topsoil brought from spoil heap not exceeding 100 m; in layers not exceeding 150 mm; grade to specified levels; remove stones over 25 mm; all by machine		
100 mm thick	m²	120.00
150 mm thick	m²	180.00
300 mm thick	m²	350.00
450 mm thick	m²	520.00
Extra to the above for imported topsoil		
Topsoil PC £24.72 m³; calculation allows for 20% settlement		
100 mm thick	100 m²	340.00
150 mm thick	100 m²	510.00
300 mm thick	100 m²	1025.00
450 mm thick	100 m²	1525.00
500 mm thick	100 m²	1700.00
600 mm thick	100 m²	2050.00
750 mm thick	100 m²	2550.00
1.00 m thick	100 m²	3400.00
Topsoil to BS3882 topsoil PC £24.72 m³; calculation allows for 20% settlement		
100 mm thick	100 m²	340.00
150 mm thick	100 m²	510.00
300 mm thick	100 m²	1025.00
450 mm thick	100 m²	1525.00
500 mm thick	100 m²	1700.00
600 mm thick	100 m²	2050.00
750 mm thick	100 m²	2550.00
1.00 m thick	100 m²	3400.00
Excavate existing topsoil and remove off site; fill with new good quality topsoil to BS3882; add 50 mm mushroom compost (delivered in 65 m³ loads) and cultivate in; grade to level removing stones		
For planting		
300 mm deep	m²	29.50
400 mm deep	m²	41.00
500 mm deep	m²	49.00

8.3.2 EXTERNAL PLANTING

Item – Overhead and Profit Included	Unit	Range £
SURFACE PREPARATIONS FOR PLANTING – cont		
Excavate existing topsoil and remove off site – cont		
Extra for incorporating mushroom compost at 50 mm/m² into the top 150 mm of topsoil (compost delivered in 20 m³) loads		
manually spread; mechanically rotavated	100 m²	245.00
mechanically spread and rotavated	100 m²	170.00
Extra for incorporating manure at 50 mm/m² into the top 150 mm of topsoil loads		
manually spread; mechanically rotavated; 20 m³ loads	100 m²	220.00
mechanically spread and rotavated; 60 m³ loads	100 m²	210.00
SURFACE PREPARATION BY HAND		
To area previously cleared and sprayed; cultivate existing topsoil to receive new ornamental planting; rotavate using pedestrian rotavator; dig over by hand 1 spit deep; rake and grade to level to receive new planting		
For ornamental planting beds	m²	0.80
Spread only and lightly consolidate topsoil brought from spoil heap in layers not exceeding 150 mm; grade to specified levels; remove stones over 25 mm; all by hand		
Excavated material alongside or within area to be treated		
100 mm thick	100 m²	640.00
150 mm thick	100 m²	970.00
300 mm thick	100 m²	1925.00
450 mm thick	100 m²	2900.00
Loading to location by barrow maximum distance 25 m; finished topsoil depth		
100 mm thick	100 m²	1425.00
150 mm thick	100 m²	2125.00
300 mm thick	100 m²	4250.00
450 mm thick	100 m²	6400.00
500 mm thick	100 m²	7100.00
600 mm thick	100 m²	8500.00
Loading to location by barrow maximum distance 100 m; finished topsoil depth		
100 mm thick	100 m²	1025.00
150 mm thick	100 m²	1550.00
300 mm thick	100 m²	3100.00
450 mm thick	100 m²	4600.00
500 mm thick	100 m²	5100.00
600 mm thick	100 m²	6200.00
Extra to the above for incorporating mushroom compost at 50 mm/m² into the top 150 mm of topsoil; by hand	100 m²	295.00
Extra to above for imported topsoil topsoil PC £24.72 m³; calculation allows for 20% settlement		
100 mm thick	100 m²	340.00
150 mm thick	100 m²	510.00
300 mm thick	100 m²	1025.00
450 mm thick	100 m²	1525.00
500 mm thick	100 m²	1700.00
600 mm thick	100 m²	2050.00
750 mm thick	100 m²	2550.00
1.00 m thick	100 m²	3400.00

8.3.2 EXTERNAL PLANTING

Item – Overhead and Profit Included	Unit	Range £
TREE PLANTING		
Specification Corresponding Tree Girth Size		
Light Standard (LS) 6–8 cm		
Standard (S) 8–10 cm		
Select Standard (SS) 10–12 cm		
Heavy Standard (HS) 12–14 cm		
Extra Heavy Standard (EHS) 14–16 cm		
Advanced Heavy Standard (AHS) 16–18 cm		
Semi mature 18–20 cm +		
Excavate tree pit by hand; fork over bottom of pit; plant tree with roots well spread out; backfill with excavated material incorporating treeplanting compost at 1 m³ per 3 m³ of soil; one tree stake and two ties; tree pits square in sizes shown		
Standard bare root tree in pit; PC £28.00		
600 × 600 mm deep	each	84.00
900 × 600 mm deep	each	98.00
Standard rootballed tree in pit; PC £37.37		
600 × 600 mm deep	each	98.00
900 × 600 mm deep	each	110.00
1.00 × 1.00 m deep	each	150.00
Selected standard bare root tree in pit; PC £49.00		
900 × 900 mm deep	each	120.00
1.00 × 1.00 m deep	each	165.00
Selected standard rootball tree in pit; PC £61.88		
900 × 900 mm deep	each	160.00
1.00 × 1.00 m deep	each	225.00
Heavy standard bare root tree in pit; PC £70.00		
900 × 900 mm deep	each	180.00
1.00 × 1.00 m deep	each	200.00
1.50 m × 750 mm deep	each	265.00
Heavy standard rootball tree in pit; PC £85.96		
900 × 900 mm deep	each	195.00
1.00 × 1.00 m deep	each	225.00
1.50 m × 750 mm deep	each	280.00
Extra heavy standard bare root tree in pit; PC £80.50		
1.00 × 1.00 m deep	each	225.00
1.20 × 1.00 m deep	each	220.00
1.50 m × 750 mm deep	each	300.00
Extra heavy standard rootball tree in pit; PC £98.16		
1.00 × 1.00 m deep	each	250.00
1.50 m × 750 mm deep	each	310.00
1.50 × 1.00 m deep	each	370.00

8.3.2 EXTERNAL PLANTING

Item – Overhead and Profit Included	Unit	Range £
TREE PLANTING – cont		
SEMI MATURE TREE PLANTING		
Trees in soft landscape environments		
Excavate tree pit deep by machine; fork over bottom of pit; plant rootballed tree Acer platanoides 'Emerald Queen' using telehandler where necessary; backfill with excavated material, incorporating Melcourt Topgrow bark/ manure mixture at 1 m³ per 3 m³ of soil; Platipus underground guying system; tree pits 1500 × 1500 × 1500 mm deep inclusive of Platimats; excavated material not backfilled to tree pit spread to surrounding area		
tree pits 1500 × 1500 × 1500 mm		
16–18 cm girth; PC £110.00	each	405.00
18–20 cm girth; PC £150.00	each	460.00
20–25 cm girth; PC £180.00	each	610.00
25–30 cm girth; PC £250.00	each	770.00
30–35 cm girth; PC £380.00	each	1075.00
tree pits 2.00 × 2.00 × 1.5 m deep		
40–45 cm girth; PC £1000.00	each	1975.00
45–50 cm girth; PC £2000.00	each	3150.00
55–60 cm girth; PC £3000.00	each	4400.00
67–70 cm girth; PC £4000.00	each	5900.00
75–80 cm girth; PC £7500.00	each	10000.00
Excavate tree pit deep by machine; fork over bottom of pit; plant rootballed tree Acer platanoides 'Emerald Queen' using telehandler where necessary; backfill with excavated material, incorporating Melcourt Topgrow bark/ manure mixture at 1 m³ per 3 m³ of soil; Double timber stake and crossbar staking; tree pits 1500 × 1500 × 1500 mm deep; excavated material not backfilled to tree pit spread to surrounding area		
16–18 cm girth; PC £110.00	each	350.00
18–20 cm girth; PC £150.00	each	405.00
20–25 cm girth; PC £180.00	each	550.00
Extra to the above for imported topsoil moved 25 m from tipping area and disposal off site of excavated material		
Tree pits 1500 × 1500 × 1500 mm deep		
16–18 cm girth	each	335.00
18–20 cm girth	each	320.00
20–25 cm girth	each	310.00
25–30 cm girth	each	300.00
30–35 cm girth	each	290.00

8.3.2 EXTERNAL PLANTING

Item – Overhead and Profit Included	Unit	Range £
Tree pit in Trafficked environments		
Excavate tree pit; remove excavated material; break up bottom of pit; line pit with root barrier; Lay drainage layer 200 mm thick; backfill tree pit with cellular load supporting cells to 7.5 tonne traffic weight limit; Plant tree with underground guying system; inclusive of root director; Add rootball irrigation system; backfill over rootball with topsoil enriched with composts and fertilizers		
Tree pit tree size		
2.00 × 2.00 × 1.00 m deep; 30–35 cm tree	nr	2300.00
2.50 × 2.50 × 1.00 m deep; 40–45 cm tree	nr	4000.00
Excavate tree pit; remove excavated material; break up bottom of pit; line pit with root barrier; Lay drainage layer 200 mm thick; backfill tree pit with cellular load supporting cells to 40 tonne traffic weight limit; Plant tree with underground guying system; inclusive of root director; Add rootball irrigation system; backfill over rootball with topsoil enriched with composts and fertilizers		
Tree pit tree size		
2.00 × 2.00 × 1.00 m deep; 30–35 cm tree	nr	9800.00
2.50 × 2.50 × 1.00 m deep; 40–45 cm tree	nr	16000.00
TREE PLANTING WITH MOBILE CRANES		
Excavate tree pit 1.50 × 1.50 × 1.00 m deep; supply and plant semi-mature trees delivered in full loads; trees lifted by crane; inclusive of backfilling tree pit with imported topsoil, compost, fertilizers and underground guying using Platipus anchors		
Self managed lift; local authority applications, health and safety, traffic management or road closures not included; tree size and distance of lift; 35 tonne crane		
25–30 cm; maximum 25 m distance	each	1075.00
30–35 cm; maximum 25 m distance	each	1375.00
35–40 cm; maximum 25 m distance	each	1950.00
55–60 cm; maximum 15 m distance	each	4400.00
80–90 cm; maximum 10 m distance	each	12500.00
Managed lift; inclusive of all local authority applications, health and safety, traffic management or road closures all by crane hire company; tree size and distance of lift; 35 tonne crane		
25–30 cm; maximum 25 m distance	each	1100.00
30–35 cm; maximum 25 m distance	each	1375.00
35–40 cm; maximum 25 m distance	each	1975.00
55–60 cm; maximum 15 m distance	each	4450.00
80–90 cm; maximum 10 m distance	each	12500.00
Self managed lift; local authority applications, health and safety, traffic management or road closures not included; tree size and distance of lift; 80 tonne crane		
25–30 cm; maximum 40 m distance	each	1075.00
30–35 cm; maximum 40 m distance	each	1375.00
35–40 cm; maximum 40 m distance	each	1975.00
55–60 cm; maximum 33 m distance	each	4400.00
80–90 cm; maximum 23 m distance	each	12500.00

8.3.2 EXTERNAL PLANTING

Item – Overhead and Profit Included	Unit	Range £
TREE PLANTING – cont		
Excavate tree pit 1.50 × 1.50 × 1.00 m deep – cont		
Managed lift; inclusive of all local authority applications, health and safety, traffic management or road closures all by crane hire company; tree size and distance of lift; 35 tonne crane		
25–30 cm; maximum 40 m distance	each	1100.00
30–35 cm; maximum 40 m distance	each	1400.00
35–40 cm; maximum 40 m distance	each	2000.00
55–60 cm; maximum 33 m distance	each	4500.00
80–90 cm; maximum 23 m distance	each	13000.00
PLANTING SHRUBS AND GROUNDCOVERS		
SHRUBS		
Excavate planting holes 250 × 250 × 300 mm deep to area previously ripped and rotavated; excavated material left alongside planting hole		
By mechanical auger		
250 mm centres (16 plants per m^2)	m^2	18.40
300 mm centres (11.11 plants per m^2)	m^2	12.80
400 mm centres (6.26 plants per m^2)	m^2	7.20
450 mm centres (4.93 plants per m^2)	m^2	5.70
500 mm centres (4 plants per m^2)	m^2	4.60
600 mm centres (2.77 plants per m^2)	m^2	3.20
750 mm centres (1.77 plants per m^2)	m^2	2.00
900 mm centres (1.23 plants per m^2)	m^2	1.40
1.00 m centres (1 plant per m^2)	m^2	1.15
1.50 m centres (0.44 plants per m^2)	m^2	0.50
Excavation by hand		
250 mm centres (16 plants per m^2)	m^2	18.40
300 mm centres (11.11 plants per m^2)	m^2	12.80
400 mm centres (6.26 plants per m^2)	m^2	7.20
450 mm centres (4.93 plants per m^2)	m^2	5.70
500 mm centres (4 plants per m^2)	m^2	4.60
600 mm centres (2.77 plants per m^2)	m^2	3.20
750 mm centres (1.77 plants per m^2)	m^2	2.00
900 mm centres (1.23 plants per m^2)	m^2	1.40
1.00 m centres (1 plant per m^2)	m^2	1.15
1.50 m centres (0.44 plants per m^2)	m^2	0.50

8.3.2 EXTERNAL PLANTING

Item – Overhead and Profit Included	Unit	Range £
Clear light vegetation from planting area and remove to dump on site; dig planting holes; plant whips with roots well spread out; backfill with excavated topsoil; including one 38 × 38 mm treated softwood stake, two tree ties and mesh guard 1.20 m high; allow for beating up once at 10% of original planting, cleaning and weeding round whips once, applying fertilizer once at 35 gm/m²; using the following mix of whips, bare rooted		
Plant bare root plants average PC £0.47 each to matrix		
1500 × 1500 mm	100 m²	435.00
Cultivate and grade shrub bed; bring top 300 mm of topsoil to a fine tilth, incorporating mushroom compost at 50 mm and Enmag slow release fertilizer; rake and bring to given levels; remove all stones and debris over 50 mm; dig planting holes average 300 × 300 × 300 mm deep; supply and plant specified shrubs in quantities as shown below; backfill with excavated material; water to field capacity and mulch 50 mm bark chips 20–40 mm size; water and weed regularly for 12 months and replace failed plants		
Shrubs 3 l PC £3.40; ground covers 9 cm PC £1.50		
100% shrub area		
300 mm centres	100 m²	6000.00
400 mm centres	100 m²	3700.00
500 mm centres	100 m²	2650.00
600 mm centres	100 m²	2075.00
100% groundcovers		
200 mm centres	100 m²	6500.00
300 mm centres	100 m²	3300.00
400 mm centres	100 m²	2200.00
500 mm centres	100 m²	1675.00
groundcover 30%/shrubs 70% at the distances shown below		
200/300 mm	100 m²	6000.00
300/400 mm	100 m²	3500.00
300/500 mm	100 m²	2800.00
400/500 mm	100 m²	2475.00
groundcover 50%/shrubs 50% at the distances shown below		
200/300 mm	100 m²	6300.00
300/400 mm	100 m²	3500.00
300/500 mm	100 m²	3000.00
400/500 mm	100 m²	2425.00
BULB PLANTING		
Cultivate ground by machine and rake to level; plant bulbs as shown; bulbs PC £25.00/100		
15 bulbs per m²	100 m²	910.00
25 bulbs per m²	100 m²	1475.00
50 bulbs per m²	100 m²	2950.00
Cultivate ground by machine and rake to level; plant bulbs as shown; bulbs PC £13.00/100		
15 bulbs per m²	100 m²	670.00
25 bulbs per m²	100 m²	1075.00
50 bulbs per m²	100 m²	2125.00

8.3.2 EXTERNAL PLANTING

Item – Overhead and Profit Included	Unit	Range £
BULB PLANTING – cont		
Form holes in grass areas and plant bulbs using bulb planter, backfill with organic manure and turf plug; bulbs PC £13.00/100		
15 bulbs per m²	100 m²	1025.00
25 bulbs per m²	100 m²	1725.00
50 bulbs per m²	100 m²	3400.00
HEDGE PLANTING		
Works by machine; excavate trench for hedge 300 mm wide × 300 mm deep; deposit spoil alongside and plant hedging plants in single row at 200 mm centres; backfill with excavated material		
Trench 300 mm × 300 mm; bare root hedging plants PC £0.55 per plant; single row		
200 mm centres	m	8.50
300 mm centres	m	7.20
400 mm centres	m	6.25
600 mm centres	m	5.30
800 mm centres	m	4.80
Trench 600 mm × 300 mm deep; bare root hedging plants PC £0.55 per plant; double staggered row		
300 mm centres	m	13.50
400 mm centres	m	5.00
500 mm centres	m	9.65
600 mm centres	m	9.20
800 mm centres	m	7.90
1.00 m centres	m	7.35
Works by hand; excavate trench for hedge 300 mm wide × 450 mm deep; deposit spoil alongside and plant hedging plants in single row at 200 mm centres; backfill with excavated material incorporating organic manure at 1 m³ per 5 m³; carry out initial cut; including delivery of plants from nursery		
Trench 300 mm × 300 mm; bare root hedging plants PC £0.55 per plant; single row		
200 mm centres	m	13.20
300 mm centres	m	11.90
400 mm centres	m	11.00
600 mm centres	m	10.00
800 mm centres	m	9.50
Trench 600 mm × 300 mm deep; bare root hedging plants PC £0.55 per plant; double staggered row		
300 mm centres	m	24.50
400 mm centres	m	22.00
500 mm centres	m	21.00
600 mm centres	m	20.00
800 mm centres	m	19.00
1.00 m centres	m	18.40

8.3.2 EXTERNAL PLANTING

Item – Overhead and Profit Included	Unit	Range £
Readyhedge Ltd; Fully preformed hedge planting excavate trench by machine 700 mm wide × 500 mm deep; add compost at 100 mm/m² mixed to excavated material; plant mature hedge plants; backfill with excavated material and compost; allow for disposal of 50% of excavated material to spoil heaps 50 m distant		
Beech or hornbeam hedging		
1.75 m × 400 mm wide at 500 mm centres	m	87.00
1.75 m × 500 mm wide at 750 mm centres	m	57.00
Yew (Taxus) hedging		
1.50 m × 500 mm wide at 500 mm centres	m	210.00
1.50 m × 500 mm wide at 750 mm centres	m	130.00
1.50 m × 500 mm wide at 1.00 m centres	m	110.00
1.75 m × 500 mm wide at 500 mm centres	m	240.00
1.75 m × 500 mm wide at 750 mm centres	m	150.00
1.75 m × 500 mm wide at 1.00 m centres	m	125.00
Laurel (Prunus) hedging		
2.00 m × 500 mm wide at 1.00 m centres	m	14.20
Box (Buxus) hedging		
800 mm/1.00 m high × 300 mm wide	m	86.00
1.00 m/1.20 m high × 300 mm wide	m	180.00
Readyhedge Ltd; pre-clipped and preformed hedges planting; excavate trench by machine 500 mm wide × 500 mm deep; add compost at 100 mm/m² mixed to excavated material; plant mature hedge plants; backfill with excavated material and compost; allow for disposal of 50% of excavated material to spoil heaps 50 m distant		
Beech or Hornbeam hedging		
1.75 m high × 300 mm wide at 400 mm centres	m	82.00
1.75 m high × 300 mm wide at 500 mm centres	m	67.00
1.75 m high × 300 mm wide at 750 mm centres	m	51.00
1.75 m high × 300 mm wide at 900 mm centres	m	42.00
2.00 m high × 300 mm wide at 400 mm centres	m	190.00
2.00 m high × 300 mm wide at 500 mm centres	m	155.00
Yew (Taxus) hedging		
700 mm high × 250 mm wide	m	210.00
1.60–1.80 m high × 350 mm wide	m	240.00
Readyhedge; strip hedge planting; excavate trench 700 mm wide × 500 mm deep by machine; add compost at 100 mm/m² mixed to excavated material; plant instant strip hedging; backfill with excavated material and compost; allow for disposal of 50% of excavated material to spoil heaps 50 m distant		
Beech hedging		
1.00 m high × 400 mm wide	m	94.00
1.60/1.80 m high × 500 mm wide	m	210.00
Yew (Taxus) hedging		
700 m high × 250 mm wide	m	210.00
Laurel (Prunus) hedging		
1.80/2.00 m high × 500 mm wide	m	190.00
Box (Buxus) hedging		
800 mm/1.00 m high × 300 mm wide	m	86.00

Approximate Estimates

8.3.2 EXTERNAL PLANTING

Item – Overhead and Profit Included	Unit	Range £
HEDGE PLANTING – cont		
Griffin Nurseries; individual feathered hedge planting; excavate trench by machine 500 mm wide × 500 mm deep; add compost at 100 mm/m² mixed to excavated material; plant mature hedge plants; backfill with excavated material and compost; allow for disposal of 50% of excavated material to spoil heaps 50 m distant		
Yew hedging		
1.0–1.20 m high × 300 mm wide at 2/m	m	77.00
1.25–1.50 m high × 300 mm wide at 3/m	m	180.00
1.50–1.75 m high at 450 mm centres	m	200.00
1.50–1.75 m high at 600 mm centres	m	150.00
Griffin nurseries; Works by machine; excavate trench for hedge 300 mm wide × 300 mm deep; deposit spoil alongside and plant hedging plants in single row; backfill with excavated material		
Box hedging; rootballed		
400–500 mm high; 200 mm centres	m	58.00
400–500 mm high; 300 mm centres	m	40.00
400–500 mm high; 500 mm centres	m	25.50
500–600 mm mm high; 200 mm centres	m	78.00
500–600 mm mm high; 300 mm centres	m	53.00
500–600 mm mm high; 600 mm centres	m	34.50
600–800 mm high; 250 mm centres	m	87.00
600–800 mm high; 400 mm centres	m	56.00
800 mm–1.0 m high; 300 mm centres	m	100.00
800 mm–1.0 m high; 500 mm centres	m	64.00
1.0–1.25 m high; 600 mm centres	m	82.00
1.0–1.25 m high; 900 mm centres	m	56.00
ANNUAL BEDDING		
BEDDING		
Spray surface with glyphosate; lift and dispose of turf when herbicide action is complete; cultivate new area for bedding plants to 400 mm deep; spread compost 100 mm deep and chemical fertilizer Enmag and rake to fine tilth to receive new bedding plants; remove all arisings		
Existing turf area		
disposal to skip	m²	8.05
disposal to compost area on site; distance 25 m	m²	8.45
Plant bedding to existing planting area; bedding planting PC £0.25 each		
Clear existing bedding; cultivate soil to 230 mm deep; incorporate compost 75 mm and rake to fine tilth; collect bedding from nursery and plant at 100 mm centres; irrigate on completion; maintain weekly for 12 weeks		
mass planted; 100 mm centres	m²	38.50
to patterns; 100 mm centres	m²	44.00
mass planted; 150 mm centres	m²	25.00
to patterns; 150 mm centres	m²	30.00
mass planted; 200 mm centres	m²	25.00
to patterns; 200 mm centres	m²	30.00

8.3.2 EXTERNAL PLANTING

Item – Overhead and Profit Included	Unit	Range £
Watering of bedding by hand held hose pipe; The prices shown here are per occasion; Bedding may require up to 40 visits during a dry season		
Flow rate 25 litres/minute; per occasion		
10 litres/m²	100 m²	0.20
15 litres/m²	100 m²	0.30
20 litres/m²	100 m²	0.40
25 litres/m²	100 m²	0.50
Flow rate 40 litres/minute		
10 litres/m²	100 m²	0.10
15 litres/m²	100 m²	0.20
20 litres/m²	100 m²	0.20
25 litres/m²	100 m²	0.30
LIVING WALL		
Living wall; Scotscape Ltd; Design and installation of planted modules with automatic irrigation systems		
Fabric based systems; indicative area rates as shown		
Wall 20 m²	m²	650.00
Wall 50 m²	m²	550.00
Wall 100 m²	m²	530.00
Wall 150 m²	m²	510.00
Annual maintenance costs for living walls		
Walls over 30–70 m²	item	5200.00
Walls under over 150 m²	item	12000.00
PLANTERS		
To brick planter; coat insides with 2 coats RIW liquid asphaltic composition; fill with 50 mm shingle and cover with geofabric; fill with screened topsoil incorporating 25% Topgrow compost and Enmag		
Planters 1.00 m deep		
1.00 × 1.00	each	150.00
1.00 × 2.00	each	250.00
1.00 × 3.00	each	355.00
Planters 1.50 m deep		
1.00 × 1.00	each	220.00
1.00 × 2.00	each	320.00
1.00 × 3.00	each	530.00
Container planting; fill with 50 mm shingle and cover with geofabric; fill with screened topsoil incorporating 25% Topgrow compost and Enmag		
Planters 1.00 m deep		
400 × 400 × 400 mm deep	each	12.30
400 × 400 × 600 mm deep	each	14.30
1.00 m × 400 mm wide × 400 mm deep	each	25.00
1.00 m × 600 mm wide × 600 mm deep	each	33.00
1.00 m × 100 mm wide × 400 mm deep	each	34.00
1.00 m × 100 mm wide × 600 mm deep	each	48.50
1.00 m × 100 mm wide × 1.00 m deep	each	79.00

8.3.2 EXTERNAL PLANTING

Item – Overhead and Profit Included	Unit	Range £
PLANTERS – cont		
Container planting – cont		
Planters 1.00 m deep – cont		
1.00 m dia. × 400 mm deep	each	27.00
1.00 m dia. × 1.00 m deep	each	63.00
2.00 m dia. × 1.00 m deep	each	240.00
Street planters		
Supply and locate in position precast concrete planters; fill with topsoil placed over 50 mm shingle and terram; plant with assorted 5 litre and 3 litre shrubs to provide instant effect		
700 mm dia. × 470 mm high; white exposed aggregate finish; Marshalls Boulevard	each	1050.00
LANDSCAPE MAINTENANCE		
AS PART OF A TERM MAINTENANCE CONTRACT		
MAINTENANCE OF ORNAMENTAL SHRUB BEDS		
Ornamental shrub beds; private gardens or business parks and resticted commercial environments		
Hand weed ornamental shrub bed during the growing season; clear leaves and litter; edge beds when necessary; prune shrubs or hedges and remove; replace mulch to 50 mm at end of each year; (mulched areas only) planting less than 2 years old		
Mulched beds; weekly visits; 12 month period planting centres shown; public areas; shopping areas		
600 mm centres	100 m^2	210.00
400 mm centres	100 m^2	190.00
300 mm centres	100 m^2	170.00
ground covers	100 m^2	150.00
Mulched beds; weekly visits; 12 month period planting centres shown; private areas; business or office parks or similar		
600 mm centres	100 m^2	190.00
400 mm centres	100 m^2	155.00
300 mm centres	100 m^2	130.00
ground covers	100 m^2	110.00
Non-mulched beds; weekly visits; planting centres		
600 mm centres	100 m^2	180.00
400 mm centres	100 m^2	170.00
300 mm centres	100 m^2	165.00
ground covers	100 m^2	160.00
Non-mulched beds; monthly visits; planting centres		
600 mm centres	100 m^2	145.00
400 mm centres	100 m^2	135.00
300 mm centres	100 m^2	130.00
ground covers	100 m^2	120.00

8.3.2 EXTERNAL PLANTING

Item – Overhead and Profit Included	Unit	Range £
Maintain rose bed		
Weed around base of plants; dead head regularly; fertilize once per annum with Enmag; prune once per annum; edge with half moon edging tool; add compost to 50 mm once per annum; clear of litter and leaves;		
private or restricted access beds	100 m²	385.00
public areas	100 m²	430.00
Remulch planting bed at the start of the planting season; top up mulch 25 mm thick; Melcourt Ltd		
Larger areas maximum distance 25 m; 80 m³ loads		
Ornamental bark mulch	100 m²	200.00
Melcourt Bark Nuggets	100 m²	200.00
Amenity Bark	100 m²	150.00
Forest biomulch	100 m²	145.00
Smaller areas; maximum distance 25 m; 25 m³ loads		
Ornamental bark mulch	100 m²	240.00
Melcourt Bark Nuggets	100 m²	235.00
Amenity Bark	100 m²	190.00
Forest biomulch	100 m²	180.00
Shrub beds in ornamental gardens		
Maintain mature shrub bed; shrubs average 1.00 m–3.00 m high with occasional mature trees in bed or in proximity; prune one third of all planting by 30% every year in rotation; fertilize; add compost; annually; add fertilizers; clear leaves; edge bed with half moon tool; water with hand held hose on 6 occasions		
massed plants; 1.00 m high	m²	590.00
plants up to 2 m high	m²	670.00
plants 2 m–3 m high	m²	870.00
VEGETATION CONTROL		
Native planting; roadside railway or forestry planted areas.		
Post planting maintenance; control of weeds and grass; herbicide spray applications; maintain weed free circles 1.00 m dia. to planting less than 5 years old in roadside, rail or forestry planting environments and the like; strim grass to 50–75 mm; prices per occasion (three applications of each operation normally required)		
Knapsack spray application; glyphosate; planting at		
1.50 m centres	ha	1625.00
1.75 m centres	ha	1025.00
2.00 m centres	ha	1100.00

8.3.2 EXTERNAL PLANTING

Item – Overhead and Profit Included	Unit	Range £
LANDSCAPE MAINTENANCE – cont		
Maintain planted areas; control of weeds and grass; maintain weed free circles 1.00 m dia. to planting less than 5 years old in roadside, rail or forestry planting environments and the like; strim surrounding grass to 50–75 mm; prices per occasion (three applications of each operation normally required)		
Herbicide spray applications; CDA (controlled droplet application) glyphosate and strimming; plants planted at the following centres		
1.50 m centres	ha	900.00
1.75 m centres	ha	1025.00
2.00 m centres	ha	1100.00
Post planting maintenance; control of weeds and grass; herbicide spray applications; CDA (controlled droplet application); Xanadu glyphosate/ diuron; maintain weed free circles 1.00 m dia. to planting less than 5 years old in roadside, rail or forestry planting environments and the like; strim grass to 50–75 mm; prices per occasion (1.5 applications of herbicide and three strim operations normally required)		
Plants planted at the following centres		
1.50 m centres	ha	950.00
1.75 m centres	ha	1075.00
2.00 m centres	ha	1150.00
MAINTENANCE AS DEFECTS LIABILITY AS PART OF A LANDSCAPE CONSTRUCTION PROJECT		
Clean and weed plant beds; fork over as necessary; water with full liability for the duration of the first growing season using moveable sprinklers; clear leaf fall; top up mulch at the end of the 12 month period		
1st year after planting		
weekly visits to high profile areas new planting areas	100 m²	510.00
monthly visits to new planting areas	100 m²	455.00

8.3.3 IRRIGATION SYSTEMS

Item – Overhead and Profit Included	Unit	Range £
IRRIGATION SYSTEMS		
LANDSCAPE IRRIGATION		
Automatic irrigation; TK Solutions Ltd		
Large garden consisting of 24 stations; 7000 m² irrigated area		
turf only	nr	13000.00
70/30% turf/shrub beds	nr	16500.00
Medium sized garden consisting of 12 stations; 3500 m² irrigated area		
turf only	nr	9000.00
70/30% turf/shrub beds	nr	13000.00
Medium sized garden consisting of 6 stations of irrigated area; 1000 m²		
turf only	nr	6000.00
70/30% turf/shrub beds	nr	6700.00
50/50% turf/shrub beds	nr	7400.00
Leaky pipe irrigation; Leaky Pipe Ltd		
Works by machine; main supply and connection to laterals; excavate trench for main or ring main 450 mm deep; supply and lay pipe; backfill and lightly compact trench		
20 mm LDPE	100 m	330.00
16 mm LDPE	100 m	295.00
Works by hand; main supply and connection to laterals; excavate trench for main or ring main 450 mm deep; supply and lay pipe; backfill and lightly compact trench		
20 mm LDPE	100 m	2175.00
16 mm LDPE	100 m	2150.00
Turf area irrigation; laterals to mains; to cultivated soil; excavate trench 150 mm deep using hoe or mattock; lay moisture leaking pipe laid 150 mm below ground at centres of 350 mm		
low leak	100 m²	660.00
high leak	100 m²	670.00
Landscape area irrigation; laterals to mains; moisture leaking pipe laid to the surface of irrigated areas at 600 mm centres		
low leak	100 m²	320.00
high leak	100 m²	325.00
Landscape area irrigation; laterals to mains; moisture leaking pipe laid to the surface of irrigated areas at 900 mm centres		
low leak	100 m²	215.00
high leak	100 m²	220.00
multi-station controller	each	430.00
solenoid valves connected to automatic controller	each	610.00

8.3.3 IRRIGATION SYSTEMS

Item – Overhead and Profit Included	Unit	Range £
RAINWATER HARVESTING		
Rainwater Harvesting Ltd; rainwater tanks; soft landscape areas installed below ground		
Excavate in soft landscape area by machine to install rainwater harvesting tank; grade base and lay compacted granular base to level; connect filters to rainwater down pipe of building 10 m distance and lay 110 mm uPVC pipe in trench. Install rainwater tank complete with filters for irrigation and submersible pump. cover with geofabric and backfill with excavated material compacting lightly; remove surplus material off site		
3600 litre	nr	2850.00
4500 litre	nr	3500.00
10000 litre	nr	5800.00
Excavate in soft landscape area by hand to install rainwater harvesting tank; grade base and lay compacted granular base to level; connect filters to rainwater down pipe of building 10 m distance and lay 110 mm uPVC pipe in trench. Install low profile rainwater tank complete with filters for irrigation and submersible pump; cover with geofabric and backfill with excavated material compacting lightly; remove surplus material off site		
Flat tank 1500 litre	nr	4800.00
Flat tank 3000 litre	nr	6000.00

8.4.1 FENCING AND RAILINGS

Item – Overhead and Profit Included	Unit	Range £
TEMPORARY FENCING		
Site fencing; supply and erect temporary protective fencing and remove at completion of works		
Cleft chestnut paling		
1.20 m high 75 mm larch posts at 3 m centres	100 m	1975.00
Heras fencing		
2 week hire period	100 m	980.00
4 week hire period	100 m	1250.00
8 week hire period	100 m	1775.00
12 week hire period	100 m	2325.00
16 week hire period	100 m	2850.00
24 week hire period	100 m	3900.00
CHAIN LINK AND WIRE FENCING		
Chain link fencing; supply and erect chain link fencing; form post holes and erect concrete posts and straining posts with struts at 50 m centres all set in 1:3:6 concrete; fix line wires		
3 mm galvanized wire 50 mm chain link fencing		
1800 mm high	m	49.00
Plastic coated 3.15 gauge galvanized wire mesh		
900 mm high	m	37.50
1200 mm high	m	41.00
1800 mm high	m	57.00
Extra for additional concrete straining posts with 1 strut set in concrete		
900 mm high	each	115.00
1200 mm high	each	125.00
1800 mm high	each	135.00
Extra for additional concrete straining posts with 2 struts set in concrete		
900 mm high	each	165.00
1200 mm high	each	175.00
1800 mm high	each	195.00
Extra for additional angle iron straining posts with 2 struts set in concrete		
900 mm high	each	105.00
1200 mm high	each	75.00
1400 mm high	each	130.00
1800 mm high	each	150.00
AGRICULTURAL FENCING		
RABBIT FENCING		
Construct rabbit-stop fencing; erect galvanized wire netting; mesh 31 mm; 900 mm above ground, 150 mm below ground turned out and buried; on 75 mm dia. treated timber posts 1.8 m long driven 700 mm into firm ground at 4.0 m centres; netting clipped to top and bottom straining wires 2.63 mm dia.; straining post 150 mm dia. × 2.3 m long driven into firm ground at 50 m intervals		
turned in 150 mm	100 m	1800.00
buried 150 mm	100 m	1950.00

8.4.1 FENCING AND RAILINGS

Item – Overhead and Profit Included	Unit	Range £
AGRICULTURAL FENCING – cont		
Deer fence		
Construct deer-stop fencing; erect five 4 mm dia. plain galvanized wires and five 2 ply galvanized barbed wires at 150 mm spacing on 45 × 45 × 5 mm angle iron posts 2.4 m long driven into firm ground at 3.0 m centres, driven into firm ground at 3.0 m centres, with timber droppers 25 × 38 mm × 1.0 m long at 1.5 m centres	100 m	2325.00
Forestry fencing		
Supply and erect forestry fencing of three lines of 3 mm plain galvanized wire tied to 1700 × 65 mm dia. angle iron posts at 2750 m centres with 1850 × 100 mm dia. straining posts and 1600 × 80 mm dia. struts at 50.0 m centres driven into firm ground		
1800 mm high; three wires	100 m	1425.00
1800 mm high; three wires including cattle fencing	100 m	1975.00
TIMBER FENCING		
Clear fence line of existing evergreen shrubs 3 m high average; grub out roots by machine chip on site and remove off site; erect closeboarded timber fence in treated softwood, pales 100 × 22 mm lapped, 150 × 22 mm gravel boards		
Concrete posts 100 × 100 mm at 3.0 m centres set into ground in 1:3:6 concrete		
1500 mm high	m	125.00
1800 mm high	m	140.00
softwood posts; three arris rails		
1350 mm high	m	125.00
1650 mm high	m	140.00
1800 mm high	m	135.00
Erect chestnut pale fencing; cleft chestnut pales; two lines galvanized wire, galvanized tying wire, treated softwood posts at 3.0 m centres and straining posts and struts at 50 m centres driven into firm ground		
900 mm high; posts 75 mm dia. × 1200 mm long	m	15.40
1200 mm high; posts 75 mm dia. × 1500 mm long	m	19.60
Construct timber rail; horizontal hit and miss type; rails 150 × 25 mm; posts 100 × 100 mm at 1.8 m centres; twice stained with coloured wood preservative; including excavation for posts and concreting into ground (C7P)		
In treated softwood		
1800 mm high	m	84.00
Construct cleft oak rail fence with rails 300 mm minimum girth tennoned both ends; 125 × 100 mm treated softwood posts double mortised for rails; corner posts 125 × 125 mm, driven into firm ground at 2.5 m centres		
3 rails	m	26.00
4 rails	m	32.00
Birdsmouth fencing; timber		
600 mm high	m	33.00
900 mm high	m	35.00

8.4.1 FENCING AND RAILINGS

Item – Overhead and Profit Included	Unit	Range £
SECURITY FENCING		
Supply and erect chain link fence; 51 × 3 mm mesh; with line wires and stretcher bars bolted to concrete posts at 3.0 m centres and straining posts at 10 m centres; posts set in concrete 450 × 450 mm × 33% of height of post deep; fit straight extension arms of 45 × 45 × 5 mm steel angle with three lines of barbed wire and droppers; all metalwork to be factory hot-dip galvanized for painting on site		
Galvanized 3 mm mesh		
1800 mm high	m	60.00
Galvanized mesh but with PVC coated 3.15 mm mesh (dia. of wire 2.5 mm)		
900 mm high	m	45.00
1200 mm high	m	51.00
1800 mm high	m	68.00
Add to fences above for base of fence to be fixed with hairpin staples cast into concrete ground beam 1:3:6 site mixed concrete; mechanical excavation disposal to on site spoil heaps		
125 × 225 mm deep	m	6.90
Add to fences above for straight extension arms of 45 × 45 × 5 mm steel angle with three lines of barbed wire and droppers	m	5.00
Supply and erect palisade security fence Jacksons Barbican 2500 mm high with rectangular hollow section steel pales at 150 mm centres on three 50 × 50 × 6 mm rails; rails bolted to 80 × 60 mm posts set in concrete 450 × 450 × 750 mm deep at 2750 mm centres; tops of pales to points and set at 45° angle; all metalwork to be hot-dip factory galvanized for painting on site	m	100.00
Supply and erect single gate to match above complete with welded hinges and lock		
1.0 m wide	each	1275.00
4.0 m wide	each	1400.00
8.0 m wide	pair	1950.00
Supply and erect Orsogril proprietary welded steel mesh panel fencing on steel posts set 750 mm deep in concrete foundations 600 × 600 mm; supply and erect proprietary single gate 2.0 m wide to match fencing		
930 mm high	100 m	8300.00
1326 mm high	100 m	10500.00
1722 mm high	100 m	13000.00
BALL STOP FENCING		
Supply and erect plastic coated 30 × 30 mm netting fixed to 60.3 mm dia. 12 mm solid bar lattice galvanized dual posts; top, middle and bottom rails with 3 horizontal rails on 60.3 mm dia. nylon coated tubular steel posts at 3.0 m centres and 60.3 mm dia. straining posts with struts at 50 m centres set 750 mm into FND2 concrete footings 300 × 300 × 600 mm deep; include framed chain link gate 900 × 1800 mm high to match complete with hinges and locking latch		
4500 mm high	100 m	14000.00
5000 mm high	100 m	17500.00
6000 mm high	100 m	20000.00

8.4.1 FENCING AND RAILINGS

Item – Overhead and Profit Included	Unit	Range £
RAILINGS		
Conservation of historic railings; Eura Conservation Ltd		
Remove railings to workshop off site; shotblast and repair mildly damaged railings; remove rust and paint with three coats; transport back to site and re-erect		
railings with finials 1.80 m high	m	650.00
railings ornate cast or wrought iron	m	1025.00
Supply and erect mild steel bar railing of 19 mm balusters at 115 mm centres welded to mild steel top and bottom rails 40 × 10 mm; bays 2.0 m long bolted to 51 × 51 mm ms hollow section posts set in C15P concrete; all metal work galvanized after fabrication		
900 mm high	m	98.00
1200 mm high	m	140.00
1500 mm high	m	150.00
GATES		
STOCK GATE		
Erect stock gate, ms tubular field gate, diamond braced 1.8 m high hung on tubular steel posts set in concrete (C7P); complete with ironmongery; all galvanized		
Width 3.00 m	each	435.00
Width 4.20 m	each	470.00

8.4.2 WALLS AND SCREENS

Item – Overhead and Profit Included	Unit	Range £
FOUNDATIONS FOR WALLS		
IN SITU CONCRETE FOUNDATIONS		
Excavate foundation trench mechanically; remove spoil off site; lay 1:3:6 site mixed concrete foundations; distance from mixer 25 m; depth of trench to be 225 mm deeper than foundation to allow for three underground brick courses priced separately		
Concrete poured to blinded exposed ground		
200 mm deep × 400 mm wide	m	36.00
300 mm deep × 500 mm wide	m	59.00
400 mm deep × 400 mm wide	m	57.00
400 mm deep × 600 mm wide	m	86.00
600 mm deep × 600 mm wide	m	120.00
Concrete poured to formwork		
200 mm deep × 400 mm wide	m	76.00
300 mm deep × 500 mm wide	m	98.00
300 mm deep × 600 mm wide	m	105.00
300 mm deep × 800 mm wide	m	130.00
400 mm deep × 400 mm wide	m	94.00
400 mm deep × 600 mm wide	m	140.00
600 mm deep × 600 mm wide	m	175.00
Excavate foundation trench by hand; remove spoil off site; lay 1:3:6 site mixed concrete foundations; distance from mixer 25 m; depth of trench to be 225 mm deeper than foundation to allow for three underground brick courses priced separately		
Disposal to spoil heap 25 m by barrow and off site by grab vehicle		
200 mm deep × 400 mm wide	m	100.00
300 mm deep × 500 mm wide	m	140.00
400 mm deep × 400 mm wide	m	140.00
400 mm deep × 600 mm wide	m	205.00
600 mm deep × 600 mm wide	m	275.00
Excavate foundation trench mechanically; remove spoil off site; lay ready mixed concrete C10 discharged directly from delivery vehicle to location; depth of trench to be 225 mm deeper than foundation to allow for three underground brick courses priced separately; foundation size		
200 mm deep × 400 mm wide	m	26.50
300 mm deep × 500 mm wide	m	41.00
400 mm deep × 400 mm wide	m	39.50
400 mm deep × 600 mm wide	m	60.00
400 mm deep × 800 mm wide	m	73.00
400 mm deep × 1000 mm wide	m	92.00
400 mm deep × 1200 mm wide	m	110.00
600 mm deep × 600 mm wide	m	84.00
Reinforced concrete wall to foundations above (site mixed concrete)		
Up to 1.00 m high × 200 mm thick	m²	99.00
Up to 1.00 m high × 300 mm thick	m²	99.00

8.4.2 WALLS AND SCREENS

Item – Overhead and Profit Included	Unit	Range £
FOUNDATIONS FOR WALLS – cont		
Reinforced concrete wall to foundations above (ready mix concrete RC35)		
1.00 m high × 200 mm thick	m^2	140.00
1.00 m high × 300 mm thick	m^2	160.00
Grading; excavate to reduce levels for concrete slab; lay waterproof membrane, 100 mm hardcore and form concrete slab reinforced with A142 mesh in 1:2:4 site mixed concrete to thickness; remove excavated material off site		
Concrete 1:2:4 site mixed		
100 mm thick	m^2	70.00
150 mm thick	m^2	82.00
250 mm thick	m^2	100.00
300 mm thick	m^2	115.00
Concrete ready mixed GEN 2		
100 mm thick	m^2	66.00
150 mm thick	m^2	75.00
250 mm thick	m^2	92.00
300 mm thick	m^2	100.00
BLOCK WALLING		
Notes: Measurements allow for works above ground only		
Concrete block walls; including excavation of foundation trench 450 mm deep; remove spoil off site; lay site mixed concrete foundations 600 × 300 mm thick; 1 block below ground		
Solid blocks 7 N/mm^2; 100 mm thick		
500 mm high	m^2	88.00
750 mm high	m^2	100.00
1.00 m high	m^2	120.00
1.25 m high	m^2	130.00
1.50 m high	m^2	145.00
1.80 m high	m^2	165.00
140 mm thick		
generally	m^2	120.00
100 mm blocks; laid on flat		
500 mm high	m^2	130.00
750 mm high	m^2	170.00
100 mm blocks laid flat; 2 courses underground		
1.00 m high	m^2	220.00
1.20 m high	m^2	250.00
1.50 m high	m^2	290.00
1.80 m high	m^2	335.00
Hollow blocks filled with concrete; 440 × 215 × 215 mm		
500 mm high	m^2	130.00
750 mm high	m^2	160.00
1.00 m high	m^2	190.00
1.25 m high	m^2	205.00
1.50 m high	m^2	245.00
1.80 m high	m^2	280.00

8.4.2 WALLS AND SCREENS

Item – Overhead and Profit Included	Unit	Range £
Concrete block retaining walls; 2 courses underground; including excavation of foundation trench 450 mm deep; remove spoil off site; lay ready mixed concrete foundations 1200 × 400 mm thick		
Solid blocks 7 N/mm^2; 100 mm thick		
1.00 m high	m	190.00
1.50 m high	m	220.00
1.80 m high	m	245.00
Double skin overall 300 mm thick; Solid blocks 7 N/mm^2; 100 mm thick with 100 cavity filled with concrete; blockwork skins tied with 200 mm stainless steel ties at 400 mm centres		
1.00 m high	m	295.00
1.50 m high	m	370.00
1.80 m high	m	410.00
100 mm blocks; laid on flat		
1.00 m high	m^2	300.00
1.50 m high	m^2	370.00
1.80 m high	m^2	410.00
Hollow blocks filled with concrete; 215 mm thick		
1.00 m high	m^2	300.00
1.50 m high	m^2	380.00
1.80 m high	m^2	410.00
Hollow blocks with steel bar cast into the foundation		
1.00 m high	m^2	385.00
1.50 m high	m^2	470.00
1.80 m high	m^2	510.00
Concrete block wall with brick face 112.5 mm thick to stretcher bond; including excavation of foundation trench 450 mm deep; remove spoil off site; lay GEN 1 concrete foundations 600 × 300 mm thick; place stainless steel ties at 4 nr/m^2 of wall face		
Solid blocks 7 N/mm^2 with stretcher bond brick face; reclaimed bricks PC £800.00/1000		
100 mm thick	m^2	250.00
140 mm thick	m^2	265.00
Concrete block wall Rendered with K- Rend; including excavation of foundation trench 450 mm deep; remove spoil off site; lay GEN 1 concrete foundations 600 × 300 mm thick; place stainless steel ties at 4 nr/m^2 of wall face		
Solid blocks 7 N/mm^2 with stretcher bond brick face; 2 coats K Rend Silicone FT		
100 mm thick	m^2	210.00
140 mm thick	m^2	225.00

Approximate Estimates

8.4.2 WALLS AND SCREENS

Item – Overhead and Profit Included	Unit	Range £
BRICK WALLING		
One brick thick wall (225 mm thick); excavation 400 mm deep; remove arisings off site; lay GEN 1 concrete foundations 450 mm wide × 250 mm thick; all in English Garden Wall bond; laid in cement: lime: sand (1:1:6) mortar with flush joints, fair face one side; DPC two courses engineering brick in cement: sand (1:3) mortar; brick on edge coping		
Wall 900 mm high above DPC		
engineering brick (class B)	m	250.00
brick PC £500.00/1000	m	280.00
brick PC £800.00/1000	m	315.00
Curved wall; 6.00 m radius 900 mm high above DPC		
engineering brick (class B)	m	295.00
brick PC £500.00/1000	m	390.00
brick PC £800.00/1000	m	420.00
Wall 1200 mm high above DPC		
engineering brick (class B)	m	320.00
brick PC £500.00/1000	m	340.00
brick PC £800.00/1000	m	390.00
Curved wall; 6.00 m radius; 1200 mm high above DPC		
engineering brick (class B)	m	385.00
brick PC £500.00/1000	m	520.00
brick PC £800.00/1000	m	760.00
Wall 1800 mm high above DPC		
engineering brick (class B)	m	430.00
brick PC £500.00/1000	m	490.00
brick PC £800.00/1000	m	570.00
Curved wall; 6.00 m radius; 1800 mm high above DPC		
engineering brick (class B)	m	550.00
brick PC £500.00/1000	m	720.00
brick PC £800.00/1000	m	770.00
One and a half brick wall; excavate 450 mm deep; remove arisings off site; lay GEN 1 concrete foundations 600 × 300 mm thick; two thick brick piers at 3.0 m centres; all in English Garden Wall bond; laid in cement: lime: sand (1:1:6) mortar with flush joints; fair face one side; DPC two courses engineering brick in cement: sand (1:3) mortar; coping of headers on edge		
Wall 900 mm high above DPC		
engineering brick (class B)	m	390.00
brick PC £500.00/1000	m	420.00
Wall 1200 mm high above DPC		
engineering brick (class B)	m	560.00
brick PC £500.00/1000	m	355.00
Wall 1800 mm high above DPC		
engineering brick (class B)	m	690.00
brick PC £500.00/1000	m	550.00
brick PC £800.00/1000	m	1525.00

8.4.2 WALLS AND SCREENS

Item – Overhead and Profit Included	Unit	Range £
BRICK WALLING WITH PIERS		
Half brick thick wall; 102.5 mm thick; with one brick piers at 2.00 m centres; excavate foundation trench 500 mm deep; remove spoil off site; lay site mixed concrete foundations 1:3:6 350 × 150 mm thick; laid in cement: lime: sand (1:1:6) mortar with flush joints; fair face one side; DPC two courses underground; engineering brick in cement: sand (1:3) mortar; coping of headers on end		
Wall 900 mm high above DPC		
engineering brick (class B)	m	290.00
brick PC £500.00/1000	m	185.00
brick PC £800.00/1000	m	235.00
One brick thick wall; 225 mm thick; with one and a half brick piers at 3.0 m centres; excavation 400 mm deep; remove arisings off site; lay GEN 1 concrete foundations 450 mm wide × 250 mm thick; all in English Garden Wall bond; laid in cement: lime: sand (1:1:6) mortar with flush joints; fair face one side; DPC two courses engineering brick in cement: sand (1:3) mortar; precast concrete coping 152 × 75 mm		
Wall 900 mm high above DPC		
engineering brick (class B)	m	490.00
brick PC £500.00/1000	m	280.00
brick PC £800.00/1000	m	700.00
Wall 1200 mm high above DPC		
engineering brick (class B)	m	550.00
brick PC £500.00/1000	m	340.00
brick PC £800.00/1000	m	3900.00
Wall 1800 mm high above DPC		
engineering brick (class B)	m	790.00
brick PC £500.00/1000	m	490.00
brick PC £800.00/1000	m	5600.00
FACED BLOCK WALLING		
Concrete block wall with brick face 112.5 mm thick to bond pattern using snap headers; including excavation of foundation trench 450 mm deep; remove spoil off site; lay GEN 1 concrete foundations 600 × 300 mm thick; place stainless steel ties at 4 nr/m² of wall face; lay coping of brick on edge in engineering brick		
Solid blocks 7 N/mm^2 with snap header brick face; bricks PC £800.00/1000		
500 mm high	m^2	185.00
750 mm high	m^2	230.00
1.00 m high	m^2	270.00
1.50 m high	m^2	365.00
1.80 m high	m^2	450.00

8.4.2 WALLS AND SCREENS

Item – Overhead and Profit Included	Unit	Range £
PIERS		
BLOCK PIERS		
Block piers; blocks 440 × 215 × 100 mm thick laid on flat piers on concrete footings 600 × 600 × 250 mm thick; inclusive of excavations and disposal off site; coping of engineering brick on edge; 2 courses underground		
Pier 450 × 450 mm; to receive cladding treatment priced separately		
500 mm high	nr	86.00
750 mm high	nr	110.00
1.00 m high	nr	150.00
1.25 m high	nr	175.00
1.50 m high	nr	200.00
1.80 m high	nr	250.00
Block piers; blocks Hollow blocks 440 × 215 × 215 mm thick on concrete footings 600 × 600 × 250 mm thick; inclusive of excavations and disposal off site; coping of engineering brick on edge; 2 courses underground		
Pier 450 × 450 mm; to receive cladding treatment priced separately		
500 mm high	nr	82.00
750 mm high	nr	110.00
1.00 m high	nr	150.00
1.25 m high	nr	175.00
1.50 m high	nr	200.00
1.80 m high	nr	250.00
BRICK PIERS		
Brick piers on concrete footings 500 × 500 × 250 mm thick; inclusive of excavations and disposal off site; coping of engineering brick on edge; 2 courses underground		
Pier 215 × 215 mm; one brick thick; brick PC £500/1000		
500 mm high	nr	79.00
750 mm high	nr	98.00
1.00 m high	nr	120.00
1.25 m high	nr	135.00
1.50 m high	nr	155.00
1.80 m high	nr	180.00
Pier 215 × 215 mm; one brick thick; brick PC £800/1000		
500 mm high	nr	86.00
750 mm high	nr	110.00
1.00 m high	nr	130.00
1.25 m high	nr	150.00
1.50 m high	nr	170.00
1.80 m high	nr	200.00
Pier 337.5 × 337.5 mm; one and a half brick thick; brick PC £500/1000		
500 mm high	nr	135.00
750 mm high	nr	175.00
1.00 m high	nr	210.00
1.25 m high	nr	250.00
1.50 m high	nr	290.00
1.80 m high	nr	340.00

8.4.2 WALLS AND SCREENS

Item – Overhead and Profit Included	Unit	Range £
Pier 337.5 × 337.5 mm; one and a half brick thick; brick PC £800/1000		
500 mm high	nr	360.00
750 mm high	nr	490.00
1.00 m high	nr	620.00
1.25 m high	nr	740.00
1.50 m high	nr	870.00
1.80 m high	nr	1025.00
Brick piers on concrete footings 800 × 800 × 600 mm thick; inclusive of excavations and disposal off site; coping of Haddonstone S150C classically moulded corbelled piercap PC £127.11; 120 mm thick overall		
Pier 450 × 450 mm; two bricks thick; brick PC £500/1000; height of brickwork above ground		
500 mm high	nr	430.00
750 mm high	nr	490.00
1.00 m high	nr	550.00
1.25 m high	nr	610.00
1.50 m high	nr	670.00
1.80 m high	nr	750.00
2.00 m high	nr	800.00
Pier 450 × 450 mm; two bricks thick; brick PC £800/1000; height of brickwork above ground		
500 mm high	nr	450.00
750 mm high	nr	520.00
1.00 m high	nr	600.00
1.25 m high	nr	670.00
1.50 m high	nr	740.00
1.80 m high	nr	820.00
2.00 m high	nr	880.00
Brick piers on concrete footings 1000 × 1000 × 600 mm thick; inclusive of excavations and disposal off site; coping Haddonstone S215C classically moulded corbelled pyramidal piercap PC £235.22		
Pier 600 × 600 mm; three bricks thick; brick PC £800/1000; height of brickwork above ground		
1.50 m high	nr	1425.00
1.80 m high	nr	1575.00
2.00 m high	nr	1700.00
IN SITU CONCRETE WALLS		
Excavate foundation trench mechanically; remove spoil off site; fix reinforcement starter bars 12 mm at 200 mm centres; lay 1:3:6 site mixed concrete foundations; distance from mixer 25 m; depth of trench to be 225 mm deeper than foundation; cast in situ concrete walls inclusive of bar reinforcement 12 mm; footings cast to blinded exposed ground		
Wall height 500 mm above ground; on foundation 400 mm × 200 mm thick		
150 thick wall; site mixed concrete 20 tonne aggregate loads	m	135.00
150 thick wall; ready mixed concrete C25	m	125.00
250 thick wall; site mixed concrete; 20 tonne aggregate loads	m	190.00
250 thick wall; ready mixed concrete C25	m	175.00

8.4.2 WALLS AND SCREENS

Item – Overhead and Profit Included	Unit	Range £
IN SITU CONCRETE WALLS – cont		
Excavate foundation trench mechanically – cont		
Wall 1.00 m high on foundation 500 mm wide × 300 mm deep		
150 thick wall; site mixed concrete; 20 tonne aggregate loads	m	220.00
150 thick wall; ready mixed concrete C25	m	200.00
250 thick wall; site mixed concrete; 20 tonne aggregate loads	m	245.00
250 thick wall; ready mixed concrete C25	m	220.00
Wall 1.50 m high on foundation 500 mm wide × 300 mm deep		
250 thick wall; site mixed concrete; 20 tonne aggregate loads	m	465.00
250 thick wall; ready mixed concrete C25	m	430.00
Wall 1.80 m high on foundation 600 mm wide × 300 mm deep		
250 thick wall; site mixed concrete; 20 tonne aggregate loads	m	590.00
250 thick wall; ready mixed concrete C25	m	550.00
Wall 1.80 m high on foundation 800 mm wide × 300 mm deep		
250 thick wall; site mixed concrete; 20 tonne aggregate loads	m	610.00
250 thick wall; ready mixed concrete C25	m	580.00
Curved walls; 6.00 m radius		
Wall 1.50 m high on foundation 500 mm wide × 300 mm deep		
250 thick wall; site mixed concrete; 20 tonne aggregate loads	m	500.00
250 thick wall; ready mixed concrete C25	m	465.00
Wall 1.80 m high on foundation 600 mm wide × 300 mm deep		
250 thick wall; site mixed concrete; 20 tonne aggregate loads	m	630.00
250 thick wall; ready mixed concrete C25	m	590.00
Curved walls; 3.00 m radius		
Wall 1.50 m high on foundation 500 mm wide × 300 mm deep		
250 thick wall; site mixed concrete; 20 tonne aggregate loads	m	550.00
250 thick wall; ready mixed concrete C25	m	520.00
Wall 1.80 m high on foundation 600 mm wide × 300 mm deep		
250 thick wall; site mixed concrete; 20 tonne aggregate loads	m	690.00
250 thick wall; ready mixed concrete C25	m	650.00

8.4.2 WALLS AND SCREENS

Item – Overhead and Profit Included	Unit	Range £
SITE MIXED CONCRETE		
Mix concrete on site; aggregates delivered in 20 tonne loads; deliver mixed concrete to location by mechanical dumper distance 25 m		
1:3:6	m³	120.00
1:2:4	m³	130.00
Ready mixed concrete; deliver mixed concrete to location by mechanical dumper distance 25 m		
10 N/mm²	m³	120.00
15 N/mm²	m³	120.00
Mix concrete on site; aggregates delivered in 20 tonne loads; deliver mixed concrete to location by barrow distance 25 m		
Aggregates delivered in 20 t loads to site		
1:3:6	m³	190.00
1:2:4	m³	210.00
Aggregates delivered in 850 kg bulk bags		
1:3:6	m³	230.00
1:2:4	m³	235.00
Ready mixed concrete		
10 N/mm²	m³	195.00
15 N/mm²	m³	200.00
BALUSTRADES		
Bottle balustrades reconstituted stone; Haddonstone Ltd; to prepared brickwork walls or bases (not included).		
Top and bottom rails to support balustrades; laid to builders work of brick block or other stonework supports (not included); under coping; pier shafts; Piers 798 mm high		
Overall 1519 mm high Balusters 457 mm high	m	680.00
Overall 1672 mm high; Balusters 610 mm high	m	750.00
Top and bottom rails to support balustrades; laid to builders work of brick block or other stonework supports (not included); under coping; pier shafts; Piers 898 mm high		
Overall 1619 mm high; Balusters 457 mm high	m	700.00
Overall 1712 mm high; Balusters 610 mm high	m	770.00

8.4.3 RETAINING WALLS

Item – Overhead and Profit Included	Unit	Range £
FOOTINGS FOR RETAINING WALLS		
Excavate trench for retaining wall; Cart material off site; pour concrete footings with L shaped T20 starter bars; total length 2.00 m at 200 mm centres; Concrete poured to 225 mm below ground level		
Ready mixed concrete		
600 mm wide × 300 mm deep	m	73.00
800 mm wide × 300 mm deep	m	86.00
1.00 m wide × 300 mm deep	m	100.00
1.20 m wide × 300 mm deep	m	130.00
1.00 m wide × 400 mm deep	m	130.00
1.50 m wide × 300 mm deep	m	150.00
1.50 m wide × 400 mm deep	m	200.00
Site mixed concrete; Aggregates in 20 tonne loads		
600 mm wide × 300 mm deep	m	81.00
800 mm wide × 300 mm deep	m	97.00
1.00 m wide × 300 mm deep	m	110.00
1.20 m wide × 300 mm deep	m	145.00
1.00 m wide × 400 mm deep	m	150.00
1.50 m wide × 300 mm deep	m	170.00
1.50 m wide × 400 mm deep	m	210.00
BLOCK RETAINING WALLS		
Note: The following models do not allow for excavation to the existing bank in preparation of the retaining system. Please amalgamate with the Grading and preparation section above.		
Concrete block retaining walls; Stepoc		
Excavate trench 750 mm deep and lay concrete foundation 750 mm wide × 400 mm deep; construct Forticrete precast hollow concrete block wall with 450 mm below ground laid all in accordance with manufacturer's instructions; fix reinforcing bar 12 mm as work proceeds; fill blocks with ready mixed concrete; Allow for two coats of bitumen based waterproofing to back of wall; backfilling against face of wall with drainage trench perforated pipe filled with gravel rejects 600 mm deep		
Walls 1.00 m high		
type 200; 400 × 225 × 200 mm	m	265.00
type 256; 400 × 225 × 256 mm	m	295.00
Walls 1.50 m high		
type 200; 400 × 225 × 200 mm	m	320.00
type 256; 400 × 225 × 256 mm	m	360.00
Excavate trench 750 mm deep and lay concrete foundation 1.80 mm wide × 450 mm deep; construct Forticrete precast hollow concrete block wall with 450 mm below ground laid all in accordance with manufacturer's instructions; fix reinforcing bar 12 mm as work proceeds; fill blocks with ready mixed concrete; Allow for two coats of bitumen based waterproofing to back of wall; backfilling against face of wall with drainage trench perforated pipe filled with gravel rejects 600 mm deep		

8.4.3 RETAINING WALLS

Item – Overhead and Profit Included	Unit	Range £
Walls 1.50 m high		
type 200; 400 × 225 × 200 mm	m	475.00
type 256; 400 × 225 × 256 mm	m	520.00
Walls 1.80 m high		
type 200; 400 × 225 × 200 mm	m	510.00
type 256; 400 × 225 × 256 mm	m	560.00
Hollow concrete block retaining wall		
Excavate trench 750 mm deep and lay concrete foundation 750 mm wide × 400 mm deep; construct hollow concrete block wall with 450 mm below ground laid all in accordance with manufacturer's instructions; fix reinforcing bar 12 mm as work proceeds; fill blocks with concrete 1:3:6 as work proceeds;Allow for two coats of bitumen based waterproofing to back of wall; backfilling against face of wall with drainage trench perforated pipe filled with gravel rejects 600 mm deep		
Walls 1.00 m high	m	250.00
Walls 1.25 m high	m	275.00
Walls 1.50 m high	m	305.00
Excavate trench 750 mm deep and lay concrete foundation 1.80 m wide × 400 mm deep; construct hollow concrete block wall with 225 mm below ground; fix reinforcing bar 12 mm as work proceeds; fill blocks with concrete 1:3:6 as work proceeds; Allow for two coats of bitumen based waterproofing to back of wall; backfilling against face of wall with drainage trench perforated pipe filled with gravel rejects 600 mm deep		
Walls 1.50 m high	m	460.00
Walls 1.80 m high	m	490.00
PRECAST CONCRETE RETAINING WALLS		
Excavate trench and lay concrete foundation 600 × 300; 1:3:6: plain concrete, supply and install Milton Precast Concrete precast concrete L shaped units, constructed all in accordance with manufacturer's instructions; backfill with approved excavated material compacted as the work proceeds		
1500 mm high × 1000 mm wide	m	345.00
2500 mm high × 1000 mm wide	m	660.00
3000 mm high × 1000 mm wide	m	760.00
GABION WALLS		
Gabion walls; excavation costs excluded; see Grading and preparation of banks above		
Gabions 500 mm high; allowance of 1.00 m working space at top of bank; backfill working space and remove balance of arisings off site; lay concrete footing; 200 mm deep × 1.50 m wide		
Height retained		
500 mm	m	175.00
1.00 m	m	285.00
1.50 m	m	410.00
2.00 m	m	520.00

8.4.3 RETAINING WALLS

Item – Overhead and Profit Included	Unit	Range £
TIMBER RETAINING WALLS		
TIMBER LOG RETAINING WALLS		
Excavate trench 300 mm wide to one third of the finished height of the retaining walls below; lay 100 mm hardcore; fix machine rounded logs set in concrete 1:3:6; remove excavated material from site; fix geofabric to rear of timber logs; backfill with previously excavated material set aside in position; all works by machine		
100 mm dia. logs		
500 mm high (constructed from 1.80 m lengths)	m	120.00
1.20 m high (constructed from 1.80 m lengths)	m	190.00
1.60 m high (constructed from 2.40 m lengths)	m	270.00
2.00 m high (constructed from 3.00 m lengths)	m	390.00
150 mm dia. logs		
1.20 m high (constructed from 1.80 m lengths)	m	330.00
1.60 m high (constructed from 2.40 m lengths)	m	380.00
Timber crib wall		
Excavate trench to receive foundation 300 mm deep; place plain concrete foundation 150 mm thick in 11.50 N/mm² concrete (sulphate-resisting cement); construct timber crib retaining wall and backfill with excavated spoil behind units		
timber crib wall system; average 5.0 m high	m	1550.00
timber crib wall system; average 4.0 m high	m	1150.00
timber crib wall system; average 2.0 m high	m	510.00
timber crib wall system; average 1.00 m high	m	260.00
Railway sleeper wall; Excavate and remove spoil off site; Pour concrete footing 500 × 250 deep; lay sleeper retaining wall		
Grade 1 hardwood sleeper 2 450 m × 250 mm wide × 150 mm high; Sleepers set to galvanized pins set in the concrete		
150 high	m	41.00
300 high	m	63.00
450 high	m	84.00
600 high	m	105.00
Grade 1 hardwood; 2 galvanized angle iron stakes per sleeper length set into concrete		
750 high	m	160.00
900 high	m	180.00
CORTEN RETAINING WALLS		
Corten Retaining Walls – Adezz; Cast strip footings 200–500 mm wide as appropriate × 250 mm deep; Fix Corten retaining wall to footings; Excavation of banks or backfilling to planters not included		
200 mm high		
straight	m	50.00
internal or external corners	m	61.00
curved – short lengths	m	110.00
curved – longer lenghts of up to 2.00 m each	m	105.00

8.4.3 RETAINING WALLS

Item – Overhead and Profit Included	Unit	Range £
400 mm high		
straight	m	87.00
internal or external corners	m	98.00
curved – short lengths	m	180.00
curved – longer lenghts of up to 2.00 m each	m	170.00
600 mm high		
straight	m	160.00
internal or external corners	m	170.00
curved – short lengths	m	305.00
curved – longer lenghts of up to 2.00 m each	m	300.00

RECYCLED PLASTIC RETAINING WALLS

HAHN RECYCLED PLASTIC RETAINING WALL SYSTEMS

Excavate trench 300 mm wide to one third of the finished height of the retaining walls below; lay 100 mm hardcore; fix recycled retaining system; concrete; 1:3:6; remove excavated material from site; fix geofabric to rear of units; backfill with previously excavated material set aside in position; all works by machine

Ogee C Section interlocking units; 100 mm dia.		
300 mm high	m	165.00
600 mm high	m	210.00
1.20 m high	m	320.00
Round palisades; 100 mm dia.; solid profile; retained height		
1.20 m high	m	280.00
1.60 m high (constructed from 2.40 m lengths)	m	455.00

8.5 EXTERNAL FIXTURES

Item – Overhead and Profit Included	Unit	Range £
8.5.1 SITE/STREET FURNITURE AND EQUIPMENT		
Bollards and access restriction		
Supply and install 10 nr cast iron Doric bollards 920 mm high above ground × 170 mm dia. bedded in concrete base 400 mm dia. × 400 mm deep	10 nr	3100.00
Benches and seating		
In grassed area excavate for base 2500 × 1575 mm and lay 100 mm hardcore, 100 mm concrete, brick pavers in stack bond bedded in 25 mm cement: lime: sand mortar; supply and fix where shown on drawing proprietary seat, hardwood slats on black powder coated steel frame, bolted down with 4 nr 24 × 90 mm recessed hex-head stainless steel anchor bolts set into concrete	set	1750.00
Cycle stand		
Supply and fix cycle stand 1250 m × 550 mm of 60.3 mm black powder coated hollow steel sections, one-piece with rounded top corners; set 250 mm into paving	each	495.00
8.5.2 ORNAMENTAL WATER FEATURES		
FAIRWATER LTD		
Excavate for small pond or lake maximum depth 1.00 m; remove arisings off site; grade and trim to shape; lay 75 mm sharp sand; line with 0.75 mm butyl liner 75 mm sharp sand and geofabric; cover over with sifted topsoil; anchor liner to anchor trench; install balancing tank and automatic top-up system		
Pond or lake of organic shape		
100 m²; perimeter 50 m	each	8900.00
250 m²; perimeter 90 m	each	13500.00
500 m²; perimeter 130 m	each	34000.00
1000 m²; perimeter 175 m	each	69000.00
Excavate for lake average depth 1.0 m, allow for bringing to specified levels; reserve topsoil; remove spoil to approved dump on site; remove all stones and debris over 75 mm; lay polythene sheet including welding all joints and seams by specialist; screen and replace topsoil 200 mm thick		
Prices are for lakes of regular shape		
500 micron sheet	1000 m²	20000.00
1000 micron sheet	1000 m²	24000.00
Extra for removing spoil to tip	m³	51.00
Extra for 25 mm sand blinding to lake bed	100 m²	210.00
Extra for screening topsoil	m²	2.20
Extra for spreading imported topsoil	100 m²	960.00
Plant aquatic plants in lake topsoil		
Representative plant prices are shown below and users should add these to to the labour rates shown		
Apongeton distachyos	nr	4.45
Acorus calamus	nr	6.20
Butomus umbellatus	nr	2.80
Typha latifolia	nr	1.70
Planting aquatics labours only	100	130.00

8.5 EXTERNAL FIXTURES

Item – Overhead and Profit Included	Unit	Range £
Formal water features		
Excavate and construct water feature of regular shape; lay 100 mm hardcore base and 150 mm concrete 1:2:4 site mixed; line base and vertical face with butyl liner 0.75 micron and construct vertical sides of reinforced blockwork; rendering two coats; anchor the liner behind blockwork; install pumps balancing tanks and all connections to mains supply		
1.00 × 1.00 × 1.00 m deep	nr	3200.00
2.00 × 1.00 × 1.00 m deep	nr	4100.00
Excavate and construct water feature of circular shape; lay 100 mm hardcore base and 150 mm concrete 1:2:4 site mixed; line base and vertical face with butyl liner 0.75 micron and construct vertical sides of reinforced blockwork; rendering two coats; anchor the liner behind blockwork; install pumps balancing tanks and all connections to mains supply		
Surround water body with ornamental fountain surround		
Architectural Heritage; Great Westwood; 6.00 m dia. 350 mm high	nr	16000.00
Haddonstone; circular pool surround; internal dia. 1780 mm; kerb features		
continuous moulding enriched with ovolvo and palmette designs; inclusive of		
plinth and integral conch shell vases flanked by dolphins	nr	8100.00
8.5.3 ORNAMENTAL BUILDINGS		
Excavate and dispose excavated material; lay concrete base 150 thick; erect garden building		
Pavillion; Cast stone; Haddonstone; floor of Yorkstone diamond sawn 6 sides		
Venetian Folly L9400; Tuscan columns, pedimented arch, quoins and optional		
balustrading; 4.70 × 3.10	nr	22000.00
Cast stone; Haddonstone temples with domed roof and stepped floor		
Small classical; 2.54 m dia.	nr	16000.00
Large classical 3.19 m dia.	nr	23000.00
Temple; Soild limestone; Architectural Heritage		
Estate temple 2.70 m dia.; floor of Yorkstone; diamond sawn 6 sides	nr	25500.00
Pergola; Solid limestone columns with oak beams; Architectural Heritage		
Estate temple 2.70 m dia.; floor of Yorkstone; diamond sawn 6 sides	nr	16000.00

8.6 EXTERNAL DRAINAGE

Item – Overhead and Profit Included	Unit	Range £
8.6.1 SURFACE WATER AND FOUL WATER DRAINAGE		
PIPE LAYING		
PVC-u pipe laying; excavate trench 300 mm wide by excavator; lay bedding; lay surface water drain pipe to depth shown; backfill to 150 mm above pipe with gravel or sand rejects; backfill with excavated material to ground level		
Trench 750 mm deep; pipe 640–590 mm deep		
110 mm	m	21.00
110 mm; short lengths	m	23.00
160 mm	m	22.00
160 mm; short lengths	m	31.50
Trench 900 mm deep; pipe 640–590 mm deep		
110 mm	m	34.50
110 mm; short lengths	m	37.00
160 mm	m	36.00
160 mm; short lengths	m	45.50
Trench 1.20 m deep; pipe 940–890 mm deep		
110 mm	m	38.00
110 mm; short lengths	m	40.50
160 mm	m	39.50
160 mm; short lengths	m	49.00
Trench 1.50 m deep; pipe 940–890 mm deep		
110 mm	m	43.00
110 mm; short lengths	m	53.00
160 mm	m	44.50
160 mm; short lengths	m	54.00
Vitrified clay pipe laying; excavate trench by excavator; backfill with excavated material to ground level; lay surface water drain pipe to depth shown; note that the use of clay pipes precludes the use of bedding materials		
Trench 750 mm deep; pipe 640–590 mm deep		
100 mm	m	29.00
100 mm; short lengths	m	31.00
160 mm	m	48.50
160 mm; short lengths	m	58.00
Trench 1.20 m deep; pipe 940–890 mm deep; bedding on reject sand		
110 mm	m	46.50
110 mm; short lengths	m	48.50
160 mm	m	63.00
160 mm; short lengths	m	76.00

8.6 EXTERNAL DRAINAGE

Item – Overhead and Profit Included	Unit	Range £
ACCESS CHAMBERS		
Excavate inspection chamber by machine; lay base of concrete 150 thick; allow for half section pipework and benchings		
Precast concrete inspection chamber		
600 × 400 × 600 mm deep	nr	620.00
600 × 400 × 900 mm deep	nr	660.00
Polypropylene inspection chamber		
Mini access chamber 600 mm deep	nr	510.00
475 mm dia. × 900 mm deep; polymer cover	nr	770.00
475 mm dia. × 900 mm deep; ductile iron cover with screw down lid	nr	830.00
GULLIES		
Gullies; vitrified clay; excavate by hand; supply and set gully in concrete (C10P); connect to drainage system with flexible joints; backfilling with excavated material to 250 mm below finished level. Lay 150 mm type 1 to receive surface treatments (not included)		
Trapped mud (dirt) gully; complete with galvanized bucket and cast iron hinged locking grate and frame		
loading to 1 tonne; 100 or 150 mm outlet	each	620.00
loading to 5 tonne; 100 mm outlet	each	710.00
loading to 5 tonne; 150 mm outlet	each	740.00
Trapped mud (dirt) gully with rodding eye; complete with galvanized bucket and cast iron hinged locking grate and frame		
100 outlet; 300 mm internal dia.; 300 mm internal depth	each	610.00
150 outlet; 400 mm internal dia.; 750 mm internal depth	each	660.00
Concrete road gully		
Excavate and lay 100 mm concrete base (1:3:6) 150 x150 mm to suit given invert level of drain; supply and connect trapped precast concrete road gully set in concrete surround; connect to vitrified clay or uPVC drainage system with flexible joints; supply and fix straight bar dished top cast iron grating and frame; bedded in cement: sand mortar (1:3)		
375 mm dia. × 750 m deep	each	570.00
450 mm dia. × 750 m deep	each	570.00
450 mm dia. × 1.05 m deep	each	610.00
Gullies PVC-u		
Excavate and lay 100 mm concrete (C20P) base 150 × 150 mm to suit given invert level of drain; connect to drainage system; backfill with DoT Type 1 granular fill; install gully; complete with cast iron grate and frame		
yard gully 300 mm dia. × 600 mm deep	each	390.00
trapped PVC-u gully; 110 mm dia. × 215 mm deep; ductile iron frame	each	215.00
bottle gully 228 × 228 × 642 mm deep	each	170.00
bottle gully 228 × 228 × 317 mm deep	each	145.00

8.6 EXTERNAL DRAINAGE

Item – Overhead and Profit Included	Unit	Range £
8.6.1 SURFACE WATER AND FOUL WATER DRAINAGE – cont		
LINEAR DRAINAGE		
Linear drainage to design sensitive areas		
Excavate trench by machine; lay Aco Brickslot channel drain on concrete base and surround to falls; all to manufacturer's specifications		
paving surround to both sides of channel	m	245.00
Linear drainage to pedestrian area		
Excavate trench by machine; lay Aco Multidrain MD polymer concrete channel drain on concrete base and surround to falls; all to manufacturer's specifications; paving surround to channel with brick paving PC £300.00/1000		
Brickslot galvanized grating; paving to both sides	m	245.00
Brickslot stainless steel grating; paving to both sides	m	450.00
slotted galvanized steel grating; paving surround to one side of channel	m	170.00
Excavate trench by machine; lay Aco Multidrain PPD recycled polypropylene channel drain on concrete base and surround to falls; all to manufacturer's specifications; paving surround to channel with brick paving PC £300.00/1000		
Heelguard composite black; paving surround to both sides of channel	m	170.00
Linear drainage to light vehicular area		
Excavate trench by machine; lay Aco Multidrain PPD recycled polypropylene channel drain on concrete base and surround to falls; all to manufacturer's specifications; paving surround to channel with brick paving PC £300.00/1000		
Heelguard composite black; paving surround to both sides of channel	m	190.00
ductile iron; paving surround to both sides of channel	m	190.00
Accessories for channel drain		
Sump unit with sediment bucket	each	180.00
End cap; inlet/outlet	each	7.30
MANHOLES		
Brick manhole; excavate pit including earthwork support and working space disposal of surplus spoil to dump on site not exceeding 100 m; lay concrete (1:2:4) base 1500 mm dia. × 200 mm thick; reinforced with mesh reinforcement; 110 mm vitrified clay channels; benching in concrete (1:3:6) allowing one outlet and two inlets for 110 mm dia. pipe; construct inspection chamber 1 brick thick walls of engineering brick Class B; backfill with excavated material; complete with 2 nr cast iron step irons		
1200 × 1200 × 1200 mm		
cover slab of precast concrete	each	1825.00
access cover; Group 2; 600 × 450 mm	each	1875.00
1200 × 1200 × 1500 mm		
access cover; Group 2; 600 × 450 mm	each	2300.00
recessed cover 5 tonne load; 600 × 450 mm; filled with block paviors	each	2475.00

8.6 EXTERNAL DRAINAGE

Item – Overhead and Profit Included	Unit	Range £
1200 × 1200 × 2000 mm; Walls 327.5 thick		
access cover; Group 2; 600 × 450 mm	each	3400.00
recessed cover 5 tonne load; 600 × 450 mm; filled with block paviors	each	3550.00
1500 × 1500 × 2500 mm; Walls 327.5 thick		
access cover; Group 2; 600 × 450 mm	each	4900.00
recessed cover 5 tonne load; 600 × 450 mm; filled with block paviors	each	5100.00
Concrete manholes; excavate pit including earthwork support and working space disposal of surplus spoil to dump on site not exceeding 100 m; lay concrete (1:2:4) base 1500 mm dia. × 200 mm thick; reinforced with mesh reinforcement; 110 mm vitrified clay channels; benching in concrete (1:3:6) allowing one outlet and two inlets for 110 mm dia. pipe; construct inspection chamber precast concrete rings; surround in concrete 150 mm thick backfill with excavated material; complete with step irons		
Manhole dia. 900 mm		
1.00 m deep	each	1375.00
Manhole dia. 1.05 m		
1.00 m deep	each	1575.00
2.00 m deep	each	2125.00
3.00 m deep	each	2600.00
Manhole dia. 1.20 m		
1.00 m deep	each	1875.00
2.00 m deep	each	2500.00
3.00 m deep	each	3100.00
Manhole dia. 1.50 m		
1.00 m deep	each	2475.00
2.00 m deep	each	3250.00
3.00 m deep	each	3950.00

8.6.4 LAND DRAINAGE

SUBSOIL DRAINAGE – BY MACHINE

Main drain; remove 150 mm topsoil and deposit alongside trench, excavate drain trench by machine and lay flexible perforated drain; lay bed of gravel rejects 100 mm; backfill with gravel rejects or similar to within 150 mm of finished ground level; complete fill with topsoil; remove surplus spoil to approved dump on site		
Main drain 160 mm supplied in 35 m lengths		
450 mm deep	100 m	670.00
600 mm deep	100 m	830.00
900 mm deep	100 m	1600.00
Extra for couplings	each	3.05
100 mm main drain supplied in 100 m lengths		
450 mm deep	100 m	1425.00
600 mm deep	100 m	1875.00
900 mm deep	100 m	3100.00
Extra for couplings	each	1.40

8.6 EXTERNAL DRAINAGE

Item – Overhead and Profit Included	Unit	Range £
8.6.4 LAND DRAINAGE – cont		
Laterals to main drains; herringbone pattern; excavation and backfilling;		
inclusive of connecting lateral to main drain		
100 mm pipe to 450 mm deep trench		
laterals at 1.0 m centres	100 m²	1425.00
laterals at 2.0 m centres	100 m²	710.00
laterals at 3.0 m centres	100 m²	470.00
laterals at 5.0 m centres	100 m²	285.00
100 mm pipe to 600 mm deep trench		
laterals at 1.0 m centres	100 m²	1875.00
laterals at 2.0 m centres	100 m²	950.00
laterals at 3.0 m centres	100 m²	620.00
laterals at 5.0 m centres	100 m²	380.00
100 mm pipe to 900 mm deep trench		
laterals at 1.0 m centres	100 m²	3100.00
laterals at 2.0 m centres	100 m²	1550.00
laterals at 3.0 m centres	100 m²	1025.00
laterals at 5.0 m centres	100 m²	620.00
80 mm pipe to 450 mm deep trench		
laterals at 1.0 m centres	100 m²	1375.00
laterals at 2.0 m centres	100 m²	690.00
laterals at 3.0 m centres	100 m²	450.00
laterals at 5.0 m centres	100 m²	275.00
80 mm pipe to 600 mm deep trench		
laterals at 1.0 m centres	100 m²	1850.00
laterals at 2.0 m centres	100 m²	920.00
laterals at 3.0 m centres	100 m²	610.00
laterals at 5.0 m centres	100 m²	370.00
80 mm pipe to 900 mm deep trench		
laterals at 1.0 m centres	100 m²	3050.00
laterals at 2.0 m centres	100 m²	1525.00
laterals at 3.0 m centres	100 m²	1000.00
laterals at 5.0 m centres	100 m²	610.00
Extra for 100/80 mm junctions connecting laterals to main drain		
laterals at 1.0 m centres	10 m	56.00
laterals at 2.0 m centres	10 m	28.00
laterals at 3.0 m centres	10 m	18.60
laterals at 5.0 m centres	10 m	11.20
SUBSOIL DRAINAGE – BY HAND		
Main drain; remove 150 mm topsoil and deposit alongside trench; excavate		
drain trench by machine and lay flexible perforated drain; lay bed of gravel		
rejects 100 mm; backfill with gravel rejects or similar to within 150 mm of		
finished ground level; complete fill with topsoil; remove surplus spoil to		
approved dump on site		
Main drain 160 mm supplied in 35 m lengths		
450 mm deep	m	38.00
600 mm deep	m	50.00
900 mm deep	m	73.00
Extra for couplings	each	3.05

8.6 EXTERNAL DRAINAGE

Item – Overhead and Profit Included	Unit	Range £
100 mm main drain supplied in 100 m lengths		
450 mm deep	m	27.00
600 mm deep	m	35.00
900 mm deep	m	50.00
Extra for couplings	each	1.40
Laterals to main drains; herringbone pattern; excavation and backfilling; inclusive of connecting lateral to main drain		
160 mm pipe to 450 mm deep trench		
laterals at 1.0 m centres	m²	38.00
laterals at 2.0 m centres	m²	27.50
160 mm pipe to 600 mm deep trench		
laterals at 1.0 m centres	m²	50.00
laterals at 2.0 m centres	m²	35.50
100 mm pipe to 450 mm deep trench		
laterals at 1.0 m centres	m²	18.90
laterals at 2.0 m centres	m²	11.80
100 mm pipe to 600 mm deep trench		
laterals at 1.0 m centres	m²	35.00
laterals at 2.0 m centres	m²	25.50
80 mm pipe to 450 mm deep trench		
laterals at 1.0 m centres	m²	18.40
laterals at 2.0 m centres	m²	11.30
80 mm pipe to 600 mm deep trench		
laterals at 1.0 m centres	m²	34.50
laterals at 2.0 m centres	m²	25.00
Extra for 100/80 mm couplings connecting laterals to main drain		
laterals at 1.0 m centres	10 m	56.00
laterals at 2.0 m centres	10 m	28.00
DRAINAGE DITCHES		
Excavate and form ditch and bank with 45° sides in light to medium soils; all widths taken at bottom of ditch		
300 mm wide × 600 mm deep	100 m	160.00
600 mm wide × 900 mm deep	100 m	550.00
1.20 m wide × 900 mm deep	100 m	1325.00
1.50 m wide × 1.20 m deep	100 m	2250.00
Clear and bottom existing ditch average 1.50 m deep, trim back vegetation and remove debris to licensed tip, lay jointed concrete pipes; including bedding, haunching and topping with 150 mm concrete; 11.50 N/mm²–40 mm aggregate; backfill with approved spoil from site		
Pipes 300 mm dia.	100 m	10000.00
Pipes 450 mm dia.	100 m	12500.00
Pipes 600 mm dia.	100 m	17000.00

8.6 EXTERNAL DRAINAGE

Item – Overhead and Profit Included	Unit	Range £
8.6.4 LAND DRAINAGE – cont		
LAND DRAINS		
Excavate trench by excavator; lay Type 2 bedding; backfill to 150 mm above pipe with gravel rejects; lay non woven geofabric and fill with topsoil to ground level		
Trench 600 deep		
160 mm PVC-u drainpipe	100 m	2600.00
110 mm PVC-u drainpipe	100 m	2475.00
150 mm vitrified clay	100 m	5300.00
100 mm vitrified clay	100 m	3300.00
SOAKAWAYS		
Construct soakaway from perforated concrete rings; excavation, casting in situ concrete ring beam base; filling with gravel 250 mm deep; placing perforated concrete rings; surrounding with geofabric and backfilling to external surround of soakaway with 250 mm wide granular surround and excavated material; step irons and cover slab; inclusive of all earthwork retention and disposal off site of surplus material		
900 mm dia.		
1.00 m deep	nr	1000.00
2.00 m deep	nr	1925.00
1200 mm dia.		
1.00 m deep	nr	1325.00
2.00 m deep	nr	2450.00
2400 mm dia.		
1.00 m deep	nr	4150.00
2.00 m deep	nr	7000.00
Aquacell Infiltration unit soakaway; 1.00 m × 500 × 400 mm; 200 litre volume each in trench		
Excavate for new soakaway; grade bottom of excavation; Lay bedding layer of 100 mm sharp sand; Install geotextile to line excavation and bedding; Install Aquacell units in bonded formation; wrap entire installation in geotextile; back filling to surround in MOT Type 2 150 mm thick. Place sharp sand over infiltration units; inclusive of connection only of in-flow pipework; Cover over 500 deep with Type 2 granular material		
8 units in trench; 1520 litres	nr	1125.00
12 units in trench; 2280 litres	nr	1850.00
16 units in trench; 3040 litres	nr	2350.00
20 units in trench; 3800 litres	nr	2900.00
30 units; in trench 5700 litres	nr	4750.00
60 units; in trench 11400 litres	nr	8600.00

8.7 EXTERNAL SERVICES

Item – Overhead and Profit Included	Unit	Range £
8.7.1 WATER MAINS SUPPLY		
Cold water		
Excavate trench for cold water pipe lay bed of sharp sand 100 mm thick; lay pipe		
1.20 m deep		
20 mm MDPE pipe	m	51.00
25 mm MDPE pipe	m	52.00
32 mm MDPE pipe	m	52.00
50 mm MDPE pipe	m	54.00
60 mm MDPE pipe	m	55.00
750 mm deep		
20 mm MDPE pipe	m	32.00
25 mm MDPE pipe	m	33.00
32 mm MDPE pipe	m	33.00
50 mm MDPE pipe	m	34.50
60 mm MDPE pipe	m	36.00
8.7.2 ELECTRICITY MAINS SUPPLY		
Power supply; Cable installation for external services; trenching and backfilling; Supply and place ring main armoured cable for supply of services for external electrical systems; Straight clear runs; 600 mm deep		
Armoured cable		
Single circuit	m	8.10
Dual circuit	m	10.80
Three circuit	m	13.40
Four circuits	m	16.10
External grade cable in ducts; laid on sand bed covered with sand and tape		
Single circuit	m	10.40
Dual circuit	m	13.00
Three circuit	m	16.40
Four circuits	m	19.70
8.7.9 EXTERNAL STREET LIGHTING SYSTEMS		
Street lighting Marshalls Bega; to parking area or precinct; Installation of lighting poles and luminaries; excludes excavation of trenches for supply cable; ducted cable		
Lighting poles to 6.00 m		
Assymetrical lighting Bega M99 515	–	1900.00
Traditional bow top pole and luminary M70–991 – single luminary	–	2650.00
Traditional bow top pole and luminary M70–910 – two luminaries	–	4100.00

Landscape Architect's Pocket Book, 2nd edition

S. Vernon *et al.*

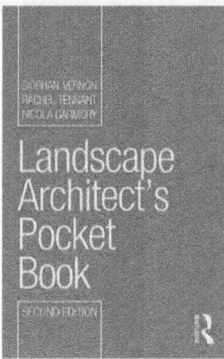

An indispensable tool for all landscape architects, this time-saving guide answers the most frequently asked questions in one pocket-sized volume. It is a concise, easy-to-read reference that gives instant access to a wide range of information needed on a daily basis, both out on site and in the office.

Covering all the major topics, including hard landscaping, soft landscaping as well as planning and legislation, the pocket book also includes a handy glossary of important terms, useful calculations and helpful contacts. Not only an essential tool for everyday queries on British standards and procedures, this is a first point of reference for those seeking more extensive, supplementary sources of information, including websites and further publications.

This new edition incorporates updates and revisions from key planning and environmental legislation, guidelines and national standards.

April 2013; 186 x 123mm, 352 pp
Pbk: 978-0-415-63084-9; £23.99
Routledge Pocket Books

To Order: Tel: +44 (0) 1235 400524 Fax: +44 (0) 1235 400525
or Post: Taylor and Francis Customer Services,
Bookpoint Ltd, Unit T1, 200 Milton Park, Abingdon, Oxon, OX14 4TA UK
Email: book.orders@tandf.co.uk

For a complete listing of all our titles visit:
www.tandf.co.uk

Prices for Measured Works

INTRODUCTION

Typical Project Profile

Contract value	£100,000.00 – £900,000.00
Labour rate (see page 7)	£28.00 per hour
Bricklayer	£36.00 per hour
Labour rate for maintenance contracts	£23.00 per hour
Number of site staff	35
Project area	6000 m^2
Project location	Outer London
Project components	50% hard landscape 50% soft landscape and planting
Access to works areas	Very good
Contract	Main contract
Delivery of materials	Full loads
Exchange rate for goods imported from the EU	We have used an exchange rate of £0.86 = €1.00

Rates shown include O&P as indicated below

Labour Type	Unit	Cost incl Overhead	Mark-up %	Rate incl. Overhead & Profit
General contracting	/man-hour	£28.00	15%	£32.20
Brickwork 1+1 man gang	/team-hour	£64.00	15%	£73.60
Brickwork 1+2 man gang	/team-hour	£92.00	15%	£105.80
Maintenance contracting	/man-hour	£23.00	15%	£23.58

Other resource categories

Materials generally	15%
Plant	15%
Subcontractors generally	10%

Some high cost material or subcontract items may be at lower rates of 5%

BIM for Landscape

Landscape Institute

BIM (Building Information Modelling) is transforming working practices across the built environment sector, as clients, professionals, contractors and manufacturers throughout the supply chain grasp the opportunities that BIM presents. The first book ever to focus on the implementation of BIM processes in landscape and external works, *BIM for Landscape* will help landscape professionals understand what BIM means for them. This book is intended to equip landscape practitioners and practices to meet the challenges and reap the rewards of working in a BIM environment - and to help professionals in related fields to understand how BIM processes can be brought into landscape projects. BIM offers significant benefits to the landscape profession, and heralds a new chapter in inter-disciplinary relationships. *BIM for Landscape* shows how BIM can enhance collaboration with other professionals and clients, streamline information processes, improve decision-making and deliver well-designed landscape projects that are right first time, on schedule and on budget.

This book looks at the organisational, technological and professional practice implications of BIM adoption. It discusses in detail the standards, structures and information processes that form BIM Level 2-compliant workflows, highlighting the role of the landscape professional within the new ways of working that BIM entails. It also looks in depth at the digital tools used in BIM projects, emphasising the 'information' in Building Information Modelling, and the possibilities that data-rich models offer in landscape design, maintenance and management. *BIM for Landscape* will be an essential companion to the landscape professional at any stage of their BIM journey.

May 2016; 246 × 174 mm, 174 pp
Hb: 978-1-138-79668-3; £48.99

To Order: Tel: +44 (0) 1235 400524 Fax: +44 (0) 1235 400525
or Post: Taylor and Francis Customer Services,
Bookpoint Ltd, Unit T1, 200 Milton Park, Abingdon, Oxon, OX14 4TA UK
Email: book.orders@tandf.co.uk

For a complete listing of all our titles visit:
www.tandf.co.uk

Taylor & Francis
Taylor & Francis Group

NEW ITEMS FOR THIS EDITION

Item – Overhead and Profit Included	PC £	Labour hours	Labour £	Plant £	Material £	Unit	Total rate £
MARKET PRICES OF AGGREGATES							
Aggregates MCM (SE) Ltd (Includes 15% mark-up)							
Aggregates in load sizes as shown; Bags in 20 tonne deliveries comprising 18–20 850 kg bags unloaded by truck mounted Hiab; Measured as compacted or settled volumes							
Type 1 primary							
20 tonne loose tipped	44.55	–	–	–	51.23	m^3	**51.23**
in bulk bags	83.65	–	–	–	96.20	m^3	**96.20**
Type 1 recycled							
20 tonne loose tipped	20.90	–	–	–	24.03	m^3	**24.03**
in bulk bags	69.88	–	–	–	80.36	m^3	**80.36**
Type 1 crushed concrete							
20 tonne loose tipped	20.90	–	–	–	24.03	m^3	**24.03**
in bulk bags	–	–	–	–	–	m^3	**–**
Type 3 primary							
20 tonne loose tipped	59.29	–	–	–	68.18	m^3	**68.18**
in bulk bags	95.29	–	–	–	109.58	m^3	**109.58**
Type 3 recycled							
20 tonne loose tipped	47.70	–	–	–	54.86	m^3	**54.86**
Ballast–all in							
20 tonne loose tipped	45.90	–	–	–	52.78	m^3	**52.78**
in bulk bags	83.65	–	–	–	96.20	m^3	**96.20**
Sand–Sharp							
20 tonne loose tipped	42.07	–	–	–	48.38	m^3	**48.38**
in bulk bags	76.68	–	–	–	88.18	m^3	**88.18**
Sand Soft building							
20 tonne loose tipped	40.80	–	–	–	46.92	m^3	**46.92**
in bulk bags	75.71	–	–	–	87.07	m^3	**87.07**
6F2 75 mm down crusher run capping material							
20 tonne loose tipped	18.00	–	–	–	20.70	m^3	**20.70**
in bulk bags	58.24	–	–	–	66.98	m^3	**66.98**
Shingle 20 mm							
20 tonne loose tipped	45.90	–	–	–	52.78	m^3	**52.78**
in bulk bags	83.65	–	–	–	96.20	m^3	**96.20**
Shingle–pea							
20 tonne loose tipped	51.30	–	–	–	58.99	m^3	**58.99**
in bulk bags	83.65	–	–	–	96.20	m^3	**96.20**
Crushed rock							
20 tonne loose tipped	107.73	–	–	–	123.89	m^3	**123.89**
Stone angular 40–20 or 20/4							
20 tonne loose tipped	72.49	–	–	–	83.36	m^3	**83.36**
in bulk bags	101.65	–	–	–	116.90	m^3	**116.90**
Grit 6–2 mm for permeable paving							
20 tonne loose tipped	–	–	–	–	72.49	m^3	**72.49**

NEW ITEMS FOR THIS EDITION

Item – Overhead and Profit Included	PC £	Labour hours	Labour £	Plant £	Material £	Unit	Total rate £
11 In situ Concrete – Bar reinforcement							
Alternative tonne rates for bar reinforcement fixing to retaining walls in external works environments; Vertical bar only; Horizontal tables shown in separate table set below							
8 mm bar. Rates per tonne based on fixing rates shown							
28.2 hrs/tonne	765.00	28.20	908.04	–	879.75	tonne	**1787.79**
10 mm bar. Rates per tonne based on fixing rates shown							
22.5 hrs/tonne	765.00	22.50	724.50	–	879.75	tonne	**1604.25**
12 mm bar. Rates per tonne based on fixing rates shown							
16.9 hrs/tonne	765.00	16.90	544.18	–	879.75	tonne	**1423.93**
16 mm bar. Rates per tonne based on fixing rates shown							
12.0 hrs/tonne	765.00	12.00	386.40	–	879.75	tonne	**1266.15**
25 mm bar. Rates per tonne based on fixing rates shown							
6.9 hrs/tonne	765.00	6.90	222.18	–	879.75	tonne	**1101.93**
32 mm bar. Rates per tonne based on fixing rates shown							
5.7 hrs/tonne	765.00	5.70	183.54	–	879.75	tonne	**1063.29**
Alternative tonne rates for bar reinforcement fixing to retaining walls in external works environments; Horizontal bar only in conjuction with vertical bar in separate table							
8 mm bar. Rates per tonne based on fixing rates shown							
7 hrs/tonne	765.00	7.00	225.40	–	879.75	tonne	**1105.15**
10 mm bar. Rates per tonne based on fixing rates shown							
5.62 hrs/tonne	765.00	5.62	180.96	–	879.75	tonne	**1060.71**
12 mm bar. Rates per tonne based on fixing rates shown							
4.22 hrs/tonne	765.00	4.22	135.88	–	879.75	tonne	**1015.63**
16 mm bar. Rates per tonne based on fixing rates shown							
3.0 hrs/tonne	765.00	3.00	96.60	–	879.75	tonne	**976.35**
25 mm bar. Rates per tonne based on fixing rates shown							
2.2 hrs/tonne	765.00	2.20	70.84	–	879.75	tonne	**950.59**
32 mm bar. Rates per tonne based on fixing rates shown							
2 hrs/tonne	765.00	2.00	64.40	–	879.75	tonne	**944.15**

NEW ITEMS FOR THIS EDITION

Item – Overhead and Profit Included	PC £	Labour hours	Labour £	Plant £	Material £	Unit	Total rate £
Reinforcement bar to concrete formwork or footings; bent							
8 mm bar							
100 ccs	2.98	0.17	10.63	–	3.43	m²	**14.06**
200 ccs	1.45	0.13	8.05	–	1.67	m²	**9.72**
300 ccs	0.99	0.08	4.83	–	1.14	m²	**5.97**
10 mm bar							
100 ccs	4.67	0.17	10.63	–	5.37	m²	**16.00**
200 ccs	2.29	0.13	8.05	–	2.63	m²	**10.68**
300 ccs	1.53	0.08	4.83	–	1.76	m²	**6.59**
12 mm bar							
100 ccs	6.73	0.17	10.63	–	7.74	m²	**18.37**
200 ccs	3.37	0.13	8.05	–	3.88	m²	**11.93**
300 ccs	2.22	0.08	4.83	–	2.55	m²	**7.38**
16 mm bar							
100 ccs	12.01	0.20	12.88	–	13.81	m²	**26.69**
200 ccs	6.04	0.17	10.63	–	6.95	m²	**17.58**
300 ccs	4.05	0.13	8.05	–	4.66	m²	**12.71**
25 mm bar							
100 ccs	29.45	0.20	12.88	–	33.87	m²	**46.75**
200 ccs	14.69	0.13	8.05	–	16.89	m²	**24.94**
300 ccs	9.79	0.13	8.05	–	11.26	m²	**19.31**
32 mm bar							
100 ccs	48.27	0.20	12.88	–	55.51	m²	**68.39**
200 ccs	24.10	0.17	10.63	–	27.72	m²	**38.35**
300 ccs	16.07	0.13	8.05	–	18.48	m²	**26.53**
6 Ground remediation – Hahn recycled retaining wall systems							
Recycled materials and Systems for External Works							
Planter retaining wall; Recycled plastic; Hahn Plastics Ltd; Hollow or Solid; inclusive of trenches, placement and surrounding with concrete; terram to rear of palisades; backfiling to height of palisade not included; brown; (grey also available)							
Ogee C Section interlocking profile palisades 100 mm dia.; hollow profile; retained height							
300 mm	99.58	0.25	24.15	–	138.21	m	**162.36**
600 mm	109.20	0.50	48.30	–	159.05	m	**207.35**
Ogee C Section interlocking profile palisades 100 mm dia.; hollow profile; retained height							
1.20 m	197.99	0.43	41.86	–	273.59	m	**315.45**

NEW ITEMS FOR THIS EDITION

Item – Overhead and Profit Included	PC £	Labour hours	Labour £	Plant £	Material £	Unit	Total rate £
6 Ground remediation – Hahn recycled retaining wall systems – cont							
Planter retaining wall – cont							
Round palisades; 100 mm dia.; hollow profile; retained height							
300 mm	40.10	0.25	24.15	–	66.58	m	**90.73**
800 mm	99.50	0.43	41.86	–	152.21	m	**194.07**
Round palisades; 100 mm dia.; solid profile; retained height							
1.00 m	99.50	0.43	41.86	–	152.21	m	**194.07**
1.20 m	155.00	0.50	48.30	–	227.41	m	**275.71**
1.50 m	193.80	1.00	96.60	–	353.54	m	**450.14**
Rectangular palisades; 120 × 160 mm; hollow profile; retained height; Grey, Brown or Black							
300 mm	65.90	0.25	24.15	–	96.25	m	**120.40**
800 mm	196.73	0.43	41.86	–	264.03	m	**305.89**
1.00 m	253.00	0.50	48.30	–	340.11	m	**388.41**
36 Site Works – Fencing – Hahn recycled fencing and barriers							
Recycled plastic; Fencing and Barriers; Hahn Plastics Ltd; Recycled and recyclable; components							
Posts; inclusive of mechanical excavation; 400 × 400 × 400 mm deep; backfilling with site mixed concrete and disposal of arisings							
Hanit Ultra – square post							
10 cm × 10 cm × 1.75 m	27.08	0.25	16.10	7.74	46.59	nr	**70.43**
Square post – reinforced							
9 cm × 9 cm × 2 m	25.40	0.25	16.10	7.74	44.65	nr	**68.49**
Round post – reinforced 8 cm dia. × 2 m	18.42	0.25	16.10	7.74	36.63	nr	**60.47**
Square post 7 cm × 7 cm × 2 m	10.52	0.25	16.10	7.74	27.54	nr	**51.38**
Round post 8 cm dia. × 2 m; in concrete inclusive of mechanical excavation and disposal	10.21	0.25	16.10	7.74	27.18	nr	**51.02**
Cross profile stake (wire fence)							
7 cm × 7 cm × 2 m; driven in	5.02	–	–	–	–	nr	–

NEW ITEMS FOR THIS EDITION

Item – Overhead and Profit Included	PC £	Labour hours	Labour £	Plant £	Material £	Unit	Total rate £
Recycled Plastic fencing; Hahn Plastics Ltd; Hanit fencing							
Fencing							
Woven Privacy Screen – Ribbon 1.50 m wide panels × 1.80 m high inclusive of recycled 9 cm × 9 cm posts set in concrete at 1.545 m centres	154.40	0.29	18.40	5.16	204.50	m	**228.06**
Picket fencing; Hahn Plastics Ltd; curved top profile; 100% recycled self-coloured hanit® material, non-rot post and panels with 20 year warranty; pre-assembled panels in grey, brown or green with either aluminium or plastic cross-bar with matching posts; hand excavation and disposal off site							
850 mm high with plastic cross-bar	117.00	0.20	12.88	–	161.58	m	**174.46**
Pale & Post fencing (palisade): Hahn Plastics Ltd; 100% recycled self-coloured hanit® Ultra material, enhanced wood grain finish, non-rot post and pales with 20 year warranty							
In black or brown woodgrain for fences; two crossrails and recycled 10 cm × 10 cm posts included; Posts at 1.80 m centres							
1.80 m high	86.70	0.40	25.76	–	107.42	m	**133.18**
2.40 m high	113.26	0.40	25.76	–	137.97	m	**163.73**
Privacy fencing: Hahn Plastics Ltd; 100% recycled self-coloured hanit®, non-rot post and tongue & grooved panels with 20 year warranty; dedicated brown plastic reinforced fence posts supplied with matching end caps with individual grey, brown or green tongue & groove panels inserted into built in recesses with a central support							
In Grey Brown or Green; Prices shown for the average priced colour; Posts at 2.00 m centres							
1.95 m high	161.90	0.20	12.88	–	193.75	m	**206.63**
Corner post or end post	117.00	0.50	32.20	–	148.64	nr	**180.84**

NEW ITEMS FOR THIS EDITION

Item – Overhead and Profit Included	PC £	Labour hours	Labour £	Plant £	Material £	Unit	Total rate £
36 Site Works – Fencing – Hahn recycled fencing and barriers – cont							
Pasture fencing: Hahn Plastics Ltd; 100% recycled self-coloured hanit®, non-rot post and tongue & grooved panels with 20 year warranty; equestrian and agricultural use							
Pointed reinforced posts driven in							
1.50 m high 3 rails (inc posts)	66.42	0.15	9.75	–	76.38	m	**86.13**
1.50 m high 5 rails (inc posts)	73.48	0.15	9.75	–	84.50	m	**94.25**
35 Site Works – Pavings – Hahn recycled Ground protection							
Ground Grid; Recycled plastic; Hahn Plastics Ltd; Interlocking; filling with angular granular material (5–15 mm)							
HDGG 8 cm thick × 40 cm wide × 60 cm long grid for HGV use; Strength: 18,600 kN/m^2 (1896 tonnes/m^2); laid to prepared base (not included)							
HDGG; Units 40 cm wide × 60 cm long × 80 mm thick; 100–500 m^2 Volume of angular gravel fill is 1.8 tonnes per 25 m^2 or 7.2 kg/m^2	19.50	0.03	3.22	2.92	26.52	m^2	**32.66**
35 Site Works – Street furniture – Hahn recycled street furniture							
Recycled plastic Street furniture; Hahn Plastics Ltd; inclusive of excavations concrete bases and anchors							
Sets							
Calero Table & Bench set							
150 cm × 180 cm × 75 cm	724.00	1.00	64.40	–	858.42	nr	**922.82**
Alton Table & Bench set							
200 cm × 205 cm × 80 cm	1520.00	1.00	64.40	–	1773.82	nr	**1838.22**
Serengeti picnic set							
200 cm × 174 cm × 76 cm	741.00	1.00	64.40	–	877.97	nr	**942.37**
Benches							
Wall Street 955 bench							
200 × 56 cm × 45 cm	899.00	0.75	48.30	–	1059.67	nr	**1107.97**
Hyde Park bench							
165 cm × 46 cm × 48 cm	206.00	0.75	48.30	–	262.72	nr	**311.02**
High line bench							
2.00 m × 480 mm × 450 mm	411.00	0.75	48.30	–	498.47	nr	**546.77**

NEW ITEMS FOR THIS EDITION

Item – Overhead and Profit Included	PC £	Labour hours	Labour £	Plant £	Material £	Unit	Total rate £
Seats							
Caserata Wall mounted seat							
150 cm × 70 cm × 65 cm	383.00	0.75	48.30	–	466.27	nr	**514.57**
Eclipse seat							
150 cm × 60 cm × 89 cm	247.00	0.75	48.30	–	309.87	nr	**358.17**
Sapo Seniors' seat							
200 cm × 61 cm × 97 cm	696.00	0.75	48.30	–	826.22	nr	**874.52**
Turon Round tree seat – octagonal							
330 cm × 305 cm × 80 cm	4303.00	3.00	193.20	–	5044.02	nr	**5237.22**
Planters; inclusive of placement and filling with planting medium and drainage; excludes planting							
Vinca 2 Planter							
108 cm × 62 cm × 44 cm	474.00	1.00	64.40	–	582.52	nr	**646.92**
Iberis 2 Planter							
110 cm × 110 cm × 47 cm	424.00	1.00	64.40	–	529.79	nr	**594.19**
Litter Bin							
Collec Litter Bin 48 cm × 66 cm; ground mounted with wire basket insert; circular; 70 litre	259.46	1.00	64.40	–	311.29	nr	**375.69**
Resco Litter Bin							
45 cm × 45 cm × 66 cm high 50 litre; square; brown	254.38	1.00	64.40	–	305.45	nr	**369.85**
35 Site Works – Play areas							
Play Areas – Lappset Ltd							
Swings							
Flat seat swing	–	–	–	–	–	nr	**2090.00**
Birds nest swing	–	–	–	–	–	nr	**3685.00**
Swing Q10144 (Estimated))	–	–	–	–	–	nr	**2365.00**
Swing with 2 × Cradle seats 137414M	–	–	–	–	–	nr	**2365.00**
Climbing equipment and play structures							
Finno Activity Tower	–	–	–	–	–	nr	**14564.00**
Finno 3D Cube Climber	–	–	–	–	–	nr	**6248.00**
Finno Hades	–	–	–	–	–	nr	**13860.00**
Castle Finno	–	–	–	–	–	nr	**30635.00**
Finno Fishing Ship	–	–	–	–	–	nr	**34050.50**
Santa Maria	–	–	–	–	–	nr	**127600.00**
Halo Climber	–	–	–	–	–	nr	**35695.00**
Halo Cubic	–	–	–	–	–	nr	**99000.00**
Flora Goblin's climbing	–	–	–	–	–	nr	**1265.00**
Flora Troll's shelter	–	–	–	–	–	nr	**7448.10**
Flora Goblin's track	–	–	–	–	–	nr	**2349.60**
Play tower Q10945	–	–	–	–	–	nr	**7364.50**
ABC Play tower Q10519	–	–	–	–	–	nr	**11110.00**
Q10259 Fire truck	–	–	–	–	–	nr	**4180.00**

NEW ITEMS FOR THIS EDITION

Item – Overhead and Profit Included	PC £	Labour hours	Labour £	Plant £	Material £	Unit	Total rate £
35 Site Works – Play areas – cont							
Play Areas – Lappset Ltd – cont							
Spinning Items							
Spinning machine	–	–	–	–	–	nr	12223.20
Waltz	–	–	–	–	–	nr	5736.50
Polka	–	–	–	–	–	nr	4435.20
Inclusive roundabout	–	–	–	–	–	nr	7200.00
Spring equipment							
010501 Horse Springer	–	–	–	–	–	nr	861.30
Tractor rocker 010515	–	–	–	–	–	nr	1958.00
Pig rocker 01516	–	–	–	–	–	nr	1023.00
Froggy 22034	–	–	–	–	–	nr	1320.00
Octopus seesaw Q10279	–	–	–	–	–	nr	1320.00
Play area Furniture							
Picnic table Accessibility adapted NF 2694	–	–	–	–	–	nr	660.00
25 Balustrades – Haddonstone **Balustrades**							
Bottle balustrades reconstituted stone; Haddonstone Ltd; to prepared brickwork walls or bases (not included)							
Top and bottom rails to support balustrades; laid to builder's work of brick block or other stonework supports (not included)							
Top Rail T110 A 900 × 30 × 86	44.40	–	–	–	51.06	m	51.06
Bottom Rail T900 A 900 × 230 × 102	44.40	–	–	–	51.06	m	51.06
Under coping							
Under Coping 900 × 377 × 76	65.49	–	–	–	75.31	m	75.31
Pier bases; 337 × 102 mm high to receive shafts							
337 × 337 × 102 mm high	31.00	0.25	16.10	–	35.65	nr	51.75
Pier shafts; at specified centres along the balustrade fixed to upstand (not included); subsequently filled with concrete							
Piers shaft 279 × 279 × 610 mm high	67.00	1.00	64.40	–	88.88	nr	153.28
Pier shafts 279 × 279 × 710 mm high	80.00	1.25	80.50	–	105.28	nr	185.78
Pier caps to Pier shafts;							
337 × 337 × 86 mm	37.00	0.25	16.10	–	45.38	nr	61.48

NEW ITEMS FOR THIS EDITION

Item – Overhead and Profit Included	PC £	Labour hours	Labour £	Plant £	Material £	Unit	Total rate £
Plinths to receive balustrade components; fixed to existing base or to low walls to receive balustrade components							
360 wide × 440 mm high	99.31	0.75	48.30	–	114.90	m	**163.20**
350 wide × 150 mm high	75.75	0.75	48.30	–	87.80	m	**136.10**
Bottle balustrades comprising reconstituted bottle balusters; fixed to plinths, top rail, bottom rail and under coping (all shown separately) with steel rods and mortar; balusters							
K457G 150 wide base × 127 top × 457 mm high; at 300 mm centres	123.21	1.00	64.40	–	144.90	m	**209.30**
K610G 140 wide base × 127 top × 610 mm high; at 300 mm centres	156.51	1.00	64.40	–	183.55	m	**247.95**
K729G 140 wide base × 127 top × 610 mm high; at 300 mm centres	183.15	1.00	64.40	–	214.19	m	**278.59**
35 Pavings – Haddonstone reconstituted stone steps							
Stone steps; Haddonstone Ltd to step base in concrete or brick substructure; Cast stone steps in Tecstone in colours Bath, Terracotta, Portland, Slate or Sienna; continuosly bedded on full mortar bed							
Steps							
Bullnose Step 1000 × 350 × 50	56.00	0.25	24.15	–	66.75	m	**90.90**
Bullnose Step 1000 × 375 × 50	65.00	0.25	24.15	–	77.25	m	**101.40**
Risers							
Step Riser 1000 × 175 × 50	27.00	0.13	12.07	–	31.13	m	**43.20**
Step Riser 1000 × 150 × 50	24.00	0.13	12.07	–	27.68	m	**39.75**
35 Decking – Millboard composite decking							
Composite decking; London Stone Ltd; Millboard composite decking Substructure for composite or timber decking; Millboard; London Stone Ltd; Pro-Plas recyled substructural elements to support timber or composite decking boards; joists and double bearers 125 × 50 mm; joists laid at 300 mm centres							
4 m² deck; 2.00 × 2.00 m	51.60	1.50	96.60	–	63.02	m²	**159.62**
10 m² deck; 3.33 × 3.33 m	53.66	1.20	77.28	–	65.39	m²	**142.67**
25 m² deck; 5.00 × 5.00 m	47.06	1.18	75.87	–	57.80	m²	**133.67**

NEW ITEMS FOR THIS EDITION

Item – Overhead and Profit Included	PC £	Labour hours	Labour £	Plant £	Material £	Unit	Total rate £
35 Decking – Millboard composite decking – cont							
Composite decking – cont Substructure for composite or timber decking – cont							
10 m² deck; 10.00 × 5.00 m	46.65	1.00	64.40	–	57.33	m²	**121.73**
Composite deck surface; Millboard; 176 mm wide; fixed with stainless screws 4.5 × 70 mm							
Enhanced grain range	45.00	0.33	21.46	–	55.71	m²	**77.17**
Weathered Oak range	57.00	0.33	21.46	–	69.51	m²	**90.97**
35 Block paving – Marshalls concrete block products							
Marshalls Plc Textured granite aggregate sett paving; on primary aggregate 6–2 mm; compacted; vibrated; joints filled with grit; Silver grey; Graphite; Charcoal; well rammed hardcore base (not included) Mistral; Mixed sizes 240 × 160 mm; 160 x160 mm; 120 × 160 mm							
60 mm thick; mixed sizes	37.43	0.25	24.15	0.18	47.79	m²	**72.12**
80 mm thick; mixed sizes	42.66	0.29	27.60	0.18	53.81	m²	**81.59**
Modal textured blocks; mixed shades and sizes; 200 × 100 mm– 400 × 200 mm; (larger sizes shown under slab paving)							
single size 80 mm thick	43.72	0.29	27.60	0.18	55.03	m²	**82.81**
mixed sizes to pattern 80 mm thick	43.72	0.33	32.20	0.18	55.03	m²	**87.41**
Modal smooth blocks; mixed shades and sizes; 200 × 100 mm– 400 × 200 mm; (larger sizes shown under slab paving)							
single size 80 mm thick	58.48	0.29	27.60	0.18	72.00	m²	**99.78**
mixed sizes to pattern 80 mm thick	58.48	0.33	32.20	0.18	72.00	m²	**104.38**
35 Flag paving – Marshalls concrete flag products Pavings; Pimple textured Buff or Natural; (Prices shown are for Buff Natural slightly less)							
450 × 450 × 50 mm	15.26	0.15	14.31	–	24.36	m²	**38.67**
600 × 300 × 50 mm	15.26	0.15	14.31	–	24.36	m²	**38.67**
400 × 400 × 65 mm	15.26	0.13	12.88	–	24.36	m²	**37.24**

NEW ITEMS FOR THIS EDITION

Item – Overhead and Profit Included	PC £	Labour hours	Labour £	Plant £	Material £	Unit	Total rate £
450 × 600 × 50 mm	15.26	0.15	14.31	–	24.36	m²	**38.67**
600 × 600 × 50 mm	15.26	0.13	12.88	–	24.36	m²	**37.24**
750 × 600 × 50 mm	15.26	0.13	12.88	–	24.36	m²	**37.24**
900 × 600 × 50 mm	15.26	0.13	12.88	–	24.36	m²	**37.24**
Precast concrete flag pavings; Marshalls Plc; on prepared subbase measured separately; bedding on 25 mm thick cement: sand mortar (1:4); pointed straight both ways Steintec jointing mixture							
Urbex paving; Practical economical paving alternative to Pimple paving; Riven or Textured in Natural or Buff							
450 × 450 × 35 mm	16.23	0.32	31.17	–	27.76	m²	**58.93**
600 × 300 × 35 mm	16.23	0.33	32.20	–	27.91	m²	**60.11**
600 × 600 × 35 mm	16.23	0.27	25.76	–	26.06	m²	**51.82**
600 × 600 × 63 mm	52.84	0.27	25.76	–	68.16	m²	**93.92**
Celestia paving; Smooth or skimmed or textured in various colour options; Luna, Nebula, Grava, Lyra with minor cost variations between colours and textures. Highest material cost used in the calculations below; Smooth pavings are generally higher priced							
450 × 450 × 50 mm	48.41	0.32	31.17	–	63.72	m²	**94.89**
600 × 600 × 50 mm	48.41	0.27	25.76	–	63.07	m²	**88.83**
600 × 200 × 50 mm	48.41	0.45	43.91	–	63.07	m²	**106.98**
900 × 300 × 50 mm	48.41	0.27	25.76	–	63.07	m²	**88.83**
Modal textured flags; mixed shades and sizes; (smaller sizes shown under slab paving); 80 mm thick							
450 × 450 mm	43.72	0.16	15.02	–	57.07	m²	**72.09**
400 × 400 mm	43.72	0.14	13.52	–	57.07	m²	**70.59**
600 × 200 mm	43.72	0.16	15.02	–	57.07	m²	**72.09**
600 × 300 mm	43.72	0.16	15.84	–	57.07	m²	**72.91**
600 × 600 mm	43.72	0.14	13.52	–	57.07	m²	**70.59**
900 × 600 mm	43.72	0.14	13.52	–	57.07	m²	**70.59**
mixed sizes to pattern	43.72	0.26	25.36	0.18	57.97	m²	**83.51**

NEW ITEMS FOR THIS EDITION

Item – Overhead and Profit Included	PC £	Labour hours	Labour £	Plant £	Material £	Unit	Total rate £
35 Flag paving – Marshalls concrete flag products – cont							
Precast concrete flag pavings; Marshalls Plc; Conservation paving; Natural granite aggregate appearance on prepared subbase measured separately; bedding on 25 mm thick cement: sand mortar (1:4); pointed straight both ways (Steintec jointing mixture)							
Conservation paving; Smooth skimmed or textured in various colour options; Silver grey, Charcoal, Heather grey, and Buff with minor cost variations between colours and textures. Highest material cost used in the calculations below; Smooth pavings are generally higher priced							
450 × 300 × 50 mm	52.84	0.33	32.20	–	69.45	m²	**101.65**
450 × 450 × 50 mm	52.84	0.32	31.17	–	68.82	m²	**99.99**
450 × 600 × 63 mm	52.84	0.15	14.31	–	68.16	m²	**82.47**
Precast concrete vehicular paving flags; Marshalls Plc; on prepared base (not included); on 50 mm compacted sharp sand bed; flags laid in 7 mm loose sand and vibrated; joints filled with sharp sand and vibrated; level and to falls only							
Trafica paving blocks;							
450 × 450 × 70 mm							
Perfecta finish; colour natural	44.42	0.17	16.10	0.18	53.74	m²	**70.02**
Perfecta finish; colour buff	48.76	0.17	16.10	0.18	58.74	m²	**75.02**
Saxon finish; colour natural	43.71	0.17	16.10	0.18	52.92	m²	**69.20**
Saxon finish; colour buff	48.36	0.17	16.10	0.18	58.27	m²	**74.55**
900 × 600 mm with additional							
central support per slab	84.66	0.31	20.13	–	97.36	m²	**117.49**

NEW ITEMS FOR THIS EDITION

Item – Overhead and Profit Included	PC £	Labour hours	Labour £	Plant £	Material £	Unit	Total rate £
28 Roof finishings – Blue roof							
Blue roof systems; ZinCo Green roof systems Ltd; (Stormwater management roof); For managment of stormwater under a green roof system; build-up of Blue roof system over roof surface with water proofing (not included) and root resistant protection							
Extensive Blue roof; Build-up height (inclusive of green roof components, priced separately: 200 mm; Saturated weight including plants 220 kg/m^2; Water storage capacity 114 l/m^2; water retention height 60 mm							
Root barrier WSF 40	2.31	0.02	1.08	–	2.66	m^2	**3.74**
Filter sheet PV; above and below the water retention unit (2 m^2/m^2)	5.39	0.03	2.14	–	6.20	m^2	**8.34**
Retention spacer RS 60	18.00	0.25	16.10	–	20.70	m^2	**36.80**
Zinco Filter Sheet SF – Rolled to drainage layer FD40-E	0.88	0.04	3.86	–	1.01	m^2	**4.87**
Optional retention spacer units							
Retention spacer RSX 100; Water retention capacity 95 l/m^2; Total water storage of system 154 l/m^2	37.00	0.29	18.40	–	42.55	m^2	**60.95**
Flow control; Adjustable flow controller to be installed on flat roofs over water outlets with a contact flange; including a suitable inspection chamber; suitable for installation on Stormwater Management Roofs (Blue Roofs) with rainwater attenuation up to ca. 85 mm. Adjustable discharge rate							
RDS 28 and inspection chamber	320.00	1.00	32.20	–	368.00	nr	**400.20**
Intensive Green roof; instead of root barrier WSF 40							
Root barrier WSB–100-PO	13.47	0.03	1.61	–	15.49	m^2	**17.10**

1 PRELIMINARIES

Item – Overhead and Profit Included	PC £	Labour hours	Labour £	Plant £	Material £	Unit	Total rate £
1.2 EMPLOYER'S REQUIREMENTS: TENDERING/ SUBLETTING/SUPPLY							
Tendering costs for an employed estimator for the acquisition of a landscape main or subcontract; inclusive of measurement, sourcing of materials and suppliers, cost and area calculations, pre- and post-tender meetings, bid compliance method statements and submission documents							
Remeasureable contract from prepared bills; contract value							
£25,000	–	–	–	–	–	nr	517.50
£50,000	–	–	–	–	–	nr	862.50
£100,000	–	–	–	–	–	nr	2070.00
£200,000	–	–	–	–	–	nr	3450.00
£500,000	–	–	–	–	–	nr	4485.00
£1,000,000	–	–	–	–	–	nr	6900.00
Lump sum contract from specifications and drawings only; contract value							
£25,000	–	–	–	–	–	nr	1380.00
£50,000	–	–	–	–	–	nr	1725.00
£100,000	–	–	–	–	–	nr	3450.00
£200,000	–	–	–	–	–	nr	6210.00
£500,000	–	–	–	–	–	nr	8625.00
£1,000,000	–	–	–	–	–	nr	13800.00
Material costs for tendering							
Drawing and plan printing and distribution costs; plans issued by employer on CD; project value							
£30,000	–	–	–	–	46.00	nr	46.00
£30,000 to £80,000	–	–	–	–	115.00	nr	115.00
£80,000 to £150,000	–	–	–	–	230.00	nr	230.00
£150,000 to £300,000	–	–	–	–	322.00	nr	322.00
£300,000 to £1,000,000	–	–	–	–	552.00	nr	552.00
Programmes; tender stage programmes							
Allow for production of an outline works programme for submission with the tenders; project value							
£30,000	–	1.50	69.00	–	–	nr	69.00
£50,000	–	2.00	92.00	–	–	nr	92.00
£75,000	–	2.50	115.00	–	–	nr	115.00

1 PRELIMINARIES

Item – Overhead and Profit Included	PC £	Labour hours	Labour £	Plant £	Material £	Unit	Total rate £
£100,000	–	3.00	138.00	–	–	nr	138.00
£200,000	–	4.00	184.00	–	–	nr	184.00
£500,000	–	4.50	207.00	–	–	nr	207.00
£1,000,000	–	6.00	276.00	–	–	nr	276.00
Method statements for private or commercial projects where award is not points based; provide detailed method statements on all aspects of the works; project value							
£30,000	–	2.50	115.00	–	–	nr	115.00
£50,000	–	3.50	161.00	–	–	nr	161.00
£75,000	–	4.00	184.00	–	–	nr	184.00
£100,000	–	5.00	230.00	–	–	nr	230.00
£200,000	–	6.00	276.00	–	–	nr	276.00
£500,000	–	7.00	322.00	–	–	nr	322.00
£1,000,000	–	8.00	368.00	–	–	nr	368.00
Method statements for local authority type projects where award is points rated; provide detailed method statements on all aspects of the works; project value							
£30,000	–	5.00	230.00	–	–	nr	230.00
£50,000	–	5.00	230.00	–	–	nr	230.00
£75,000	–	6.00	276.00	–	–	nr	276.00
£100,000	–	7.00	322.00	–	–	nr	322.00
£200,000	–	8.00	368.00	–	–	nr	368.00
£500,000	–	9.00	414.00	–	–	nr	414.00
£1,000,000	–	16.00	736.00	–	–	nr	736.00
Health and safety Produce health and safety file including preliminary meeting and subsequent progress meetings with external planning officer in connection with health and safety; project value							
£35,000	–	8.00	368.00	–	–	nr	368.00
£75,000	–	12.00	552.00	–	–	nr	552.00
£100,000	–	16.00	736.00	–	–	nr	736.00
£200,000 to £500,000	–	40.00	1840.00	–	–	nr	1840.00
Maintain health and safety file for project duration; project value							
£35,000	–	4.00	184.00	–	–	week	184.00
£75,000	–	4.00	184.00	–	–	week	184.00
£100,000	–	8.00	368.00	–	–	week	368.00
£200,000 to £500,000	–	8.00	368.00	–	–	week	368.00

1 PRELIMINARIES

Item – Overhead and Profit Included	PC £	Labour hours	Labour £	Plant £	Material £	Unit	Total rate £
1.2 EMPLOYER'S REQUIREMENTS: TENDERING/ SUBLETTING/SUPPLY – cont							
Health and safety – cont							
Produce written risk assessments on all areas of operations within the scope of works of the contract; project value							
£35,000	–	2.00	92.00	–	–	nr	**92.00**
£75,000	–	3.00	138.00	–	–	nr	**138.00**
£100,000	–	5.00	230.00	–	–	nr	**230.00**
£200,000 to £500,000	–	8.00	368.00	–	–	nr	**368.00**
Produce COSHH assessments on all substances to be used in connection with the contract; project value							
£35,000	–	1.50	69.00	–	–	nr	**69.00**
£75,000	–	2.00	92.00	–	–	nr	**92.00**
£100,000	–	3.00	138.00	–	–	nr	**138.00**
£200,000 to £500,000	–	3.00	138.00	–	–	nr	**138.00**
1.2.4 EMPLOYER'S REQUIREMENTS: SECURITY/ SAFETY/PROTECTION							
Temporary security fence; HSS Hire; mesh framed unclimbable fencing; including precast concrete supports and couplings							
Weekly hire; 2.85 × 2.00 m high							
weekly hire rate	–	–	–	1.35	–	m	**1.35**
erection of fencing; labour only	–	0.10	3.22	–	–	m	**3.22**
removal of fencing loading to collection vehicle	–	0.07	2.15	–	–	m	**2.15**
delivery charge	–	–	–	0.99	–	m	**0.99**
return haulage charge	–	–	–	0.74	–	m	**0.74**

1 PRELIMINARIES

Item – Overhead and Profit Included	PC £	Labour hours	Labour £	Plant £	Material £	Unit	Total rate £
Tree Root zone protection; Greenfix Ltd Carefully hand dig around base of tree but away from trunk to required depth; Lay Treetex geofabric to subgrade; Lay Geoweb cellular root protection to required depth and backfill with angular material							
To graded and compacted substrate (not included); on 50 mm sharp sand base filled with angular drainage aggregate 20–5 mm; not mechanical traffic on the area of the rootzone permitted; Excavated material loaded outside the root protection zone to spoil heaps							
Geoweb 75 mm depth	7.26	0.07	6.44	1.16	16.87	m^2	**24.47**
Geoweb 100 mm depth	8.63	0.08	7.24	1.35	20.38	m^2	**28.97**
Geoweb 150 mm depth	10.55	0.10	9.18	1.46	26.09	m^2	**36.73**
Geoweb 200 mm depth	13.06	0.10	10.00	2.20	32.48	m^2	**44.68**
Geoweb 300 mm depth	19.72	0.13	12.72	3.51	47.12	m^2	**63.35**
Lay ply boards to protect surfaces							
To horizontal surfaces	–	0.01	0.70	–	5.85	m^2	**6.55**
To vertical surfaces	–	0.07	3.74	–	5.85	m^2	**9.59**
1.2.6 SUBMISSION OF SAMPLES							
Set up sample panels for the works inclusive of arranging delivery of sample materials; paving samples on 100 mm base							
Costs of labours only							
brick paving panel (pointed)	–	6.00	168.00	–	–	m^2	**168.00**
brick paving panel (butt jointed)	–	4.00	112.00	–	–	m^2	**112.00**
block paving panel	–	4.00	112.00	–	–	m^2	**112.00**
stone slab paving panel	–	6.00	168.00	–	–	m^2	**168.00**
cladding panel	–	7.00	196.00	–	–	m^2	**196.00**
brick wall panel	–	5.00	140.00	–	–	m^2	**140.00**
render panel	–	3.50	98.00	–	–	m^2	**98.00**
paint panel	–	3.00	84.00	–	–	m^2	**84.00**

1 PRELIMINARIES

Item – Overhead and Profit Included	PC £	Labour hours	Labour £	Plant £	Material £	Unit	Total rate £
1.2.7 CONTRACTOR'S GENERAL COST ITEMS: MECHANICAL PLANT							
Southern Conveyors; moving only of granular material by belt conveyor; conveyors fitted with troughed belts, receiving hopper, front and rear undercarriages and driven by electrical and air motors; support work for conveyor installation (scaffolding), delivery, collection and installation all excluded							
Delivery collection haulage and setup charges							
Conveyors up to 10 m long	–	–	–	575.00	–	nr	**575.00**
Conveyors up to 20 m long	–	–	–	1035.00	–	nr	**1035.00**
Conveyors up to 50 m long	–	–	–	2875.00	–	nr	**2875.00**
Conveyor belt width 400 mm; mechanically loaded and removed at offload point; conveyor length							
up to 5 m	–	1.50	48.30	19.60	–	m^3	**67.90**
10 m	–	1.50	48.30	20.18	–	m^3	**68.48**
12.5 m	–	1.50	48.30	20.87	–	m^3	**69.17**
15 m	–	1.50	48.30	21.37	–	m^3	**69.67**
20 m	–	1.50	48.30	22.55	–	m^3	**70.85**
25 m	–	1.50	48.30	24.03	–	m^3	**72.33**
30 m	–	1.50	48.30	25.52	–	m^3	**73.82**
Conveyor belt width 600 mm; mechanically loaded and removed at offload point; conveyor length							
up to 5 m	–	1.00	32.21	13.40	–	m^3	**45.61**
10 m	–	1.00	32.21	14.05	–	m^3	**46.26**
12.5 m	–	1.00	32.21	14.52	–	m^3	**46.73**
15 m	–	1.00	32.21	15.04	–	m^3	**47.25**
20 m	–	1.00	32.21	16.03	–	m^3	**48.24**
25 m	–	1.00	32.21	17.68	–	m^3	**49.89**
30 m	–	1.00	32.21	19.00	–	m^3	**51.21**
Conveyor belt width 400 mm; mechanically loaded and removed by hand at offload point; conveyor length							
up to 5 m	–	3.19	102.73	6.95	–	m^3	**109.68**
10 m	–	3.19	102.73	7.74	–	m^3	**110.47**
12.5 m	–	3.19	102.73	8.66	–	m^3	**111.39**
15 m	–	3.19	102.73	9.31	–	m^3	**112.04**
20 m	–	3.19	102.73	10.90	–	m^3	**113.63**
25 m	–	3.19	102.73	12.88	–	m^3	**115.61**
30 m	–	3.19	102.73	14.85	–	m^3	**117.58**

1 PRELIMINARIES

Item – Overhead and Profit Included	PC £	Labour hours	Labour £	Plant £	Material £	Unit	Total rate £
Terranova Cranes Ltd; crane hire; materials handling and lifting; telescopic cranes supply and management; exclusive of roadway management or planning applications; prices below illustrate lifts of 1 tonne at maximum crane reach; lift cycle of 0.25 hours; mechanical filling of material skip if appropriate							
35 tonne mobile crane; lifting capacity of 1 tonne at 26 m; C.P.A hire (all management by hirer)							
granular materials including concrete	–	0.75	24.15	24.25	–	tonne	**48.40**
palletized or packed materials	–	0.50	16.10	21.56	–	tonne	**37.66**
35 tonne mobile crane; lifting capacity of 1 tonne at 26 m; contract lift							
granular materials or concrete	–	0.75	24.15	57.32	–	tonne	**81.47**
palletized or packed materials	–	0.50	16.10	54.63	–	tonne	**70.73**
50 tonne mobile crane; lifting capacity of 1 tonne at 34 m; C.P.A hire (all management by hirer)							
granular materials including concrete	–	0.75	24.15	27.13	–	tonne	**51.28**
palletized or packed materials	–	1.00	32.20	24.44	–	tonne	**56.64**
50 tonne mobile crane; lifting capacity of 1 tonne at 34 m; contract lift							
granular materials or concrete	–	0.75	24.15	68.82	–	tonne	**92.97**
palletized or packed materials	–	0.50	16.10	66.13	–	tonne	**82.23**
80 tonne mobile crane; lifting capacity of 1 tonne at 44 m; C.P.A. hire (all management by hirer)							
granular materials including concrete	–	0.75	24.15	42.94	–	tonne	**67.09**
palletized or packed materials	–	0.50	16.10	40.25	–	tonne	**56.35**
80 tonne mobile crane; lifting capacity of 1 tonne at 44 m; contract lift							
granular materials or concrete	–	0.75	24.15	80.32	–	tonne	**104.47**
palletized or packed materials	–	0.50	16.10	77.63	–	tonne	**93.73**
Other Mechanical or hired Plant required to manage a landscape contract							
Hired in plant							
General purpose Excavator and dumper – non productive time, or not task allocated; operated	–	–	–	80.59	–	hour	**80.59**
Telehandler; operated	–	–	–	37.46	–	week	**37.46**
Site skips	–	–	–	234.00	–	nr	**234.00**
Fuel bowsers	–	–	–	70.00	–	week	**70.00**

1 PRELIMINARIES

Item – Overhead and Profit Included	PC £	Labour hours	Labour £	Plant £	Material £	Unit	Total rate £
1.2.8 CONTRACTOR'S GENERAL COST ITEMS: TEMPORARY WORKS							
LiveTrakway; portable roadway systems Temporary roadway system laid directly onto existing surface or onto PVC matting to protect existing surface; most systems are based on a weekly hire charge with transportation, installation and recovery charges included							
Heavy Duty Trakpanel; per panel (3.05 × 2.59 m); per week	–	–	–	–	–	m²	5.57
Outrigger Mats for use in conjunction with Heavy Duty Trakpanels; per set of 4 mats per week	–	–	–	–	–	set	94.60
Medium Duty Trakpanel; per panel (2.44 × 3.00 m); per week	–	–	–	–	–	m²	6.01
LD20 Evolution; Light Duty Trakway; roll out system minimum delivery 50 m	–	–	–	–	–	m²	7.33
Terraplas Walkways; turf protection system; per section (1 × 1 m); per week	–	–	–	–	–	m²	5.50
1.5 TENDER AND CONTRACT DOCUMENTS							
Note: The preliminary requirements are different for all projects. Many are time related and many are set costs. Others vary depending on the project requirements. We have attempted to provide a guide to the cost related items in Section A of the National Building Specification. These costs are a guide only. Users of the book should evaluate each project separately. There are other preliminary cost items which may be added in to the preliminary costs of any project which may not be featured here.							

1 PRELIMINARIES

Item – Overhead and Profit Included	PC £	Labour hours	Labour £	Plant £	Material £	Unit	Total rate £
Method statements							
Provide detailed method statements on all aspects of the works; project value							
£30,000	–	2.50	100.00	–	–	nr	**100.00**
£50,000	–	3.50	140.00	–	–	nr	**140.00**
£75,000	–	4.00	160.00	–	–	nr	**160.00**
£100,000	–	5.00	200.00	–	–	nr	**200.00**
£200,000	–	6.00	240.00	–	–	nr	**240.00**
Tender and contract documents							
Drawing and plan printing and distribution costs; plans issued by employer on CD; project value							
£30,000	–	–	–	–	40.00	nr	**40.00**
£30,000 to £80,000	–	–	–	–	100.00	nr	**100.00**
£80,000 to £150,000	–	–	–	–	200.00	nr	**200.00**
£150,000 to £300,000	–	–	–	–	280.00	nr	**280.00**
£300,000 to £1,000,000	–	–	–	–	480.00	nr	**480.00**
Contract drawings for project management and distribution to suppliers, subcontractors and site staff							
£30,000	–	–	–	–	80.00	nr	**80.00**
£30,000 to £80,000	–	–	–	–	200.00	nr	**200.00**
£80,000 to £150,000	–	–	–	–	400.00	nr	**400.00**
£150,000 to £300,000	–	–	–	–	560.00	nr	**560.00**
£300,000 to £1,000,000	–	–	–	–	960.00	nr	**960.00**
1.5 THE CONTRACT/ SUBCONTRACT							
Contract/subcontract evaluation							
Due diligence on evaluation or examination of clauses in contract documents; evaluation and report by a suitably qualified quantity surveyor or legal advisor where necessary							
minor works contract	–	1.00	40.00	–	–	nr	**40.00**
JCLI	–	–	–	–	70.00	nr	**70.00**
JCT intermediate contract or subcontract	–	1.00	40.00	–	140.00	nr	**180.00**

1 PRELIMINARIES

Item – Overhead and Profit Included	PC £	Labour hours	Labour £	Plant £	Material £	Unit	Total rate £
1.6 DESIGN COSTS FOR LANDSCAPE WORKS							
Landscape design; Landscape architects; mixed hard and soft landscape; indicative prices for design project stages; brief; planning; detailed design and supervision during construction stage; the complexity ratings shown refer to the Landscape Institute tables published in this book							
Commercial urban development; complexity rating 2; landscape construction value							
£100,000.00	–	–	–	–	–	nr	**10000.00**
£250,000.00	–	–	–	–	–	nr	**20000.00**
£500,000.00	–	–	–	–	–	nr	**35000.00**
£1,000,000.00	–	–	–	–	–	nr	**65000.00**
Hospitals and education; complexity rating 3; landscape construction value							
£100,000.00	–	–	–	–	–	nr	**11000.00**
£250,000.00	–	–	–	–	–	nr	**23000.00**
£500,000.00	–	–	–	–	–	nr	**40000.00**
£1,000,000.00	–	–	–	–	–	nr	**74000.00**
Public urban open space; complexity rating 4; landscape construction value							
£100,000.00	–	–	–	–	–	nr	**12000.00**
£250,000.00	–	–	–	–	–	nr	**26000.00**
£500,000.00	–	–	–	–	–	nr	**47500.00**
£1,000,000.00	–	–	–	–	–	nr	**85000.00**
Countryside or leisure open space; complexity rating 1							
£100,000.00	–	–	–	–	–	nr	**9000.00**
£250,000.00	–	–	–	–	–	nr	**18000.00**
£500,000.00	–	–	–	–	–	nr	**32000.00**
£1,000,000.00	–	–	–	–	–	nr	**60000.00**
Reclamation project; complexity rating 4							
£100,000.00	–	–	–	–	–	nr	**12000.00**
£250,000.00	–	–	–	–	–	nr	**26000.00**
£500,000.00	–	–	–	–	–	nr	**45000.00**
£1,000,000.00	–	–	–	–	–	nr	**80000.00**
Housing; complexity rating 3							
£100,000.00	–	–	–	–	–	nr	**11000.00**
£250,000.00	–	–	–	–	–	nr	**23000.00**
£500,000.00	–	–	–	–	–	nr	**40000.00**
£1,000,000.00	–	–	–	–	–	nr	**74000.00**

1 PRELIMINARIES

Item – Overhead and Profit Included	PC £	Labour hours	Labour £	Plant £	Material £	Unit	Total rate £
1.7 EMPLOYER'S REQUIREMENTS: PROVISION CONTENT AND USE OF DOCUMENTS							
Provision, content and use of documents							
Supply as built drawings for elements of the project that may have carried from the original design drawings							
£35,000	–	2.00	87.40	–	–	nr	**87.40**
£75,000	–	2.00	87.40	–	–	nr	**87.40**
£100,000	–	4.00	174.80	–	–	nr	**174.80**
£200,000 to £500,000	–	5.00	218.50	–	–	nr	**218.50**
Fees for Considerate Contractors Scheme (CCS)							
Project value							
up to £100,000	–	–	–	–	115.00	nr	**115.00**
£100,000–£500,000	–	–	–	–	270.25	nr	**270.25**
£500,000–£5,000,000	–	–	–	–	540.50	nr	**540.50**
over £5,000,000	–	–	–	–	810.75	nr	**810.75**
Climatic conditions – keep records of temperature and rainfall; delays due to weather including descriptions of weather							
daily cost	–	0.08	3.83	–	–	day	**3.83**
weekly cost	–	0.42	19.16	–	–	day	**19.16**
Programmes; master programme for the contract works							
Allow for production of works programmes prior to the start of the works; project value							
£30,000	–	3.00	138.00	–	–	nr	**138.00**
£50,000	–	6.00	276.00	–	–	nr	**276.00**
£75,000	–	8.00	368.00	–	–	nr	**368.00**
£100,000	–	10.00	460.00	–	–	nr	**460.00**
£200,000	–	14.00	644.00	–	–	nr	**644.00**
£500,000	–	15.00	690.00	–	–	nr	**690.00**
£1,000,000	–	18.00	828.00	–	–	nr	**828.00**
Indicative staff resource chart							
Project value							
up to £100,000	–	1.00	46.00	–	–	nr	**46.00**
up to £500,000	–	1.50	69.00	–	–	nr	**69.00**
up to £1,000,000	–	2.00	92.00	–	–	nr	**92.00**

1 PRELIMINARIES

Item – Overhead and Profit Included	PC £	Labour hours	Labour £	Plant £	Material £	Unit	Total rate £
1.7 EMPLOYER'S REQUIREMENTS: PROVISION CONTENT AND USE OF DOCUMENTS – cont							
Curriculum vitae of staff							
Prepare and submit curriculum vitae of pre-construction and construction phase staff							
per staff member	–	0.75	34.50	–	–	nr	**34.50**
1.7 EMPLOYER'S REQUIREMENTS: MANAGEMENT OF THE WORKS							
Allow for updating the works programme during the course of the works; project value							
£30,000	–	1.00	46.00	–	–	nr	**46.00**
£50,000	–	1.50	69.00	–	–	nr	**69.00**
£75,000	–	2.00	92.00	–	–	nr	**92.00**
£100,000	–	3.00	138.00	–	–	nr	**138.00**
£200,000	–	5.00	230.00	–	–	nr	**230.00**
Setting out							
Setting out for external works operations comprising hard and soft works elements; placing of pegs and string lines to Landscape Architect's drawings; surveying levels and placing level pegs; obtaining approval from the Landscape Architect to commence works; areas of entire site							
1000 m^2	–	5.00	161.00	–	9.66	nr	**170.66**
2500 m^2	–	8.00	257.60	–	19.32	nr	**276.92**
5000 m^2	–	8.00	257.60	–	19.32	nr	**276.92**
10000 m^2	–	32.00	1030.40	–	48.30	nr	**1078.70**
Setting out engineer							
per half day	–	–	–	–	–	nr	**210.00**
per day	–	–	–	–	–	nr	**350.00**

1 PRELIMINARIES

Item – Overhead and Profit Included	PC £	Labour hours	Labour £	Plant £	Material £	Unit	Total rate £
1.11 CONTRACTOR'S GENERAL COST ITEMS: SITE ACCOMMODATION							
General							
The following items are instances of the commonly found preliminary costs associated with external works contracts. The assumption is made that the external works contractor is subcontracted to a main contractor.							
Elliot Hire; erect temporary accommodation and storage on concrete base measured separately							
Prefabricated office hire; jackleg; open plan							
3.6 × 2.4 m	–	–	–	33.88	–	week	**33.88**
4.8 × 2.4 m	–	–	–	39.01	–	week	**39.01**
Armoured store; erect temporary secure storage container for tools and equipment							
3.0 × 2.4 m	–	–	–	13.97	–	week	**13.97**
3.6 × 2.4 m	–	–	–	13.97	–	week	**13.97**
Delivery and collection charges on site offices							
delivery charge	–	–	–	234.60	–	load	**234.60**
collection charge	–	–	–	234.60	–	load	**234.60**
Toilet facilities							
HSS Hire; serviced self-contained toilet delivered to and collected from site; maintained by toilet supply company							
single chemical toilet including wash hand basin and water	–	–	–	40.25	–	week	**40.25**
delivery and collection; each way	–	–	–	34.50	–	nr	**34.50**
1.11 THE SITE/EXISTING BUILDINGS							
Existing mains and services; mark positions of existing mains and services; locate and mark; site area							
less than 500 m²	–	1.50	48.30	–	–	nr	**48.30**
up to 1000 m²	–	3.00	96.60	–	–	nr	**96.60**
up to 2000 m²	–	8.00	257.60	–	–	nr	**257.60**
up to 4000 m²	–	16.00	515.20	–	–	nr	**515.20**

1 PRELIMINARIES

Item – Overhead and Profit Included	PC £	Labour hours	Labour £	Plant £	Material £	Unit	Total rate £
1.11 THE SITE/EXISTING BUILDINGS – cont							
Access to the site							
For pedestrians							
security kiosk	–	–	–	212.75	–	week	**212.75**
security guard	–	40.00	552.00	–	–	week	**552.00**
protected walkways; Heras fencing on 2 sides	–	–	–	2.69	–	m	**2.69**
Parking							
Parking expenses where vehicles do not park on the site area; per vehicle							
metropolitan area; city centre	–	–	–	–	–	week	**230.00**
metropolitan area; outer areas	–	–	–	–	–	week	**184.00**
suburban restricted parking areas	–	–	–	–	–	week	**46.00**
Congestion charging							
London only	–	–	–	–	–	week	**66.13**
Site visit; pre-tender for purposes of understanding site and tender requirements; prices below are based on a single senior manager attending site							
City location; average distance of travel 20 miles inclusive of travel and parking costs							
less than 500 m²	–	4.00	184.00	–	45.42	nr	**229.42**
up to 1000 m²	–	5.00	230.00	–	47.72	nr	**277.72**
up to 2000 m²	–	6.00	276.00	–	54.63	nr	**330.63**
up to 4000 m²	–	8.00	368.00	–	63.82	nr	**431.82**
Town or rural location; average distance of travel 20 miles inclusive of travel and parking costs							
less than 500 m²	–	3.00	138.00	–	25.30	nr	**163.30**
up to 1000 m²	–	4.00	184.00	–	26.45	nr	**210.45**
up to 2000 m²	–	5.00	230.00	–	27.60	nr	**257.60**
up to 4000 m²	–	6.00	276.00	–	27.60	nr	**303.60**

2 OFF SITE MANUFACTURED MATERIALS, COMPONENTS AND BUILDINGS

Item – Overhead and Profit Included	PC £	Labour hours	Labour £	Plant £	Material £	Unit	Total rate £
PREFABRICATED BUILDINGS/ STRUCTURES/UNITS							
Cast stone buildings; Haddonstone Ltd; ornamental garden buildings in Portland Bath or Terracotta finished cast stone; prices for stonework and facades only; excavations, foundations, reinforcement, concrete infill, roofing and floors all priced separately							
Pavilion Venetian Folly L9400; Tuscan columns, pedimented arch, quoins and optional balustrading							
4184 mm high × 4728 mm wide × 3147 mm deep	12663.73	175.00	5635.00	115.00	14760.12	nr	**20510.12**
Pavilion L9300; Tuscan columns 3496 mm high × 3634 mm wide	7686.08	144.00	4636.80	115.00	9035.83	nr	**13787.63**
Small Classical Temple L9250; 6 column with fibreglass lead effect finish dome roof							
overall height 3610 mm; dia. 2540 mm	6652.87	130.00	4186.00	57.50	7847.63	nr	**12091.13**
Large Classical Temple L9100; 8 column with fibreglass lead effect finish dome roof							
overall height 4664 mm; dia. 3190 mm	12412.92	165.00	5313.00	57.50	14471.69	nr	**19842.19**
Stepped floors to temples							
single step; Small Classical Temple	1218.01	24.00	772.80	–	1488.46	nr	**2261.26**
single step; Large Classical Temple	1719.62	26.00	837.20	–	2082.86	nr	**2920.06**
Stone structures; Architectural Heritage Ltd; hand carved from solid natural limestone with wrought iron domed roof, decorated frieze and base and integral seats; supply and erect only; excavations and concrete bases priced separately							
The Park Temple; 5 columns; 3500 mm high × 1650 mm dia.	10000.00	96.00	3091.20	108.66	11500.00	nr	**14699.86**
The Estate Temple; 6 columns; 4000 mm high × 2700 mm dia.	17600.00	120.00	3864.00	217.33	20240.00	nr	**24321.33**

2 OFF SITE MANUFACTURED MATERIALS, COMPONENTS AND BUILDINGS

Item – Overhead and Profit Included	PC £	Labour hours	Labour £	Plant £	Material £	Unit	Total rate £
PREFABRICATED BUILDINGS/ STRUCTURES/UNITS – cont							
Stone structures; Architectural Heritage Ltd; hand carved from solid natural limestone with solid oak trelliage; prices for stonework and facades only; supply and erect only; excavations and concrete bases priced separately The Pergola; 2240 mm high × 2640 mm wide × 6990 mm long	7600.00	96.00	3091.20	434.64	8784.23	nr	**12310.07**
Ornamental stone structures; Architectural Heritage Ltd; setting to bases or plinths (not included) The Small Obelisk; classic natural stone obelisk; tapering square form on panelled square base; 1860 mm high × 360 mm square	1800.00	2.00	64.40	–	2070.00	nr	**2134.40**
The Inverted Sundial Pedestal with Griffin Armillary Sphere	2200.00	2.00	64.40	–	2530.00	nr	**2594.40**
Large obelisk; 2.40 m high × 460 mm square; hand carved natural limestone obelisk of tapering square form raised upon a panelled square base supported by four spheres. Reproduction.	2400.00	1.00	32.20	–	2760.00	nr	**2792.20**

3 DEMOLITIONS

Item – Overhead and Profit Included	PC £	Labour hours	Labour £	Plant £	Material £	Unit	Total rate £
DEMOLITIONS/CLEARANCE							
Demolish existing structures; disposal off site; mechanical demolition; with 3 tonne excavator and dumper							
Brick wall							
112.5 mm thick	–	0.07	2.15	3.21	7.43	m²	**12.79**
225 mm thick	–	0.08	2.68	4.01	14.85	m²	**21.54**
337.5 mm thick	–	–	–	11.65	24.91	m²	**36.56**
450 mm thick	–	–	–	15.52	29.69	m²	**45.21**
Demolish existing structures; disposal off site mechanically loaded; all other works by hand							
Brick wall							
112.5 mm thick	–	0.33	10.73	–	7.43	m²	**18.16**
225 mm thick	–	0.50	16.10	–	14.85	m²	**30.95**
337.5 mm thick	–	0.67	21.47	–	24.91	m²	**46.38**
450 mm thick	–	1.00	32.20	–	29.69	m²	**61.89**
Demolish existing structures; disposal off site mechanically loaded; by diesel or electric breaker; all other works by hand							
Brick wall							
112.5 mm thick	–	0.17	5.37	0.29	7.43	m²	**13.09**
225 mm thick	–	0.20	6.44	0.35	14.85	m²	**21.64**
337.5 mm thick	–	0.25	8.05	0.44	22.02	m²	**30.51**
450 mm thick	–	0.33	10.73	0.59	29.69	m²	**41.01**
Break out concrete footings associated with free-standing walls; inclusive of all excavation and backfilling with excavated material; disposal off site mechanically loaded							
By mechanical breaker; diesel or electric							
plain concrete	–	1.50	48.30	18.32	39.94	m³	**106.56**
reinforced concrete	–	2.50	80.50	28.67	65.99	m³	**175.16**
Remove existing free-standing buildings; demolition by hand							
Timber building with suspended timber floor; hardstanding or concrete base not included; disposal off site							
shed 6.0 m²	–	2.00	64.40	–	68.74	nr	**133.14**
shed 10.0 m²	–	3.00	96.61	–	123.73	nr	**220.34**
shed 15.0 m²	–	3.50	112.71	–	192.47	nr	**305.18**

3 DEMOLITIONS

Item – Overhead and Profit Included	PC £	Labour hours	Labour £	Plant £	Material £	Unit	Total rate £
DEMOLITIONS/CLEARANCE – cont							
Remove existing free-standing buildings – cont							
Timber building; insulated; with timber or concrete posts set in concrete, felt covered timber or tiled roof; internal walls cladding with timber or plasterboard; load arisings to self loaded muck-away							
timber structure 6.0 m²	–	2.50	80.50	–	296.94	nr	**377.44**
timber structure 12.0 m²	–	3.50	112.71	–	415.16	nr	**527.87**
timber structure 20.0 m²	–	8.00	257.60	–	439.91	nr	**697.51**
Demolition of free-standing brick buildings with tiled or sheet roof; concrete foundations measured separately; mechanical demolition; maximum distance to stockpile 25 m; inclusive for all access scaffolding and the like; maximum height of roof 4.0 m; inclusive of all doors, windows, guttering and down pipes; including disposal by grab							
Half brick thick							
10 m²	–	8.00	257.60	206.36	224.49	nr	**688.45**
20 m²	–	16.00	515.20	372.39	288.72	nr	**1176.31**
1 brick thick							
10 m²	–	10.00	322.00	332.13	448.98	nr	**1103.11**
20 m²	–	16.00	515.20	498.32	577.42	nr	**1590.94**
Cavity wall with blockwork inner skin and brick outer skin; insulated							
10 m²	–	12.00	386.55	458.06	623.23	nr	**1467.84**
20 m²	–	20.00	644.00	624.31	802.92	nr	**2071.23**
Extra over to the above for disconnection of services							
Electrical							
disconnection	–	2.00	82.80	–	–	nr	**82.80**
grub out cables and dispose; backfilling; by machine	–	–	–	0.40	1.10	m	**1.50**
grub out cables and dispose; backfilling; by hand	–	0.50	16.10	–	1.10	m	**17.20**
Water supply, foul or surface water drainage							
disconnection; capping off	–	1.00	32.20	–	79.20	nr	**111.40**
grub out pipes and dispose; backfilling; by machine	–	–	–	0.40	1.10	m	**1.50**
grub out pipes and dispose; backfilling; by hand	–	0.50	16.10	–	1.10	m	**17.20**

3 DEMOLITIONS

Item – Overhead and Profit Included	PC £	Labour hours	Labour £	Plant £	Material £	Unit	Total rate £
Demolish existing fence; remove stakes or grub out posts as appropriate; remove to stock pile for removal off site; (not included)							
Mechanical demolition; clear fence line							
chain link fence 1.20–1.50 m high	–	0.01	0.64	0.70	–	m	1.34
closeboard or timber panel fence 1.80 m high	–	0.03	1.61	1.61	–	m	3.22
Mechanical demolition; light shrubs or creepers in chain links							
chain link fence 1.20–1.50 m high	–	0.01	0.81	0.90	–	m	1.71
closeboard or timber panel fence 1.80 m high	–	0.03	2.14	2.16	–	m	4.30
Mechanical demolition; heavy shrubs or creepers bramble and the like requiring clearance to enable removal							
fence 1.20–1.50 m high	–	0.05	3.22	2.27	–	m	5.49
closeboard or timber panel fence 1.80 m high	–	0.05	3.22	3.42	–	m	6.64
Demolition and site transport by hand; clear fence lines							
fence 1.20–1.50 m high	–	0.03	2.14	–	–	m	2.14
closeboard or timber panel fence 1.80 m high	–	0.07	4.30	–	–	m	4.30
Demolition and site transport by hand; light vegetation and creepers							
fence 1.20–1.50 m high	–	0.05	3.22	–	–	m	3.22
closeboard or timber panel fence 1.80 m high	–	0.13	8.05	–	–	m	8.05
Demolition and site transport by hand; heavy shrubs or creepers bramble and the like requiring clearance to enable removal							
fence 1.20–1.50 m high	–	0.13	8.05	–	–	m	8.05
closeboard or timber panel fence 1.80 m high	–	0.17	10.74	–	–	m	10.74
Demolition of posts and straining posts							
Break out straining post; grub out concrete footings; remove to stockpile on site							
single straining post by machine	–	0.25	8.05	16.55	–	nr	24.60
double straining post by machine	–	0.30	9.66	18.98	–	nr	28.64
single straining post by hand	–	1.00	64.40	–	–	nr	64.40
double straining post by hand	–	1.25	80.50	–	–	nr	80.50
Break out gatepost; concrete steel or timber; girth not exceeding 150 mm; grub out concrete footings							
by machine	–	0.25	8.05	16.55	–	nr	24.60
by hand	–	1.00	32.20	–	–	nr	32.20

3 DEMOLITIONS

Item – Overhead and Profit Included	PC £	Labour hours	Labour £	Plant £	Material £	Unit	Total rate £
SITE PREPARATION							
Topsoil stripping for preservation; stripping to subsoil layer 300 mm deep							
Larger sites with 21 tonne excavators							
moving to stockpile	–	–	–	2.46	–	m²	**2.46**
spreading locally	–	–	–	3.37	–	m²	**3.37**
Tree felling							
Felling and leaving on site							
girth 600 mm–1.50 m (95–240 mm trunk dia.)	–	1.00	96.60	–	–	nr	**96.60**
girth 1.50–3.00 m (240–475 mm trunk dia.)	–	4.00	386.40	–	–	nr	**386.40**
girth 3.00–4.00 m (475–630 mm trunk dia.)	–	6.00	579.60	–	–	nr	**579.60**
Felling and removing trees off site; logging to arboriculturalists store; chipped material to free green waste site							
girth 600 mm–1.50 m (95–240 mm trunk dia.)	–	2.50	241.50	59.89	–	nr	**301.39**
girth 1.50–3.00 m (240–475 mm trunk dia.)	–	6.00	579.60	119.80	–	nr	**699.40**
girth 3.00–4.00 m (475–630 mm trunk dia.)	–	8.00	772.80	191.67	–	nr	**964.47**
Removing tree stumps							
girth 600 mm–1.50 m	–	2.00	64.40	119.53	–	nr	**183.93**
girth 1.50–3.00 m	–	7.00	225.40	418.35	–	nr	**643.75**
girth over 3.00 m	–	12.00	386.40	717.16	–	nr	**1103.56**
Stump grinding; disposing to spoil heaps							
girth 600 mm–1.50 m	–	2.00	64.40	126.50	–	nr	**190.90**
girth 1.50–3.00 m	–	2.50	80.50	253.00	–	nr	**333.50**
girth over 3.00 m	–	4.00	128.80	221.38	–	nr	**350.18**
Clearing site vegetation							
mechanical clearance	–	0.25	8.05	14.71	–	100 m²	**22.76**
hand clearance	–	2.00	64.40	–	–	100 m²	**64.40**
Lifting turf for preservation							
sod cutter machine lift and stack	–	0.75	24.15	16.10	–	100 m²	**40.25**
hand lift and stack	–	8.33	268.32	–	–	100 m²	**268.32**
Lifting turf for disposal (disposal not included)							
by excavator; 5 tonne	–	–	–	43.84	–	100 m²	**43.84**
by excavator; 8 tonne	–	–	–	31.15	–	100 m²	**31.15**
by excavator; 8 tonne; working with two 3 tonne dumpers	–	–	–	35.16	–	100 m²	**35.16**
by excavator; 8 tonne; working with two 5 tonne dumpers	–	–	–	28.87	–	100 m²	**28.87**
hand lift and stack	–	8.33	268.32	–	–	100 m²	**268.32**

3 DEMOLITIONS

Item – Overhead and Profit Included	PC £	Labour hours	Labour £	Plant £	Material £	Unit	Total rate £
Site clearance; by machine; clear site of mature shrubs from existing cultivated beds; dig out roots by machine							
Mixed shrubs in beds; planting centres 500 mm average							
height less than 1 m	–	0.03	0.81	1.10	–	m²	**1.91**
1.00–1.50 m	–	0.04	1.29	1.77	–	m²	**3.06**
1.50–2.00 m; pruning to ground level by hand	–	0.10	3.22	4.44	–	m²	**7.66**
2.00–3.00 m; pruning to ground level by hand	–	0.10	3.22	8.87	–	m²	**12.09**
3.00–4.00 m; pruning to ground level by hand	–	0.20	6.44	14.79	–	m²	**21.23**
Site clearance; by hand; clear site of mature shrubs from existing cultivated beds; dig out roots							
Mixed shrubs in beds; planting centres 500 mm average							
height less than 1 m	–	0.33	10.73	–	–	m²	**10.73**
1.00–1.50 m	–	0.50	16.10	–	–	m²	**16.10**
1.50–2.00 m	–	1.00	32.20	–	–	m²	**32.20**
2.00–3.00 m	–	2.00	64.40	–	–	m²	**64.40**
3.00–4.00 m	–	3.00	96.61	–	–	m²	**96.61**
Spraying of vegetation; Glyphosate at 5 l/ha by tractor drawn spray equipment	–	–	–	5.06	1.91	100 m²	**6.97**
Excavated material; on site							
In spoil heaps							
average 25 m distance	–	–	–	3.14	–	m³	**3.14**
average 50 m distance	–	–	–	3.62	–	m³	**3.62**
average 100 m distance (1 dumper)	–	–	–	4.54	–	m³	**4.54**
average 100 m distance (2 dumpers)	–	–	–	5.53	–	m³	**5.53**
average 200 m distance	–	–	–	5.47	–	m³	**5.47**
average 200 m distance (2 dumpers)	–	–	–	6.66	–	m³	**6.66**

5 EXCAVATING AND FILLING

Item – Overhead and Profit Included	PC £	Labour hours	Labour £	Plant £	Material £	Unit	Total rate £
CLARIFICATION NOTES ON LABOUR COSTS IN THIS SECTION							
General groundworks team Generally a three man team is used in this section; The column 'Labour hours' reports team hours. The column 'Labour £' reports the total cost of the team for the unit of work shown							
3 man team	–	1.00	96.60	–	–	hr	**96.60**
banksman	–	1.00	32.20	–	–	hr	**32.20**
MACHINE SELECTION							
Machine volumes for excavating/ filling only and placing excavated material alongside or to a dumper; no bulkages are allowed for in the material volumes; these rates should be increased by user-preferred percentages to suit prevailing site conditions; the figures in the next section for 'Excavation mechanical' and filling allow for the use of banksmen within the rates shown below							
1.5 tonne excavators: digging volume							
1 cycle/minute; $0.04\,m^3$	–	0.42	13.42	2.66	–	m^3	**16.08**
2 cycles/minute; $0.08\,m^3$	–	0.21	6.70	1.67	–	m^3	**8.37**
3 cycles/minute; $0.12\,m^3$	–	0.14	4.47	1.36	–	m^3	**5.83**
3 tonne excavators; digging volume							
1 cycle/minute; $0.13\,m^3$	–	0.13	4.13	1.64	–	m^3	**5.77**
2 cycles/minute; $0.26\,m^3$	–	0.06	2.06	2.40	–	m^3	**4.46**
3 cycles/minute; $0.39\,m^3$	–	0.04	1.38	1.07	–	m^3	**2.45**
5 tonne excavators; digging volume							
1 cycle/minute; $0.28\,m^3$	–	–	–	2.89	–	m^3	**2.89**
2 cycles/minute; $0.56\,m^3$	–	–	–	1.45	–	m^3	**1.45**
3 cycles/minute; $0.84\,m^3$	–	–	–	0.97	–	m^3	**0.97**
8 tonne excavators; supplied with operator; digging volume							
1 cycle/minute; $0.28\,m^3$	–	–	–	3.23	–	m^3	**3.23**
2 cycles/minute; $0.56\,m^3$	–	–	–	1.62	–	m^3	**1.62**
3 cycles/minute; $0.84\,m^3$	–	–	–	1.08	–	m^3	**1.08**

5 EXCAVATING AND FILLING

Item – Overhead and Profit Included	PC £	Labour hours	Labour £	Plant £	Material £	Unit	Total rate £
21 tonne excavators; supplied with operator; digging volume							
1 cycle/minute; 1.21 m³	–	–	–	1.23	–	m³	**1.23**
2 cycles/minute; 2.42 m³	–	–	–	0.51	–	m³	**0.51**
3 cycles/minute; 3.63 m³	–	–	–	0.35	–	m³	**0.35**
Note: All volumes below are based on excavated 'earth' moist at 1,997 kg/m³ solid or 1,598 kg/m³ loose; a 25% bulkage factor has been used; the weight capacities below exceed the volume capacities of the machine in most cases; see the memorandum section at the back of this book for further weights of materials.							
Dumpers							
1 tonne high tip skip loader; volume 0.485 m³ (775 kg)							
5 loads per hour	–	0.41	13.28	1.69	–	m³	**14.97**
7 loads per hour	–	0.29	9.49	1.25	–	m³	**10.74**
10 loads per hour	–	0.21	6.64	0.93	–	m³	**7.57**
3 tonne dumper; excavated material volume 1.3 m³							
4 loads per hour	–	–	–	7.53	–	m³	**7.53**
5 loads per hour	–	–	–	6.03	–	m³	**6.03**
7 loads per hour	–	–	–	5.36	–	m³	**5.36**
10 loads per hour	–	–	–	2.93	–	m³	**2.93**
6 tonne dumper; maximum volume 3.40 m³ (5.4 t); available volume 3.77 m³							
4 loads per hour	–	0.07	2.14	0.32	–	m³	**2.46**
5 loads per hour	–	0.05	1.71	0.26	–	m³	**1.97**
7 loads per hour	–	0.04	1.22	0.21	–	m³	**1.43**
10 loads per hour	–	0.03	0.85	0.16	–	m³	**1.01**
EXCAVATING							
Market prices of topsoil; prices shown include for 20% settlement							
Multiple source screened topsoil	–	–	–	–	29.66	m³	**29.66**
Single source topsoil; British Sugar PLC	–	–	–	–	29.66	m³	**29.66**
High grade topsoil for planting	–	–	–	–	80.34	m³	**80.34**

5 EXCAVATING AND FILLING

Item – Overhead and Profit Included	PC £	Labour hours	Labour £	Plant £	Material £	Unit	Total rate £
EXCAVATING – cont							
Note: The figures in this section relate to the machine capacities shown earlier in this section. The figures below however allow for dig efficiency based on depth. A banksman is allowed for in all excavation build and disposal costs shown. Bulkages of 25% allowed; adjustments should be made for different soil types.							
Excavating; mechanical; topsoil for preservation							
3 tonne tracked excavator (bucket volume 0.13 m³)							
average depth 100 mm	–	–	–	1.29	–	m²	**1.29**
average depth 150 mm	–	–	–	1.45	–	m²	**1.45**
average depth 200 mm	–	–	–	1.93	–	m²	**1.93**
average depth 250 mm	–	–	–	2.15	–	m²	**2.15**
average depth 300 mm	–	–	–	2.58	–	m²	**2.58**
8 tonne excavator							
average depth 100 mm	–	1.00	32.20	59.77	–	100 m²	**91.97**
average depth 150 mm	–	1.25	40.25	74.70	–	100 m²	**114.95**
average depth 200 mm	–	1.78	57.32	106.38	–	100 m²	**163.70**
average depth 250 mm	–	1.90	61.18	113.55	–	100 m²	**174.73**
average depth 300 mm	–	2.00	64.40	119.53	–	100 m²	**183.93**
Excavating; mechanical; to reduce levels							
5 tonne excavator (bucket volume 0.28 m³)							
maximum depth not exceeding 0.25 m	–	0.07	2.25	3.39	–	m³	**5.64**
maximum depth not exceeding 1.00 m	–	0.05	1.53	2.31	–	m³	**3.84**
maximum depth not exceeding 2.00 m	–	0.06	1.92	2.89	–	m³	**4.81**
8 tonne tracked excavator (bucket volume 0.28 m³)							
maximum depth not exceeding 1.00 m	–	0.06	2.01	3.39	–	m³	**5.40**
maximum depth not exceeding 2.00 m	–	0.07	2.30	3.88	–	m³	**6.18**
maximum depth not exceeding 3.00 m	–	0.09	2.93	4.93	–	m³	**7.86**

5 EXCAVATING AND FILLING

Item – Overhead and Profit Included	PC £	Labour hours	Labour £	Plant £	Material £	Unit	Total rate £
21 tonne 360 tracked excavator (bucket volume 1.21 m^3)							
maximum depth not exceeding 1.00 m	–	0.01	0.36	0.82	–	m^3	**1.18**
maximum depth not exceeding 2.00 m	–	0.02	0.53	1.22	–	m^3	**1.75**
maximum depth not exceeding 3.00 m	–	0.03	1.07	2.46	–	m^3	**3.53**
Pits; 3 tonne tracked excavator							
maximum depth not exceeding 0.25 m	–	0.33	10.73	15.05	–	m^3	**25.78**
maximum depth not exceeding 1.00 m	–	0.25	8.05	11.29	–	m^3	**19.34**
maximum depth not exceeding 2.00 m	–	0.40	12.88	18.07	–	m^3	**30.95**
Pits; 8 tonne tracked excavator							
maximum depth not exceeding 2.00 m	–	0.40	12.88	21.73	–	m^3	**34.61**
maximum depth not exceeding 3.00 m	–	0.50	16.10	27.16	–	m^3	**43.26**
Trenches; width not exceeding 0.30 m; 3 tonne excavator							
maximum depth not exceeding 0.25 m	–	1.00	32.20	5.39	–	m^3	**37.59**
maximum depth not exceeding 1.00 m	–	0.69	22.06	3.69	–	m^3	**25.75**
maximum depth not exceeding 2.00 m	–	0.60	19.32	3.23	–	m^3	**22.55**
Trenches; width exceeding 0.30 m; 3 tonne excavator							
maximum depth not exceeding 0.25 m	–	0.60	19.32	3.23	–	m^3	**22.55**
maximum depth not exceeding 1.00 m	–	0.50	16.10	2.69	–	m^3	**18.79**
maximum depth not exceeding 2.00 m	–	0.38	12.39	2.07	–	m^3	**14.46**
Extra over any types of excavating irrespective of depth for breaking out existing materials; Excavator with breaker attachment							
hard rock	–	0.50	16.10	138.31	–	m^3	**154.41**
concrete	–	0.50	16.10	51.87	–	m^3	**67.97**
reinforced concrete	–	1.00	32.20	80.19	–	m^3	**112.39**
brickwork, blockwork or stonework	–	0.25	8.05	51.87	–	m^3	**59.92**

5 EXCAVATING AND FILLING

Item – Overhead and Profit Included	PC £	Labour hours	Labour £	Plant £	Material £	Unit	Total rate £
EXCAVATING – cont							
Excavating – cont							
Extra over any types of excavating irrespective of depth for breaking out existing hard pavings; Excavator with breaker attachment							
concrete; 100 mm thick	–	–	–	3.81	–	m²	**3.81**
concrete; 150 mm thick	–	–	–	6.34	–	m²	**6.34**
concrete; 200 mm thick	–	–	–	7.60	–	m²	**7.60**
concrete; 300 mm thick	–	–	–	11.41	–	m²	**11.41**
reinforced concrete; 100 mm thick	–	0.08	2.68	4.15	–	m²	**6.83**
reinforced concrete; 150 mm thick	–	0.08	2.42	5.81	–	m²	**8.23**
reinforced concrete; 200 mm thick	–	0.10	3.22	7.74	–	m²	**10.96**
reinforced concrete; 300 mm thick	–	0.15	4.83	11.61	–	m²	**16.44**
tarmacadam; 75 mm thick	–	–	–	3.81	–	m²	**3.81**
tarmacadam and hardcore; 150 mm thick	–	–	–	6.08	–	m²	**6.08**
Extra over any types of excavating irrespective of depth for taking up							
precast concrete paving slabs	–	0.07	2.15	0.97	–	m²	**3.12**
natural stone paving	–	0.10	3.22	1.44	–	m²	**4.66**
cobbles	–	0.13	4.03	1.81	–	m²	**5.84**
brick paviors	–	0.13	4.03	1.81	–	m²	**5.84**
Excavating; hand							
Topsoil for preservation; loading to barrows							
average depth 100 mm	–	0.08	7.65	–	–	m²	**7.65**
average depth 150 mm	–	0.12	11.48	–	–	m²	**11.48**
average depth 200 mm	–	0.19	18.37	–	–	m²	**18.37**
average depth 250 mm	–	0.24	22.95	–	–	m²	**22.95**
average depth 300 mm	–	0.29	27.54	–	–	m²	**27.54**
Excavating; hand							
Topsoil to reduce levels							
maximum depth not exceeding 0.25 m	–	0.79	76.51	–	–	m³	**76.51**
maximum depth not exceeding 1.00 m	–	1.03	99.46	–	–	m³	**99.46**
Pits							
maximum depth not exceeding 0.25 m	–	0.88	85.01	–	–	m³	**85.01**
maximum depth not exceeding 1.00 m	–	1.14	110.51	–	–	m³	**110.51**
maximum depth not exceeding 2.00 m (includes earthwork support)	–	2.29	221.02	–	–	m³	**290.59**

5 EXCAVATING AND FILLING

Item – Overhead and Profit Included	PC £	Labour hours	Labour £	Plant £	Material £	Unit	Total rate £
Trenches; width not exceeding 0.30 m							
maximum depth not exceeding 0.25 m	–	0.94	91.08	–	–	m³	**91.08**
maximum depth not exceeding 1.00 m	–	1.23	118.61	–	–	m³	**118.61**
maximum depth not exceeding 2.00 m (includes earthwork support)	–	1.23	118.61	–	–	m³	**153.40**
Trenches; width exceeding 0.30 m							
maximum depth not exceeding 0.25 m	–	0.94	91.08	–	–	m³	**91.08**
maximum depth not exceeding 1.00 m	–	1.32	127.51	–	–	m³	**127.51**
maximum depth not exceeding 2.00 m (includes earthwork support)	–	1.98	191.27	–	–	m³	**260.84**
Extra over any types of excavating irrespective of depth for breaking out existing materials; hand held pneumatic breaker							
rock	–	1.65	159.39	48.07	–	m³	**207.46**
concrete	–	0.82	79.69	24.03	–	m³	**103.72**
reinforced concrete	–	1.32	127.51	43.98	–	m³	**171.49**
brickwork, blockwork or stonework	–	0.50	47.82	14.42	–	m³	**62.24**
Earthwork Retention							
Manhole boxes to excavations (not included); Timber plywood with timber braces to shore excavations to 1.00 m deep; opposing faces distance less than 1.00 m							
1.00 m deep	–	0.50	32.20	–	16.19	m²	**48.39**
2.00 m deep	–	1.00	64.40	–	16.19	m²	**80.59**
Trench boxes; A-Plant							
Trench box soil retention; mechanical placement of hydraulic trenchboxes	–	0.50	48.30	191.61	–	week	**239.91**
Manhole protection system to 3.0 m deep to enable earthwork retention; maximum distance between opposing faces 1.8 m	–	0.50	48.30	259.46	–	week	**307.76**

5 EXCAVATING AND FILLING

Item – Overhead and Profit Included	PC £	Labour hours	Labour £	Plant £	Material £	Unit	Total rate £
FILLING							
Filling to make up levels; mechanical							
Arising from the excavations							
average thickness not exceeding 0.25 m	–	0.07	2.37	3.99	–	m³	**6.36**
average thickness less than 500 mm	–	0.06	2.01	3.39	–	m³	**5.40**
average thickness 1.00 m	–	0.05	1.50	2.52	–	m³	**4.02**
Obtained from on site spoil heaps; average 25 m distance; multiple handling							
average thickness less than 250 mm	–	–	–	10.80	–	m³	**10.80**
average thickness less than 500 mm	–	–	–	8.90	–	m³	**8.90**
average thickness 1.00 m	–	–	–	7.57	–	m³	**7.57**
Obtained off site; planting quality topsoil PC £24.72/m³							
average thickness less than 250 mm	–	–	–	10.80	34.11	m³	**44.91**
average thickness less than 500 mm	–	–	–	8.90	34.11	m³	**43.01**
average thickness 1.00 m	–	–	–	7.57	34.11	m³	**41.68**
Obtained off site; hardcore; PC £22.50/m³							
average thickness less than 250 mm	–	–	–	12.60	31.05	m³	**43.65**
average thickness less than 500 mm	–	–	–	9.46	31.05	m³	**40.51**
average thickness 1.00 m	–	–	–	7.96	31.05	m³	**39.01**
Filling to make up levels; hand							
Arising from the excavations							
average thickness exceeding 0.25 m; depositing in layers 150 mm maximum thickness	–	0.60	19.32	–	–	m³	**19.32**
Obtained from on site spoil heaps; average 25 m distance; multiple handling							
average thickness exceeding 0.25 m thick; depositing in layers 150 mm maximum thickness	–	1.00	32.20	–	–	m³	**32.20**

5 EXCAVATING AND FILLING

Item – Overhead and Profit Included	PC £	Labour hours	Labour £	Plant £	Material £	Unit	Total rate £
DISPOSAL							
Note							
Most commercial site disposal is carried out by 20 tonne, 8 wheeled vehicles. It has been customary to calculate disposal from construction sites in terms of full 15 m³ loads. Spon's research has found that based on weights of common materials such as clean hardcore and topsoil, vehicles could not load more than 12 m³ at a time. Most hauliers do not make it apparent that their loads are calculated by weight and not by volume. The rates below reflect these lesser volumes which are limited by the 20 tonne limit. The volumes shown below are based on volumes 'in the solid'. Weight and bulking factors have been applied. For further information please see the weights of typical materials in the Earthworks section of the Memoranda at the back of this book.							
Disposal; mechanical (all rates include banksman)							
Excavated material; off site; to tip; mechanically loaded (360 Excavator)							
hardcore	–	0.02	0.67	1.31	21.78	m³	**23.76**
inert (clean landfill, rubble concrete); loose	–	0.03	0.90	–	39.94	m³	**40.84**
inert (clean excavated material); compacted	–	0.03	0.90	–	49.92	m³	**50.82**
slightly contaminated – non hazardous	–	0.03	0.90	1.31	68.75	m³	**70.96**
clean concrete	–	0.03	0.90	1.31	17.82	m³	**20.03**
macadam	–	0.02	0.67	1.31	22.77	m³	**24.75**
broken out compacted materials such as road bases and the like	–	0.03	0.90	–	59.91	m³	**60.81**
Light soils and loams (bulking factor of 1.25); 40 tonnes (2 loads per hour)							
soil (sandy and loam); dry	–	0.03	0.90	–	39.94	m³	**40.84**
soil (sandy and loam); wet	–	0.03	0.90	–	48.50	m³	**49.40**
soil (clay); dry	–	0.03	0.90	–	55.91	m³	**56.81**
soil (clay); wet	–	0.03	0.90	–	59.91	m³	**60.81**

5 EXCAVATING AND FILLING

Item – Overhead and Profit Included	PC £	Labour hours	Labour £	Plant £	Material £	Unit	Total rate £
DISPOSAL – cont							
Disposal – cont							
Other materials							
rubbish (mixed loads)	–	0.03	1.07	2.09	53.16	m³	**56.32**
green waste	–	0.03	1.07	2.09	9.90	m³	**13.06**
Costs if allowing for 3 loads (36 m³, 60 tonne) per hour loaded and removed							
inert material	–	0.03	0.90	1.75	79.08	m³	**81.73**
Excavated material; off site to tip; mechanically loaded by grab; capacity of load 13 m³ (18 tonne)							
inert material	–	0.50	16.10	–	43.56	m³	**59.66**
hardcore	–	0.50	16.10	–	58.74	m³	**74.84**
macadam	–	0.50	16.10	–	37.40	m³	**53.50**
clean concrete	–	0.50	16.10	–	20.90	m³	**37.00**
soil (sandy and loam); dry	–	0.50	16.10	–	35.20	m³	**51.30**
soil (sandy and loam); wet	–	0.50	16.10	–	46.20	m³	**62.30**
broken out compacted materials such as road bases and the like	–	0.50	16.10	–	48.40	m³	**64.50**
soil (clay); dry	–	0.50	16.10	–	46.20	m³	**62.30**
soil (clay); wet	–	0.50	16.10	–	58.74	m³	**74.84**
rubbish (mixed loads)	–	0.50	16.10	–	50.35	m³	**66.45**
green waste unchipped; compacted	–	–	–	–	–	m³	**42.90**
Disposal by skip; 8 yd³ (6.1 m³)							
Excavated material loaded to skip							
by machine	–	–	–	62.27	–	m³	**62.27**
by hand	–	3.00	96.60	58.91	–	m³	**155.51**
Excavated material; on site							
In spoil heaps							
average 25 m distance	–	–	–	3.14	–	m³	**3.14**
average 50 m distance	–	–	–	3.62	–	m³	**3.62**
average 100 m distance (1 dumper)	–	–	–	4.54	–	m³	**4.54**
average 100 m distance (2 dumpers)	–	–	–	5.53	–	m³	**5.53**
average 200 m distance	–	–	–	5.47	–	m³	**5.47**
average 200 m distance (2 dumpers)	–	–	–	6.66	–	m³	**6.66**
Spreading on site							
average 25 m distance	–	0.02	0.72	7.23	–	m³	**7.95**
average 50 m distance	–	0.04	1.25	8.07	–	m³	**9.32**
average 100 m distance	–	0.06	2.00	9.82	–	m³	**11.82**
average 200 m distance	–	0.12	3.86	9.02	–	m³	**12.88**

5 EXCAVATING AND FILLING

Item – Overhead and Profit Included	PC £	Labour hours	Labour £	Plant £	Material £	Unit	Total rate £
Disposal; hand							
Excavated material; on site; in spoil heaps							
average 25 m distance	–	2.40	77.28	–	–	m³	**77.28**
average 50 m distance	–	2.64	85.01	–	–	m³	**85.01**
average 100 m distance	–	3.00	96.60	–	–	m³	**96.60**
average 200 m distance	–	3.60	115.92	–	–	m³	**115.92**
Excavated material; spreading on site							
average 25 m distance	–	2.64	85.01	–	–	m³	**85.01**
average 50 m distance	–	3.00	96.60	–	–	m³	**96.60**
average 100 m distance	–	3.60	115.92	–	–	m³	**115.92**
average 200 m distance	–	4.20	135.24	–	–	m³	**135.24**
Disposal of material from site clearance operations							
Shrubs and groundcovers less than 1.00 m height; disposal by 15 m³ self loaded truck							
deciduous shrubs; not chipped; winter	–	0.05	1.61	1.05	8.25	m²	**10.91**
deciduous shrubs; chipped; winter	–	0.05	1.61	3.30	2.07	m²	**6.98**
evergreen or deciduous shrubs; chipped; summer	–	0.08	2.42	2.25	8.25	m²	**12.92**
Shrubs 1.00–2.00 m height; disposal by 15 m³ self loaded truck							
deciduous shrubs; not chipped; winter	–	0.17	5.37	1.05	11.55	m²	**17.97**
deciduous shrubs; chipped; winter	–	0.25	8.05	2.25	8.25	m²	**18.55**
evergreen or deciduous shrubs; chipped; summer	–	0.30	9.66	2.64	16.50	m²	**28.80**
Shrubs or hedges 2.00–3.00 m height; disposal by 15 m³ self loaded truck							
deciduous plants non-woody growth; not chipped; winter	–	0.25	8.05	1.05	26.40	m²	**35.50**
deciduous shrubs; chipped; winter	–	0.50	16.10	2.25	20.63	m²	**38.98**
evergreen or deciduous shrubs non-woody growth; chipped; summer	–	0.25	8.05	2.25	8.25	m²	**18.55**
evergreen or deciduous shrubs woody growth; chipped; summer	–	0.67	21.47	3.45	24.75	m²	**49.67**
Shrubs and groundcovers less than 1.00 m height; disposal to spoil heap							
deciduous shrubs; not chipped; winter	–	0.05	1.61	–	–	m²	**1.61**
deciduous shrubs; chipped; winter	–	0.05	1.61	1.20	–	m²	**2.81**
evergreen or deciduous shrubs; chipped; summer	–	0.08	2.42	1.20	–	m²	**3.62**

5 EXCAVATING AND FILLING

Item – Overhead and Profit Included	PC £	Labour hours	Labour £	Plant £	Material £	Unit	Total rate £
DISPOSAL – cont							
Disposal of material from site clearance operations – cont							
Shrubs 1.00–2.00 m height; disposal to spoil heap							
deciduous shrubs; not chipped; winter	–	0.17	5.37	–	–	m²	**5.37**
deciduous shrubs; chipped; winter	–	0.25	8.05	1.20	–	m²	**9.25**
evergreen or deciduous shrubs; chipped; summer	–	0.30	9.66	1.60	–	m²	**11.26**
Shrubs or hedges 2.00–3.00 m height; disposal to spoil heaps							
deciduous plants non-woody growth; not chipped; winter	–	0.25	8.05	–	–	m²	**8.05**
deciduous shrubs; chipped; winter	–	0.50	16.10	1.20	–	m²	**17.30**
evergreen or deciduous shrubs non-woody growth; chipped; summer	–	0.67	21.47	2.39	–	m²	**23.86**
evergreen or deciduous shrubs woody growth; chipped; summer	–	1.50	48.30	4.80	–	m²	**53.10**

6 GROUND REMEDIATION AND SOIL STABILIZATION

Item – Overhead and Profit Included	PC £	Labour hours	Labour £	Plant £	Material £	Unit	Total rate £
CLARIFICATION NOTES ON LABOUR COSTS IN THIS SECTION							
General groundworks team Generally a three man team is used in this section; The column Labour hours reports team hours. The column Labour £ reports the total cost of the team for the unit of work shown							
3 man team	–	1.00	96.60	–	–	hr	**96.60**
RETAINING WALLS AND REVETMENTS							
Preamble Earth retaining and stabilization can take numerous forms depending on the height, soil type and loadings. The main types include gravity walls, reinforced soil and soil nailing. Prices for these items depend on quantity, access for the installation and the strength of the soils. Estimates should be obtained from the manufacturer when site conditions and layout have been determined.							
Corten prefabricated retaining walls; Adezz limited 200 mm high on prepared concrete base (not included); top edge of 50 × 25 mm; bolted together internally with 30 mm M8 bolts							
straight sections 1.00 m long	26.10	0.20	12.88	–	30.02	m	**42.90**
straight sections 1.50 m long	25.75	0.17	10.74	–	29.61	m	**40.35**
straight sections 2.00 m long	25.32	0.13	8.05	–	29.12	m	**37.17**
inside corner or outside corner 500 mm × 500 mm	25.06	0.25	16.10	–	28.82	m	**44.92**
inside or outside corner 1.00 × 1.00 m	24.80	0.30	19.32	–	28.52	m	**47.84**
inside or outside curve 500 mm	72.04	0.20	12.88	–	82.85	m	**95.73**
inside or outside curve; 1.00 m lengths	72.04	0.25	16.10	–	82.85	m	**98.95**
inside or outside curve; 1.50 m lengths	72.03	0.17	10.74	–	82.83	m	**93.57**
inside or outside curve; 2.00 m lengths	72.03	0.14	9.20	–	82.83	m	**92.03**

6 GROUND REMEDIATION AND SOIL STABILIZATION

Item – Overhead and Profit Included	PC £	Labour hours	Labour £	Plant £	Material £	Unit	Total rate £
RETAINING WALLS AND REVETMENTS – cont							
Corten prefabricated retaining walls – cont							
400 mm high; on prepared concrete base (not included); top edge of 50 × 25 mm; bolted together internally with 30 mm M8 bolts							
straight sections 1.00 m long	46.98	0.22	14.17	–	54.03	m	**68.20**
straight sections 1.50 m long	46.29	0.18	11.81	–	53.23	m	**65.04**
straight sections 2.00 m long	45.67	0.14	8.86	–	52.52	m	**61.38**
inside corner or outside corner 500 mm × 500 mm	44.89	0.28	17.71	–	51.62	m	**69.33**
inside or outside corner 1.00 × 1.00 m	44.63	0.33	21.25	–	51.32	m	**72.57**
inside or outside curve 500 mm	120.06	0.22	14.17	–	138.07	m	**152.24**
inside or outside curve; 1.00 m lengths	120.06	0.28	17.71	–	138.07	m	**155.78**
inside or outside curve; 1.50 m lengths	131.71	0.18	11.81	–	151.47	m	**163.28**
inside or outside curve; 2.00 m lengths	119.80	0.16	10.12	–	137.77	m	**147.89**
600 mm high; on prepared concrete base (not included); top edge of 50 × 25 mm; bolted together internally with 30 mm M8 bolts							
straight sections 1.00 m long	90.31	0.24	15.58	–	103.86	m	**119.44**
straight sections 1.50 m long	88.74	0.20	13.00	–	102.05	m	**115.05**
straight sections 2.00 m long	88.22	0.15	9.74	–	101.45	m	**111.19**
inside corner or outside corner 500 mm × 500 mm	86.13	0.30	19.48	–	99.05	m	**118.53**
inside or outside corner 1.00 × 1.00 m	85.87	0.36	23.38	–	98.75	m	**122.13**
inside or outside curve 500 mm	210.88	0.24	15.58	–	242.51	m	**258.09**
inside or outside curve; 1.00 m lengths	210.89	0.30	19.48	–	242.52	m	**262.00**
inside or outside curve; 1.50 m lengths	210.89	0.20	13.00	–	242.52	m	**255.52**
inside or outside curve; 2.00 m lengths	210.90	0.17	11.13	–	242.54	m	**253.67**

6 GROUND REMEDIATION AND SOIL STABILIZATION

Item – Overhead and Profit Included	PC £	Labour hours	Labour £	Plant £	Material £	Unit	Total rate £
Retaining walls; FP McCann Concrete Retaining walls of units with plain concrete finish; prices based on 24 tonne loads but other quantities available (excavation, temporary shoring, foundations and backfilling not included)							
1000 mm wide × 1000 mm high	145.95	0.50	32.20	29.88	220.17	m	**282.25**
1000 mm wide × 1500 mm high	185.85	0.42	27.24	29.88	278.13	m	**335.25**
1000 mm wide × 2000 mm high	309.75	0.42	27.28	29.88	446.34	m	**503.50**
1000 mm wide × 2500 mm high	399.00	0.42	27.27	29.88	582.96	m	**640.11**
1000 mm wide × 3000 mm high	432.60	0.42	27.24	62.97	637.84	m	**728.05**
Retaining walls; FP McCann Ltd; Easi-Bloc; (Lego Block); precast self-locking concrete gravity wall system; storage bays, partitions retaining walls, revetments; to firm and levelled base; concrete or granular; min. 800 mm wide (not included) Blocks 450 mm high laid stretcher bond							
1.20 m × 600 mm (750 kg per block)	172.86	0.46	44.72	25.15	198.79	m²	**268.66**
600 mm × 600 mm (375 kg per block)	207.46	0.50	48.30	27.16	238.58	m²	**314.04**
Retaining walls; Tensar International Tensartech TW1 retaining wall system; modular dry laid concrete blocks; 220 × 400 mm long × 150 mm high connected to Tensar RE geogrid with proprietary connectors; geogrid laid horizontally within the fill at 300 mm centres; on 150 × 450 mm concrete foundation; filling with imported granular material							
1.00 m high	126.69	1.00	64.40	14.94	154.96	m²	**234.30**
2.00 m high	126.69	1.33	85.86	14.94	154.96	m²	**255.76**
3.00 m high	126.69	1.50	96.60	14.94	154.96	m²	**266.50**

6 GROUND REMEDIATION AND SOIL STABILIZATION

Item – Overhead and Profit Included	PC £	Labour hours	Labour £	Plant £	Material £	Unit	Total rate £
RETAINING WALLS AND REVETMENTS – cont							
Retaining walls; Grass Concrete Ltd Betoflor precast concrete landscape retaining walls including soil filling to pockets (excavation, concrete foundations, backfilling stones to rear of walls and planting not included)							
Betoflor interlocking units; 250 mm long × 250 mm wide × 200 mm modular deep; in walls 250 mm wide	–	–	–	–	–	m²	**135.99**
extra over Betoflor interlocking units for colours	–	–	–	–	–	m²	**18.11**
Betoatlas earth retaining walls; 250 mm long × 500 mm wide × 200 mm modular deep; in walls 500 mm wide	–	–	–	–	–	m²	**210.39**
extra over Betoatlas interlocking units for colours	–	–	–	–	–	m²	**24.67**
Retaining walls; Forticrete Ltd Keystone precast concrete block retaining wall; geogrid included for walls over 1.00 m high; excavation, concrete foundation, stone backfill to rear of wall all measured separately; includes design							
1.0 m high	122.45	0.80	77.28	–	140.82	m²	**218.10**
2.0 m high	122.45	0.80	77.28	2.88	161.37	m²	**241.53**
3.0 m high	122.45	0.80	77.28	3.83	175.65	m²	**256.76**
4.0 m high	122.45	0.80	77.28	4.60	186.71	m²	**268.59**
Retaining walls; Forticrete Ltd Stepoc Blocks; interlocking blocks; 10 mm reinforcing laid loose horizontally to preformed notches and vertical reinforcing nominal size 10 mm fixed to starter bars; infilling with concrete; foundations and starter bars measured separately							
type 325; 400 × 225 × 325 mm	62.86	0.47	45.08	–	102.78	m²	**147.86**
type 256; 400 × 225 × 256 mm	59.20	0.40	38.64	–	94.07	m²	**132.71**
type 200; 400 × 225 × 200 mm	49.47	0.33	32.20	–	78.38	m²	**110.58**

6 GROUND REMEDIATION AND SOIL STABILIZATION

Item – Overhead and Profit Included	PC £	Labour hours	Labour £	Plant £	Material £	Unit	Total rate £
Timber log retaining walls; timber posts are kiln dried redwood with 15 year guarantee; Machine rounded softwood logs to trenches priced separately; disposal of excavated material priced separately; inclusive of 75 mm hardcore blinding to trench and backfilling trench with site mixed concrete 1:3:6; geofabric pinned to rear of logs; heights of logs above ground							
100 mm dia.							
500 mm (constructed from 1.80 m lengths)	48.55	0.50	48.30	–	69.01	m	**117.31**
1.20 m (constructed from 1.80 m lengths)	97.10	0.43	41.86	–	142.05	m	**183.91**
1.60 m (constructed from 2.70 m lengths)	129.60	0.83	80.47	–	187.06	m	**267.53**
2.00 m (constructed from 3.00 m lengths)	197.50	1.17	112.69	–	272.79	m	**385.48**
150 mm dia.							
500 mm	62.38	0.83	80.47	–	84.92	m	**165.39**
1.20 m	207.73	0.58	56.35	–	269.26	m	**325.61**
1.60 m	207.93	1.00	96.60	–	277.15	m	**373.75**
RECYCLED MATERIALS AND SYSTEMS FOR EXTERNAL WORKS							
Planter retaining wall; Recycled plastic; Hahn Plastics Ltd; Hollow or Solid; inclusive of trenches, placement and surrounding with concrete; terram to rear of palisades; backfiling to height of palisade not included; brown; (grey also available)							
Ogee C Section interlocking profile palisades 100 mm dia.; hollow profile; retained height							
300 mm	99.58	0.25	24.15	–	138.21	m	**162.36**
600 mm	109.20	0.50	48.30	–	159.05	m	**207.35**
Ogee C Section interlocking profile palisades 100 mm dia.; hollow profile; retained height							
1.20 m	197.99	0.43	41.86	–	273.59	m	**315.45**
Round palisades; 100 mm dia.; hollow profile; retained height							
300 mm	40.10	0.25	24.15	–	66.58	m	**90.73**
800 mm	99.50	0.43	41.86	–	152.21	m	**194.07**

6 GROUND REMEDIATION AND SOIL STABILIZATION

Item – Overhead and Profit Included	PC £	Labour hours	Labour £	Plant £	Material £	Unit	Total rate £
RETAINING WALLS AND REVETMENTS – cont							
Planter retaining wall – cont							
Round palisades; 100 mm dia.; solid profile; retained height							
1.00 m	99.50	0.43	41.86	–	152.21	m	**194.07**
1.20 m	155.00	0.50	48.30	–	227.41	m	**275.71**
1.50 m	193.80	1.00	96.60	–	353.54	m	**450.14**
Rectangular palisades; 120 × 160 mm; hollow profile; retained height; Grey, Brown or Black							
300 mm	65.90	0.25	24.15	–	96.25	m	**120.40**
800 mm	196.73	0.43	41.86	–	264.03	m	**305.89**
1.00 m	253.00	0.50	48.30	–	340.11	m	**388.41**
Railway sleeper walls; retaining wall from railway sleepers; fixed with steel galvanized pins 12 mm driven into the ground; sleepers laid flat							
Grade 1 hardwood; 2590 × 250 × 150 mm							
150 mm; 1 sleeper high	4.22	–	–	–	–	m	**–**
300 mm; 2 sleepers high	8.43	0.33	32.20	–	11.10	m	**43.30**
450 mm; 3 sleepers high	12.54	0.50	48.30	–	16.27	m	**64.57**
600 mm; 4 sleepers high	16.90	0.67	64.40	–	21.29	m	**85.69**
Grade 1 hardwood; 2590 × 250 × 150 mm with 2 nr galvanized angle iron stakes set into concrete internally and screwed to the inside face of the sleepers							
750 mm; 5 sleepers high	20.91	0.83	80.50	–	62.25	m²	**142.75**
900 mm; 6 sleepers high	25.02	0.92	88.55	–	71.67	m²	**160.22**
New pine softwood; 2.40 m × 250 mm × 125 mm							
125 mm; 1 sleeper high	8.05	0.17	16.10	–	9.96	m	**26.06**
250 mm; 2 sleepers high	16.10	0.33	32.20	–	19.92	m	**52.12**
375 mm; 3 sleepers high	24.14	0.50	48.30	–	29.61	m	**77.91**
500 mm; 4 sleepers high	32.19	0.67	64.40	–	38.87	m	**103.27**
New pine softwood 2.40 m × 250 mm × 125 mm with 2 nr galvanized angle iron stakes set into concrete internally and screwed to the inside face of the sleepers							
625 mm; 5 sleepers high	40.24	0.83	80.50	–	84.48	m²	**164.98**
775 mm; 6 sleepers high	48.29	0.92	88.55	–	98.43	m²	**186.98**

6 GROUND REMEDIATION AND SOIL STABILIZATION

Item – Overhead and Profit Included	PC £	Labour hours	Labour £	Plant £	Material £	Unit	Total rate £
New oak hardwood							
2400 × 200 × 100 mm							
100 mm; 1 sleeper high	10.66	0.17	16.09	–	12.96	m	**29.05**
200 mm; 2 sleepers high	21.32	0.33	32.20	–	25.92	m	**58.12**
300 mm; 3 sleepers high	31.20	0.75	72.45	–	37.73	m	**110.18**
400 mm; 4 sleepers high	43.16	1.00	96.60	–	51.49	m	**148.09**
New oak hardwood							
2400 × 200 × 100 mm but with 2 nr galvanized angle iron stakes set into concrete internally and screwed to the inside face of the sleepers							
500 mm; 5 sleepers high	54.08	0.83	80.50	–	102.74	m²	**183.24**
600 mm; 6 sleepers high	65.00	1.10	106.26	–	117.64	m²	**223.90**
Excavate foundation trench; set railway sleepers vertically on end in concrete 1:3:6 continuous foundation to 33.3% of their length to form retaining wall							
Grade 1 hardwood; finished height above ground level							
300 mm	33.85	1.00	96.60	2.24	46.48	m	**145.32**
500 mm	60.94	1.00	96.60	2.24	77.63	m	**176.47**
600 mm	73.12	1.00	96.60	2.24	97.40	m	**196.24**
750 mm	81.25	1.17	112.70	2.24	106.75	m	**221.69**
1.00 m	81.25	1.25	120.75	2.24	107.90	m	**230.89**
Excavate and place vertical steel universal beams 165 mm wide in concrete base at 2.59 m centres; fix railway sleepers set horizontally between beams to form horizontal fence or retaining wall							
Grade 1 hardwood; bay length 2.59 m							
400 mm high; 2 sleepers	21.42	0.83	80.50	–	53.74	bay	**134.24**
600 mm high; 3 sleepers	31.48	0.87	83.72	–	79.40	bay	**163.12**
800 mm high; 4 sleepers	42.19	1.00	96.60	–	101.49	bay	**198.09**
1.00 m high; 5 sleepers	53.87	1.17	112.70	–	132.39	bay	**245.09**
1.20 m high; 6 sleepers	63.28	1.00	96.60	–	156.83	bay	**253.43**
1.40 m high; 7 sleepers	73.66	1.00	96.60	–	181.23	bay	**277.83**

6 GROUND REMEDIATION AND SOIL STABILIZATION

Item – Overhead and Profit Included	PC £	Labour hours	Labour £	Plant £	Material £	Unit	Total rate £
RETAINING WALLS AND REVETMENTS – cont							
Waterside revetments; Willowbank Services Ltd; natural engineering solutions; ecological sustainable waterside stabilization systems Live willow spiling; 1 m high uprights at 0.50 m centres; woven geotextile backed; included filling with site based topsoil behind spiling to top of bank							
less than 50 m	–	–	–	–	–	m	**186.91**
50–100 m	–	–	–	–	–	m	**146.01**
Brushwood faggot and pre-planted coir revetments; 900 mm high revetment; 300 mm dia. faggot; 300 mm dia. pre-planted coir roll; staked in position with 1.8 m tanalized posts; 2 posts per m; faggots and coir fixed to posts with wire							
less than 50 m	–	–	–	–	–	m	**167.49**
50–100 m	–	–	–	–	–	m	**144.19**
Brushwood faggot revetments; 900 mm high revetment; 2 nr 300 mm dia. faggot staked in position with 1.8 m tanalized posts; 2 posts per m; faggots and coir fixed to posts with wire							
less than 50 m	–	–	–	–	–	m	**143.77**
50–100 m	–	–	–	–	–	m	**99.60**

6 GROUND REMEDIATION AND SOIL STABILIZATION

Item – Overhead and Profit Included	PC £	Labour hours	Labour £	Plant £	Material £	Unit	Total rate £
GENERALLY							
Soil stabilization – General							
Preamble: Earth-retaining and							
stabilizing materials are often							
specified as part of the earth-forming							
work in landscape contracts,							
therefore this section lists a number							
of products specially designed for							
large-scale earth control. There are							
two types: rigid units for structural							
retention of earth on steep slopes;							
and flexible meshes and sheets for							
control of soil erosion where							
structural strength is not required.							
Prices for these items depend on							
quantity, difficulty of access to the							
site and availability of suitable filling							
material; estimates should be							
obtained from the manufacturer							
when the site conditions have been							
determined.							
GRADING AND SURFACE PREPARATION							
Grading operations; surface previously excavated to reduce levels to prepare to receive subsequent treatments; grading to accurate levels and falls 20 mm tolerances							
Clay or heavy soils or hardcore							
5 tonne excavator	–	0.04	1.29	0.22	–	m²	**1.51**
13 tonne excavator	–	0.01	0.32	0.63	–	m²	**0.95**
21 tonne 360 tracked excavator	–	0.01	0.22	0.49	–	m²	**0.71**
by hand	–	0.10	3.22	–	–	m²	**3.22**
Loamy topsoils							
5 tonne excavator	–	0.03	0.85	0.14	–	m²	**0.99**
13 tonne excavator	–	0.01	0.22	0.43	–	m²	**0.65**
21 tonne 360 tracked excavator	–	0.01	0.24	0.55	–	m²	**0.79**
by hand	–	0.05	1.61	–	–	m²	**1.61**
Sand or graded granular materials							
5 tonne excavator	–	0.02	0.64	0.10	–	m²	**0.74**
13 tonne excavator	–	0.01	0.22	0.43	–	m²	**0.65**
21 tonne 360 tracked excavator	–	0.01	0.18	0.43	–	m²	**0.61**
by hand	–	0.03	1.07	–	–	m²	**1.07**

6 GROUND REMEDIATION AND SOIL STABILIZATION

Item – Overhead and Profit Included	PC £	Labour hours	Labour £	Plant £	Material £	Unit	Total rate £
GRADING AND SURFACE PREPARATION – cont							
Grading operations to bottoms of excavations or trenches to receive foundations or bases; surface previously excavated to reduce levels to prepare to receive subsequent treatments; grading to accurate levels and falls 20 mm tolerances							
Trenches							
5 tonne excavator	–	0.32	18.25	0.72	–	m²	**18.97**
by hand	–	0.33	21.46	–	–	m²	**21.46**
Excavated areas to receive bases or formwork							
5 tonne excavator	–	0.13	7.51	0.36	–	m²	**7.87**
by hand	–	0.17	10.74	–	–	m²	**10.74**
Grading operations; surface recently filled to raise levels to prepare to receive subsequent treatments; grading to accurate levels and falls 20 mm tolerances							
Clay or heavy soils or hardcore							
5 tonne excavator	–	0.03	0.97	0.16	–	m²	**1.13**
21 tonne 360 tracked excavator	–	0.01	0.22	0.49	–	m²	**0.71**
by hand	–	0.08	2.68	–	–	m²	**2.68**
Loamy topsoils							
5 tonne excavator	–	0.02	0.67	0.12	–	m²	**0.79**
21 tonne 360 tracked excavator	–	0.01	0.22	0.49	–	m²	**0.71**
by hand	–	0.04	1.29	–	–	m²	**1.29**
Sand or graded granular materials							
5 tonne excavator	–	0.02	0.48	0.08	–	m²	**0.56**
13 tonne excavator	–	0.01	0.24	0.47	–	m²	**0.71**
21 tonne 360 tracked excavator	–	0.01	0.16	0.37	–	m²	**0.53**
by hand	–	0.03	0.92	–	–	m²	**0.92**
Surface grading; Agripower Ltd; using laser controlled equipment							
Maximum variation of existing ground level							
75 mm; laser box grader and tractor	–	–	–	–	–	ha	**2194.52**
100 mm; laser controlled low ground pressure bulldozer	–	–	–	–	–	ha	**3599.01**

6 GROUND REMEDIATION AND SOIL STABILIZATION

Item – Overhead and Profit Included	PC £	Labour hours	Labour £	Plant £	Material £	Unit	Total rate £
Fraise mowing; scarification of existing turf layer to rootzone level; Agripower Ltd							
Cart arisings to tip on site							
average thickness 50–75 mm	–	–	–	–	–	ha	3937.60
Cultivating							
Ripping up subsoil; using approved subsoiling machine; minimum depth 250 mm below topsoil; at 1.20 m centres; in							
gravel or sandy clay	–	–	–	5.20	–	100 m²	5.20
soil compacted by machines	–	–	–	5.78	–	100 m²	5.78
clay	–	–	–	8.67	–	100 m²	8.67
chalk or other soft rock	–	–	–	8.67	–	100 m²	8.67
Extra for subsoiling at 1 m centres	–	–	–	3.38	–	100 m²	3.38
Breaking up existing ground; using pedestrian operated tine cultivator or rotavator; loam or sandy soil							
100 mm deep	–	0.22	7.08	1.41	–	100 m²	8.49
150 mm deep	–	0.28	8.86	1.76	–	100 m²	10.62
200 mm deep	–	0.37	11.80	2.35	–	100 m²	14.15
Breaking up existing ground; using pedestrian operated tine cultivator or rotavator; heavy clay or wet soils							
100 mm deep	–	0.44	14.17	2.82	–	100 m²	16.99
150 mm deep	–	0.66	21.25	4.23	–	100 m²	25.48
200 mm deep	–	0.82	26.57	5.29	–	100 m²	31.86
Breaking up existing ground; using tractor drawn tine cultivator or rotavator							
100 mm deep	–	–	–	1.30	–	100 m²	1.30
150 mm deep	–	–	–	1.63	–	100 m²	1.63
200 mm deep	–	–	–	2.16	–	100 m²	2.16
400 mm deep	–	–	–	6.51	–	100 m²	6.51
Final levelling to ploughed and rotavated ploughed ground; using disc, drag, or chain harrow							
4 passes	–	–	–	1.18	–	100 m²	1.18
Rolling cultivated ground lightly; using self-propelled agricultural roller	–	0.06	1.79	0.70	–	100 m²	2.49
Importing and storing selected and approved topsoil; 20 tonne load = 11.88 m³ average							
20 tonne loads	29.66	–	–	–	34.11	m³	34.11
Surface treatments							
Compacting							
bottoms of excavations	–	0.01	0.16	0.04	–	m²	0.20

6 GROUND REMEDIATION AND SOIL STABILIZATION

Item – Overhead and Profit Included	PC £	Labour hours	Labour £	Plant £	Material £	Unit	Total rate £
GRADING AND SURFACE PREPARATION – cont							
Surface treatments – cont							
Surface preparation							
Trimming surfaces of cultivated ground to final levels; removing roots, stones and debris exceeding 50 mm in any direction to tip off site; slopes less than 15°							
clean ground with minimal stone content	–	0.25	8.05	–	–	100 m²	**8.05**
slightly stony; 0.5 kg stones per m²	–	0.33	10.72	–	0.02	100 m²	**10.74**
very stony; 1.0–3.0 kg stones per m²	–	0.50	16.10	–	0.05	100 m²	**16.15**
clearing mixed, slightly contaminated rubble; inclusive of roots and vegetation	–	0.50	16.10	–	0.10	100 m²	**16.20**
clearing brick-bats, stones and clean rubble	–	0.60	19.32	–	0.10	100 m²	**19.42**

10 CRIB WALLS, GABIONS AND REINFORCED EARTH

Item – Overhead and Profit Included	PC £	Labour hours	Labour £	Plant £	Material £	Unit	Total rate £
CRIB WALLS, GABIONS AND REINFORCED EARTH							
Crib walls; Phi Group							
Permacrib (timber) or Andacrib (concrete) crib walling system; machine filled infill with crushed rock; inclusive of concrete footing and rear wall drain; excluding excavation and backfill material							
retaining walls up to 2.0 m	–	–	–	–	–	m²	**254.10**
retaining walls up to 4.0 m	–	–	–	–	–	m²	**284.35**
retaining walls up to 6.0 m	–	–	–	–	–	m²	**308.55**
Retaining walls							
Wire mesh gabions; galvanized mesh 80 × 100 mm; filling with broken stones 125–200 mm size; wire down securely to manufacturer's instructions; filling front face by hand							
2 × 1 × 0.50 m	29.87	0.67	64.40	19.92	158.24	nr	**242.56**
2 × 1 × 1.0 m	42.66	1.33	128.80	39.85	296.84	nr	**465.49**
PVC coated gabions							
2 × 1 × 0.5 m	40.08	0.67	64.40	19.92	169.98	nr	**254.30**
2 × 1 × 1.0 m	55.33	1.33	128.80	39.85	311.41	nr	**480.06**
Mattress gabions							
6 × 2 × 0.17 m	42.73	1.00	96.60	29.88	301.88	nr	**428.36**
6 × 2 × 0.30 m	47.59	2.00	193.20	44.83	500.73	nr	**738.76**
Embankments; Tensar International							
Embankments; reinforced with Tensar Geogrid RE520; geogrid laid horizontally within fill to 100% of vertical height of slope at 1.00 m centres; filling with excavated material							
slopes less than 45°; Tensartech Natural Green System	3.19	0.61	59.18	5.09	3.67	m³	**67.94**
slopes exceeding 45°; Tensartech Greenslope System	6.86	0.06	5.97	8.90	7.89	m³	**22.76**
extra over to slopes exceeding 45° for temporary shuttering; shuttering measured per m² of vertical elevation	–	–	–	–	15.40	m²	**15.40**

10 CRIB WALLS, GABIONS AND REINFORCED EARTH

Item – Overhead and Profit Included	PC £	Labour hours	Labour £	Plant £	Material £	Unit	Total rate £
CRIB WALLS, GABIONS AND REINFORCED EARTH – cont							
Embankments – cont							
Embankments; reinforced with Tensar Geogrid RE540; geogrid laid horizontally within fill to 100% of vertical height of slope at 1.00 m centres; filling with excavated material							
slopes less than 45°; Tensartech Natural Green System	3.19	0.61	59.18	62.15	3.67	m³	**125.00**
slopes exceeding 45°; Tensartech Greenslope System	–	0.06	5.98	131.55	–	m³	**137.53**
extra over to slopes exceeding 45° for temporary shuttering; shuttering measured per m² of vertical elevation	–	–	–	–	15.40	m²	**15.40**
Extra for Tensar Mat 400; erosion control mats; to faces of slopes of 45° or less; filling with 20 mm fine topsoil; seeding	4.33	0.01	0.63	0.30	6.73	m²	**7.66**
Embankments; reinforced with Tensar Geogrid RE540; geogrid laid horizontally within fill to 50% of horizontal length of slopes at specified centres; filling with imported fill 6F2 PC £18.00/m³							
300 mm centres; slopes up to 45°; Tensartech Natural Green System	10.52	0.08	7.52	100.88	36.94	m³	**145.34**
600 mm centres; slopes up to 45°; Tensartech Natural Green System	5.29	0.06	5.98	56.25	30.92	m³	**93.15**
extra over to slopes exceeding 45° for temporary shuttering; shuttering measured per m² of vertical elevation	–	–	–	–	15.40	m²	**15.40**
Extra for Tensar Mat 400; erosion control mats; to faces of slopes of 45° or less; filling with 20 mm fine topsoil; seeding	4.33	0.01	0.63	0.30	6.73	m²	**7.66**

10 CRIB WALLS, GABIONS AND REINFORCED EARTH

Item – Overhead and Profit Included	PC £	Labour hours	Labour £	Plant £	Material £	Unit	Total rate £
Embankments; reinforced with Tensar Geogrid RE540; geogrid laid horizontally within fill; filling with excavated material; wrapping around at faces							
300 mm centres; slopes exceeding 45°; Tensartech Natural Green System	11.67	0.06	5.97	195.06	13.42	m³	**214.45**
600 mm centres; slopes exceeding 45°; Tensartech Greenslope System	6.35	0.06	5.97	99.80	7.30	m³	**113.07**
extra over to slopes exceeding 45° for temporary shuttering; shuttering measured per m² of vertical elevation	–	–	–	–	15.40	m²	**15.40**
Extra over embankments for bagwork face supports for slopes exceeding 45°	–	0.17	16.09	–	17.77	m²	**33.86**
Extra over embankments for seeding of bags	–	0.03	3.22	–	0.38	m²	**3.60**
Extra over embankments for Bodkin joints	–	–	–	–	1.48	m	**1.48**
Green faced reinforced soil; Phi Group Textomur reinforced soil system embankments; to reinforced soil slopes 55–70° to the horizontal; including facing and geogrid; excluding backfill material							
retaining walls up to 3.0 m	–	–	–	–	–	m²	**175.45**
retaining walls up to 6.0 m	–	–	–	–	–	m²	**181.50**
Split face concrete blocks reinforced soil; Phi Group Modular blocks reinforced soil system embankments; to reinforced soil slopes near vertical; including geogrid; excluding backfill material							
retaining walls up to 3.0 m	–	–	–	–	–	m²	**254.10**
retaining walls up to 6.0 m	–	–	–	–	–	m²	**308.55**
Embankments; Grass Concrete Ltd; In situ grass concrete; Fully installed by subcontractor to cleared and graded embankments Grasscrete; in situ reinforced concrete surfacing; to 20 mm thick sand blinding layer (not included); including soiling and seeding							
GC1; 100 mm thick	–	–	–	–	47.23	m²	**47.23**
GC2; 150 mm thick	–	–	–	–	61.57	m²	**61.57**

10 CRIB WALLS, GABIONS AND REINFORCED EARTH

Item – Overhead and Profit Included	PC £	Labour hours	Labour £	Plant £	Material £	Unit	Total rate £
CRIB WALLS, GABIONS AND REINFORCED EARTH – cont							
Embankments – cont							
Grassblock 103; solid matrix precast concrete blocks; to 20 mm thick sand blinding layer; excluding edge restraint; including soiling and seeding							
406 × 406 × 103 mm; fully interlocking	–	–	–	–	57.59	m^2	**57.59**
Flexible sheet materials; Tensar International							
Tensar Mat 400 erosion mats; 3.0–4.5 m wide; securing with Tensar pegs; lap rolls 100 mm; anchors at top and bottom of slopes; in trenches	4.89	0.01	0.53	–	5.62	m^2	**6.15**
Topsoil filling to Tensar Mat; including brushing and raking	0.77	0.01	0.53	0.10	0.89	m^2	**1.52**
Tensar Bi-axial Geogrid; to graded compacted base; filling with 200 mm granular fill; compacting (turf or paving to surfaces not included); 400 mm laps							
TriAx 150; 39 × 39 mm mesh	1.58	–	0.44	0.60	6.79	m^2	**7.83**
TriAx 160; 39 × 39 mm mesh	1.93	–	0.44	0.60	7.19	m^2	**8.23**
TriAx 170; 39 × 39 mm mesh	1.58	–	0.44	0.60	6.79	m^2	**7.83**
Flexible sheet materials; Terram Ltd							
Terram synthetic fibre filter fabric; to graded base (not included)							
Terram 1000; 0.70 mm thick; mean water flow 50 l/m^2/s	0.49	0.01	0.53	–	0.56	m^2	**1.09**
Terram 2000; 1.00 m thick; mean water flow 33 l/m^2/s	1.37	0.01	0.53	–	1.58	m^2	**2.11**
Terram Minipack	0.86	–	0.10	–	0.99	m^2	**1.09**
Flexible sheet materials; Greenfix Ltd							
Geojute lightweight biodegradeable mats for erosion control to pre-seeded slopes; 10–15 mm thick; fixing with 4 nr crimped pins in accordance with manufacturer's instructions; to graded surface (not included)							
Geojute 500; 500 kg/m^2	137.50	0.67	64.40	–	173.07	100 m^2	**237.47**
Greenfix C–100-P Unseeded mat	181.50	0.67	64.40	–	223.68	100 m^2	**288.08**
Greenfix C–10-J Unseeded mat	217.70	0.67	64.40	–	265.30	100 m^2	**329.70**

10 CRIB WALLS, GABIONS AND REINFORCED EARTH

Item – Overhead and Profit Included	PC £	Labour hours	Labour £	Plant £	Material £	Unit	Total rate £
Coir matting							
Geocoir 400	159.60	0.67	64.40	–	198.49	100 m²	**262.89**
Geocoir 900	278.25	0.67	64.40	–	334.94	100 m²	**399.34**
Bioroll; 300 mm dia.; to river banks and revetments	19.50	0.04	4.03	–	27.19	m	**31.22**
Extra over Greenfix erosion control mats for fertilizer applied at 70 g/m²	7.81	0.07	6.44	–	8.98	100 m²	**15.42**
Extra over Greenfix erosion control mats for laying to slopes exceeding 30°	–	–	–	–	–	25%	**–**
Extra for the following operations							
Spreading 25 mm approved topsoil							
by machine	0.74	–	0.21	0.18	0.85	m²	**1.24**
by hand	0.74	0.01	1.13	–	0.85	m²	**1.98**
Grass seed; PC £4.50/kg; spreading in two operations; by hand							
35 g/m²	15.75	0.06	5.37	–	18.11	100 m²	**23.48**
50 g/m²	22.50	0.06	5.37	–	25.88	100 m²	**31.25**
70 g/m²	31.50	0.06	5.37	–	36.22	100 m²	**41.59**
100 g/m²	45.00	0.07	6.44	–	51.75	100 m²	**58.19**
125 g/m²	56.25	0.67	64.39	–	64.69	100 m²	**129.08**
Extra over seeding by hand for slopes over 30° (allowing for the actual area but measured in plan)							
35 g/m²	2.34	–	0.06	–	2.69	100 m²	**2.75**
50 g/m²	3.38	–	0.06	–	3.89	100 m²	**3.95**
70 g/m²	4.72	–	0.06	–	5.43	100 m²	**5.49**
100 g/m²	6.75	–	0.06	–	7.76	100 m²	**7.82**
125 g/m²	8.41	–	0.06	–	9.67	100 m²	**9.73**
Grass seed; PC £4.50/kg; spreading in two operations; by machine							
35 g/m²	15.75	–	–	1.35	18.11	100 m²	**19.46**
50 g/m²	22.50	–	–	1.35	25.88	100 m²	**27.23**
70 g/m²	31.50	–	–	1.35	36.22	100 m²	**37.57**
100 g/m²	45.00	–	–	1.35	51.75	100 m²	**53.10**
125 kg/ha	562.50	–	–	134.23	646.88	ha	**781.11**
150 kg/ha	675.00	–	–	134.23	776.25	ha	**910.48**
200 kg/ha	900.00	–	–	134.23	1035.00	ha	**1169.23**
250 kg/ha	1125.00	–	–	134.23	1293.75	ha	**1427.98**
300 kg/ha	1350.00	–	–	134.23	1552.50	ha	**1686.73**
350 kg/ha	1575.00	–	–	134.23	1811.25	ha	**1945.48**
400 kg/ha	1800.00	–	–	134.23	2070.00	ha	**2204.23**
500 kg/ha	2250.00	–	–	134.23	2587.50	ha	**2721.73**
700 kg/ha	3150.00	–	–	134.23	3622.50	ha	**3756.73**
1400 kg/ha	6300.00	–	–	134.23	7245.00	ha	**7379.23**

10 CRIB WALLS, GABIONS AND REINFORCED EARTH

Item – Overhead and Profit Included	PC £	Labour hours	Labour £	Plant £	Material £	Unit	Total rate £
CRIB WALLS, GABIONS AND REINFORCED EARTH – cont							
Extra for the following operations – cont							
Extra over seeding by machine for slopes over 30° (allowing for the actual area but measured in plan)							
35 g/m²	2.36	–	–	0.20	2.71	100 m²	**2.91**
50 g/m²	3.38	–	–	0.20	3.89	100 m²	**4.09**
70 g/m²	4.72	–	–	0.20	5.43	100 m²	**5.63**
100 g/m²	6.75	–	–	0.20	7.76	100 m²	**7.96**
125 kg/ha	84.38	–	–	20.14	97.04	ha	**117.18**
150 kg/ha	101.25	–	–	20.14	116.44	ha	**136.58**
200 kg/ha	135.00	–	–	20.14	155.25	ha	**175.39**
250 kg/ha	168.75	–	–	20.14	194.06	ha	**214.20**
300 kg/ha	202.50	–	–	20.14	232.88	ha	**253.02**
350 kg/ha	236.25	–	–	20.14	271.69	ha	**291.83**
400 kg/ha	270.00	–	–	20.14	310.50	ha	**330.64**
500 kg/ha	337.50	–	–	20.14	388.13	ha	**408.27**
700 kg/ha	472.50	–	–	20.14	543.38	ha	**563.52**
1400 kg/ha	945.00	–	–	20.14	1086.75	ha	**1106.89**

11 IN SITU CONCRETE WORKS

Item – Overhead and Profit Included	PC £	Labour hours	Labour £	Plant £	Material £	Unit	Total rate £
CLARIFICATION NOTES ON LABOUR COSTS IN THIS SECTION							
In situ concrete team Generally a three man team is used in this section; The column 'Labour hours' reports team hours. The column 'Labour £' reports the total cost of the team for the unit of work shown							
3 man team	–	1.00	96.60	–	–	hr	**96.60**
2 man team	–	1.00	64.40	–	–	hr	**64.40**
CONCRETE MIX INFORMATION							
General The industry standard refences to concrete are Designed, Standard, Prescribed and Designated mixes. Most Items in this section refer to Standard mixes							
Designed mix User specifies the performance of the concrete. Producer is responsible for selecting the appropriate mix. Strength testing is essential.							
Prescribed mix User specifies the mix constituents and is responsible for ensuring that the concrete meets performance requirements. Accurate mix proportions are essential.							
Standard mix Made with a restricted range of materials. Specification to include the proposed use of the material as well as the standard mix reference, type of cement, type and size of aggregate, and slump (workability). Quality assurance is required.							

11 IN SITU CONCRETE WORKS

Item – Overhead and Profit Included	PC £	Labour hours	Labour £	Plant £	Material £	Unit	Total rate £
CONCRETE MIX INFORMATION – cont							
Designated mix Producer to hold current product conformity certification and quality approval. Quality assurance is essential. The mix must not be modified.							
CONCRETE MIXES Note: The mixes below are all mixed in 113 litre mixers.							
Concrete mixes at cost; mixed on site; costs for producing concrete; prices for commonly used mixes for various types of work; based on bulk load 20 tonne rates for aggregates Roughest type mass concrete such as footings and road haunchings; 300 mm thick							
1:3:6	101.20	–	–	–	101.20	m^3	**101.20**
Concrete mixes; mixed on site; costs for producing concrete; prices for commonly used mixes for various types of work; materials delivered in 10 tonne loads Roughest type mass concrete such as footings and road haunchings; 300 mm thick							
1:3:6	125.37	–	–	–	125.37	m^3	**125.37**
Concrete mixes; mixed on site; costs for producing concrete; prices for commonly used mixes for various types of work; materials delivered in 850 kg bulk bags Roughest type mass concrete such as footings and road haunchings; 300 mm thick							
1:3:6	133.27	–	–	–	133.27	m^3	**133.27**

11 IN SITU CONCRETE WORKS

Item – Overhead and Profit Included	PC £	Labour hours	Labour £	Plant £	Material £	Unit	Total rate £
Most ordinary use of concrete such as mass walls above ground, road slabs, etc. and general reinforced concrete work							
1:2:4; aggregates delivered in 20 tonne ballast loads	–	–	–	–	113.66	m³	**113.66**
1:2:4; aggregates delivered in 10 tonne ballast loads	130.76	–	–	–	130.76	m³	**130.76**
1:2:4; aggregates delivered in 850 kg bulk bags	138.66	–	–	–	138.66	m³	**138.66**
Watertight floors, pavements, walls, tanks, pits, steps, paths, surface of two course roads; reinforced concrete where extra strength is required							
1:1.5:3; aggregates delivered in 20 tonne ballast loads	100.36	1.20	33.52	–	100.36	m³	**133.88**
1:1.5:3; aggregates delivered in 10 tonne ballast loads	117.46	1.20	33.52	–	117.46	m³	**150.98**
1:1.5:3; aggregates delivered in 850 kg bulk bags	125.36	1.20	33.52	–	125.36	m³	**158.88**
IN SITU CONCRETE							
Plain in situ concrete; site mixed; 10 N/mm²–40 mm aggregate (1:3:6) (aggregate delivery indicated)							
Foundations							
ordinary Portland cement; 20 tonne ballast loads	103.73	0.33	31.88	–	119.29	m³	**151.17**
ordinary Portland cement; 10 tonne ballast loads	128.50	0.33	31.88	–	147.77	m³	**179.65**
ordinary Portland cement; 850 kg bulk bags	136.60	0.33	31.88	–	157.09	m³	**188.97**
Foundations; poured on or against earth or unblinded hardcore							
ordinary Portland cement; 20 tonne ballast loads	106.26	0.33	31.88	–	122.20	m³	**154.08**
ordinary Portland cement; 10 tonne ballast loads	131.64	0.33	31.88	–	151.39	m³	**183.27**
ordinary Portland cement; 850 kg bulk bags	139.93	0.33	31.88	–	160.92	m³	**192.80**
Isolated foundations							
ordinary Portland cement; 20 tonne ballast loads	103.73	0.33	31.88	–	119.29	m³	**151.17**
ordinary Portland cement; 850 kg bulk bags	136.60	0.33	31.88	–	157.09	m³	**188.97**

11 IN SITU CONCRETE WORKS

Item – Overhead and Profit Included	PC £	Labour hours	Labour £	Plant £	Material £	Unit	Total rate £
IN SITU CONCRETE – cont							
Plain in situ concrete; site mixed; 21 N/mm^2–20 mm aggregate (1:2:4)							
Foundations							
ordinary Portland cement; 20 tonne ballast loads	116.50	0.33	31.88	–	133.98	m^3	**165.86**
ordinary Portland cement; 850 kg bulk bags	142.12	0.33	31.88	–	163.44	m^3	**195.32**
Foundations; poured on or against earth or unblinded hardcore							
ordinary Portland cement; 20 tonne ballast loads	119.34	0.33	31.88	–	137.24	m^3	**169.12**
ordinary Portland cement; 850 kg bulk bags	145.59	0.33	31.88	–	167.43	m^3	**199.31**
Isolated foundations							
ordinary Portland cement; 20 tonne ballast loads	116.50	0.33	31.88	–	133.98	m^3	**165.86**
ordinary Portland cement; 850 kg bulk bags	142.12	0.33	31.88	–	163.44	m^3	**195.32**
Ready mix concrete; concrete mixed on site; Euromix							
Ready mix concrete mixed on site; placed by barrow not more than 25 m distance (16 barrows/m^3)							
concrete 1:3:6	97.50	0.25	24.15	–	112.13	m^3	**136.28**
concrete 1:2:4	98.29	0.25	24.15	–	113.03	m^3	**137.18**
concrete C30	105.00	0.25	24.15	–	120.75	m^3	**144.90**
Reinforced in situ concrete; site mixed; 21 N/mm^2–20 mm aggregate (1:2:4); aggregates delivered in 10 tonne loads							
Foundations							
ordinary Portland cement	116.50	0.67	64.33	–	133.98	m^3	**198.31**
Foundations; poured on or against earth or unblinded hardcore							
ordinary Portland cement	116.50	0.67	64.33	–	133.98	m^3	**198.31**
Isolated foundations							
ordinary Portland cement	116.50	0.83	80.47	–	133.98	m^3	**214.45**
Plain in situ concrete; ready mixed; 10 N/mm mixes; suitable for mass concrete fill and blinding							
Foundations							
C10; 10 N/mm^2	87.00	0.20	19.32	–	100.05	m^3	**119.37**
C25; 25 N/mm^2	91.35	0.20	19.32	–	105.05	m^3	**124.37**

11 IN SITU CONCRETE WORKS

Item – Overhead and Profit Included	PC £	Labour hours	Labour £	Plant £	Material £	Unit	Total rate £
Foundations; poured on or against earth or unblinded hardcore							
C10; 10 N/mm^2	87.00	0.20	19.32	–	100.05	m^3	**119.37**
C20; 20 N/mm^2	91.00	0.20	19.32	–	104.65	m^3	**123.97**
Isolated foundations							
C10; 10 N/mm^2	87.00	0.67	64.33	–	100.05	m^3	**164.38**
C25; 25 N/mm^2	91.50	0.67	64.33	–	105.22	m^3	**169.55**
Plain in situ concrete; ready mixed; 15 N/mm mixes; suitable for oversite below suspended slabs and strip footings in non aggressive soils							
Foundations							
C15; 15 N/mm^2	88.00	0.20	19.32	–	101.20	m^3	**120.52**
C20; 20 N/mm^2	91.00	0.20	19.32	–	104.65	m^3	**123.97**
Foundations; poured on or against earth or unblinded hardcore							
C15; 15 N/mm^2	88.00	0.20	19.32	–	101.20	m^3	**120.52**
ST 3; 15 N/mm^2	91.00	0.20	19.32	–	104.65	m^3	**123.97**
Isolated foundations							
C15; 15 N/mm^2	88.00	0.67	64.33	–	101.20	m^3	**165.53**
C20; 20 N/mm^2	–	0.67	64.33	–	104.65	m^3	**168.98**
Plain in situ concrete; ready mixed; air entrained mixes suitable for paving							
Beds or slabs; house drives, parking and external paving							
PAV 1; 35 N/mm^2; Designated mix 100 mm thick	103.37	0.25	24.15	–	118.88	m^3	**143.03**
Beds or slabs; house drives, parking and external paving							
PAV 1; 35 N/mm^2; Designated mix	103.37	0.25	24.15	–	118.88	m^3	**143.03**
Reinforced in situ concrete; ready mixed; 35 N/mm^2 mix; suitable for foundations in class 2 sulphate conditions							
Foundations							
RC 35; Designated mix	97.65	0.20	19.32	–	112.30	m^3	**131.62**
Foundations; poured on or against earth or unblinded hardcore							
RC 35; Designated mix	99.97	0.20	19.32	–	114.97	m^3	**134.29**
Isolated foundations							
RC 35; Designated mix	93.00	0.67	64.33	–	106.95	m^3	**171.28**

11 IN SITU CONCRETE WORKS

Item – Overhead and Profit Included	PC £	Labour hours	Labour £	Plant £	Material £	Unit	Total rate £
IN SITU CONCRETE – cont							
Reinforced in situ concrete; 21 N/mm²–20 mm aggregate (1:2:4); aggregates delivered in 20 tonne loads							
Site mixed concrete; aggregates in 20 tonne loads; Walls; thickness not exceeding 150 mm							
1:2:4: 21 N/mm²; 20 mm							
aggregate	116.50	0.67	64.33	4.60	133.98	m³	**202.91**
Ready mixed concrete							
C10 Concrete	87.00	0.33	32.20	4.60	100.05	m³	**136.85**
C15 Concrete	92.40	0.33	32.20	4.60	106.26	m³	**143.06**
C20 Concrete	95.55	0.33	32.20	4.60	109.88	m³	**146.68**
C25 Concrete	96.08	0.33	32.20	4.60	110.49	m³	**147.29**
RC 35; Designated mix	97.65	0.33	32.20	4.60	112.30	m³	**149.10**
Walls; thickness not exceeding 150–450 mm							
Site mixed concrete; aggregates in 20 tonne loads 1:2:4: 21 N/mm²; 20 mm aggregate	116.50	0.50	48.30	4.60	133.98	m³	**186.88**
Ready mixed concrete							
C10 Concrete	87.00	0.20	19.32	4.60	100.05	m³	**123.97**
C15 Concrete	88.00	0.20	19.32	4.60	101.20	m³	**125.12**
C20 Concrete	91.00	0.20	19.32	4.60	104.65	m³	**128.57**
RC 35; Designated mix	93.00	0.20	19.32	4.60	106.95	m³	**130.87**
Adjustments for site mixed concrete; add the following amounts for aggregates delivered in different load sizes							
850 kg bulk bags	–	–	–	–	28.75	m³	**28.75**
10 tonne loads	–	–	–	–	19.67	m³	**19.67**
PLAIN VERTICAL FORMWORK							
Plain vertical formwork; basic finish							
Sides of foundations							
height not exceeding 250 mm	–	0.25	16.10	–	2.17	m	**18.27**
height 250–500 mm	–	0.25	16.10	–	2.46	m	**18.56**
height 500 mm–1.00 m	–	0.33	21.46	–	6.89	m	**28.35**
height exceeding 1.00 m	–	0.50	32.20	–	8.87	m²	**41.07**
Sides of foundations; left in							
height not exceeding 250 mm	–	0.20	12.88	–	5.19	m	**18.07**
height 250–500 mm	–	0.20	12.88	–	8.48	m	**21.36**
height 500 mm–1.00 m	–	0.25	16.10	–	16.96	m	**33.06**
height over 1.00 m	–	0.40	25.76	–	10.23	m²	**35.99**

11 IN SITU CONCRETE WORKS

Item – Overhead and Profit Included	PC £	Labour hours	Labour £	Plant £	Material £	Unit	Total rate £
Walls							
height not exceeding 250 mm	–	0.31	20.13	–	7.92	m	**28.05**
height 250–500 mm	–	0.33	21.46	–	11.06	m	**32.52**
height 500 mm–1.00 m	–	0.42	26.83	–	24.10	m	**50.93**
height exceeding 1.00 m	–	0.75	48.30	–	33.20	m²	**81.50**
Walls curved to 6.00 radius							
height not exceeding 250 mm	–	0.38	24.15	–	6.00	m	**30.15**
height 250–500 mm	–	0.40	25.76	–	11.24	m	**37.00**
height 500 mm–1.00 m	–	0.50	32.19	–	24.44	m	**56.63**
height exceeding 1.00 m	–	0.90	57.96	–	33.53	m²	**91.49**
Walls curved to 3.00 radius							
height not exceeding 250 mm	–	0.47	30.19	–	6.16	m	**36.35**
height 250–500 mm	–	0.50	32.20	–	11.06	m	**43.26**
height 500 mm–1.00 m	–	0.62	40.24	–	24.10	m	**64.34**
height exceeding 1.00 m	–	1.13	72.45	–	33.20	m²	**105.65**
BAR REINFORCEMENT							
The rates for reinforcement shown							
for steel bars below are based on							
prices which would be supplied on a							
typical landscape contract. The steel							
prices shown have been priced on a							
selection of steel delivered to site							
where the total order quantity is in							
the region of 2 tonnes. The							
assumption is that should larger							
quantities be required, the work							
would fall outside the scope of the							
typical landscape contract defined in							
the front of this book. Keener rates							
can be obtained for larger orders.							
Reinforcement bars; hot rolled							
plain round mild steel; straight or							
bent							
Bars							
8 mm nominal size	765.00	13.50	869.40	–	879.75	tonne	**1749.15**
10 mm nominal size	765.00	13.00	837.20	–	879.75	tonne	**1716.95**
12 mm nominal size	765.00	12.50	805.00	–	879.75	tonne	**1684.75**
16 mm nominal size	765.00	12.00	772.80	–	879.75	tonne	**1652.55**
20 mm nominal size	765.00	11.50	740.60	–	879.75	tonne	**1620.35**
Weights of bar reinforcement							
(m/tonne) – kg/m							
6 mm (4505)	–	–	–	–	–	0.22 kg	-
8 mm (2534)	–	–	–	–	–	0.40 kg	-
10 mm (1622)	–	–	–	–	–	0.62 kg	-
12 mm (1126)	–	–	–	–	–	0.89 kg	-
16 mm (634)	–	–	–	–	–	1.58 kg	-
20 mm (405)	–	–	–	–	–	2.47 kg	-
25 mm (260)	–	–	–	–	–	3.85 kg	-
32 mm (158)	–	–	–	–	–	6.31 kg	-

11 IN SITU CONCRETE WORKS

Item – Overhead and Profit Included	PC £	Labour hours	Labour £	Plant £	Material £	Unit	Total rate £
BAR REINFORCEMENT – cont							
Reinforcement bar to concrete formwork straight; Slabs and foundations; Straight bar							
8 mm bar							
100 ccs	2.98	0.17	10.63	–	3.43	m²	**14.06**
200 ccs	1.45	0.13	8.05	–	1.67	m²	**9.72**
300 ccs	0.99	0.08	4.83	–	1.14	m²	**5.97**
tonne rate	765.00	11.25	724.50	–	879.75	tonne	**1604.25**
10 mm bar							
100 ccs	4.67	0.17	10.63	–	5.37	m²	**16.00**
200 ccs	2.29	0.13	8.05	–	2.63	m²	**10.68**
300 ccs	1.53	0.08	4.83	–	1.76	m²	**6.59**
tonne rate	765.00	11.25	724.50	–	879.75	tonne	**1604.25**
12 mm bar							
100 ccs	6.73	0.17	10.63	–	7.74	m²	**18.37**
200 ccs	3.37	0.13	8.05	–	3.88	m²	**11.93**
300 ccs	2.22	0.08	4.83	–	2.55	m²	**7.38**
tonne rate	765.00	8.45	544.18	–	879.75	tonne	**1423.93**
16 mm bar							
100 ccs	12.01	0.20	12.88	–	13.81	m²	**26.69**
200 ccs	6.04	0.17	10.63	–	6.95	m²	**17.58**
300 ccs	4.05	0.13	8.05	–	4.66	m²	**12.71**
tonne rate	765.00	6.00	386.40	–	879.75	tonne	**1266.15**
25 mm bar							
100 ccs	29.45	0.20	12.88	–	33.87	m²	**46.75**
200 ccs	14.69	0.17	10.63	–	16.89	m²	**27.52**
300 ccs	9.79	0.17	10.63	–	11.26	m²	**21.89**
tonne rate	765.00	3.45	222.18	–	879.75	tonne	**1101.93**
32 mm bar							
100 ccs	48.27	0.20	12.88	–	55.51	m²	**68.39**
200 ccs	24.10	0.17	10.63	–	27.72	m²	**38.35**
300 ccs	16.07	0.13	8.05	–	18.48	m²	**26.53**
tonne rate	765.00	2.85	183.54	–	879.75	tonne	**1063.29**
Alternative tonne rates for bar reinforcement fixing to retaining walls in external works environments; Vertical bar only; Horizontal tables shown in separate table set below							
8 mm bar. Rates per tonne based on fixing rates shown							
28.2 hrs/tonne	765.00	28.20	908.04	–	879.75	tonne	**1787.79**
10 mm bar. Rates per tonne based on fixing rates shown							
22.5 hrs/tonne	765.00	22.50	724.50	–	879.75	tonne	**1604.25**

11 IN SITU CONCRETE WORKS

Item – Overhead and Profit Included	PC £	Labour hours	Labour £	Plant £	Material £	Unit	Total rate £
12 mm bar. Rates per tonne based on fixing rates shown							
16.9 hrs/tonne	765.00	16.90	544.18	–	879.75	tonne	**1423.93**
16 mm bar. Rates per tonne based on fixing rates shown							
12.0 hrs/tonne	765.00	12.00	386.40	–	879.75	tonne	**1266.15**
25 mm bar. Rates per tonne based on fixing rates shown							
6.9 hrs/tonne	765.00	6.90	222.18	–	879.75	tonne	**1101.93**
32 mm bar. Rates per tonne based on fixing rates shown							
5.7 hrs/tonne	765.00	5.70	183.54	–	879.75	tonne	**1063.29**
Alternative tonne rates for bar reinforcement fixing to retaining walls in external works environments; Horizontal bar only in conjuction with vertical bar in separate table							
8 mm bar. Rates per tonne based on fixing rates shown							
7 hrs/tonne	765.00	7.00	225.40	–	879.75	tonne	**1105.15**
10 mm bar. Rates per tonne based on fixing rates shown							
5.62 hrs/tonne	765.00	5.62	180.96	–	879.75	tonne	**1060.71**
12 mm bar. Rates per tonne based on fixing rates shown							
4.22 hrs/tonne	765.00	4.22	135.88	–	879.75	tonne	**1015.63**
16 mm bar. Rates per tonne based on fixing rates shown							
3.0 hrs/tonne	765.00	3.00	96.60	–	879.75	tonne	**976.35**
25 mm bar. Rates per tonne based on fixing rates shown							
2.2 hrs/tonne	765.00	2.20	70.84	–	879.75	tonne	**950.59**
32 mm bar. Rates per tonne based on fixing rates shown							
2 hrs/tonne	765.00	2.00	64.40	–	879.75	tonne	**944.15**
Reinforcement bar to concrete formwork or footings; bent							
8 mm bar							
100 ccs	2.98	0.17	10.63	–	3.43	m²	**14.06**
200 ccs	1.45	0.13	8.05	–	1.67	m²	**9.72**
300 ccs	0.99	0.08	4.83	–	1.14	m²	**5.97**
10 mm bar							
100 ccs	4.67	0.17	10.63	–	5.37	m²	**16.00**
200 ccs	2.29	0.13	8.05	–	2.63	m²	**10.68**
300 ccs	1.53	0.08	4.83	–	1.76	m²	**6.59**

11 IN SITU CONCRETE WORKS

Item – Overhead and Profit Included	PC £	Labour hours	Labour £	Plant £	Material £	Unit	Total rate £
BAR REINFORCEMENT – cont							
Reinforcement bar to concrete formwork or footings – cont							
12 mm bar							
100 ccs	6.73	0.17	10.63	–	7.74	m²	**18.37**
200 ccs	3.37	0.13	8.05	–	3.88	m²	**11.93**
300 ccs	2.22	0.08	4.83	–	2.55	m²	**7.38**
16 mm bar							
100 ccs	12.01	0.20	12.88	–	13.81	m²	**26.69**
200 ccs	6.04	0.17	10.63	–	6.95	m²	**17.58**
300 ccs	4.05	0.13	8.05	–	4.66	m²	**12.71**
25 mm bar							
100 ccs	29.45	0.20	12.88	–	33.87	m²	**46.75**
200 ccs	14.69	0.13	8.05	–	16.89	m²	**24.94**
300 ccs	9.79	0.13	8.05	–	11.26	m²	**19.31**
32 mm bar							
100 ccs	48.27	0.20	12.88	–	55.51	m²	**68.39**
200 ccs	24.10	0.17	10.63	–	27.72	m²	**38.35**
300 ccs	16.07	0.13	8.05	–	18.48	m²	**26.53**
Reinforcement fabric; lapped; in beds or suspended slabs							
Fabric							
A98 (1.54 kg/m²)	1.08	0.02	1.40	–	1.24	m²	**2.64**
A142 (2.22 kg/m²)	3.55	0.02	1.40	–	4.08	m²	**5.48**
A193 (3.02 kg/m²)	4.09	0.03	1.89	–	4.70	m²	**6.59**
A252 (3.95 kg/m²)	5.16	0.03	1.86	–	5.93	m²	**7.79**
A393 (6.16 kg/m²)	6.88	0.03	2.14	–	7.91	m²	**10.05**

14 MASONRY

Item – Overhead and Profit Included	PC £	Labour hours	Labour £	Plant £	Material £	Unit	Total rate £
MARKET PRICES OF MATERIALS							
Cement; Builder Centre							
Portland cement	–	–	–	–	3.50	25 kg	**3.50**
Sulphate-resistant cement	–	–	–	–	3.89	25 kg	**3.89**
White cement	–	–	–	–	12.53	25 kg	**12.53**
Sand; Builder Centre							
Building sand							
loose	47.70	–	–	–	47.70	m³	**47.70**
850 kg bulk bags	–	–	–	–	33.60	nr	**33.60**
850 kg bulk bags	–	–	–	–	65.52	m³	**65.52**
Sharp sand							
loose	–	–	–	–	42.07	m³	**42.07**
850 kg bulk bags	–	–	–	–	33.60	nr	**33.60**
850 kg bulk bags	–	–	–	–	65.52	m³	**65.52**
Sand; Yeoman Aggregates Ltd							
Sharp sand	–	–	–	–	25.50	tonne	**25.50**
Sharp sand	–	–	–	–	40.80	m³	**40.80**
Soft sand	–	–	–	–	26.50	tonne	**26.50**
Soft sand	–	–	–	–	40.80	m³	**40.80**
CLARIFICATION NOTES ON LABOUR COSTS IN THIS SECTION							
Brick/Block walling team							
Generally a three man team is used in this section; The column 'Labour hours' reports team hours. The column 'Labour £' reports the total cost of the team for the unit of work shown							
3 man team	–	1.00	96.60	–	–	hr	**96.60**
2 man team	–	1.00	64.40	–	–	hr	**64.40**
Brick/Block walling Data							
Note: Batching quantities for these mortar mixes may be found in the Tables and Memoranda section of this book.							

14 MASONRY

Item – Overhead and Profit Included	PC £	Labour hours	Labour £	Plant £	Material £	Unit	Total rate £
Brick/Block walling Data – cont							
Mortar mixes for brickwork and pavings; all mortars mixed on site and priced at delivery to waiting wheelbarrows at the cement mixer; cement and lime in 25 kg bags; aggregates as described							
Mortar grade 1:3; aggregates delivered in 20 tonne load; cement: sand volumes shown below (bags: litres)							
85 litre mixer (1.84:93.5)	–	0.75	24.15	–	138.55	m³	**162.70**
150 litre mixer (3.24:165)	–	0.50	16.10	–	138.55	m³	**154.65**
Mortar grade 1:3; aggregates delivered in 10 tonne load; cement: sand volumes shown below (bags: litres)							
85 litre mixer (1.84:93.5)	–	0.75	24.15	–	166.63	m³	**190.78**
150 litre mixer (3.24:165)	–	0.50	16.10	–	166.63	m³	**182.73**
Mortar grade 1:3; aggregates delivered in 850 kg bulk bags; cement: sand volumes shown below (bags: litres)							
85 litre mixer (1.84:93.5)	–	0.75	24.15	–	166.92	m³	**191.07**
150 litre mixer (3.24:165)	–	0.50	16.10	–	166.92	m³	**183.02**
Mortar grade 1:4; aggregates delivered in 20 tonne load; cement: sand volumes shown below (bags: litres)							
85 litre mixer (1.36:102)	–	0.75	24.15	–	120.70	m³	**144.85**
150 litre mixer (2.4:180)	–	0.50	16.10	–	120.70	m³	**136.80**
Mortar grade 1:4; aggregates delivered in 10 tonne load; cement: sand volumes shown below (bags: litres)							
85 litre mixer (1.36:102)	–	0.75	24.15	–	151.34	m³	**175.49**
150 litre mixer (2.4:180)	–	0.50	16.10	–	151.34	m³	**167.44**
Mortar grade 1:4; aggregates delivered in 850 kg bulk bags; cement: sand volumes shown below (bags: litres)							
85 litre mixer (1.36:102)	–	0.75	24.15	–	151.73	m³	**175.88**
150 litre mixer (2.4:180)	–	0.50	16.10	–	151.73	m³	**167.83**
Mortar grade 1:1:6; aggregates delivered in 20 tonne load; cement: lime: sand volumes shown below (bags: bags: litres)							
85 litre mixer (0.9:0.44:93.5)	–	0.75	24.15	–	137.36	m³	**161.51**
150 litre mixer (1.6: 0.8: 165)	–	0.50	16.10	–	137.36	m³	**153.46**

14 MASONRY

Item – Overhead and Profit Included	PC £	Labour hours	Labour £	Plant £	Material £	Unit	Total rate £
Mortar grade 1:1:6; aggregates delivered in 10 tonne load; cement: lime: sand volumes shown below (bags: bags: litres)							
85 litre mixer (0.9:0.44:93.5)	–	0.75	24.15	–	172.68	m³	196.83
150 litre mixer (1.6:0.8:165)	–	0.50	16.10	–	172.68	m³	188.78
Mortar grade 1:1:6; aggregates delivered in 850 kg bulk bags; cement: lime: sand volumes shown below (bags: bags: litres)							
85 litre mixer (0.9:0.44:93.5)	–	0.75	24.15	–	165.74	m³	189.89
150 litre mixer (1.6:0.8:165)	–	0.50	16.10	–	165.74	m³	181.84
Mortar grade 1:2:9; aggregates delivered in 20 tonne load; cement: lime: sand volumes shown below (bags: bags: litres)							
85 litre mixer (0.7:0.5:102)	–	0.75	24.15	–	137.29	m³	161.44
150 litre mixer (1.2:0.9:180)	–	0.50	16.10	–	132.59	m³	148.69
Mortar grade 1:2:9; aggregates delivered in 10 tonne load; cement: lime: sand volumes shown below (bags: bags: litres)							
85 litre mixer (0.7:0.5:102)	–	0.75	24.15	–	167.92	m³	192.07
150 litre mixer (1.2:0.9:180)	–	0.50	16.10	–	167.92	m³	184.02
Mortar grade 1:2:9; aggregates delivered in 850 kg bulk bags; cement: lime: sand volumes shown below (bags: bags: litres)							
85 litre mixer (0.7:0.5:102)	–	0.75	24.15	–	168.31	m³	192.46
150 litre mixer (1.2:0.9:180)	–	0.50	16.10	–	167.92	m³	184.02
Variation in brick prices (area × price difference × value shown below)							
Add or subtract the following amounts for every £1.00/1000 difference in the PC price of the measured items below							
half brick thick	–	–	–	–	7.24	m²	7.24
one brick thick	–	–	–	–	14.49	m²	14.49
one and a half brick thick	–	–	–	–	21.73	m²	21.73
two brick thick	–	–	–	–	28.98	m²	28.98
Mortar (1:3) required per m² of brickwork; brick size 215 × 102.5 × 65 mm							
Half brick wall (103 mm); 58 bricks/m²							
no frog	–	–	–	–	3.82	m²	3.82
single frog	–	–	–	–	4.40	m²	4.40
double frog	–	–	–	–	5.21	m²	5.21

14 MASONRY

Item – Overhead and Profit Included	PC £	Labour hours	Labour £	Plant £	Material £	Unit	Total rate £
Brick/Block walling Data – cont							
Mortar (1:3) required per m² of brickwork – cont							
2 × half brick cavity wall (270 mm); 116 bricks/m²							
no frog	–	–	–	–	7.61	m²	**7.61**
single frog	–	–	–	–	9.03	m²	**9.03**
double frog	–	–	–	–	11.03	m²	**11.03**
One brick wall (215 mm); 116 bricks/m²							
no frog	–	–	–	–	9.21	m²	**9.21**
single frog	–	–	–	–	11.03	m²	**11.03**
double frog	–	–	–	–	12.82	m²	**12.82**
One and a half brick wall (328 mm) 174 bricks/m²							
no frog	–	–	–	–	12.63	m²	**12.63**
single frog	–	–	–	–	14.82	m²	**14.82**
double frog	–	–	–	–	17.63	m²	**17.63**
Mortar (1:3) required per m² of blockwork; blocks 440 × 215 mm							
Block thickness							
100 mm	–	–	–	–	1.41	m²	**1.41**
140 mm	–	–	–	–	1.82	m²	**1.82**
hollow blocks 440 × 215 mm	–	–	–	–	0.46	m²	**0.46**
Movement of materials							
Loading to wheelbarrows and transporting to location; per 215 mm thick walls; maximum distance 25 m	–	0.42	13.42	–	–	m²	**13.42**
Adjustments for mortars in the brick prices below							
Difference for cement: mortar (1:3) in lieu of gauged mortar							
half brick thick	–	–	–	–	–0.14	m²	**–0.14**
one brick thick	–	–	–	–	–0.35	m²	**–0.35**
one and a half brick thick	–	–	–	–	–0.46	m²	**–0.46**
two brick thick	–	–	–	–	–0.68	m²	**–0.68**

14 MASONRY

Item – Overhead and Profit Included	PC £	Labour hours	Labour £	Plant £	Material £	Unit	Total rate £
BRICK WALLING AND BRICK PIERS							
Note: frog types and mortar quantities used in the calculations below are based on bricks with single frogs. Please make adjustments for mortar volumes used with varying brick frog formats. See data above for mortar quantities for various frog types.							
Delivery volumes of aggregates for mortars are based on 10 tonne deliveries. The figures below show the approximate areas of walling which would be constructed mixed with building sand from a 10 tonne (6.25 m³) load of building sand							
half brick wall = 256 m²							
one brick wall = 124 m²							
Class B engineering bricks; PC £358.00/1000; double Flemish bond in cement: mortar (1:3)							
Offloading mechanically; loading to wheelbarrows; transporting to location maximum 25 m distance per 215 mm thick walls	–	0.42	13.42	–	–	m²	**13.42**
Construct walls							
half brick thick	22.58	0.60	63.48	–	30.16	m²	**93.64**
one brick thick	45.16	1.20	126.96	–	65.37	m²	**192.33**
one and a half brick thick	67.74	1.80	190.44	–	93.43	m²	**283.87**
two brick thick	90.32	2.40	253.82	–	126.95	m²	**380.77**
Construct walls; curved; mean radius 6 m							
half brick thick	22.53	0.90	95.22	–	30.11	m²	**125.33**
one brick thick	45.16	1.80	190.44	–	62.42	m²	**252.86**
Construct walls; curved; mean radius 1.50 m							
half brick thick	22.58	1.20	126.95	–	30.16	m²	**157.11**
one brick thick	45.16	3.60	264.96	–	62.42	m²	**327.38**
Construct walls; tapering; one face battering; average							
one and a half brick thick	67.74	2.22	234.49	–	92.01	m²	**326.50**
two brick thick	90.32	2.96	312.64	–	124.86	m²	**437.50**
Construct walls; battering (retaining)							
one and a half brick thick	67.74	2.22	234.49	–	92.01	m²	**326.50**
two brick thick	90.32	2.96	312.64	–	124.86	m²	**437.50**

14 MASONRY

Item – Overhead and Profit Included	PC £	Labour hours	Labour £	Plant £	Material £	Unit	Total rate £
BRICK WALLING AND BRICK PIERS – cont							
Class B engineering bricks – cont							
Projections; vertical							
one brick × half brick	5.02	2.33	246.84	–	6.44	m	**253.28**
one brick × one brick	10.03	0.47	49.30	–	12.87	m	**62.17**
one and a half brick × one brick	15.05	0.70	74.06	–	19.31	m	**93.37**
two brick by one brick	20.07	0.77	81.04	–	25.75	m	**106.79**
Walls; half brick thick							
in honeycomb bond	14.69	0.60	63.48	–	20.71	m²	**84.19**
in quarter bond	22.04	0.56	58.72	–	28.21	m²	**86.93**
Facing bricks; PC £500.00/1000; English garden wall bond; in gauged mortar (1:1:6); facework one side							
Mechanically offloading; maximum 25 m distance; loading to wheelbarrows and transporting to location; walls	–	0.42	13.42	–	–	m²	**13.42**
Construct walls							
half brick thick	–	0.60	63.38	–	42.29	m²	**105.67**
half brick thick (using site cut snap headers to form bond)	–	0.80	85.07	–	42.29	m²	**127.36**
one brick thick	–	1.20	126.96	–	84.58	m²	**211.54**
one and a half brick thick	–	1.80	190.44	–	120.66	m²	**311.10**
two brick thick	–	2.40	253.92	–	163.10	m²	**417.02**
Walls; curved; mean radius 6 m							
half brick thick	–	0.90	95.22	–	40.56	m²	**135.78**
one brick thick	–	1.80	190.44	–	180.71	m²	**371.15**
Walls; curved; mean radius 1.50 m							
half brick thick	–	1.20	126.96	–	41.42	m²	**168.38**
one brick thick	–	2.40	253.92	–	182.44	m²	**436.36**
Walls; tapering; one face battering; average							
one and a half brick thick	–	2.20	232.76	–	128.42	m²	**361.18**
two brick thick	–	3.00	317.40	–	173.45	m²	**490.85**
Walls; battering (retaining)							
one and a half brick thick	–	2.00	211.60	–	128.42	m²	**340.02**
two brick thick	–	2.70	285.66	–	163.61	m²	**449.27**
Projections; vertical							
one brick × half brick	–	0.23	24.33	–	8.74	m	**33.07**
one brick × one brick	–	0.50	52.90	–	17.47	m	**70.37**
one and a half brick × one brick	–	0.70	74.06	–	26.22	m	**100.28**
two brick × one brick	–	0.80	84.64	–	36.13	m	**120.77**
Brickwork fair faced both sides; facing bricks in gauged mortar (1:1:6)							
extra for fair face both sides; flush, struck, weathered, or bucket-handle pointing	–	0.67	21.46	–	–	m²	**21.46**

14 MASONRY

Item – Overhead and Profit Included	PC £	Labour hours	Labour £	Plant £	Material £	Unit	Total rate £
Bricks; PC £800.00/1000; English garden wall bond; in gauged mortar (1:1:6)							
Walls							
half brick thick (stretcher bond)	–	0.60	63.48	–	60.92	m²	**124.40**
half brick thick (using site cut snap headers to form bond)	–	0.80	84.64	–	60.92	m²	**145.56**
one brick thick	–	1.20	126.96	–	126.71	m²	**253.67**
one and a half brick thick	–	1.80	190.44	–	185.14	m²	**375.58**
two brick thick	–	2.40	253.92	–	251.05	m²	**504.97**
Walls; curved; mean radius 6 m							
half brick thick	–	0.90	95.22	–	62.30	m²	**157.52**
one brick thick	–	2.40	253.92	–	127.73	m²	**381.65**
Walls; curved; mean radius 1.50 m							
half brick thick	–	0.53	56.07	–	67.82	m²	**123.89**
one brick thick	–	1.07	112.88	–	138.77	m²	**251.65**
Walls; stretcher bond; wall ties at 450 mm centres vertically and horizontally							
one brick thick	0.76	1.09	115.32	–	126.40	m²	**241.72**
two brick thick	2.28	1.64	173.51	–	253.67	m²	**427.18**
Brickwork fair faced both sides; facing bricks in gauged mortar (1:1:6)							
extra for fair face both sides; flush, struck, weathered or bucket-handle pointing	–	0.67	21.46	–	–	m²	**21.46**
Brick copings							
Copings; all brick headers-on-edge; two angles rounded 53 mm radius; flush pointing top and both sides as work proceeds; one brick wide; horizontal							
machine-made specials	48.05	0.16	17.35	–	56.44	m	**73.79**
handmade specials	61.80	0.16	17.35	–	72.25	m	**89.60**
Extra over copings for two courses machine-made tile creasings; projecting 25 mm each side; 260 mm wide copings; horizontal	6.24	0.17	17.99	–	7.76	m	**25.75**
Copings; all brick headers-on-edge; flush pointing top and both sides as work proceeds; one brick wide; horizontal							
engineering bricks PC £358.00/ 1000	5.02	0.16	17.35	–	5.77	m	**23.12**
facing bricks PC £500.00/1000	7.16	0.16	17.35	–	8.23	m	**25.58**
facing bricks PC £800.00/1000	10.66	0.16	17.35	–	12.26	m	**29.61**

14 MASONRY

Item – Overhead and Profit Included	PC £	Labour hours	Labour £	Plant £	Material £	Unit	Total rate £
BRICK WALLING AND BRICK PIERS – cont							
Isolated brick piers in English bond; in gauged mortar 1:1:6							
Engineering brick PC £358.00/1000							
one brick thick (225 mm)	–	2.33	246.84	–	63.74	m²	**310.58**
one and a half brick thick (337.5 mm)	–	3.00	317.40	–	95.61	m²	**413.01**
two brick thick (450 mm)	–	3.33	352.64	–	129.46	m²	**482.10**
three brick thick (675 mm)	–	4.13	437.28	–	191.22	m²	**628.50**
Engineering brick PC £358.00/1000 (not SMM)							
one brick (225 mm)	–	0.54	57.62	–	14.64	m	**72.26**
one and a half brick thick (337.5 mm)	–	1.00	105.80	–	42.53	m	**148.33**
two brick thick (450 mm)	–	1.50	158.70	–	68.39	m	**227.09**
three brick thick (675 mm)	–	2.79	295.18	–	140.60	m	**435.78**
Brick PC £500.00/1000							
one brick thick (225 mm)	–	2.33	246.84	–	84.26	m²	**331.10**
one and a half brick thick (337.5 mm)	–	3.00	317.40	–	126.38	m²	**443.78**
two brick thick (450 mm)	–	3.33	352.64	–	170.49	m²	**523.13**
three brick thick (675 mm)	–	4.13	437.28	–	252.78	m²	**690.06**
Brick PC £500.00/1000 (not SMM)							
one brick (225 mm)	–	0.54	57.56	–	19.43	m	**76.99**
one and a half brick thick (337.5 mm)	–	1.00	105.80	–	52.91	m	**158.71**
two brick thick (450 mm)	–	1.50	158.70	–	86.86	m	**245.56**
three brick thick (675 mm)	–	2.79	295.18	–	182.14	m	**477.32**
Brick PC £800.00/1000							
one brick thick (225 mm)	–	2.33	246.84	–	127.73	m²	**374.57**
one and a half brick thick (337.5 mm)	–	3.00	317.40	–	191.59	m²	**508.99**
two brick thick (450 mm)	–	3.33	352.64	–	257.43	m²	**610.07**
three brick thick (675 mm)	–	4.13	437.28	–	383.19	m²	**820.47**
Brick PC £800.00/1000 (not SMM)							
one brick (225 mm)	–	0.54	57.56	–	29.58	m	**87.14**
one and a half brick thick (337.5 mm)	–	1.00	105.80	–	74.92	m	**180.72**
two brick thick (450 mm)	–	1.50	158.70	–	125.98	m	**284.68**
three brick thick (675 mm)	–	2.79	295.18	–	270.17	m	**565.35**

14 MASONRY

Item – Overhead and Profit Included	PC £	Labour hours	Labour £	Plant £	Material £	Unit	Total rate £
RESTORATION BRICKWORK							
Carefully dismantle existing coping to walls; Salvage copings (50%); Stack for reuse and dispose of balance off site							
Walls over 1.5 m high to 2.40 m high							
one brick thick	–	0.50	36.80	4.60	0.44	m	**41.84**
one and a half brick thick	–	0.75	55.20	4.60	0.65	m	**60.45**
two brick thick	–	1.00	73.60	4.60	0.87	m	**79.07**
Walls under 1.5 m high							
one brick thick	–	0.50	36.80	–	0.44	m	**37.24**
one and a half brick thick	–	0.75	55.20	–	0.65	m	**55.85**
two brick thick	–	1.00	73.60	–	0.87	m	**74.47**
Dismantle existing wall; salvage bricks and dispose damaged material off site (25%)							
half brick thick	–	0.40	29.44	0.74	1.50	m	**31.68**
one brick thick	–	0.70	51.52	0.74	3.00	m	**55.26**
one and a half brick thick	–	1.30	95.68	0.74	4.50	m	**100.92**
two brick thick	–	1.80	132.48	0.74	6.01	m	**139.23**
Cut out damaged brickwork in isolated areas							
half brick thick	–	0.17	5.37	–	–	brick	**5.37**
one brick thick to 2 bricks thick	–	0.14	4.60	–	–	brick	**4.60**
Isolated areas under 0.5 m^2							
half brick thick	–	0.50	16.10	–	–	m^2	**16.10**
one brick thick to 2 bricks thick per side	–	0.40	12.88	–	–	m^2	**12.88**
Isolated areas over 1.0 m^2							
half brick thick	–	1.00	32.20	–	–	m^2	**32.20**
one brick thick to 2 bricks thick	–	0.80	25.76	–	–	m^2	**25.76**
Cut out existing damaged pointing and replace with new mortar pointing							
half brick thick walls in strecher bond							
in standard mortar	0.28	0.42	13.42	–	0.32	m^2	**13.74**
in lime mortar	2.63	0.45	14.49	–	3.02	m^2	**17.51**
one brick thick walls in English garden wall bond							
in standard mortar	0.61	0.58	18.78	–	0.70	m^2	**19.48**
in lime mortar	5.25	0.58	18.78	–	6.04	m^2	**24.82**
flemish bond							
in standard mortar	0.65	0.69	22.08	–	0.75	m^2	**22.83**
in lime mortar	5.25	0.69	22.08	–	6.04	m^2	**28.12**

14 MASONRY

Item – Overhead and Profit Included	PC £	Labour hours	Labour £	Plant £	Material £	Unit	Total rate £
RESTORATION BRICKWORK – cont							
Replace damaged brickwork in existing wall; Cut out damaged bricks and replace with new bricks to match;reclaimed bricks PC £600.00/1000							
One brick thick; one side fair faced							
per brick	0.60	0.27	8.58	–	0.78	nr	9.36
larger areas	72.00	2.50	142.60	–	100.35	m²	242.95
One brick thick both sides fair faced							
per brick	0.60	0.28	9.12	–	0.78	nr	9.90
larger areas	72.00	3.50	216.20	–	100.35	m²	316.55
One and a half brick thick; one side fair faced							
per brick	1.20	0.57	18.23	–	1.58	nr	19.81
larger areas	108.00	5.50	322.00	–	150.52	m²	472.52
Construct new wall sections to match existing walls in lime mortar; Facing bricks; PC £1000.00/1000; English garden wall bond; in lime mortar							
Facing bricks; PC £1000.00/1000; English garden wall bond;							
half brick thick	31.71	0.64	67.88	–	107.19	m²	175.07
half brick thick (using site cut snap headers to form bond)	31.71	0.84	89.32	–	107.19	m²	196.51
one brick thick	58.17	1.26	133.31	–	208.35	m²	341.66
one and a half brick thick	84.63	1.89	199.96	–	309.50	m²	509.46
two brick thick	111.09	2.52	266.62	–	410.65	m²	677.27
Walls; curved; mean radius 6 m							
half brick thick	31.71	0.94	99.98	–	108.92	m²	208.90
one brick thick	58.17	1.89	199.96	–	211.80	m²	411.76
Walls; curved; mean radius 1.50 m							
half brick thick	31.71	1.26	133.31	–	110.64	m²	243.95
one brick thick	58.17	2.52	266.62	–	215.25	m²	481.87
Walls; tapering; one face battering; average							
one and a half brick thick	84.63	2.31	244.40	–	325.02	m²	569.42
two brick thick	111.09	3.15	333.27	–	431.35	m²	764.62
Walls; battering (retaining)							
one and a half brick thick	84.63	2.10	222.18	–	325.02	m²	547.20
two brick thick	84.63	2.83	299.94	–	400.92	m²	700.86
Projections; vertical							
one brick × half brick	11.13	0.24	25.55	–	28.90	m	54.45
one brick × one brick	17.01	0.53	55.54	–	51.75	m	107.29
one and a half brick × one brick	22.89	0.73	77.76	–	74.62	m	152.38
two brick × one brick	28.77	0.84	88.87	–	97.49	m	186.36

14 MASONRY

Item – Overhead and Profit Included	PC £	Labour hours	Labour £	Plant £	Material £	Unit	Total rate £
Brickwork fair faced both sides; facing bricks in gauged mortar (1:1:6)							
extra for fair face both sides; flush, struck, weathered, or bucket-handle pointing	5.25	0.70	22.54	–	6.04	m²	**28.58**
Brick copings; lime mortar bedding and jointing							
Copings; all brick headers-on-edge; two angles rounded 53 mm radius; flush pointing top and both sides as work proceeds; one brick wide; horizontal							
machine-made specials PC £3.50 each	54.52	0.17	18.22	–	62.70	m	**80.92**
handmade specials PC £4.50 each	72.93	0.17	18.22	–	83.87	m	**102.09**
Copings; all brick headers-on-edge; flush pointing top and both sides as work proceeds; one brick wide; horizontal							
reclaimed bricks PC £1000.00 /1000	6.27	0.17	18.22	–	22.93	m	**41.15**
Isolated brick piers in English bond; in gauged mortar 1:1:6							
Brick PC £1000.00/1000							
one brick thick (225 mm)	58.17	2.45	259.18	–	211.80	m²	**470.98**
one and a half brick thick (337.5 mm)	17.01	3.15	333.27	–	236.91	m²	**570.18**
two brick thick (450 mm)	111.09	3.50	370.27	–	417.55	m²	**787.82**
three brick thick (675 mm)	164.01	4.34	459.13	–	623.31	m²	**1082.44**
Brick PC £1000.00/1000 (not SMM)							
one brick (225 mm)	15.90	0.57	60.43	–	52.11	m	**112.54**
one and a half brick thick (337.5 mm)	95.88	1.05	111.09	–	110.26	m	**221.35**
two brick thick (450 mm)	54.72	1.57	166.63	–	193.34	m	**359.97**
three brick thick (675 mm)	117.79	2.93	309.94	–	428.88	m	**738.82**

14 MASONRY

Item – Overhead and Profit Included	PC £	Labour hours	Labour £	Plant £	Material £	Unit	Total rate £
BLOCK WALLING AND BLOCK PIERS							
Dense aggregate concrete blocks; Tarmac Topblock or other equal and approved; in gauged mortar (1:2:9); one course underground Walls							
Solid blocks 7 N/mm^2							
440 × 215 × 100 mm thick	12.80	0.40	42.32	–	16.25	m^2	**58.57**
440 × 215 × 140 mm thick	20.83	0.44	46.87	–	25.67	m^2	**72.54**
Solid blocks 7 N/mm^2 laid flat							
440 × 100 × 215 mm thick	26.11	1.07	113.21	–	31.75	m^2	**144.96**
Hollow concrete blocks							
440 × 215 × 215 mm thick	27.40	0.43	45.82	–	33.22	m^2	**79.04**
Filling of hollow concrete blocks with concrete as work proceeds; tamping and compacting							
440 × 215 × 215 mm thick	9.29	0.25	26.45	–	10.68	m^2	**37.13**
Isolated blockwork piers; to receive facing treatments (not included) Solid blocks; 440 × 100 × 215 mm thick; laid on-flat; 7 N/mm^2							
450 × 450 mm (SMM measurement)	73.86	2.00	147.20	–	86.27	m^2	**233.47**
450 × 450 mm (not SMM)	33.27	0.90	66.24	–	38.87	m	**105.11**
NATURAL STONE WALLING							
Expert stonework team Generally a three man team is used in this section; The column Labour hours reports team hours. The column Labour £ reports the total cost of the team for the unit of work shown							
2 man team	–	1.00	64.40	–	–	hr	**64.40**

14 MASONRY

Item – Overhead and Profit Included	PC £	Labour hours	Labour £	Plant £	Material £	Unit	Total rate £
Dry stone walling – General							
Preamble: In rural areas where natural stone is a traditional material, it may be possible to use dry stone walling or dyking as an alternative to fences or brick walls. Many local authorities are willing to meet the extra cost of stone walling in areas of high landscape value, and they may hold lists of available craftsmen. DSWA Office, Westmorland County Showground, Lane Farm, Crooklands, Milnthorpe, Cumbria, LA7 7NH; Tel: 01539 567953; E-mail: information@dswa.org.uk Note: Traditional walls are not built on concrete foundations.							
Dry stone wall; wall on concrete foundation (not included); dry stone coursed wall inclusive of locking stones and filling to wall with broken stone or rubble; walls up to 1.20 m high; battered; 2 sides fair faced							
Yorkstone	–	4.00	257.60	–	123.16	m²	**380.76**
Cotswold stone	–	4.00	257.60	–	62.09	m²	**319.69**
Purbeck	–	4.00	257.60	–	94.19	m²	**351.79**
Drystone style walling; mortared facing stone to blockwork wall; (not included); stainless steel ties between blockwork and stone; stone wall mortar raked subsequently raked out to provide drystone impression							
Yorkstone	–	3.25	209.30	–	127.29	m²	**336.59**
Cotswold stone	–	3.25	209.30	–	66.23	m²	**275.53**
Purbeck	–	3.25	209.30	–	98.31	m²	**307.61**
Portland stone walling							
Portland stone walling to external wall; Portland stone 300 × 100 mm × 250 mm thick; laid to White cement mortar; jointed with white cement mortar							
fair faced 1 side	330.00	1.00	96.60	–	386.39	m²	**482.99**
fair faced 2 sides	380.00	1.33	128.48	–	443.89	m²	**572.37**
Portland stone walling to external wall; 440 × 300 mm × 225 mm thick; laid to white cement mortar; jointed with white cement mortar							
fair faced 1 side	768.77	1.25	120.75	–	890.97	m²	**1011.72**
fair faced 2 sides	881.52	1.54	148.61	–	1020.64	m²	**1169.25**

14 MASONRY

Item – Overhead and Profit Included	PC £	Labour hours	Labour £	Plant £	Material £	Unit	Total rate £
NATURAL STONE WALLING – cont							
Portland stone walling – cont							
Coping Portland stone							
Coping 380 mm wide × 50 mm thick; Bull nosed single side; double drip	224.47	0.50	32.20	–	264.53	m	**296.73**
RECONSTITUTED STONE WALLING							
Haddonstone Ltd; cast stone piers; ornamental masonry in Portland Bath or Terracotta finished cast stone							
Gate Pier S120; to foundations and underground work measured separately; concrete infill							
S120G base unit to pier; 699 × 699 × 172 mm	181.76	0.25	24.15	–	215.41	nr	**239.56**
S120F/F shaft base unit; 533 × 533 × 280 mm	147.31	0.33	32.17	–	176.02	nr	**208.19**
S120E/E main shaft unit; 533 × 533 × 280 mm; nr of units required dependent on height of pier	147.31	0.33	32.17	–	176.02	nr	**208.19**
S120D/D top shaft unit; 33 × 533 × 280 mm	147.31	0.33	32.17	–	176.02	nr	**208.19**
S120C pier cap unit; 737 × 737 × 114 mm	171.07	0.17	16.03	–	203.34	nr	**219.37**
S120B pier block unit; base for finial; 533 × 533 × 64 mm	74.84	0.11	10.63	–	86.31	nr	**96.94**
Pier blocks; flat to receive gate finial							
S100B; 440 × 440 × 63 mm	48.71	0.11	10.63	–	56.57	nr	**67.20**
S120B; 546 × 546 × 64 mm	74.84	0.11	10.63	–	86.62	nr	**97.25**
S150B; 330 × 330 × 51 mm	26.14	0.11	10.63	–	30.61	nr	**41.24**
Pier caps; part weathered							
S100C; 915 × 915 × 150 mm	407.48	0.25	24.15	–	469.71	nr	**493.86**
S120C; 737 × 737 × 114 mm	171.07	0.17	16.03	–	197.28	nr	**213.31**
S150C; 584 × 584 × 120 mm	127.12	0.17	16.03	–	146.74	nr	**162.77**
Pier caps; weathered							
S230C; 1029 × 1029 × 175 mm	533.42	0.17	16.03	–	613.98	nr	**630.01**
S215C; 687 × 687 × 175 mm	235.22	0.17	16.03	–	271.05	nr	**287.08**
S210C; 584 × 584 × 175 mm	149.69	0.17	16.03	–	172.70	nr	**188.73**
Pier strings							
S100S; 800 × 800 × 55 mm	139.00	1.66	160.36	–	160.18	nr	**320.54**
S120S; 555 × 555 × 44 mm	64.75	0.17	16.03	–	74.80	nr	**90.83**
S150S; 457 × 457 × 48 mm	40.39	0.17	16.03	–	46.78	nr	**62.81**

14 MASONRY

Item – Overhead and Profit Included	PC £	Labour hours	Labour £	Plant £	Material £	Unit	Total rate £
Balls and bases							
E150 A Ball 535 mm and E150C collared base	414.00	0.17	16.03	–	476.43	nr	**492.46**
E120 A Ball 330 mm and E120C collared base	157.00	0.17	16.03	–	180.88	nr	**196.91**
E110 A Ball 230 mm and E110C collared base	105.00	0.17	16.03	–	121.08	nr	**137.11**
E100 A Ball 170 mm and E100B plain base	90.00	0.17	16.03	–	103.83	nr	**119.86**
Haddonstone Ltd; cast stone copings; ornamental masonry in Portland Bath or Terracotta finished cast stone							
Copings for walls; bedded, jointed and pointed in approved coloured cement-lime mortar 1:2:9							
T100 weathered coping; 102 mm high × 178 mm wide × 914 mm	76.30	0.11	10.72	–	88.22	m	**98.94**
T140 weathered coping; 102 mm high × 337 mm wide × 914 mm	73.25	0.11	10.72	–	84.71	m	**95.43**
T200 weathered coping; 127 mm high × 508 mm wide × 750 mm	109.87	0.11	10.72	–	126.82	m	**137.54**
T170 weathered coping; 108 mm high × 483 mm wide × 914 mm	117.72	0.11	10.72	–	135.85	m	**146.57**
T340 raked coping; 75–100 mm high × 290 mm wide × 900 mm	66.71	0.11	10.72	–	77.19	m	**87.91**
T310 raked coping; 76–89 mm high × 381 mm wide × 914 mm	78.48	0.11	10.72	–	90.72	m	**101.44**
Slope-Loc – Anderton Concrete Ltd							
Dry stacked random sized units 150–400 mm long × 150 mm high cast stone wall mechanically interlocked with fibreglass pins; constructed to levelling pad of coarse compacted granular material (not included) back filled behind the elevation with 300 mm wide granular drainage material; walls to 5.00 m high (retaining walls over heights shown below require individual design)							
Slope-loc 300 × 245 × 125 mm; gravity wall up to 850 mm;	56.56	1.00	64.40	48.53	71.25	m²	**184.18**
Slope-loc 300 × 245 × 125 mm; reinforced wall; inclusive of design and reinforcement grids; retaining up to 1.70 m	117.49	1.00	64.40	48.53	141.32	m²	**254.25**

14 MASONRY

Item – Overhead and Profit Included	PC £	Labour hours	Labour £	Plant £	Material £	Unit	Total rate £
RECONSTITUTED STONE WALLING – cont							
Retaining wall; but reinforced with Tensar Geogrid RE520 laid between every two courses horizontally into the face of the excavation (excavation not included)							
Near vertical; 250 mm thick; 1.50 m of geogrid length							
up to 1.20 m high; 2 layers of geogrid	56.56	0.50	48.30	–	75.52	m²	**123.82**
1.20–1.50 m high; 3 layers of geogrid	56.56	0.67	64.33	–	80.76	m²	**145.09**
1.50–1.80 m high; 4 layers of geogrid	56.56	0.75	72.45	–	86.00	m²	**158.45**
Battered; maximum 1:3 slope							
up to 1.20 m high; 3 layers of geogrid	56.56	0.83	80.47	–	80.76	m²	**161.23**
1.20–1.50 m high; 4 layers of geogrid	56.56	0.83	80.47	–	86.00	m²	**166.47**
1.50–1.80 m high; 5 layers of geogrid	56.56	1.00	96.60	–	90.37	m²	**186.97**
COPINGS AND PIER CAPS							
Copings extra over to all copings for access over 1.50 m							
Bandstand	–	–	–	57.50	–	week	**57.50**
Portland stone copings; Albion stone; laid to 10 mm mortar bed; stainless steel dowels at 4 nr/m drilled and set to wall and coping							
Coping Portland stone							
Coping 380 mm wide × 50 mm thick; Bull nosed single side; double drip	224.47	0.50	32.20	–	267.98	m	**300.18**
Granite copings; CED Ltd; on 10 mm mortar bed with 4 nr. 12 mm stainless steel dowels/lm drilled to wall and coping							
Coping Granite; Bush hammered top and exposed sides with 30 mm single bullnose to long edge and exposed sides; straight copings							
310 × 100 mm	24.81	0.13	8.05	–	32.57	m	**40.62**
380 × 100 mm	30.36	0.10	6.44	–	39.05	m	**45.49**

14 MASONRY

Item – Overhead and Profit Included	PC £	Labour hours	Labour £	Plant £	Material £	Unit	**Total rate £**
Coping Granite; Bush hammered top and exposed sides with 30 mm double bullnose; straight copings							
110 × 50 mm	9.10	0.08	5.36	–	10.46	m	**15.82**
500 × 50 mm	29.29	0.17	10.74	–	33.68	m	**44.42**
500 × 100 mm	43.79	0.33	21.46	–	50.36	m	**71.82**
430 × 100 mm	42.12	0.29	18.40	–	48.44	m	**66.84**
680 × 100 mm	41.45	0.36	23.41	–	47.67	m	**71.08**
780 × 100 mm	66.49	0.40	25.76	–	76.46	m	**102.22**
Coping Granite; Bush hammered top and exposed sides with 30 mm double bullnose; curved							
310 × 100 mm; external radius of 3.00–4.00 m	32.76	0.25	16.10	–	37.67	m	**53.77**
310 × 100 mm; external radius of 15.00 m	32.76	0.10	6.44	–	37.67	m	**44.11**
430 × 100 mm; curved to external radius of 8.00–9.00 m	42.47	0.17	10.74	–	48.84	m	**59.58**
430 × 100 mm; curved to external radius of 12.00 m	42.47	0.13	8.05	–	48.84	m	**56.89**
Copings; London Stone; on 10 mm mortar bed with 4 nr. 12 mm stainless steel dowels/lm drilled to wall and coping; 300 wide × 40 mm thick							
Sandstone; Rock faced and with double pencil edge and drip							
Autumn brown/Kandla grey/Mint/ Heath sandstones	20.28	0.13	8.05	–	27.36	m	**35.41**
Sandstone; Sawn and flamed with double pencil edge and drip							
Beige/Heath sandstone	42.18	0.13	8.05	–	52.54	m	**60.59**
Sandstone and Black basalt; Sawn with double pencil edge and drip							
Black basalt/Buff sandstone/ Harvest sandstone	–	0.13	8.05	–	52.54	m	**60.59**

14 MASONRY

Item – Overhead and Profit Included	PC £	Labour hours	Labour £	Plant £	Material £	Unit	Total rate £
COPINGS AND PIER CAPS – cont							
Copings; Haddonstone; Cast stone in Portland, Bath or Terracotta; on 10 mm mortar bed with 4 nr. 12 mm stainless steel dowels/lm drilled to wall and coping; 300 wide × 40 mm thick							
Plain ended coping stones for walls up to 248 mm thick							
T130; flat topped 337 mm wide at top to 248 mm at base; overall 89 mm thick	77.83	0.10	6.44	–	89.50	m	**95.94**
T140; twice weathered 337 mm wide at top to 248 mm wide at base; overall 102 mm thick	80.48	0.10	6.44	–	92.55	m	**98.99**
T150; twice weathered 368 mm wide at top to 279 mm at base; 102 mm thick overall	83.10	0.10	6.44	–	95.56	m	**102.00**
Plain ended coping stones for walls up to 394 mm thick							
T440; twice weathered 394 mm wide at top to 356 mm at base; overall 76 mm thick	81.85	0.14	9.20	–	94.13	m	**103.33**
T470; twice weathered 410 mm wide at top to 250 mm wide at base; overall 114 mm thick	117.34	0.17	10.74	–	134.94	m	**145.68**
Precast concrete coping							
Copings; twice weathered; twice grooved							
600 × 300 mm, 50–30 mm thick	12.24	0.40	12.88	–	14.42	m	**27.30**
600 × 300 mm Bevelled edge	19.20	0.40	12.88	–	22.55	m	**35.43**
600 × 600 mm, 75–50 mm thick	21.00	0.50	16.10	–	24.91	m	**41.01**
Pier caps; four sides weathered							
295 × 295 mm	12.00	1.00	32.20	–	14.03	nr	**46.23**
380 × 380 mm	18.00	1.00	32.20	–	20.93	nr	**53.13**
520 × 520 mm	28.80	1.20	38.64	–	33.35	nr	**71.99**

19 WATERPROOFING

Item – Overhead and Profit Included	PC £	Labour hours	Labour £	Plant £	Material £	Unit	Total rate £
COLD APPLIED BITUMINOUS EMULSION							
Tanking and damp-proofing; 'Blackjack DPM'; cold applied bituminous emulsion waterproof coating to smooth finished concrete or screeded slabs; flat; blinding with sand							
two coats	2.57	0.22	7.15	–	3.20	m²	**10.35**
three coats	3.86	0.31	9.98	–	4.68	m²	**14.66**
Blackjack DPM; to fair faced brickwork with flush joints, rendered brickwork or smooth finished concrete walls; vertical							
two coats	2.83	0.29	9.20	–	3.50	m²	**12.70**
three coats	4.50	0.40	12.88	–	5.42	m²	**18.30**
SELF ADHESIVE							
Tanking and damp-proofing; Grace Construction Products Bitu-thene 3000; 1.50 mm thick; overlapping and bonding; including sealing all edges							
to concrete slabs; flat	7.70	0.25	8.05	–	9.60	m²	**17.65**
to brick/concrete walls; vertical	7.70	0.40	12.88	–	9.60	m²	**22.48**

21 CLADDING AND COVERING

Item – Overhead and Profit Included	PC £	Labour hours	Labour £	Plant £	Material £	Unit	Total rate £
NATURAL STONE SLAB CLADDING/FEATURES							
Tier natural stone cladding; CED Ltd; Tier cladding system; real stone panelling system, natural stone finish; interlocking tiers, fitting seamlessly together. For both interior and exterior walls; multiple finishes							
To exterior blockwork brickwork or concrete walls; inclusive of mechanical fixings and adhesive; Girth exceeding 400 mm							
multiple colours and stone types	59.46	3.00	193.20	–	89.85	m²	**283.05**
To exterior blockwork brickwork or concrete walls; inclusive of mechanical fixings and adhesive; Girth not exceeding 400 mm							
multiple colours and stone types	62.29	4.00	257.60	–	95.09	m²	**352.69**
Sawn Yorkstone cladding; Johnsons Wellfield Quarries Ltd							
Six sides sawn stone; rubbed face; sawn and jointed edges; fixed to blockwork (not included) with stainless steel fixings; Ancon Ltd; grade 304 stainless steel frame cramp and dowel 7 mm; cladding units drilled 4 × to receive dowels							
440 × 200 × 50 mm thick	108.09	1.87	60.21	–	131.10	m²	**191.31**
Portland stone cladding							
Cladding 40 mm thick fixed with 6 mm stainless steel dowell supports to blockwork; fixings at 200 mm centres; load support of 30 mm angle every 4 courses;							
Single size 400 × 400 mm	140.00	3.13	201.25	–	197.29	m²	**398.54**
Coursed – 3 sizes 400 mm × 400 mm –							
400 mm × 200 mm	140.00	4.00	257.60	–	205.56	m²	**463.16**
Coping Portland stone							
Coping 380 mm wide × 50 mm thick; Bull nosed single side;							
double drip	177.00	0.50	32.20	–	209.94	m	**242.14**

21 CLADDING AND COVERING

Item – Overhead and Profit Included	PC £	Labour hours	Labour £	Plant £	Material £	Unit	Total rate £
CAST STONE SLAB CLADDING/ FEATURES							
Cast stone cladding; Haddonstone Ltd Reconstituted stone in Portland Bath or Terracotta; fixed to blockwork or concrete (not included) with stainless steel fixings; stainless steel frame cramp and dowel M6 mm; cladding units drilled 4 × to receive dowels							
1000 × 300 × 50 mm thick	45.00	1.87	60.21	–	58.55	m²	**118.76**

25 BALUSTRADES

Item – Overhead and Profit Included	PC £	Labour hours	Labour £	Plant £	Material £	Unit	Total rate £
GLASS BALUSTRADES							
Glass balustrades; Elite Balustrades Ltd; Glass 17 mm laminated toughened with radius corners							
Installed to stainless steel uprights with bolt fixings to concrete base; Cranage or porting of units not included							
1.10 m high	365.06	0.33	21.46	–	419.82	m	**441.28**
1.25 m high	379.23	0.33	21.46	–	436.11	m	**457.57**
1.81 m high	477.47	0.50	32.20	–	549.09	m	**581.29**
1.95 m high	507.77	0.50	32.20	–	583.94	m	**616.14**
2.15 m high	641.45	2.00	128.80	–	737.67	m	**866.47**
Gates to glass balustrades; toughened 17 mm; with secure locking mechanisms							
1.81 m high	948.00	1.00	64.40	–	1090.20	m	**1154.60**
1.90 m high	980.00	1.20	77.28	–	1127.00	m	**1204.28**
2.05 m high	988.00	1.33	85.65	–	1136.20	m	**1221.85**
RECONSTITUTED STONE BALUSTRADES							
Bottle balustrades reconstituted stone; Haddonstone Ltd; to prepared brickwork walls or bases (not included)							
Top and bottom rails to support balustrades; laid to builders work of brick block or other stonework supports (not included)							
Top Rail T110 A 900 × 30 × 86	44.40	–	–	–	51.06	m	**51.06**
Bottom Rail T900 A							
900 × 230 × 102	44.40	–	–	–	51.06	m	**51.06**
Under coping							
Under Coping 900 × 377 × 76	65.49	–	–	–	75.31	m	**75.31**
Pier bases; 337 × 102 mm high to receive shafts							
337 × 337 × 102 mm high	31.00	0.25	16.10	–	35.65	nr	**51.75**
Pier shafts; at specified centres along the balustrade fixed to upstand (not included); subsequently filled with concrete							
Piers shaft 279 × 279 × 610 mm high	67.00	1.00	64.40	–	88.88	nr	**153.28**
Pier shafts 279 × 279 × 710 mm high	80.00	1.25	80.50	–	105.28	nr	**185.78**

25 BALUSTRADES

Item – Overhead and Profit Included	PC £	Labour hours	Labour £	Plant £	Material £	Unit	Total rate £
Pier caps to Pier shafts;							
337 × 337 × 86 mm	37.00	0.25	16.10	–	45.38	nr	**61.48**
Plinths to receive balustrade components; fixed to existing base or to low walls to receive balustrade components							
360 wide × 440 mm high	99.31	0.75	48.30	–	114.90	m	**163.20**
350 wide × 150 mm high	75.75	0.75	48.30	–	87.80	m	**136.10**
Bottle balustrades comprising reconstituted bottle balusters; fixed to plinths, top rail, bottom rail and under coping (all shown separately) with steel rods and mortar; balusters							
K457G 150 mm wide base x 127 mm top × 457 mm high; at 300 mm centres	123.21	1.00	64.40	–	144.90	m	**209.30**
K610G 140 mm wide base x 127 mm top × 610 mm high; at 300 mm centres	156.51	1.00	64.40	–	183.55	m	**247.95**
K729G 140 mm wide base x 127 mm top × 610 mm high; at 300 mm centres	183.15	1.00	64.40	–	214.19	m	**278.59**

28 FLOOR, WALL, CEILING FINISHINGS

Item – Overhead and Profit Included	PC £	Labour hours	Labour £	Plant £	Material £	Unit	Total rate £
PLASTERED/RENDERED/ ROUGHCAST COATINGS							
Specialist renders; K-Rend – London Stone; 2 coats to blockwork or in situ concrete or brickwork walls							
Silicone FT; scraped texture render, incorporating silicone technology low-maintenance, water-repellent finish; six colours; single or double coat 8–12 mm thick each							
Primer	5.14	0.10	6.44	–	38.91	m²	**45.35**
Single coat 2.2 kg/m²	5.14	0.10	6.44	–	38.91	m²	**45.35**
2 coats; 4.4 kg/m²	51.37	0.67	42.94	–	59.08	m²	**102.02**
Beads and stops for K-Rend							
K Bead Stop Bead White 2.5 m × 4 mm	1.09	0.07	4.30	–	1.25	m	**5.55**
K Bead Stop Bead Ivory 2.5 m × 4 mm	1.32	0.07	4.30	–	1.52	m	**5.82**
K Bead Feather Edge Angle Ivory 3 m × 4 mm	1.84	0.07	4.30	–	2.12	m	**6.42**
Cement: lime: sand (1:1:6); 19 mm thick; two coats; wood floated finish							
Walls							
width exceeding 300 mm; to brickwork or blockwork base	–	–	–	–	–	m²	**15.68**
Extra over cement: sand: lime (1:1:6) coatings for decorative texture finish with water repellent cement							
combed or floated finish	–	–	–	–	–	m²	**4.18**
STONE/CONCRETE/QUARRY/ CERAMIC TILING/MOSAIC							
Ceramic tiles; unglazed slip-resistant; various colours and textures; jointing							
Floors							
level or to falls only not exceeding 15° from horizontal; 150 × 150 × 8 mm thick	21.00	1.00	32.20	–	27.97	m²	**60.17**
level or to falls only not exceeding 15° from horizontal; 150 × 150 × 12 mm thick	26.25	1.00	32.20	–	34.01	m²	**66.21**

28 FLOOR, WALL, CEILING FINISHINGS

Item – Overhead and Profit Included	PC £	Labour hours	Labour £	Plant £	Material £	Unit	Total rate £
Clay tiles – General							
Preamble: Typical specification – Clay tiles shall be reasonably true to shape, flat, free from flaws, frost resistant and true to sample approved by the Landscape Architect prior to laying. Quarry tiles (or semi-vitrified tiles) shall be of external quality, either heather brown or blue, to size specified, laid on 1:2:4 concrete, with 20 mm maximum aggregate 100 mm thick, on 100 mm hardcore. The hardened concrete should be well wetted and the surplus water taken off. Clay tiles shall be thoroughly wetted immediately before laying and then drained and shall be bedded to 19 mm thick cement: sand (1:3) screed. Joints should be approximately 4 mm (or 3 mm for vitrified tiles) grouted in cement: sand (1:2) and cleaned off immediately.							
Quarry tiles; external quality; including bedding; jointing							
Floors							
level or to falls only not exceeding 15° from horizontal; 150 × 150 × 12.5 mm thick; heather brown; PC £15.00/m²	–	0.80	25.76	–	21.93	m²	**47.69**
level or to falls only not exceeding 15° from horizontal; 225 × 225 × 29 mm thick; heather brown PC £25.00/m²	–	0.67	21.46	–	34.01	m²	**55.47**
level or to falls only not exceeding 15° from horizontal; 150 × 150 × 12.5 mm thick; blue/black PC £16.00/m²	–	1.00	32.20	–	23.14	m²	**55.34**
level or to falls only not exceeding 15° from horizontal; 194 × 194 × 12.5 mm thick; heather brown PC £15.00/m²	–	0.80	25.76	–	21.93	m²	**47.69**

28 ROOF FINISHINGS – GREEN ROOFS

Item – Overhead and Profit Included	PC £	Labour hours	Labour £	Plant £	Material £	Unit	Total rate £
GREEN ROOFS							
Preamble: A variety of systems are available which address all the varied requirements for a successful Green Roof. For installation by approved contractors only. The prices shown are for budgeting purposes only, as each installation is site specific and may incorporate some or all of the resources shown. Specifiers should verify that the systems specified include for design liability and inspections by the suppliers. The systems below assume commercial insulation levels are required to the space below the proposed Green Roof. Extensive Green Roofs are those of generally lightweight construction with low maintenance planting and shallow soil designed for aesthetics only; Intensive Green Roofs are designed to allow use for recreation and trafficking. They require more maintenance and allow a greater variety of surfaces and plant types. The following systems which are compliant with European Technical Approval ETA 13/0668 (Kits for Green roofs) are indicated as such. The rates below assume that all elements are in position in the works location; no allowances for access, lifting or materials movement to the roof areas.							
Root barriers; Zinco Green Roof Systems Ltd							
Where waterproofing does not contain a root barrier							
WSB 100-PO Weldable reinforced flexible polyolefin. 1.1 mm thick. Root resistant tested according to FLL	13.47	–	–	–	–	m²	–
WSF 40 High pressure polyethylene 0.34 mm thick	2.31	0.02	1.93	–	2.66	m²	**4.59**

28 ROOF FINISHINGS – GREEN ROOFS

Item – Overhead and Profit Included	PC £	Labour hours	Labour £	Plant £	Material £	Unit	Total rate £
VOID FORMERS							
Void former; Filcor; lightweight expanded polystyrene; to roof or planter voids to make up levels. Sheets 2400 × 1200 mm; exclusive of carriage or cranage to point of laying; thickness							
Filcor 20							
165 mm	8.32	0.02	1.13	–	9.57	m²	**10.70**
200 mm	10.08	0.02	1.29	–	11.59	m²	**12.88**
245 mm	12.35	0.02	1.33	–	14.20	m²	**15.53**
300 mm	13.75	0.03	1.77	–	15.81	m²	**17.58**
360 mm	18.15	0.03	1.74	–	20.87	m²	**22.61**
450 mm	22.69	0.03	1.84	–	26.09	m²	**27.93**
Filcor 45							
165 mm	9.58	0.02	1.13	–	11.02	m²	**12.15**
200 mm	11.61	0.02	1.29	–	13.35	m²	**14.64**
245 mm	14.22	0.02	1.33	–	16.35	m²	**17.68**
300 mm	17.41	0.03	1.61	–	20.02	m²	**21.63**
360 mm	20.90	0.03	1.74	–	24.03	m²	**25.77**
450 mm	26.12	0.03	1.84	–	30.04	m²	**31.88**
EXTENSIVE GREEN ROOFS							
Extensive Sedum type Green Roof System; Zinco Green Roof Systems Ltd. System for planting with Sedum plugs or overlaying with pre-grown cultivated Sedum Carpet. (ETA 13/0668)							
Flat roofs with no standing water up to 10° or for Inverted roofs and as an alternative to single layer systems. Drainage, water retention and protection mat with attached filter sheet. 20 mm thick.							
Zinco Drainage Element Fixodrain XD20; 3 l/m² reservoir	4.80	0.04	3.86	–	5.52	m²	**9.38**
Zinco System Substrate Growing medium 'Sedum Carpet' – laid to Fixodrain XD20 at min. 70 mm thickness	100.00	0.67	64.40	–	115.00	m³	**179.40**
Zinco Premium Sedum Carpet Mat – Laid over System substrate	20.00	0.07	6.44	–	23.00	m²	**29.44**

28 ROOF FINISHINGS – GREEN ROOFS

Item – Overhead and Profit Included	PC £	Labour hours	Labour £	Plant £	Material £	Unit	Total rate £
EXTENSIVE GREEN ROOFS – cont							
Extensive Rockery type Green Roof System; Zinco Green Roof Systems Ltd. For planting with Sedum plugs or rockery type plants; flat roofs with no standing water up to 10°. (Not suitable for inverted roofs); Suitable system depth 90 mm. System saturated weight 110 kg/m^2; system water storage capacity: 36 l/m2. (ETA 13/ 0668)							
Non-rotting synthetic fibre mat for protection of the waterproofing layer and provision of additional water reservoir of 5 l/m^2							
Zinco Moisture mat/protection fleece SSM45.	1.95	0.04	3.86	–	2.24	m^2	**6.10**
Pressure resistant drainage and water storage element; recycled polyolefin; laid to SSM45; 25 mm thick with diffusion openings; water storage capacity 3 l/m^2.							
Zinco Drainage Element Fixodrain FD25-E	6.00	0.04	3.86	–	6.90	m^2	**10.76**
Zinco Filter Sheet SF – Rolled to drainage layer FD25-E	0.80	0.04	3.86	–	0.92	m^2	**4.78**
Zinco System Substrate Growing medium 'Rockery Type Plants' laid to Filter Fleece SF min. 70 mm thick	100.00	0.67	64.40	–	115.00	m^3	**179.40**
Planting to substrates							
Zinco Premium Sedum Carpet Mat	22.00	0.04	3.86	–	25.30	m^2	**29.16**
Zinco Seed Mix – Meadow Scents; 15 g/m^2	0.66	0.01	0.97	–	0.76	m^2	**1.73**

28 ROOF FINISHINGS – GREEN ROOFS

Item – Overhead and Profit Included	PC £	Labour hours	Labour £	Plant £	Material £	Unit	Total rate £
Extensive Rockery type Green Roof System for ZERO degree roofs; Zinco Green Roof Systems Ltd; for Zero degree roofs where isolated standing water may be present; planting with Sedum plugs or rockery type plants; system depth 150 mm; saturated weight: 105 kg/m²; water storage capacity; 33 l/m²							
Non rotting synthetic fibre mat for protection of the waterproofing layer and provision of additional water reservoir of 3 l/m²							
Zinco Moisture mat/protection fleece TSM32	1.76	0.04	3.86	–	2.02	m²	**5.88**
Drainage Element; laid to TSM32; ensures that a gap is maintained between any standing water and the soil substrate and the necessary aeration of the root zone; expanded polystyrene; 50 mm thick; lightweight (0.6 kg/m²); water storage capacity of 3 l/m²							
Zinco Drainage Element; Floraset FS50	7.59	0.04	3.86	–	8.73	m²	**12.59**
Zinco Filter Sheet SF – Rolled to drainage layer FS50	0.80	0.01	0.64	–	0.92	m²	**1.56**
Zinco System Substrate Growing medium 'Rockery Type Plants' – laid to filter fleece SF; min 70 mm thick	100.00	0.67	64.40	–	115.00	m³	**179.40**
Planting to substrates							
Zinco Premium Sedum Carpet Mat	22.00	0.04	3.86	–	25.30	m²	**29.16**
Zinco Seed Mix – Meadow Scents; 15 g/m²	0.66	0.01	0.97	–	0.76	m²	**1.73**

28 ROOF FINISHINGS – GREEN ROOFS

Item – Overhead and Profit Included	PC £	Labour hours	Labour £	Plant £	Material £	Unit	Total rate £
EXTENSIVE GREEN ROOF PITCH SYSTEMS							
Extensive Green Roof System for Pitched roofs; for pitched roofs with slopes from 10° up to 25°; Zinco Green Roof Systems Ltd; suitable for planting with Sedum plugs or overlaying with pre-grown cultivated Sedum Carpet. System depth – 130 mm; system saturated weight – 115 kg/m^2; system water storage capacity – 38 l/m^2; Waterproofing layer must contain a root barrier as a separate root barrier cannot be used on pitched roofs non rotting synthetic fibre mat for protection of the waterproofing layer and provision of additional water reservoir of 7 l/m^2; laid to water proofing layer							
Protection Mat BSM64	2.55	0.10	9.66	–	2.93	m^2	**12.59**
Lightweight (1 kg/m^2) drainage element of expanded polystyrene. 75 mm thick laid with large studs upwards; laid to BSM64							
Zinco Drainage Element Floraset FS75	10.25	0.10	9.66	–	11.79	m^2	**21.45**
Zinco System Substrate Growing medium infilled to drainage board and to a level of 50 mm above the large studs							
'Rockery Type Plants' – laid to FS75	8.25	0.10	9.66	–	9.49	m^2	**19.15**
Erosion net; laid over substrate to aid prevention of wind erosion							
Zinco Jute Anti Erosion Net JEG	0.90	0.07	6.44	–	1.04	m^2	**7.48**
Vegetation layer; laid over System substrate							
Zinco Premium Sedum Carpet Mat	20.00	0.07	6.44	–	23.00	m^2	**29.44**

28 ROOF FINISHINGS – GREEN ROOFS

Item – Overhead and Profit Included	PC £	Labour hours	Labour £	Plant £	Material £	Unit	Total rate £
Extensive Green Roof System for Steep Pitched roofs with slopes from 25° up to 35°; Zinco Green Roof Systems Ltd. Pitched roofs with slopes from 25° up to 35°; Suitable for overlay with pre-grown cultivated Sedum Carpet. System height – 120 mm; system saturated weight – 155 kg/m^2; water storage capacity – 64 l/m^2 (depending on slope); Waterproofing layer must contain a root barrier as a separate root barrier cannot be used on pitched roofs							
Protection mat; non rotting needle fleeced synthetic mat for protection of the waterproofing layer and provision of additional water reservoir of 12 l/m^2							
WSM150 – laid to waterproofing layer	5.95	0.10	9.66	–	6.84	m^2	**16.50**
Drainage element; Stable Grid element 100 mm high made from recycled polyethylene. Grid dimension 625 mm with integrated connecting plug							
Zinco Drainage Element Georaster – Laid to WSM150	14.65	0.13	12.88	–	16.85	m^2	**29.73**
Zinco System Substrate Growing medium 'Heather with Lavender' filled to Georaster							
Zinco Premium Sedum Carpet Mat – Laid over Georaster	20.00	0.20	19.32	–	23.00	m^2	**42.32**
Shear barriers and brackets for pitched roofs; Zinco Green Roof Systems Ltd. Elements in pitched roof constructions transfer shear forces into the roof construction. Structural calculations required to establish the exact location and fixing of the shear barriers and brackets							
Zinco Eaves Profile and support brackets							
Eaves Profile TRP140 E; Stainless steel angle 140 mm high × 3.00 m long	15.47	0.08	8.05	–	17.79	m	**25.84**
Support bracket TSH100	22.20	0.50	48.30	–	25.53	nr	**73.83**
Shear fix bracket LF300	83.00	0.50	48.30	–	95.45	nr	**143.75**

28 ROOF FINISHINGS – GREEN ROOFS

Item – Overhead and Profit Included	PC £	Labour hours	Labour £	Plant £	Material £	Unit	Total rate £
INTENSIVE GREEN ROOFS							
Flat roofs up to 10°; Intensive Simple and Wildflower Green Roof System; Zinco Green Roof Systems Ltd; (not suitable for Inverted roofs); for planting with small shrubs, blooming perennials and wildflowers; can be combined with patios and walkways if required. System depth – 145 mm–200 mm; saturated weight – 145 kg/m^2– 245 kg/m^2; water storage capacity – 62 l/m^2 to 80 l/m^2. (ETA 13/0668)							
Non-rotting synthetic fibre mat for protection of the waterproofing layer and provision of additional water reservoir of 5 l/m^2							
Moisture mat/protection fleece SSM45	1.95	0.04	3.86	–	2.24	m^2	**6.10**
Pressure resistant drainage and water storage element made from recycled polyolefin laid to SSM45; 40 mm thick with diffusion openings and water storage capacity of 6 l/m^2.							
Fixodrain FD40-E	7.26	0.04	3.86	–	8.35	m^2	**12.21**
Zinco Filter Sheet SF – Rolled to drainage layer FD40-E	0.88	0.04	3.86	–	1.01	m^2	**4.87**
System Substrate; growing medium 'Heather with Lavender'; laid to Filter Fleece SF at 100 mm thickness up to 200 mm	125.00	0.20	19.32	–	143.75	m^3	**163.07**

28 ROOF FINISHINGS – GREEN ROOFS

Item – Overhead and Profit Included	PC £	Labour hours	Labour £	Plant £	Material £	Unit	Total rate £
Flat roofs up to 10°; Intensive Simple and Wildflower Green Roof System; Zinco Green Roof Systems Ltd; (suitable for Inverted roofs); for planting with small shrubs, blooming perennials and wildflowers; can be combined with patios and walkways if required. System depth – 145 mm–200 mm; saturated weight – 145 kg/m^2– 245 kg/m^2; water storage capacity – 62 l/m^2 to 80 l/m^2. Water repellent, air permeable membrane used as a diffusionable separation layer; laid above waterproofing and separation membrane; rolled over the insulation.							
Separation Layer; TGV21	0.82	0.04	3.86	–	0.94	m^2	**4.80**
Pressure resistant drainage and water storage element made from recycled polyolefin laid to SSM45; 40 mm thick with diffusion openings and water storage capacity of 6 l/m^2.							
Fixodrain FD40-E; laid to TGV21	6.60	0.04	3.86	–	7.59	m^2	**11.45**
Zinco Filter Sheet SF – Rolled to drainage layer FD40-E	0.88	0.04	3.86	–	1.01	m^2	**4.87**
System Substrate; growing medium 'Heather with Lavender'; laid to Filter Fleece SF at 100 mm thickness up to 200 mm	125.00	0.67	64.40	–	143.75	m^3	**208.15**
Flat roofs up to 10°; Intensive Simple and Wildflower Green Roof System; Zinco Green Roof Systems Ltd; (suitable for Inverted roofs); for planting with lawn, shrubs, small trees, and perennials; can be combined with patios and walkways if required. System depth – min 160 mm; saturated weight – 342 kg/m^2; water storage capacity – 110 l/m^2. Water repellent, air permeable membrane used as a diffusionable separation layer; laid above waterproofing and separation membrane; rolled over the insulation.							
Separation Layer; TGV21	0.82	0.04	3.86	–	0.94	m^2	**4.80**

28 ROOF FINISHINGS – GREEN ROOFS

Item – Overhead and Profit Included	PC £	Labour hours	Labour £	Plant £	Material £	Unit	Total rate £
INTENSIVE GREEN ROOFS – cont							
Flat roofs up to 10° – cont							
Pressure resistant drainage and water storage element made from thermoformed hard plastic; laid to TGV21; 60 mm thick filled with Zincolit Plus mineral aggregate							
Zincolit Plus; mineral aggregate infill; 27 l/m^2	2.97	0.03	2.42	–	3.42	m^2	**5.84**
Filter Sheet SF – Rolled to drainage layer FD 60	0.88	0.04	3.86	–	1.01	m^2	**4.87**
System Substrate Growing medium; 'Roof Garden' laid to Filter Fleece SF min. 200 mm thick dependent on planting required and depth capacities	125.00	1.00	96.60	–	143.75	m^3	**240.35**
Flat roofs up to 10°; Intensive Simple and Wildflower Green Roof System; Zinco Green Roof Systems Ltd; (suitable for Inverted roofs); for planting with lawn, shrubs, small trees, and perennials; System depth – min 270 mm; saturated weight – 365 kg/m^2; water storage capacity – 135 l/m^2. (ETA 13/0668)							
Moisture and protection mat; Synthetic fibre resistant to mechanical stress resistant for protection of waterproofing layer and provision of additional extra water reservoir of 4 l/m^2							
Zinco Moisture mat/protection fleece ISM50.	5.61	0.04	3.86	–	6.45	m^2	**10.31**
Zinco Drainage Element FD 60 – Laid to ISM50	13.64	0.04	3.86	–	15.69	m^2	**19.55**
Zincolit Plus mineral aggregate infill 27 l/m^2	2.97	0.03	2.42	–	3.42	m^2	**5.84**
Filter Sheet SF – Rolled to drainage layer FD 60	0.88	0.04	3.86	–	1.01	m^2	**4.87**
System Substrate Growing medium; 'Roof Garden' laid to Filter Fleece SF min. 200 mm thick dependent on planting required and depth capacities	125.00	0.67	64.40	–	143.75	m^3	**208.15**

28 ROOF FINISHINGS – GREEN ROOFS

Item – Overhead and Profit Included	PC £	Labour hours	Labour £	Plant £	Material £	Unit	Total rate £
LAWN Green Roof System; Zinco Green Roof Systems Ltd. Lawns on roofs where build-up depth is restricted; Flat roofs – 0° up to 5°; Lawn surface treatment; System depth 150 mm system saturated weight min 165 kg/m². System water storage capacity; from 65 l/m²							
Protection Layer – polypropylene mat, mechanical stress resistant for protection of the waterproofing layer							
Zinco Filter sheet PV	2.69	0.04	3.86	–	3.09	m²	6.95
Drainage and distribution element; Water storage, distribution and drainage element made from thermoformed recycled hard plastic; 45 mm thick and preformed with clipping points for Zinco Irrigation Dripperline 100-L1							
Zinco Aquatec 45 – laid to Filter Sheet PV	18.30	0.07	6.44	–	21.05	m²	27.49
Drippers; clipped into Aquatec 45 and linked to Zinco irrigation manager BM–4							
Zinco Dripperline 100-L1	2.69	0.07	6.44	–	3.09	m²	9.53
Programmable Irrigation management unit in lockable steel box for automated irrigation of lawn and green roofs; linked to dripperlines 100-L1							
Zinco Irrigation Manager BM–4	1527.92	2.00	193.20	–	1977.11	nr	2170.31
Water distributing polyester fleece with capillary effective fibres; laid over Aquatec 45 containing dripperlines 100-L1							
Zinco Wicking Mat DV 40	9.85	0.03	2.76	–	11.33	m²	14.09
System Substrate Growing medium; Minimum thickness 100 mm up to 150 mm laid over Wicking mat DV 40							
Zinco LAWN	18.75	0.07	6.44	–	21.56	m²	28.00
Top dressing; 15 mm layer laid over LAWN substrate							
ZINCOHUM	1.43	0.01	1.29	–	1.64	m²	2.93

28 ROOF FINISHINGS – GREEN ROOFS

Item – Overhead and Profit Included	PC £	Labour hours	Labour £	Plant £	Material £	Unit	Total rate £
LANDSCAPE OPTIONS TO GREEN ROOF SYSTEMS							
Green roof drainage							
Outlet inspection chambers							
150 mm deep	–	–	–	–	–	nr	77.00
250 mm deep	–	–	–	–	–	nr	93.50
350 mm deep	–	–	–	–	–	nr	104.50
Green roof drainage; HydroBlox; Plastic composite solution sheets with high permeability laid to green roof surface to drain water to drainage outlets (not included) on roof surface;							
laid in conjuction with reservoir boards							
45 mm thick; 1003 l/m permeability; units 1.00 m × 220 mm wide × 45 mm thick (4.55 planks /m²)	145.60	0.08	5.36	–	167.44	m²	172.80
Adjustable pedestals to rooftops for decking and pavings							
Decking; 300 mm joists at 1 pedestal per linear metre (3.3/m²) along the length of the joists;							
17 mm high	2.81	0.08	5.31	–	3.23	m²	8.54
28 mm high	3.79	0.08	5.31	–	4.36	m²	9.67
175–285 mm high	23.03	0.08	5.43	–	26.48	m²	31.91
285–400 mm high	39.86	0.08	5.43	–	45.84	m²	51.27
355–515 mm high	49.01	0.08	5.43	–	56.36	m²	61.79
465–625 mm high	56.69	0.09	5.69	–	65.19	m²	70.88
545–740 mm high	65.83	0.09	5.85	–	75.70	m²	81.55
645–850 mm high	73.52	0.09	5.85	–	84.55	m²	90.40
720–960 mm high	82.67	0.09	5.85	–	95.07	m²	100.92
830–1070 mm high	90.35	0.09	5.85	–	103.90	m²	109.75
Decking; 400 mm joists at 1 pedestal per linear metre (2.5/m²) along the length of the joists;							
17 mm high	2.13	0.07	4.30	–	2.45	m²	6.75
28 mm high	2.88	0.07	4.30	–	3.31	m²	7.61
175–285 mm high	17.45	0.07	4.60	–	20.07	m²	24.67
285–400 mm high	30.20	0.07	4.60	–	34.73	m²	39.33
355–515 mm high	37.13	0.07	4.60	–	42.70	m²	47.30
465–625 mm high	42.95	0.08	4.96	–	49.39	m²	54.35
545–740 mm high	49.88	0.08	4.96	–	57.36	m²	62.32
645–850 mm high	55.70	0.08	4.96	–	64.06	m²	69.02
720–960 mm high	62.63	0.08	4.96	–	72.02	m²	76.98
830–1070 mm high	68.45	0.08	4.96	–	78.72	m²	83.68

28 ROOF FINISHINGS – GREEN ROOFS

Item – Overhead and Profit Included	PC £	Labour hours	Labour £	Plant £	Material £	Unit	Total rate £
Adjustable pedestals to rooftops for decking and pavings; Note other heights available up to 1.07 m							
Paving slabs; laid at uniform sizes with one pedestal per corner of each paving slab; paving slab sizes; laid on adjustable pedestal 28 mm high above the slab							
300 × 300 mm	19.55	0.50	32.20	–	22.48	m²	**54.68**
400 × 400 mm	11.50	0.50	32.20	–	13.23	m²	**45.43**
450 × 450 mm	8.63	0.50	32.20	–	9.92	m²	**42.12**
600 × 450 mm with additional central support per slab	7.59	0.33	21.46	–	8.73	m²	**30.19**
900 × 600 mm with additional central support per slab	4.37	0.25	16.10	–	5.03	m²	**21.13**
Paving slabs; laid at uniform sizes with one pedestal per corner of each paving slab; paving slab sizes; laid on adjustable pedestal 285–400 mm high above the slab							
300 × 300 mm	205.36	0.63	40.25	–	236.16	m²	**276.41**
400 × 400 mm	120.80	0.63	40.25	–	138.92	m²	**179.17**
450 × 450 mm	90.60	0.63	40.25	–	104.19	m²	**144.44**
600 × 450 mm with additional central support per slab	79.73	0.42	26.83	–	91.69	m²	**118.52**
900 × 600 mm with additional central support per slab	45.90	0.31	20.13	–	52.78	m²	**72.91**
Paving slabs; laid at uniform sizes with one pedestal per corner of each paving slab; paving slab sizes; laid on adjustable pedestal 645–850 mm high above the slab							
300 × 300 mm	378.76	0.63	40.25	–	435.57	m²	**475.82**
400 × 400 mm	222.80	0.63	40.25	–	256.22	m²	**296.47**
450 × 450 mm	167.10	0.63	40.25	–	192.16	m²	**232.41**
600 × 450 mm with additional central support per slab	147.05	0.42	26.83	–	169.11	m²	**195.94**
900 × 600 mm with additional central support per slab	84.66	0.31	20.13	–	97.36	m²	**117.49**

28 ROOF FINISHINGS – GREEN ROOFS

Item – Overhead and Profit Included	PC £	Labour hours	Labour £	Plant £	Material £	Unit	Total rate £
BLUE ROOFS							
Blue roof systems; ZinCo Green roof systems Ltd; (Stormwater management roof); For managment of stormwater under a green roof system; build-up of Blue roof system over roof surface with water proofing (not included) and root resistant protection							
Extensive Blue roof; Build-up height (inclusive of green roof components, priced separately: 200 mm; Saturated weight including plants 220 kg/m^2; Water storage capacity 114 l/m^2; water retention height 60 mm							
Root barrier WSF 40	2.31	0.02	1.08	–	2.66	m^2	**3.74**
Filter sheet PV; above and below the water retention unit (2 m^2/m^2)	5.39	0.03	2.14	–	6.20	m^2	**8.34**
Retention spacer RS 60	18.00	0.25	16.10	–	20.70	m^2	**36.80**
Zinco Filter Sheet SF – Rolled to drainage layer FD40-E	0.88	0.04	3.86	–	1.01	m^2	**4.87**
Optional retention spacer units							
Retention spacer RSX 100; Water retention capacity 95 l/m^2; Total water storage of system 154 l/m^2	37.00	0.29	18.40	–	42.55	m^2	**60.95**
Flow control; Adjustable flow controller to be installed on flat roofs over water outlets with a contact flange; including a suitable inspection chamber; suitable for installation on Stormwater Management Roofs (Blue Roofs) with rainwater attenuation up to ca. 85 mm. Adjustable discharge rate							
RDS 28 and inspection chamber	320.00	1.00	32.20	–	368.00	nr	**400.20**
Intensive Green roof; instead of root barrier WSF 40							
Root barrier WSB–100-PO	13.47	0.03	1.61	–	15.49	m^2	**17.10**

29 DECORATION

Item – Overhead and Profit Included	PC £	Labour hours	Labour £	Plant £	Material £	Unit	Total rate £
PAINTING/CLEAR FINISHING							
Prepare; touch up primer; two undercoats and one finishing coat of gloss oil paint; on metal surfaces							
General surfaces							
girth exceeding 300 mm	3.27	0.67	21.47	–	3.76	m²	**25.23**
isolated surfaces; girth not exceeding 300 mm	9.91	0.40	12.88	–	11.40	m	**24.28**
isolated areas not exceeding 0.50 m² irrespective of girth	6.54	0.67	21.47	–	7.52	nr	**28.99**
Ornamental railings; each side measured separately							
girth exceeding 300 mm	3.27	0.57	18.40	–	3.76	m²	**22.16**
Prepare; one coat primer; two finishing coats of gloss paint; on wood surfaces							
New wood surfaces							
girth exceeding 300 mm	2.67	0.80	25.76	–	3.07	m²	**28.83**
isolated surfaces; girth not exceeding 300 mm	0.89	1.00	32.20	–	1.02	m	**33.22**
isolated areas not exceeding 0.50 m² irrespective of girth	3.22	0.50	16.10	–	3.70	nr	**19.80**
Previously painted wood surfaces							
girth exceeding 300 mm	1.41	0.50	16.10	–	1.62	m²	**17.72**
isolated surfaces; girth not exceeding 300 mm	0.47	0.67	21.47	–	0.54	m	**22.01**
isolated areas not exceeding 0.50 m² irrespective of girth	2.82	0.40	12.88	–	3.24	nr	**16.12**
Fences and sheds; prepare; two coats of Protek wood preserver on wood surfaces							
Planed surfaces							
girth exceeding 300 mm	0.45	0.07	2.15	–	0.52	m²	**2.67**
isolated surfaces; girth not exceeding 300 mm	0.15	0.17	5.37	–	0.17	m	**5.54**
isolated areas not exceeding 0.50 m² irrespective of girth	0.45	0.25	8.05	–	0.52	nr	**8.57**
Sawn surfaces							
girth exceeding 300 mm	0.17	0.10	3.22	–	0.20	m²	**3.42**
isolated surfaces; girth not exceeding 300 mm	2.27	0.17	5.37	–	2.61	m	**7.98**
isolated areas not exceeding 0.50 m² irrespective of girth	0.68	0.25	8.05	–	0.78	nr	**8.83**

29 DECORATION

Item – Overhead and Profit Included	PC £	Labour hours	Labour £	Plant £	Material £	Unit	Total rate £
PAINTING/CLEAR FINISHING – cont							
Fences and sheds; prepare; two coats of Sadolin wood preserver on wood surfaces							
Planed surfaces							
girth exceeding 300 mm	2.43	0.07	2.15	–	2.79	m²	**4.94**
isolated surfaces; girth not exceeding 300 mm	0.81	0.17	5.37	–	0.93	m	**6.30**
isolated areas not exceeding 0.50 m² irrespective of girth	1.21	0.25	8.05	–	1.39	nr	**9.44**
Sawn surfaces							
girth exceeding 300 mm	1.09	0.10	3.22	–	1.25	m²	**4.47**
isolated surfaces; girth not exceeding 300 mm	1.29	0.17	5.37	–	1.48	m	**6.85**
isolated areas not exceeding 0.50 m² irrespective of girth	1.94	0.25	8.05	–	2.23	nr	**10.28**
Prepare; proprietary solution primer; two coats of dark stain; on wood surfaces							
General surfaces							
girth exceeding 300 mm	2.24	0.17	5.37	–	2.58	m²	**7.95**
isolated surfaces; girth not exceeding 300 mm	0.75	0.13	4.03	–	0.86	m	**4.89**
Two coats Weathershield; to clean, dry surfaces; in accordance with manufacturer's instructions							
Brick or block walls							
girth exceeding 300 mm	0.66	0.40	12.88	–	0.76	m²	**13.64**
Cement render or concrete walls							
girth exceeding 300 mm	0.66	0.33	10.73	–	0.76	m²	**11.49**

33 DRAINAGE ABOVE GROUND

Item – Overhead and Profit Included	PC £	Labour hours	Labour £	Plant £	Material £	Unit	Total rate £
DRAINAGE TO ROOF DECKS AND PLANTERS							
Leca (light expanded clay aggregate); drainage aggregate to roofdecks and planters							
Placed mechanically to planters; average 100 mm thick; by mechanical plant tipped into planters							
aggregate size 10–20 mm; delivered in 30 m³ loads	50.47	0.40	12.88	6.23	58.04	m³	**77.15**
aggregate size 10–20 mm; delivered in 70 m³ loads	39.14	0.20	6.44	8.86	45.01	m³	**60.31**
Placed by light aggregate blower (maximum 40 m)							
aggregate size 10–20 mm; delivered in 30 m³ loads	50.47	0.14	4.60	–	63.47	m³	**68.07**
aggregate size 10–20 mm; delivered in 55 m³ loads on blower vehicle	42.72	0.14	4.60	–	49.13	m³	**53.73**
aggregate size 10–20 mm; delivered in 70 m³ loads	39.14	0.14	4.60	–	50.44	m³	**55.04**
By hand							
aggregate size 10–20 mm; delivered in 30 m³ loads	50.47	1.33	42.93	–	58.04	m³	**100.97**
aggregate size 10–20 mm; delivered in 70 m³ loads	39.14	1.33	42.93	–	45.01	m³	**87.94**
Drainage boards laid to insulated slabs on roof decks; boards laid below growing medium and granulated drainage layer and geofabric (all not included) to collect and channel water to drainage outlets (not included)							
Zinco Floradrain; polyethylene irrigation/drainage layer; inclusive of geofabric laid over the surface of the drainage board							
Floradrain FD40; 0.96 × 2.08 m panels	11.00	0.05	1.61	–	13.98	m²	**15.59**
Floradrain FD60; 1.00 × 2.00 m panel	17.39	0.07	2.15	–	21.33	m²	**23.48**

34 DRAINAGE BELOW GROUND

Item – Overhead and Profit Included	PC £	Labour hours	Labour £	Plant £	Material £	Unit	Total rate £
CLARIFICATION ON LABOUR COSTS IN THIS SECTION							
General groundworks team Generally a three man team is used in this section; The column Labour hours reports team hours. The column Labour £ reports the total cost of the team for the unit of work shown							
3 man team	–	1.00	96.60	–	–	hr	**96.60**
TRENCHES/PIPEWAYS/PITS FOR BURIED ENGINEERING							
Excavating trenches; using 3 tonne tracked excavator; to receive pipes; grading bottoms; earthwork support; filling with excavated material to within 150 mm of finished surfaces and compacting; completing fill with topsoil; disposal of surplus soil Services not exceeding 200 mm nominal size							
average depth of run not exceeding 0.50 m	1.33	0.12	3.86	1.20	1.53	m	**6.59**
average depth of run not exceeding 0.75 m	1.33	0.16	5.24	1.66	1.53	m	**8.43**
average depth of run not exceeding 1.00 m	1.33	0.28	9.12	2.90	1.53	m	**13.55**
average depth of run not exceeding 1.25 m	1.33	0.38	12.34	3.90	1.53	m	**17.77**
Excavating trenches; using 3 tonne tracked excavator; to receive pipes; grading bottoms; earthwork support; filling with imported granular material (6F2) and compacting; disposal of surplus soil Services not exceeding 200 mm nominal size							
average depth of run not exceeding 0.50 m	3.24	0.09	2.79	0.84	9.73	m	**13.36**
average depth of run not exceeding 0.75 m	4.86	0.11	3.48	1.05	14.58	m	**19.11**
average depth of run not exceeding 1.00 m	6.48	0.14	4.50	1.37	19.43	m	**25.30**
average depth of run not exceeding 1.25 m	8.10	0.23	7.36	2.32	24.29	m	**33.97**

34 DRAINAGE BELOW GROUND

Item – Overhead and Profit Included	PC £	Labour hours	Labour £	Plant £	Material £	Unit	Total rate £
Excavating trenches; using 3 tonne tracked excavator; to receive pipes; grading bottoms; earthwork support; filling with lean mix concrete; disposal of surplus soil							
Services not exceeding 200 mm nominal size							
average depth of run not exceeding 0.50 m	13.05	0.11	3.43	0.44	22.20	m	**26.07**
average depth of run not exceeding 0.75 m	19.57	0.13	4.19	0.54	33.29	m	**38.02**
average depth of run not exceeding 1.00 m	26.10	0.17	5.37	0.72	44.40	m	**50.49**
average depth of run not exceeding 1.25 m	32.63	0.23	7.24	1.08	55.49	m	**63.81**
Earthwork support; providing support to opposing faces of excavation; moving along as work proceeds							
Maximum depth not exceeding 2.00 m							
trenchbox; distance between opposing faces not exceeding 2.00 m	–	0.80	25.76	35.39	–	m	**61.15**
timber; distance between opposing faces not exceeding 500 mm	–	0.50	32.20	–	2.67	m	**34.87**
CHANNELS							
Best quality vitrified clay half section channels; Hepworth Plc; bedding and jointing in cement: mortar (1:2)							
Channels; straight							
100 mm	11.80	0.80	25.76	–	17.17	m	**42.93**
150 mm	19.62	1.00	32.20	–	26.16	m	**58.36**
225 mm	44.05	1.35	43.47	–	54.26	m	**97.73**
300 mm	90.44	1.80	57.96	–	107.61	m	**165.57**
Bends; 15, 30, 45 or 90°							
100 mm bends	10.61	0.75	24.15	–	14.01	nr	**38.16**
150 mm bends	18.35	0.90	28.98	–	23.82	nr	**52.80**
225 mm bends	71.09	1.20	38.64	–	85.35	nr	**123.99**
300 mm bends	144.93	1.10	35.42	–	171.18	nr	**206.60**

34 DRAINAGE BELOW GROUND

Item – Overhead and Profit Included	PC £	Labour hours	Labour £	Plant £	Material £	Unit	Total rate £
CHANNELS – cont							
Best quality vitrified clay channels; Hepworth Plc; bedding and jointing in cement: mortar (1:2)							
Branch bends; 15, 30, 45 or 90°; left or right hand							
100 mm	10.61	0.75	24.15	–	14.01	nr	**38.16**
150 mm	18.35	0.90	28.98	–	23.82	nr	**52.80**
EXCAVATION FOR DRAINAGE							
Machine excavation							
Excavating pits; starting from ground level; works exclude for earthwork retention							
maximum depth not exceeding 1.00 m	–	0.50	16.10	22.57	–	m³	**38.67**
maximum depth not exceeding 2.00 m	–	0.50	16.10	36.40	–	m³	**52.50**
maximum depth not exceeding 4.00 m	–	0.50	16.10	89.64	–	m³	**105.74**
Disposal of excavated material; depositing on site in permanent spoil heaps; average 50 m	–	0.04	1.35	2.70	–	m³	**4.05**
Filling to excavations; obtained from on site spoil heaps; average thickness not exceeding 0.25 m	–	0.13	4.29	7.97	–	m³	**12.26**
Hand excavation							
Excavating pits; starting from ground level; works exclude for earthwork retention							
maximum depth not exceeding 1.00 m	–	1.20	77.28	–	–	m³	**77.28**
maximum depth not exceeding 2.00 m	–	1.50	96.60	–	–	m³	**96.60**
Works to pits or drainage excavations							
Surface treatments; compacting; bottoms of excavations	–	0.05	1.61	–	–	m²	**1.61**
Earthwork support							
Manhole boxes to excavations (not included); Timber plywood with timber braces to shore excavations to 1.00 m deep; opposing faces distance less than 1.00 m							
1.00 m deep	–	0.50	32.20	–	16.19	m²	**48.39**
2.00 m deep	–	1.00	64.40	–	16.19	m²	**80.59**

34 DRAINAGE BELOW GROUND

Item – Overhead and Profit Included	PC £	Labour hours	Labour £	Plant £	Material £	Unit	Total rate £
Trench boxes; A-Plant							
Manhole box soil retention Placement of manhole protection system to 3.0 m deep to enable earthwork retention; maximum distance between opposing faces	–	0.50	48.30	191.61	–	week	**239.91**
1.8 m	–	0.50	48.30	259.46	–	week	**307.76**
GULLIES AND INTERCEPTION TRAPS							
Intercepting traps; Hepworth Plc Vitrified clay; inspection arms; brass stoppers; iron levers; chains and staples; galvanized; staples cut and pinned to brickwork; cement: mortar (1:2) joints to vitrified clay pipes and channels; bedding and surrounding in concrete; 11.50 N/mm^2–40 mm aggregate; cutting and fitting brickwork; making good facings							
100 mm inlet; 100 mm outlet	127.61	3.00	96.60	–	164.90	nr	**261.50**
150 mm inlet; 150 mm outlet	184.00	2.00	64.40	–	241.65	nr	**306.05**
GULLIES; CONCRETE; FP MCCANN							
Concrete road gullies; trapped; cement: mortar (1:2) joints to concrete pipes; bedding and surrounding in concrete; 11.50 N/mm^2–40 mm aggregate; rodding eye; stoppers							
375 mm dia. × 750 mm deep	56.09	3.00	193.20	–	92.18	nr	**285.38**
375 mm dia. × 900 mm deep	58.39	3.00	193.20	–	94.83	nr	**288.03**
450 mm dia. × 750 mm deep	58.41	3.00	193.20	–	94.85	nr	**288.05**
450 mm dia. × 900 mm deep	63.28	3.00	193.20	–	100.45	nr	**293.65**
450 mm dia. × 1.05 m deep	68.21	3.50	225.40	–	106.12	nr	**331.52**
450 mm dia. × 1.20 m deep	87.72	4.00	257.60	–	128.56	nr	**386.16**
Accessories for clay gullies							
coverslab	24.96	0.25	16.10	–	28.70	nr	**44.80**
seating rings	26.38	–	–	–	30.34	nr	**30.34**

34 DRAINAGE BELOW GROUND

Item – Overhead and Profit Included	PC £	Labour hours	Labour £	Plant £	Material £	Unit	Total rate £
GULLIES AND INTERCEPTION TRAPS – cont							
Gullies; vitrified clay; Hepworth Plc; bedding in concrete; 11.50 N/mm²–40 mm aggregate							
Yard gullies (mud); trapped; domestic duty (up to 1 tonne)							
100 mm outlet; 100 mm dia.; 225 mm internal width; 585 mm internal depth	200.51	3.50	112.70	–	231.39	nr	**344.09**
150 mm outlet; 100 mm dia.; 225 mm internal width; 585 mm internal depth	200.51	3.50	112.70	–	231.39	nr	**344.09**
Yard gullies (mud); trapped; medium duty (up to 5 tonnes)							
100 mm outlet; 100 mm dia.; 225 mm internal width; 585 mm internal depth	283.47	3.50	112.70	–	326.80	nr	**439.50**
150 mm outlet; 100 mm dia.; 225 mm internal width; 585 mm internal depth	310.59	3.50	112.70	–	357.98	nr	**470.68**
Combined filter and silt bucket for yard gullies							
225 mm dia.	72.51	–	–	–	83.39	nr	**83.39**
Road gullies; trapped with rodding eye							
100 mm outlet; 300 mm internal dia.; 600 mm internal depth	–	3.50	112.70	–	222.58	nr	**335.28**
150 mm outlet; 300 mm internal dia.; 600 mm internal depth	197.49	3.50	112.70	–	227.92	nr	**340.62**
150 mm outlet; 400 mm internal dia.; 750 mm internal depth	229.05	3.50	112.70	–	264.21	nr	**376.91**
150 mm outlet; 450 mm internal dia.; 900 mm internal depth	309.89	3.50	112.70	–	357.18	nr	**469.88**
Hinged gratings and frames for gullies; alloy							
193 mm for 150 mm dia. gully	–	–	–	–	55.27	nr	**55.27**
120 × 120 mm	–	–	–	–	19.67	nr	**19.67**
150 × 150 mm	–	–	–	–	35.55	nr	**35.55**
230 × 230 mm	–	–	–	–	65.00	nr	**65.00**
316 × 316 mm	–	–	–	–	172.32	nr	**172.32**
Hinged gratings and frames for gullies; cast iron							
265 mm for 225 mm dia. gully	–	–	–	–	110.42	nr	**110.42**
150 × 150 mm	–	–	–	–	35.55	nr	**35.55**
230 × 230 mm	–	–	–	–	65.00	nr	**65.00**
316 × 316 mm	–	–	–	–	172.32	nr	**172.32**

34 DRAINAGE BELOW GROUND

Item – Overhead and Profit Included	PC £	Labour hours	Labour £	Plant £	Material £	Unit	Total rate £
Universal gully trap Plastech Southern; bedding in concrete; 11.50 N/mm²–40 mm aggregate							
Universal gully fitting; comprising gully trap only							
110 mm outlet; 110 mm dia.; 205 mm internal depth	11.28	3.50	112.70	–	13.78	nr	**126.48**
Vertical inlet hopper; c/w plastic grate							
272 × 183 mm	16.84	0.25	8.05	–	19.37	nr	**27.42**
Sealed access hopper							
110 × 110 mm	35.64	0.25	8.05	–	40.99	nr	**49.04**
Universal gully PVC-u; Wavin Plastics Ltd; OsmaDrain system; accessories to universal gully trap							
Hoppers; backfilling with clean granular material; tamping; surrounding in lean mix concrete							
plain hopper; with 110 mm spigot; 150 mm long	12.39	0.40	12.88	–	14.74	nr	**27.62**
vertical inlet hopper; with 110 mm spigot; 150 mm long	16.84	0.40	12.88	–	19.37	nr	**32.25**
sealed access hopper; with 110 mm spigot; 150 mm long	35.64	0.40	12.88	–	40.99	nr	**53.87**
plain hopper; solvent weld to trap	9.40	0.40	12.88	–	10.81	nr	**23.69**
vertical inlet hopper; solvent wed to trap	16.16	0.40	12.88	–	18.58	nr	**31.46**
sealed access cover; PVC-u	18.15	0.10	3.22	–	20.87	nr	**24.09**
Gullies PVC-u; Wavin Plastics Ltd; OsmaDrain system; bedding in concrete; 11.50 N/mm²–40 mm aggregate							
Bottle gully; providing access to the drainage system for cleaning							
bottle gully; 228 × 228 × 317 mm deep	33.55	0.50	16.10	–	39.28	nr	**55.38**
sealed access cover; PVC-u; 217 × 217 mm	23.39	0.10	3.22	–	26.90	nr	**30.12**
grating; ductile iron; 215 × 215 mm	18.03	0.10	3.22	–	20.73	nr	**23.95**
bottle gully riser; 325 mm	4.29	0.50	16.10	–	6.44	nr	**22.54**
Yard gully; trapped; 300 mm dia. × 600 mm deep; including catchment bucket and ductile iron cover and frame; medium duty loading							
305 mm dia. × 600 mm deep	186.71	2.50	80.50	–	219.33	nr	**299.83**

34 DRAINAGE BELOW GROUND

Item – Overhead and Profit Included	PC £	Labour hours	Labour £	Plant £	Material £	Unit	Total rate £
GULLIES AND INTERCEPTION TRAPS – cont							
Kerbs to gullies							
One course Class B engineering bricks to four sides; rendering in cement: mortar (1:3); dished to gully gratings							
150 × 150 mm	1.79	0.33	10.73	–	2.54	nr	**13.27**
ACCESS COVERS AND FRAMES							
Load Classes for Access Covers							
FACTA (Fabricated Access Cover Trade Association) class:							
A – 0.5 tonne maximum slow moving wheel load							
AA – 1.5 tonne maximum slow moving wheel load							
AAA – 2.5 tonne maximum slow moving wheel load							
B – 5 tonne maximum slow moving wheel load							
C – 6.5 tonne maximum slow moving wheel load							
D – 11 tonne maximum slow moving wheel load							
Access covers and frames; solid top; galvanized; Steelway Brickhouse; Bristeel; bedding frame in cement: mortar (1:3); cover in grease and sand; clear opening sizes; base size shown in brackets (50 mm depth)							
FACTA AA; single seal							
450 × 450 mm (520 × 520 mm)	87.34	1.50	48.30	–	114.08	nr	**162.38**
600 × 450 mm (670 × 520 mm)	96.94	1.80	57.96	–	126.64	nr	**184.60**
600 × 600 mm (670 × 670 mm)	103.64	2.00	64.40	–	137.37	nr	**201.77**
FACTA AA; double seal							
450 × 450 mm (560 × 560 mm)	141.89	1.50	48.30	–	176.81	nr	**225.11**
600 × 450 mm (710 × 560 mm)	155.90	1.80	57.96	–	194.44	nr	**252.40**
600 × 600 mm (710 × 710 mm)	170.10	2.00	64.40	–	213.80	nr	**278.20**
FACTA AAA; single seal							
450 × 450 mm (520 × 520 mm)	134.52	1.50	48.30	–	168.34	nr	**216.64**
600 × 450 mm (670 × 520 mm)	145.18	1.80	57.96	–	182.11	nr	**240.07**
600 × 600 mm (670 × 670 mm)	154.68	2.00	64.40	–	196.06	nr	**260.46**

34 DRAINAGE BELOW GROUND

Item – Overhead and Profit Included	PC £	Labour hours	Labour £	Plant £	Material £	Unit	Total rate £
FACTA AAA; double seal							
450 mm × 450 mm (560 × 560)	171.61	1.50	48.30	–	210.99	nr	**259.29**
600 mm × 450 mm (710 × 560)	187.11	1.80	57.96	–	230.33	nr	**288.29**
600 mm × 600 mm (710 × 710)	208.38	2.00	64.40	–	257.82	nr	**322.22**
FACTA B; single seal							
450 × 450 mm (520 × 520 mm)	112.91	1.50	48.30	–	143.49	nr	**191.79**
600 × 450 mm (670 × 520 mm)	140.97	1.80	57.96	–	177.27	nr	**235.23**
600 × 600 mm (670 × 670 mm)	163.87	2.00	64.40	–	206.63	nr	**271.03**
FACTA B; double seal							
450 × 450 mm (560 × 560 mm)	162.94	1.50	48.30	–	201.02	nr	**249.32**
600 × 450 mm (710 × 560 mm)	181.39	1.80	57.96	–	223.76	nr	**281.72**
600 × 600 mm (710 × 710 mm)	217.83	2.00	64.40	–	268.69	nr	**333.09**
Access covers and frames; recessed; galvanized; Steelway Brickhouse; Bripave; bedding frame in cement: mortar (1:3); cover in grease and sand; clear opening sizes; base size shown in brackets (for block depths 50 mm, 65 mm, 80 mm or 100 mm)							
FACTA AA							
450 × 450 mm (562 × 562 mm)	253.18	1.50	48.30	–	304.80	nr	**353.10**
600 × 450 mm (712 × 562 mm)	277.72	1.80	57.96	–	334.53	nr	**392.49**
600 × 600 mm (712 × 712 mm)	287.86	2.00	64.40	–	348.55	nr	**412.95**
750 × 600 mm (862 × 712 mm)	321.19	2.40	77.28	–	389.07	nr	**466.35**
FACTA B							
450 × 450 mm (562 × 562 mm)	253.61	1.50	48.30	–	305.29	nr	**353.59**
600 × 450 mm (712 × 562 mm)	294.82	1.80	57.96	–	354.20	nr	**412.16**
600 × 600 mm (712 × 712 mm)	298.49	2.00	64.40	–	361.45	nr	**425.85**
750 × 600 mm (862 × 712 mm)	299.45	2.40	77.28	–	364.07	nr	**441.35**
FACTA D							
450 × 450 mm (562 × 562 mm)	284.23	1.50	48.30	–	340.50	nr	**388.80**
600 × 450 mm (712 × 562 mm)	301.99	1.80	57.96	–	362.45	nr	**420.41**
600 × 600 mm (712 × 712 mm)	312.01	2.00	64.40	–	376.99	nr	**441.39**
750 × 600 mm (862 × 712 mm)	379.71	2.40	77.28	–	456.37	nr	**533.65**
Extra over manhole frames and covers for							
filling recessed manhole covers with brick paviors; PC £500.00/ 1000	–	1.00	32.20	–	23.58	m^2	**55.78**
filling recessed manhole covers with vehicular paving blocks; PC £11.76/m^2	11.76	0.75	24.15	–	13.52	m^2	**37.67**
filling recessed manhole covers with concrete paving flags; PC £17.97/m^2	17.97	0.35	11.27	–	20.67	m^2	**31.94**

34 DRAINAGE BELOW GROUND

Item – Overhead and Profit Included	PC £	Labour hours	Labour £	Plant £	Material £	Unit	Total rate £
INSPECTION CHAMBERS AND MANHOLES							
Inspection chambers; in situ concrete; excavations; earthwork reinforcement all not included							
Beds; plain in situ concrete; 11.50 N/mm²–40 mm aggregate							
thickness not exceeding 150 mm	–	2.00	64.40	–	130.71	m³	**195.11**
thickness 150–450 mm	–	1.75	56.35	–	130.71	m³	**187.06**
Benchings in bottoms; plain in situ concrete; 25.50 N/mm²–20 mm aggregate							
thickness 150–450 mm	–	2.00	64.40	–	130.71	m³	**195.11**
Isolated cover slabs; reinforced In situ concrete; 21.00 N/mm²–20 mm aggregate							
thickness not exceeding 150 mm	113.66	4.00	128.80	–	130.71	m³	**259.51**
Fabric reinforcement; A193 (3.02 kg/ m²) in cover slabs	4.09	0.06	2.01	–	4.70	m²	**6.71**
Formwork to reinforced in situ concrete; isolated cover slabs							
soffits; horizontal	–	3.28	105.62	–	5.27	m²	**110.89**
height not exceeding 250 mm	–	0.97	31.23	–	3.12	m	**34.35**
Inspection chambers (House – HIC); precast concrete units; to excavations, earthwork reinforcement, in situ bases benchings all priced separately; includes cover slab							
Precast concrete inspection chamber units; FP McCann Ltd; bedding, jointing and pointing in cement: mortar (1:3); 600 × 450 mm internally							
600 mm deep	58.94	6.50	209.30	–	75.36	nr	**284.66**
900 mm deep	78.50	7.00	225.40	–	96.00	nr	**321.40**
Drainage chambers; FP McCann Ltd; 1200 × 750 mm; no base unit; depth of invert							
1.00 m deep	216.46	3.00	289.80	–	258.46	nr	**548.26**
1.50 m deep	276.87	3.67	354.20	–	333.66	nr	**687.86**
2.00 m deep	337.29	4.17	402.50	–	406.96	nr	**809.46**

34 DRAINAGE BELOW GROUND

Item – Overhead and Profit Included	PC £	Labour hours	Labour £	Plant £	Material £	Unit	Total rate £
Bases to chamber ring manholes; in situ concrete; to excavated ground (not included); on 50 mm C10 concrete blinding on 225 mm mass concrete to receive bottom manhole ring							
Dia. of manhole							
900 mm	–	4.00	257.60	–	41.35	nr	**298.95**
1.05 m	–	4.00	257.60	–	50.05	nr	**307.65**
1.20 m	–	4.50	289.80	–	69.90	nr	**359.70**
1.50 m	–	5.50	354.20	–	105.87	nr	**460.07**
Bottom manhole chamber ring; FP McCann Ltd; including pipe entrances to bases (not included); forming pipe entrances to manhole ring; seating and sealing to base; connecting pipes (not included); depth 1.00 m; dia.							
900 mm	91.62	3.00	289.80	108.66	114.91	nr	**513.37**
1.05 m	96.61	3.25	313.95	108.66	120.65	nr	**543.26**
1.20 m	117.95	4.00	386.40	135.83	145.19	nr	**667.42**
1.50 m	205.93	4.50	434.70	162.99	246.35	nr	**844.04**
Benchings to manholes; 400 mm concrete; dia. of manhole							
900 mm	–	0.75	48.30	–	50.43	nr	**98.73**
1.05 m	–	0.75	48.30	–	70.61	nr	**118.91**
1.20 m	–	0.90	57.96	–	90.78	nr	**148.74**
1.50 m	–	1.10	70.84	–	143.26	nr	**214.10**
Easi-Base Polypropylene liner; FP McCann Ltd; complete with prefabricated benching and channels as one unit; Easi-Base dia.							
1200 mm	499.26	1.00	64.40	54.33	574.15	nr	**692.88**
1500 mm	998.54	1.00	64.40	54.33	1148.32	nr	**1267.05**
Concrete manhole chamber rings; FP McCann Ltd; standard ring units to excavated ground (not included); sealed with bitumen tape; tongue and groove units installed to 1.00 m deep base units (not included); surrounding in concrete min 150 thick and backfilling; Note: The final depths shown below show operations onto the 1.00 m deep base rings above							
Internal dia. 900 mm; final manhole depth							
1.00 m deep	–	1.00	96.60	54.33	104.27	nr	**255.20**

34 DRAINAGE BELOW GROUND

Item – Overhead and Profit Included	PC £	Labour hours	Labour £	Plant £	Material £	Unit	Total rate £
INSPECTION CHAMBERS AND MANHOLES – cont							
Concrete manhole chamber rings – cont							
Internal dia. 1.05 mm; final manhole depth							
1.00 m deep	–	1.00	96.60	108.66	104.27	nr	**309.53**
2.00 m deep	96.61	2.50	241.50	190.15	310.11	nr	**741.76**
3.00 m deep	193.22	4.00	386.40	217.33	514.44	nr	**1118.17**
Internal dia. 1200 mm; final manhole depth							
1.00 m deep	–	1.00	96.60	162.99	104.27	nr	**363.86**
2.00 m deep	117.95	3.00	289.80	217.33	334.65	nr	**841.78**
3.00 m deep	235.90	4.00	386.40	325.98	563.52	nr	**1275.90**
Internal dia. 1500 mm							
1.00 m deep	–	2.00	193.20	162.99	104.27	nr	**460.46**
2.00 m deep	205.93	4.00	386.40	217.33	435.83	nr	**1039.56**
3.00 m deep	411.86	5.00	483.00	325.98	765.88	nr	**1574.86**
Concrete manhole chamber rings; Wide wall; FP McCann Ltd; 130 mm thick precast manholes; (concrete surround on backfill not required) to excavated ground (not included); Watertight, complete with sealant; to prepared base (or Easi-Base priced elsewhere)							
Internal dia. 1200 mm							
1.00 m deep	198.50	3.00	289.80	162.99	228.27	nr	**681.06**
2.00 m deep	397.00	4.00	386.40	217.33	456.55	nr	**1060.28**
3.00 m deep	595.50	6.00	579.60	271.65	684.83	nr	**1536.08**
Internal dia. 1500 mm							
1.00 m deep	312.79	3.00	289.80	217.33	369.25	nr	**876.38**
2.00 m deep	625.58	5.00	483.00	325.98	728.96	nr	**1537.94**
3.00 m deep	938.37	7.00	676.20	380.32	1088.66	nr	**2145.18**
Cover slabs for chambers or shaft sections; FP McCann Ltd; heavy duty							
900 mm dia. internally	93.09	0.67	21.45	–	107.05	nr	**128.50**
1050 mm dia. internally	99.91	2.00	64.40	19.92	114.90	nr	**199.22**
1200 mm dia. internally	121.17	1.00	32.20	19.92	139.35	nr	**191.47**
1500 mm dia. internally	220.53	1.00	32.20	19.92	253.61	nr	**305.73**
1800 mm dia. internally	362.09	2.00	64.40	44.83	416.40	nr	**525.63**

34 DRAINAGE BELOW GROUND

Item – Overhead and Profit Included	PC £	Labour hours	Labour £	Plant £	Material £	Unit	Total rate £
BRICKWORK							
Walls to manholes; engineering bricks; in cement: mortar (1:3)							
one brick thick	45.16	3.00	96.60	–	63.38	m²	**159.98**
one and a half brick thick	67.74	4.00	128.80	–	95.07	m²	**223.87**
two brick thick projection of footing or the like	90.32	4.80	154.56	–	126.75	m²	**281.31**
Extra over common or engineering bricks in any mortar for fair face; flush pointing as work proceeds; English bond walls or the like	–	0.13	4.29	–	–	m²	**4.29**
In situ finishings; cement: sand mortar (1:3); steel trowelled; 13 mm one coat work to manhole walls; to brickwork or blockwork base; over 300 mm wide	–	0.80	25.76	–	3.82	m²	**29.58**
Building into brickwork; ends of pipes; making good facings or renderings							
small	–	0.20	6.44	–	–	nr	**6.44**
large	–	0.30	9.66	–	–	nr	**9.66**
extra large	–	0.40	12.88	–	–	nr	**12.88**
extra large; including forming ring arch cover	–	0.50	16.10	–	–	nr	**16.10**
Inspection chambers; polypropylene; Hepworth Plc							
Mini access chamber; up to 600 mm deep; including cover and frame							
300 mm dia. × 600 mm deep; three 100/110 mm inlets	212.09	3.00	96.60	–	247.40	nr	**344.00**
Up to 1200 mm deep; including polymer cover and frame with screw down lid							
475 mm dia. × 940 mm deep; five 100/110 mm inlets; supplied with four stoppers in inlets	313.56	4.00	128.80	–	480.67	nr	**609.47**
Extra for square ductile iron cover and frame to 1 tonne load; screw down lid	154.54	–	–	–	61.15	nr	**61.15**
Step irons; drainage systems; malleable cast iron; galvanized; building into joints							
General purpose pattern; for one brick walls	5.05	0.17	5.47	–	5.81	nr	**11.28**

34 DRAINAGE BELOW GROUND

Item – Overhead and Profit Included	PC £	Labour hours	Labour £	Plant £	Material £	Unit	Total rate £
INSPECTION CHAMBERS AND MANHOLES – cont							
Step irons – cont							
Accessories in PVC-u							
110 mm screwed access cover	4.58	–	–	–	5.27	nr	**5.27**
110 mm rodding eye	13.25	0.50	16.10	–	21.06	nr	**37.16**
Kerbs; to gullies; in one course Class B engineering bricks; to four sides; rendering in cement: mortar (1:3); dished to gully gratings	3.01	1.00	32.20	–	4.61	nr	**36.81**
CAST IRON DRAINAGE							
Cast iron pipes and fittings; PAM; Timesaver							
100 mm cast iron pipes double spiggot; in trenches (trenches not included)							
laid straight	63.25	0.11	7.15	–	72.74	m	**79.89**
short runs under 3.00 m	85.53	0.17	10.74	–	98.36	m	**109.10**
Extra over 100 mm cast iron pipes for							
bends; medium radius – 87.5°–60°	128.04	0.33	10.63	–	147.25	nr	**157.88**
bends; medium radius – 45°–10°	131.40	0.33	10.63	–	151.11	nr	**161.74**
bends; long radius	169.01	0.25	8.05	–	194.36	nr	**202.41**
plain branch	155.27	0.25	8.05	–	178.56	nr	**186.61**
150 mm cast iron pipes; in trenches (trenches not included)							
laid straight	109.98	0.13	8.05	–	126.48	m	**134.53**
short runs under 3.00 m	136.96	0.25	16.10	–	157.50	m	**173.60**
Extra over 150 mm cast iron pipes for							
bends; medium radius – 87.5°–60°	152.17	0.33	10.63	–	175.00	nr	**185.63**
bends; medium radius – 45°–10°	196.79	0.33	10.63	–	226.31	nr	**236.94**
225 mm cast iron pipes; in trenches (trenches not included)							
laid straight	381.58	0.20	12.88	–	438.82	m	**451.70**
short runs under 3.00 m	381.58	0.50	32.20	–	438.82	m	**471.02**
Extra over 225 mm cast iron pipes for							
bends; medium radius – 87.5°–10°	482.54	0.50	16.10	–	554.92	nr	**571.02**

34 DRAINAGE BELOW GROUND

Item – Overhead and Profit Included	PC £	Labour hours	Labour £	Plant £	Material £	Unit	Total rate £
Cast iron inspection chambers; PAM; Timesaver; drainage systems bolted flat covers; bedding in cement: mortar (1:3); mechanical coupling joints							
100 × 100 mm							
one branch each side	297.81	1.50	96.60	–	421.19	nr	**517.79**
two branches each side	586.27	2.50	161.00	–	906.51	nr	**1067.51**
100 × 150 mm							
one branch each side	533.56	2.00	128.80	–	613.97	nr	**742.77**
two branch each side	1012.26	2.50	161.00	–	1164.48	nr	**1325.48**
150 × 150 mm							
one branch each side	489.64	2.50	161.00	–	749.58	nr	**910.58**
two branches each side	977.72	3.00	193.20	–	1403.93	nr	**1597.13**
Accessories in Cast iron							
Gullies; including concrete bed and surround							
Gully trap with surface access							
100 mm × 87.5°	279.55	1.00	64.40	–	359.08	nr	**423.48**
Gully trap with 225 mm inlet	246.36	1.00	64.40	–	320.91	nr	**385.31**
Trapless gully	274.68	1.00	64.40	–	353.48	nr	**417.88**
Gully trap square 225 × 225 mm							
with sediment trap and grate	544.17	1.00	64.40	–	663.39	nr	**727.79**
Deans trap gully 600 mm deep							
inclusive of sediment pan	637.98	1.00	64.40	–	771.27	nr	**835.67**
Road Gully including hinged grate							
and frame	734.13	3.00	193.20	–	881.84	nr	**1075.04**
LINEAR DRAINAGE							
Marshalls Plc; Mini Beany combined kerb and channel drainage system; to trenches (not included)							
Precast concrete drainage channel base; 185–385 mm deep; bedding, jointing and pointing in cement: mortar (1:3); on 150 mm deep concrete (ready mixed) foundation; including haunching with in situ concrete; 11.50 N/mm^2–40 mm aggregate one side; channels 250 mm wide × 1.00 m long							
straight; 1.00 m long	37.38	1.00	32.20	–	47.69	m	**79.89**
straight; 500 mm long	43.97	1.05	33.81	–	55.27	m	**89.08**
radial; 30/10 or 9/6 internal or							
external	40.62	1.33	42.93	–	51.42	m	**94.35**
angles 45 or 90°	117.29	1.00	32.20	–	139.59	nr	**171.79**

34 DRAINAGE BELOW GROUND

Item – Overhead and Profit Included	PC £	Labour hours	Labour £	Plant £	Material £	Unit	Total rate £
LINEAR DRAINAGE – cont							
Marshalls Plc – cont							
Mini Beany Top Block; perforated kerb unit to drainage channel above; natural grey							
straight	37.51	0.33	10.73	–	44.88	m	**55.61**
radial; 30/10 or 9/6 internal or external	28.94	0.50	16.10	–	35.03	m	**51.13**
angles 45 or 90°	116.76	0.50	16.10	–	136.02	nr	**152.12**
Mini Beany; outfalls; two section concrete trapped outfall with Mini Beany cast iron access cover and frame; to concrete foundation							
high capacity outfalls; silt box 150/225 mm outlet; two section trapped outfall silt box and cast iron access cover	358.62	1.00	32.20	–	858.57	nr	**890.77**
inline side or end outlet; outfall 150 mm; 2 section concrete trapped outfall; cast iron Mini Beany access cover and frame	309.92	1.00	32.20	–	357.36	nr	**389.56**
Ancillaries to Mini Beany							
end cap	23.44	0.25	8.05	–	26.96	nr	**35.01**
end cap outlets	61.00	0.25	8.05	–	70.15	nr	**78.20**
Slot and Channel drains							
Loadings for slot drains							
A15; 1.5 tonne; pedestrian							
B125; 12.5 tonne; domestic use							
C250; 25 tonne; car parks, supermarkets, industrial units							
D400; 40 tonne; highways							
E600; 60 tonne; forklifts							
Fully galvanized hexagonal format slot drains; Gatic Ltd; for paving or asphalt use; 269 mm depth							
'Facade drain' fixed to building facade to collect water falling from the building facade							
115 mm maximum flow rate 19.6 l/sec	49.00	0.33	21.46	–	56.35	m	**77.81**

34 DRAINAGE BELOW GROUND

Item – Overhead and Profit Included	PC £	Labour hours	Labour £	Plant £	Material £	Unit	Total rate £
Slotdrain channels; Hexagonally profiled fully galvanized slotdrain system; In trenches (not included) surrounded by concrete							
Pavslot 10 mm throat 100 mm deep	54.61	0.33	21.46	–	66.30	m	87.76
Pavslot 10 mm throat 150 mm deep	62.40	0.33	21.46	–	75.43	m	96.89
Gatic Access points at 1nr per 30 m	5.33	0.25	16.10	–	9.63	m	25.73
Gatic outlets	170.00	1.00	64.40	–	199.00	nr	263.40
Slot drains; ACO Technologies; laid to concrete bed C25 on compacted granular base on 200 mm deep concrete bed; haunched with 200 mm concrete surround; all in 750 × 430 mm wide trench with compacted 200 mm granular base surround (excavation and subbase not included)							
ACO MultiDrain M100PPD; recycled polypropylene drainage channel; range of gratings to complement installations which require discreet slot drainage							
142 mm wide × 150 mm deep	80.15	0.83	26.83	–	106.40	m	133.23
Accessories for M100PPD							
connectors; vertical outlet 110 mm	15.75	–	–	–	18.11	nr	18.11
connectors; vertical outlet 160 mm	13.44	–	–	–	15.46	nr	15.46
sump units; 110 mm	92.72	1.50	48.30	–	132.77	nr	181.07
universal gully and bucket; 440 × 440 × 1315 mm deep	380.26	3.00	96.60	–	486.31	nr	582.91
Channel drains; ACO Technologies; ACO MultiDrain MD polymer concrete channel drainage system; traditional channel and grate drainage solution							
ACO MultiDrain MD Brickslot; offset galvanized slot drain grating for M100PPD; load class C250							
brickslot; galvanized steel; 1.00 m	78.25	–	–	–	89.99	m	89.99
ACO MultiDrain MD Brickslot; offset slot drain grating; load class C250–400							
brickslot; galvanized steel; 1.00 m	109.25	1.20	38.64	–	125.64	m	164.28
brickslot; stainless steel; 1.00 m	168.26	1.20	38.64	–	387.00	m	425.64

34 DRAINAGE BELOW GROUND

Item – Overhead and Profit Included	PC £	Labour hours	Labour £	Plant £	Material £	Unit	Total rate £
LINEAR DRAINAGE – cont							
Channel drains – cont							
Accessories for MultiDrain MD							
sump unit; complete with							
sediment bucket and access unit	92.72	1.00	32.20	–	190.08	nr	**222.28**
end cap; closing piece	7.29	–	–	–	8.38	nr	**8.38**
end cap; inlet/outlet	6.37	–	–	–	7.33	nr	**7.33**
ACO Drainlock Gratings for **M100PPD and M100D system**							
A15 loading							
slotted galvanized steel	11.46	0.15	4.83	–	13.18	m	**18.01**
perforated galvanized steel	20.18	0.15	4.83	–	23.21	m	**28.04**
C250 loading							
Heelguard composite black;							
500 mm long with security locking	31.16	–	–	–	35.83	m	**35.83**
Intercept; ductile iron; 500 mm							
long	31.16	–	–	–	35.83	m	**35.83**
slotted galvanized steel; 1.00 m							
long	20.21	–	–	–	23.24	m	**23.24**
perforated galvanized steel;							
1.00 m long	49.34	–	–	–	56.74	m	**56.74**
mesh galvanized steel; 1.00 m							
long	44.60	–	–	–	51.29	m	**51.29**
Wade Ltd; stainless steel channel **drains; specialized applications;** **bespoke manufacture**							
Drain in stainless steel; c/w tie in lugs and inbuilt falls to 100 mm spigot outlet; stainless steel gratings							
NE channel; Ref 12430; secured							
gratings	432.60	1.20	38.64	–	511.72	m	**550.36**
PIPE LAYING							
Excavating trenches; to receive **pipes; grading bottoms to falls;** **backfilling with excavated** **material and compacting;** **disposal of surplus material off** **site; volumes allow for bedding** **materials which are priced** **separately below**							
Trenches 300 mm wide							
depth of pipe 750 mm	–	–	–	12.89	6.81	m³	**19.70**
depth of pipe 900 mm	–	–	–	14.73	8.60	m³	**23.33**
depth of pipe 1.20 m	–	–	–	15.47	12.20	m³	**27.67**
depth of pipe 1.50 m	–	–	–	17.19	15.80	m³	**32.99**
depth of pipe 2.00 m	–	–	–	10.79	21.79	m³	**32.58**

34 DRAINAGE BELOW GROUND

Item – Overhead and Profit Included	PC £	Labour hours	Labour £	Plant £	Material £	Unit	Total rate £
Trenches 300 mm wide							
depth of pipe 750 mm	–	–	–	4.30	4.49	m	**8.79**
depth of pipe 900 mm	–	–	–	5.22	5.39	m	**10.61**
depth of pipe 1.20 m	–	–	–	6.96	7.19	m	**14.15**
depth of pipe 1.50 m	–	–	–	8.71	8.99	m	**17.70**
depth of pipe 2.00 m	–	–	–	10.79	21.79	m	**32.58**
Earthwork support; providing support to opposing faces of excavation; moving along as work proceeds							
Maximum depth not exceeding 2.00 m							
distance between opposing faces not exceeding 2.00 m	–	0.80	25.76	32.13	–	m	**57.89**
Excavating trenches; using 3 tonne tracked excavator; to receive pipes; grading bottoms; earthwork support; filling with excavated material to within 150 mm of finished surfaces and compacting; completing fill with topsoil; disposal of surplus soil							
Services not exceeding 200 mm nominal size							
average depth of run not exceeding 0.50 m	1.33	0.12	3.86	1.20	1.53	m	**6.59**
average depth of run not exceeding 0.75 m	1.33	0.16	5.24	1.66	1.53	m	**8.43**
average depth of run not exceeding 1.00 m	1.33	0.28	9.12	2.90	1.53	m	**13.55**
average depth of run not exceeding 1.25 m	1.11	0.38	12.34	3.90	1.28	m	**17.52**
Granular beds to trenches; lay granular material to trenches excavated separately; to receive pipes (not included)							
300 mm wide × 100 mm thick							
reject sand	–	0.05	1.61	0.26	4.40	m	**6.27**
reject gravel	–	0.05	1.61	0.26	4.35	m	**6.22**
shingle 40 mm aggregate	–	0.05	1.61	0.26	5.54	m	**7.41**
sharp sand	4.63	0.05	1.61	0.26	5.32	m	**7.19**
300 mm wide × 150 mm thick							
reject sand	–	0.08	2.42	0.40	6.60	m	**9.42**
reject gravel	–	0.08	2.42	0.40	6.52	m	**9.34**
shingle 40 mm aggregate	–	0.08	2.42	0.40	8.31	m	**11.13**
sharp sand	6.94	0.08	2.42	0.40	7.98	m	**10.80**

34 DRAINAGE BELOW GROUND

Item – Overhead and Profit Included	PC £	Labour hours	Labour £	Plant £	Material £	Unit	Total rate £
PIPE LAYING – cont							
Excavating trenches; using 3 tonne tracked excavator; to receive pipes; grading bottoms; earthwork support; filling with imported granular material type 2 and compacting; disposal of surplus soil							
Services not exceeding 200 mm nominal size							
average depth of run not exceeding 0.50 m	3.24	0.09	2.79	0.84	10.92	m	**14.55**
average depth of run not exceeding 0.75 m	4.86	0.11	3.48	1.05	16.37	m	**20.90**
average depth of run not exceeding 1.00 m	6.48	0.14	4.50	1.37	21.83	m	**27.70**
average depth of run not exceeding 1.25 m	8.10	0.23	7.36	2.32	27.28	m	**36.96**
Excavating trenches; using 3 tonne tracked excavator; to receive pipes; grading bottoms; earthwork support; filling with concrete, ready mixed C10; disposal of surplus soil							
Services not exceeding 200 mm nominal size							
average depth of run not exceeding 0.50 m	13.38	0.11	3.43	0.44	22.58	m	**26.45**
average depth of run not exceeding 0.75 m	20.06	0.13	4.19	0.54	33.85	m	**38.58**
average depth of run not exceeding 1.00 m	26.75	0.17	5.37	0.72	45.14	m	**51.23**
average depth of run not exceeding 1.25 m	33.44	0.23	7.24	1.08	56.43	m	**64.75**
Earthwork support; providing support to opposing faces of excavation; moving along as work proceeds							
Maximum depth not exceeding 2.00 m							
trenchbox; distance between opposing faces not exceeding 2.00 m	–	0.80	25.76	32.13	–	m	**57.89**
timber; distance between opposing faces not exceeding 500 mm	–	0.20	12.88	–	0.03	m	**12.91**

34 DRAINAGE BELOW GROUND

Item – Overhead and Profit Included	PC £	Labour hours	Labour £	Plant £	Material £	Unit	Total rate £
Clay pipes and fittings; Hepworth Plc; Supersleve							
100 mm clay pipes; polypropylene slip coupling; in trenches (trenches not included)							
laid straight	9.27	0.25	8.05	–	15.61	m	**23.66**
short runs under 3.00 m	9.27	0.31	10.06	–	15.61	m	**25.67**
Extra over 100 mm clay pipes for							
bends; 15–90°; single socket	19.87	0.25	8.05	–	22.85	nr	**30.90**
junction; 45 or 90°; double socket	41.89	0.25	8.05	–	48.17	nr	**56.22**
slip couplings; polypropylene	6.88	0.08	2.68	–	7.91	nr	**10.59**
gully with P trap; 100 mm;							
154 × 154 mm plastic grating	48.36	1.00	32.20	–	89.22	nr	**121.42**
150 mm clay pipes; polypropylene slip coupling; in trenches (trenches not included)							
laid straight	29.20	0.30	9.66	–	33.58	m	**43.24**
short runs under 3.00 m	29.20	0.60	19.32	–	33.58	m	**52.90**
Extra over 150 mm clay pipes for							
bends; 15–90°	25.00	0.28	9.02	–	42.62	nr	**51.64**
junction; 45 or 90°; 100 × 150 mm	33.46	0.40	12.88	–	66.21	nr	**79.09**
junction; 45 or 90°; 150 × 150 mm	36.74	0.40	12.88	–	69.97	nr	**82.85**
slip couplings; polypropylene	12.05	0.05	1.61	–	13.86	nr	**15.47**
tapered pipe; 100–150 mm	37.84	0.50	16.10	–	43.52	nr	**59.62**
tapered pipe; 150–225 mm	96.54	0.50	16.10	–	111.02	nr	**127.12**
socket adaptor; connection to							
traditional pipes and fittings	25.36	0.33	10.63	–	43.02	nr	**53.65**
Accessories in clay							
access pipe; 150 mm	94.26	–	–	–	136.13	nr	**136.13**
rodding eye; 150 mm	90.17	0.50	16.10	–	109.51	nr	**125.61**
gully with P traps; 150 mm;							
154 × 154 mm plastic grating	106.38	0.80	25.76	–	138.53	nr	**164.29**
PVC-u pipes and fittings; Wavin Plastics Ltd; OsmaDrain; laid to and covered over with imported granular material							
110 mm PVC-u pipes; in trenches (trenches not included)							
laid straight	6.84	0.08	2.58	2.25	10.57	m	**15.40**
short runs under 3.00 m	7.69	0.12	3.86	2.25	11.56	m	**17.67**
Extra over 110 mm PVC-u pipes for							
bends; short radius	14.78	0.25	8.05	–	17.00	nr	**25.05**
bends; long radius	24.92	0.25	8.05	–	28.66	nr	**36.71**
junctions; equal; double socket	15.86	0.25	8.05	–	18.24	nr	**26.29**
slip couplings	7.70	0.25	8.05	–	8.86	nr	**16.91**
adaptors to clay	14.14	0.50	16.10	–	16.26	nr	**32.36**

34 DRAINAGE BELOW GROUND

Item – Overhead and Profit Included	PC £	Labour hours	Labour £	Plant £	Material £	Unit	Total rate £
PIPE LAYING – cont							
PVC-u pipes and fittings – cont							
160 mm PVC-u pipes; in trenches (trenches not included)							
laid straight	8.01	0.08	2.58	2.25	11.93	m	**16.76**
short runs under 3.00 m	15.22	0.12	3.86	2.25	20.22	m	**26.33**
Extra over 160 mm PVC-u pipes for							
bends; short radius	18.68	0.25	8.05	–	38.48	nr	**46.53**
bends; long radius	64.72	0.25	8.05	–	74.43	nr	**82.48**
junctions; single	75.90	0.33	10.73	–	87.29	nr	**98.02**
pipe coupler	9.34	0.05	1.61	–	10.74	nr	**12.35**
slip couplings PVC-u	9.34	0.05	1.61	–	10.74	nr	**12.35**
adaptors to clay	40.75	0.50	16.10	–	46.86	nr	**62.96**
level invert reducer	15.33	0.50	16.10	–	17.63	nr	**33.73**
spiggot	24.67	0.20	6.44	–	28.37	nr	**34.81**
CATCHWATER OR FRENCH DRAINS							
Catchwater or French drains							
Excavation; disposal to spoil heaps 100 m max; lining trench; wrapping trench with filter fabric; backfilling with shingle							
depth to 1.5 m	47.59	0.17	10.74	11.91	54.73	m^3	**77.38**
Pipe perforated to trench excavated prepared and backfilled separately							
80 mm Ø	0.73	0.01	0.32	–	0.84	m	**1.16**
100 mm Ø	1.17	0.01	0.32	–	1.35	m	**1.67**
160 mm Ø	0.71	0.01	0.32	–	0.82	m	**1.14**

34 DRAINAGE BELOW GROUND

Item – Overhead and Profit Included	PC £	Labour hours	Labour £	Plant £	Material £	Unit	Total rate £
HydroBlox surface and subsurface drains; Recycled plastic; water conveyance plank format as a French/land drain alternative and/or retrofit solution for existing drainage installations Sheet material to trenches providing 4560 litre/min/m² water permeability to trenches by trenching machine backfilled with excavated material							
Single unit 1.00 m × 220 mm deep × 45 mm thick × providing 1003 l/min permeability; backfilled with excavated material	32.00	0.13	7.51	0.36	36.80	m	44.67
Two units 1.00 m × 220 mm deep × 90 mm thick × providing 2006 l/min permeability; backfilled with excavated material	64.00	0.16	9.12	0.36	73.60	m	83.08
Four units 1.00 m × 220 mm deep × 180 mm thick × providing 4012 l/min permeability; backfilled with excavated material	128.00	0.20	11.81	0.36	147.20	m	159.37
DITCHING							
Ditching; clear silt and bottom ditch not exceeding 1.50 m deep; strim back vegetation; disposing to spoil heaps; by machine							
Up to 1.50 m wide at top	–	6.00	193.20	32.35	–	100 m	225.55
1.50–2.50 m wide at top	–	8.00	257.60	43.13	–	100 m	300.73
2.50–4.00 m wide at top	–	9.00	289.80	48.52	–	100 m	338.32
Ditching; clear only vegetation from ditch not exceeding 1.50 m deep; disposing to spoil heaps; by strimmer							
Up to 1.50 m wide at top	–	5.00	161.00	11.50	–	100 m	172.50
1.50–2.50 m wide at top	–	6.00	193.20	20.70	–	100 m	213.90
2.50–4.00 m wide at top	–	7.00	225.40	32.20	–	100 m	257.60
Ditching; clear silt from ditch not exceeding 1.50 m deep; trimming back vegetation; disposing to spoil heaps; by hand							
Up to 1.50 m wide at top	–	15.00	483.00	11.50	–	100 m	494.50
1.50–2.50 m wide at top	–	27.00	869.40	20.70	–	100 m	890.10
2.50–4.00 m wide at top	–	42.00	1352.40	32.20	–	100 m	1384.60

34 DRAINAGE BELOW GROUND

Item – Overhead and Profit Included	PC £	Labour hours	Labour £	Plant £	Material £	Unit	Total rate £
DITCHING – cont							
Ditching; excavating and forming ditch and bank to given profile (normally 45°); in loam or sandy loam; by machine							
Width 300 mm							
depth 600 mm	–	3.70	119.14	39.89	–	100 m	**159.03**
depth 900 mm	–	5.20	167.44	56.06	–	100 m	**223.50**
depth 1200 mm	–	7.20	231.84	77.63	–	100 m	**309.47**
depth 1500 mm	–	9.40	302.68	88.68	–	100 m	**391.36**
Width 600 mm							
depth 600 mm	–	–	–	370.53	–	100 m	**370.53**
depth 900 mm	–	–	–	549.83	–	100 m	**549.83**
depth 1200 mm	–	–	–	747.04	–	100 m	**747.04**
depth 1500 mm	–	–	–	1075.74	–	100 m	**1075.74**
Width 900 mm							
depth 600 mm	–	–	–	570.47	–	100 m	**570.47**
depth 900 mm	–	–	–	1005.11	–	100 m	**1005.11**
depth 1200 mm	–	–	–	1350.65	–	100 m	**1350.65**
depth 1500 mm	–	–	–	1673.38	–	100 m	**1673.38**
Width 1200 mm							
depth 600 mm	–	–	–	896.45	–	100 m	**896.45**
depth 900 mm	–	–	–	1331.09	–	100 m	**1331.09**
depth 1200 mm	–	–	–	1763.02	–	100 m	**1763.02**
depth 1500 mm	–	–	–	2259.06	–	100 m	**2259.06**
Width 1500 mm							
depth 600 mm	–	–	–	1105.62	–	100 m	**1105.62**
depth 900 mm	–	–	–	1685.32	–	100 m	**1685.32**
depth 1200 mm	–	–	–	2241.13	–	100 m	**2241.13**
depth 1500 mm	–	–	–	2808.88	–	100 m	**2808.88**
Extra for ditching in clay	–	–	–	–	–	20%	**–**
Ditching; excavating and forming ditch and bank to given profile (normal 45°); in loam or sandy loam; by hand							
Width 300 mm							
depth 600 mm	–	36.00	1159.20	–	–	100 m	**1159.20**
depth 900 mm	–	42.00	1352.40	–	–	100 m	**1352.40**
depth 1200 mm	–	56.00	1803.20	–	–	100 m	**1803.20**
depth 1500 mm	–	70.00	2254.00	–	–	100 m	**2254.00**
Width 600 mm							
depth 600 mm	–	56.00	1803.20	–	–	100 m	**1803.20**
depth 900 mm	–	84.00	2704.80	–	–	100 m	**2704.80**
depth 1200 mm	–	112.00	3606.40	–	–	100 m	**3606.40**
depth 1500 mm	–	140.00	4508.00	–	–	100 m	**4508.00**
Width 900 mm							
depth 600 mm	–	84.00	2704.80	–	–	100 m	**2704.80**
depth 900 mm	–	126.00	4057.20	–	–	100 m	**4057.20**
depth 1200 mm	–	168.00	5409.60	–	–	100 m	**5409.60**
depth 1500 mm	–	210.00	6762.00	–	–	100 m	**6762.00**

34 DRAINAGE BELOW GROUND

Item – Overhead and Profit Included	PC £	Labour hours	Labour £	Plant £	Material £	Unit	Total rate £
Width 1200 mm							
depth 600 mm	–	112.00	3606.40	–	–	100 m	**3606.40**
depth 900 mm	–	168.00	5409.60	–	–	100 m	**5409.60**
depth 1200 mm	–	224.00	7212.80	–	–	100 m	**7212.80**
depth 1500 mm	–	280.00	9016.00	–	–	100 m	**9016.00**
Extra for ditching in clay	–	–	–	–	–	50%	**–**
Extra for ditching in compacted soil	–	–	–	–	–	90%	**–**
Piped ditching							
Jointed concrete pipes; FP McCann Ltd; including bedding, haunching and topping with 150 mm concrete; 11.50 N/mm^2–40 mm aggregate; to existing ditch							
300 mm dia.	25.52	0.67	21.46	11.95	53.73	m	**87.14**
450 mm dia.	38.02	0.67	21.46	11.95	80.30	m	**113.71**
600 mm dia.	61.08	0.67	21.46	11.95	122.18	m	**155.59**
900 mm dia.	161.89	1.00	32.20	17.93	214.38	m	**264.51**
Jointed concrete pipes; FP McCann Ltd; including bedding, haunching and topping with 150 mm concrete; 11.50 N/mm^2–40 mm aggregate; to existing ditch							
1200 mm dia.	277.84	1.00	32.20	17.93	356.43	m	**406.56**
extra over jointed concrete pipes for bends to 45°	255.10	0.67	21.46	14.94	304.62	nr	**341.02**
extra over jointed concrete pipes for single junctions 300 mm dia.	174.77	0.67	21.46	14.94	212.24	nr	**248.64**
extra over jointed concrete pipes for single junctions 450 mm dia.	259.08	0.67	21.46	13.58	309.20	nr	**344.24**
extra over jointed concrete pipes for single junctions 600 mm dia.	419.93	0.67	21.46	13.58	494.18	nr	**529.22**
extra over jointed concrete pipes for single junctions 900 mm dia.	709.85	0.67	21.46	13.58	827.59	nr	**862.63**
extra over jointed concrete pipes for single junctions 1200 mm dia.	1084.40	0.67	21.46	13.58	1258.32	nr	**1293.36**
Outfalls							
Reinforced concrete outfalls to water course; flank walls; for 150 mm drain outlets; overall dimensions							
Small headwall FP McCann Ltd	745.90	4.00	386.40	54.33	941.59	m	**1382.32**
LAND DRAINS							
Market prices of backfilling materials							
Sand	41.76	–	–	–	41.76	m^3	**41.76**
Gravel rejects	39.60	–	–	–	39.60	m^3	**39.60**
Imported topsoil (allowing for 20% settlement)	29.66	–	–	–	29.66	m^3	**29.66**

34 DRAINAGE BELOW GROUND

Item – Overhead and Profit Included	PC £	Labour hours	Labour £	Plant £	Material £	Unit	Total rate £
LAND DRAINS – cont							
Land drainage; calculation table							
For calculation of drainage per hectare the following table can be used; rates show the lengths of drains per unit and not the value							
Lateral drains							
10.00 m centres	–	–	–	–	–	m/ha	**1000.00**
15.00 m centres	–	–	–	–	–	m/ha	**650.00**
25.00 m centres	–	–	–	–	–	m/ha	**400.00**
30.00 m centres	–	–	–	–	–	m/ha	**330.00**
Main drains							
1 nr (at 100 m centres)	–	–	–	–	–	m/ha	**100.00**
2 nr (at 50 m centres)	–	–	–	–	–	m/ha	**200.00**
3 nr (at 33.3 m centres)	–	–	–	–	–	m/ha	**300.00**
4 nr (at 25 m centres)	–	–	–	–	–	m/ha	**400.00**
Land drainage; excavating							
Removing 150 mm depth of topsoil; 300 mm wide; depositing beside trench; by machine	–	1.50	48.30	16.17	–	100 m	**64.47**
Removing 150 mm depth of topsoil; 300 mm wide; depositing beside trench; by hand	–	8.00	257.60	–	–	100 m	**257.60**
Excavated material; disposal on site							
In spoil heaps							
average 25 m distance	–	–	–	3.14	–	m³	**3.14**
average 50 m distance	–	–	–	3.62	–	m³	**3.62**
average 100 m distance (1 dumper)	–	–	–	4.54	–	m³	**4.54**
average 100 m distance (2 dumpers)	–	–	–	5.53	–	m³	**5.53**
average 200 m distance	–	–	–	5.47	–	m³	**5.47**
average 200 m distance (2 dumpers)	–	–	–	6.66	–	m³	**6.66**
Removing excavated material from site to tip; mechanically loaded							
excavated material and clean hardcore rubble	–	–	–	2.27	38.11	m³	**40.38**

34 DRAINAGE BELOW GROUND

Item – Overhead and Profit Included	PC £	Labour hours	Labour £	Plant £	Material £	Unit	Total rate £
Land drainage; Agripower Ltd; excavating trenches (with minimum project size of 1000 m) by trenching machine; spreading arisings on site; laying perforated pipe; as shown below; backfilling with shingle to 350 mm from surface							
Width 150 mm; 80 mm perforated pipe							
depth 750 mm	–	–	–	–	–	m	**10.20**
Width 175 mm; 100 mm perforated pipe							
depth 800 mm	–	–	–	–	–	m	**10.59**
Width 250 mm; 150 mm perforated pipe							
depth 900 mm	–	–	–	–	–	m	**14.42**
Land drainage; excavating for drains; by 5 tonne mini-excavator; including disposing spoil to spoil heaps not exceeding 100 m; boning to levels by laser							
Width 150–225 mm							
depth 450 mm	–	9.67	311.27	391.87	–	100 m	**703.14**
depth 600 mm	–	13.00	418.60	587.80	–	100 m	**1006.40**
depth 700 mm	–	15.50	499.10	734.75	–	100 m	**1233.85**
depth 900 mm	–	23.00	740.60	1175.60	–	100 m	**1916.20**
Width 300 mm							
depth 450 mm	–	9.67	311.27	391.87	–	100 m	**703.14**
depth 700 mm	–	14.00	450.80	646.58	–	100 m	**1097.38**
depth 900 mm	–	23.00	740.60	1175.60	–	100 m	**1916.20**
depth 1000 mm	–	25.00	805.00	1293.16	–	100 m	**2098.16**
depth 1200 mm	–	25.00	805.00	1469.50	–	100 m	**2274.50**
depth 1500 mm	–	31.00	998.20	1645.85	–	100 m	**2644.05**
Land drainage; excavating for drains; by 8 tonne tracked excavator; including disposing spoil to spoil heaps not exceeding 100 m							
Width 225 mm							
depth 450 mm	–	7.00	225.40	260.61	–	100 m	**486.01**
depth 600 mm	–	7.17	230.77	271.48	–	100 m	**502.25**
depth 700 mm	–	7.35	236.60	283.28	–	100 m	**519.88**
depth 900 mm	–	7.76	249.93	310.26	–	100 m	**560.19**
depth 1000 mm	–	8.00	257.60	325.77	–	100 m	**583.37**
depth 1200 mm	–	8.26	266.08	342.92	–	100 m	**609.00**

34 DRAINAGE BELOW GROUND

Item – Overhead and Profit Included	PC £	Labour hours	Labour £	Plant £	Material £	Unit	Total rate £
LAND DRAINS – cont							
Land drainage – cont							
Width 300 mm							
depth 450 mm	–	7.35	236.60	283.28	–	100 m	**519.88**
depth 600 mm	–	7.76	249.93	310.26	–	100 m	**560.19**
depth 700 mm	–	8.26	266.08	342.92	–	100 m	**609.00**
depth 900 mm	–	9.25	297.85	407.22	–	100 m	**705.07**
depth 1000 mm	–	10.14	326.60	465.39	–	100 m	**791.99**
depth 1200 mm	–	11.00	354.20	521.24	–	100 m	**875.44**
Width 600 mm							
depth 450 mm	–	13.00	418.60	651.54	–	100 m	**1070.14**
depth 600 mm	–	14.11	454.38	723.94	–	100 m	**1178.32**
depth 700 mm	–	16.33	525.93	868.72	–	100 m	**1394.65**
depth 900 mm	–	17.29	556.60	930.78	–	100 m	**1487.38**
depth 1000 mm	–	18.38	591.99	1002.37	–	100 m	**1594.36**
depth 1200 mm	–	21.18	682.05	1184.61	–	100 m	**1866.66**
Land drainage; excavating for drains; by hand, including disposing spoil to spoil heaps not exceeding 100 m							
Width 150 mm							
depth 450 mm	–	0.44	14.19	–	–	m	**14.19**
depth 600 mm	–	0.59	18.92	–	–	m	**18.92**
depth 700 mm	–	0.69	22.06	–	–	m	**22.06**
depth 900 mm	–	0.88	28.37	–	–	m	**28.37**
Width 225 mm							
depth 450 mm	–	0.66	21.29	–	–	m	**21.29**
depth 600 mm	–	0.88	28.37	–	–	m	**28.37**
depth 700 mm	–	1.03	33.11	–	–	m	**33.11**
depth 900 mm	–	1.32	42.57	–	–	m	**42.57**
Width 300 mm							
depth 450 mm	–	0.88	28.37	–	–	m	**28.37**
depth 600 mm	–	1.18	37.83	–	–	m	**37.83**
depth 700 mm	–	1.37	44.14	–	–	m	**44.14**
Width 375 mm							
depth 450 mm	–	1.10	35.47	–	–	m	**35.47**
depth 600 mm	–	1.47	47.30	–	–	m	**47.30**
depth 700 mm	–	1.71	55.18	–	–	m	**55.18**

34 DRAINAGE BELOW GROUND

Item – Overhead and Profit Included	PC £	Labour hours	Labour £	Plant £	Material £	Unit	Total rate £
LAND DRAINAGE; PIPE LAYING							
Wavin Plastics Ltd; flexible plastic perforated pipes in trenches (not included); to a minimum depth of 450 mm (couplings not included)							
OsmaDrain; flexible plastic perforated pipes in trenches (not included); to a minimum depth of 450 mm (couplings not included)							
80 mm dia.; available in 100 m coil	74.61	2.00	64.40	–	85.80	100 m	**150.20**
100 mm dia.; available in 100 m coil	120.09	2.00	64.40	–	138.10	100 m	**202.50**
160 mm dia.; available in 35 m coil	70.72	2.00	64.40	–	81.33	100 m	**145.73**
WavinCoil; plastic pipe junctions							
100 × 80 mm	3.47	0.05	1.61	–	3.99	nr	**5.60**
160 × 160 mm	8.67	0.05	1.61	–	9.97	nr	**11.58**
WavinCoil; couplings for flexible pipes							
100 mm dia.	0.27	0.03	1.07	–	0.31	nr	**1.38**
160 mm dia.	1.72	0.03	1.07	–	1.98	nr	**3.05**
Land drainage; backfilling trench after laying pipes with gravel rejects or similar; blind filling with ash or sand; topping with 150 mm imported topsoil from dumps not exceeding 100 m; by machine							
Width 150 mm							
depth 450 mm	–	3.30	106.26	96.16	255.23	100 m	**457.65**
depth 600 mm	–	4.30	138.46	126.04	355.58	100 m	**620.08**
depth 900 mm	–	6.30	202.86	185.81	534.68	100 m	**923.35**
Width 225 mm							
depth 600 mm	–	6.45	207.69	189.06	533.58	100 m	**930.33**
depth 900 mm	–	9.45	304.29	278.71	813.03	100 m	**1396.03**
Land drainage; backfilling trench after laying pipes with gravel rejects or similar, blind filling with ash or sand; topping with 150 mm topsoil from dumps not exceeding 100 m; by hand							
Width 150 mm							
depth 450 mm	–	0.18	5.80	–	2.40	m	**8.20**
depth 600 mm	–	24.84	799.85	–	355.58	100 m	**1155.43**
depth 750 mm	–	31.05	999.81	–	346.52	100 m	**1346.33**
depth 900 mm	–	37.26	1199.77	–	534.68	100 m	**1734.45**
Width 225 mm							
depth 450 mm	–	27.94	899.67	–	394.06	100 m	**1293.73**
depth 600 mm	–	37.26	1199.77	–	533.58	100 m	**1733.35**
depth 750 mm	–	46.57	1499.55	–	673.51	100 m	**2173.06**
depth 900 mm	–	55.89	1799.66	–	813.03	100 m	**2612.69**

34 DRAINAGE BELOW GROUND

Item – Overhead and Profit Included	PC £	Labour hours	Labour £	Plant £	Material £	Unit	Total rate £
LAND DRAINS – cont							
Land drainage – cont							
Width 375 mm							
depth 450 mm	–	46.57	1499.55	–	656.49	100 m	**2156.04**
depth 600 mm	–	61.10	1967.42	–	889.16	100 m	**2856.58**
depth 750 mm	–	77.63	2499.69	–	1122.24	100 m	**3621.93**
depth 900 mm	–	93.15	2999.43	–	1354.91	100 m	**4354.34**
MARKET PRICES OF BACKFILLING MATERIALS							
Market prices of commonly used materials used for backfilling							
The prices below show the market price with standard settlement factors allowed for							
Sharp sand	–	–	–	–	44.17	m³	**44.17**
Gravel (washed river or pit)	–	–	–	–	43.12	m³	**43.12**
Shingle	–	–	–	–	47.05	m³	**47.05**
Topsoil	–	–	–	–	29.66	m³	**29.66**
Gravel rejects	–	–	–	–	36.90	m³	**36.90**
Selected granular material 6F2	–	–	–	–	21.60	m³	**21.60**
SAND SLITTING/GROOVING							
Sand slitting; Agripower Ltd							
Drainage slits; at 1.00 m centres; using spinning disc trenching machine; backfilling to 100 mm of surface with pea gravel; finished surface with medium grade sand 100; arisings to be loaded, hauled and tipped on site; minimum pitch size 6000 m²							
250 mm depth	–	–	–	–	–	m	**2.82**
300 mm depth	–	–	–	–	–	m	**3.10**
400 mm depth	–	–	–	–	–	m	**4.67**
Sand grooving or banding; Agripower Ltd							
Drainage bands at 260 mm centres; using Blec Sandmaster; to existing grass at 260 mm centres; minimum pitch size 4000 m²							
sand banding or sand grooving	–	–	–	–	–	m²	**1.68**

34 DRAINAGE BELOW GROUND

Item – Overhead and Profit Included	PC £	Labour hours	Labour £	Plant £	Material £	Unit	Total rate £
SOAKAWAYS							
Preamble: Flat rate hourly rainfall = 50 mm/hr and assumes 100 impermeability of the run-off area. A storage capacity of the soakaway should be ⅓; of the hourly rainfall. Formulae for calculating soakaway depths are provided in the publications mentioned below and in the Tables and Memoranda section of this publication. The design of soakaways is dependent on, amongst other factors, soil conditions, permeability, groundwater level and run-off. The definitive documents for design of soakaways are CIRIA 156 and BRE Digest 365 dated September 1991. The suppliers of the systems below will assist through their technical divisions. Excavation earthwork support of pits and disposal not included.							
Soakaway excavation							
Pits; 3 tonne tracked excavator							
maximum depth not exceeding 1.00 m	–	0.25	8.05	11.29	–	m³	**19.34**
maximum depth not exceeding 2.00 m	–	0.40	12.88	18.07	–	m³	**30.95**
Pits; 8 tonne tracked excavator							
maximum depth not exceeding 2.00 m	–	0.40	12.88	21.73	–	m³	**34.61**
maximum depth not exceeding 3.00 m	–	0.50	16.10	27.16	–	m³	**43.26**
Disposal							
Excavated material; off site; to tip; mechanically loaded (8 tonne excavator							
inert	–	–	–	–	–	m³	**39.94**
In situ concrete ring beam foundations to base of soakaway; 300 mm wide × 250 mm deep; poured on or against earth or unblinded hardcore							
Internal dia.s of rings							
900 mm	23.61	2.00	128.80	–	27.15	nr	**155.95**
1200 mm	31.47	2.25	144.90	–	36.19	nr	**181.09**
1500 mm	39.34	2.50	161.00	–	45.24	nr	**206.24**
2400 mm	62.95	2.75	177.10	–	72.39	nr	**249.49**

34 DRAINAGE BELOW GROUND

Item – Overhead and Profit Included	PC £	Labour hours	Labour £	Plant £	Material £	Unit	Total rate £
SOAKAWAYS – cont							
Concrete manhole rings; Milton Pipes Ltd; perforations and step irons to concrete rings at manufacturer's recommended centres; placing of concrete ring manholes to in situ concrete ring beams (1:3:6) (not included); filling and surrounding base with gravel 225 mm deep (not included)							
Ring dia. 900 mm							
1.00 m deep; volume 636 litres	112.40	0.63	40.25	23.77	299.18	nr	**363.20**
1.50 m deep; volume 954 litres	168.60	0.82	53.13	62.76	477.73	nr	**593.62**
2.00 m deep; volume 1272 litres	224.80	0.90	57.96	62.76	634.18	nr	**754.90**
Ring dia. 1200 mm							
1.00 m deep; volume 1131 litres	146.84	1.50	96.60	57.05	389.52	nr	**543.17**
1.50 m deep; volume 1696 litres	220.26	2.25	144.90	85.57	576.82	nr	**807.29**
2.00 m deep; volume 2261 litres	293.68	2.50	161.00	85.57	764.12	nr	**1010.69**
2.50 m deep; volume 2827 litres	367.10	3.00	193.20	114.09	874.06	nr	**1181.35**
Ring dia. 1500 mm							
1.00 m deep; volume 1767 litres	236.83	2.25	144.90	57.05	545.62	nr	**747.57**
1.50 m deep; volume 2651 litres	355.25	3.00	193.20	79.87	806.76	nr	**1079.83**
2.00 m deep; volume 3534 litres	473.66	3.50	225.40	79.87	1067.91	nr	**1373.18**
2.50 m deep; volume 4418 litres	367.10	3.75	241.50	142.62	1070.34	nr	**1454.46**
Ring dia. 2400 mm							
1.00 m deep; volume 4524 litres	909.68	3.00	193.20	57.05	1418.86	nr	**1669.11**
1.50 m deep; volume 6786 litres	1364.52	3.75	241.50	79.87	2111.10	nr	**2432.47**
2.00 m deep; volume 9048 litres	1819.36	4.00	257.60	79.87	2814.40	nr	**3151.87**
2.50 m deep; volume 11310 litres	2274.20	4.50	289.80	142.62	3506.64	nr	**3939.06**
Extra over for							
250 mm depth chamber ring	–	–	–	–	–	100%	–
500 mm depth chamber ring	–	–	–	–	–	50%	–
Cover slabs to soakaways							
Heavy duty precast concrete							
900 mm dia.	185.91	0.50	48.30	28.52	213.80	nr	**290.62**
1200 mm dia.	208.40	0.50	48.30	28.52	239.66	nr	**316.48**
1500 mm dia.	337.65	0.50	48.30	28.52	388.30	nr	**465.12**
2400 mm dia.	1211.24	0.50	48.30	28.52	1392.93	nr	**1469.75**
Step irons to concrete chamber rings	44.36	–	–	–	51.01	m	**51.01**
Extra over soakaways for filter wrapping with a proprietary filter membrane							
900 mm dia. × 1.00 m deep	1.74	0.50	48.30	–	2.00	nr	**50.30**
900 mm dia. × 2.00 m deep	3.47	0.75	72.45	–	3.99	nr	**76.44**
1050 mm dia. × 1.00 m deep	2.03	0.75	72.45	–	2.33	nr	**74.78**
1050 mm dia. × 2.00 m deep	4.05	1.00	96.60	–	4.66	nr	**101.26**
1200 mm dia. × 1.00 m deep	2.31	1.00	96.60	–	2.66	nr	**99.26**
1200 mm dia. × 2.00 m deep	4.63	1.25	120.75	–	5.32	nr	**126.07**
1500 mm dia. × 1.00 m deep	2.89	1.25	120.75	–	3.32	nr	**124.07**

34 DRAINAGE BELOW GROUND

Item – Overhead and Profit Included	PC £	Labour hours	Labour £	Plant £	Material £	Unit	Total rate £
1500 mm dia. × 2.00 m deep	5.77	1.25	120.75	–	6.64	nr	**127.39**
1800 mm dia. × 1.00 m deep	3.47	1.50	144.90	–	3.99	nr	**148.89**
1800 mm dia. × 2.00 m deep	6.94	1.63	156.98	–	7.98	nr	**164.96**
Gravel surrounding to concrete ring soakaway 40 mm aggregate backfilled to vertical face of soakaway wrapped with geofabric (not included)							
250 mm thick	45.90	0.10	3.22	18.89	52.78	m³	**74.89**
Backfilling to face of soakaway; carefully compacting as work proceeds Arising from the excavations average thickness exceeding 0.25 m; depositing in layers							
150 mm maximum thickness	–	0.03	1.07	6.30	–	m³	**7.37**
Aquacell soakaway; Wavin Plastics Ltd; preformed polypropylene soakaway infiltration crate units; to trenches; surrounded by geotextile and 40 mm aggregate laid 100 mm thick in trenches (excavation, disposal and backfilling not included) 1.00 m × 500 × 400 mm; internal volume 190 litres							
4 crates; 2.00 m × 1.00 m × 400 mm; 760 litres	179.29	0.80	51.52	25.10	274.60	nr	**351.22**
8 crates; 2.00 m × 1.00 m × 800 mm; 1520 litres	358.58	1.60	103.04	32.27	497.62	nr	**632.93**
12 crates; 6.00 m × 500 mm × 800 mm; 2280 litres	537.88	2.40	154.56	66.93	785.14	nr	**1006.63**
16 crates; 4.00 m × 1.00 m × 800 mm; 3040 litres	717.17	3.20	206.08	59.77	974.50	nr	**1240.35**
20 crates; 5.00 m × 1.00 m × 800 mm; 3800 litres	896.46	4.00	257.60	59.17	1188.79	nr	**1505.56**
30 crates; 15.00 m × 1.00 m × 400 mm; 5700 litres	1344.69	6.00	386.40	161.36	1944.06	nr	**2491.82**
60 crates; 15.00 m × 1.00 m × 800 mm; 11400 litres	2689.38	10.00	644.00	210.97	3625.89	nr	**4480.86**

34 DRAINAGE BELOW GROUND

Item – Overhead and Profit Included	PC £	Labour hours	Labour £	Plant £	Material £	Unit	Total rate £
SOAKAWAYS – cont							
Geofabric surround to Aquacell units; Terram Ltd							
Terram synthetic fibre filter fabric; to face of concrete rings (not included); anchoring whilst backfilling (not included)							
Terram 1000; 0.70 mm thick; mean water flow 50 l/m^2/s	0.59	0.03	1.61	–	0.68	m^2	**2.29**
CULVERTS							
Culverts; precast box sections FP McCann Ltd; For directing watercourses and underpasses; to trenches or prepared bases (not included) on 200 mm prepared granular or concrete base blinded with 50 mm sand							
Plain box culvert; 800 mm high; width							
1.20 m	671.61	3.00	289.80	86.16	781.88	m	**1157.84**
1.50 m	706.65	3.50	338.10	107.70	822.18	m	**1267.98**
Plain box culvert; 1.00 m high; width							
1.20 m	700.81	4.00	386.40	129.24	815.47	m	**1331.11**
1.50 m	817.61	4.50	434.70	150.78	949.78	m	**1535.26**
1.80 m	905.22	6.00	579.60	215.41	1050.54	m	**1845.55**
Plain box culvert; 1.20 m high; width							
1.50 m	864.34	6.00	579.60	215.41	1003.52	m	**1798.53**
1.80 m	934.42	8.00	772.80	258.49	1084.12	m	**2115.41**
Extra over for cantilevered animal ledges	233.60	–	–	–	268.64	m	**268.64**
HEADWALLS							
Headwalls; FP McCann Ltd; Precast concrete headwalls to outflows; Placing to prepared graded and blinded 300 mm thick (not included); connection of outfall pipe; 150 mm thick wall with non standard opening; excludes standard grating							
size on plan 1.618 m tapering to 889 mm × 1.32 m × 663 mm high							
Small Headwall; pipe size up to 300 mm	745.90	4.00	386.40	54.33	941.59	nr	**1382.32**

34 DRAINAGE BELOW GROUND

Item – Overhead and Profit Included	PC £	Labour hours	Labour £	Plant £	Material £	Unit	Total rate £
size on plan 1.878 m tapering to 1.026 m × 1.37 m × 998 mm high							
Medium Headwall; pipe size up to 450 mm	1215.09	7.00	676.20	162.99	1528.52	nr	**2367.71**
size on plan 2.351 m tapering to 1.656 m × 2.150 m × 1.258 m high							
Large Headwall; pipe size up to 900 mm	1792.56	8.00	772.80	325.98	2061.44	nr	**3160.22**
SPORTS AND AMENITY DRAINAGE							
Sports or amenity drainage; Agripower Ltd; excavating trenches (with minimum project size of 1000 m) by trenching machine; arisings to spoil heap on site maximum 100 m; backfill with shingle to within 125 mm from surface and with rootzone to ground level							
Width 145 mm; 80 mm perforated pipe							
depth 550 mm	–	–	–	–	–	m	**11.91**
Width 145 mm; 100 mm perforated pipe							
depth 650 mm	–	–	–	–	–	m	**12.86**
Width 145 mm; 150 mm perforated pipe							
depth 700 mm	–	–	–	–	–	100 m	**16.14**

35 SITE WORKS – BASES

Item – Overhead and Profit Included	PC £	Labour hours	Labour £	Plant £	Material £	Unit	Total rate £
HARDCORE BASES							
Hardcore bases; obtained off site; PC £32.40/m³							
By machine							
100 mm thick	3.24	0.05	1.61	1.48	3.73	m²	**6.82**
150 mm thick	4.86	0.07	2.15	1.93	5.59	m²	**9.67**
200 mm thick	6.48	0.08	2.58	2.01	7.45	m²	**12.04**
300 mm thick	9.72	0.08	2.68	2.15	11.18	m²	**16.01**
exceeding 300 mm thick	38.88	0.17	5.36	4.95	44.71	m³	**55.02**
By hand							
100 mm thick	3.24	0.20	6.44	0.15	3.73	m²	**10.32**
150 mm thick	5.83	0.30	9.67	0.15	6.70	m²	**16.52**
200 mm thick	7.78	0.40	12.88	0.24	8.95	m²	**22.07**
300 mm thick	11.66	0.60	19.32	0.15	13.41	m²	**32.88**
exceeding 300 mm thick	38.88	2.00	64.40	0.51	44.71	m³	**109.62**
Hardcore; difference for each £1.00 increase/decrease in PC price per m³; price will vary with type and source of hardcore							
average 75 mm thick	–	–	–	–	0.09	m²	**0.09**
average 100 mm thick	–	–	–	–	0.12	m²	**0.12**
average 150 mm thick	–	–	–	–	0.17	m²	**0.17**
average 200 mm thick	–	–	–	–	0.23	m²	**0.23**
average 250 mm thick	–	–	–	–	0.29	m²	**0.29**
average 300 mm thick	–	–	–	–	0.35	m²	**0.35**
exceeding 300 mm thick	–	–	–	–	1.27	m³	**1.27**
TYPE 1 BASES							
Type 1 granular fill base; PC £24.75/tonne (£55.45/m³ compacted)							
By machine							
100 mm thick	5.45	0.02	1.77	0.75	6.27	m²	**8.79**
150 mm thick	8.17	0.03	2.66	0.75	9.40	m²	**12.81**
200 mm thick	10.89	0.04	3.54	1.48	12.52	m²	**17.54**
250 mm thick	13.61	0.05	4.44	1.48	15.65	m²	**21.57**
over 250 mm thick	54.45	0.18	17.72	4.58	62.62	m³	**84.92**
By hand (mechanical compaction)							
100 mm thick	3.63	0.07	6.44	0.05	4.17	m²	**10.66**
150 mm thick	8.17	0.08	8.05	0.08	9.40	m²	**17.53**
250 mm thick	13.61	0.13	12.07	0.13	15.65	m²	**27.85**
over 250 mm thick	27.23	0.20	19.32	0.13	31.31	m³	**50.76**

35 SITE WORKS – BASES

Item – Overhead and Profit Included	PC £	Labour hours	Labour £	Plant £	Material £	Unit	Total rate £
Type 1 (crushed concrete) granular fill base; PC £9.50/tonne (£20.90/m³ compacted)							
By machine							
100 mm thick	2.09	0.02	1.77	0.49	2.40	m²	**4.66**
150 mm thick	3.13	0.03	2.66	0.75	3.60	m²	**7.01**
200 mm thick	5.22	0.04	3.54	1.24	6.00	m²	**10.78**
250 mm thick	5.22	0.05	4.44	1.24	6.00	m²	**11.68**
over 250 mm thick	9.50	0.18	17.72	4.58	10.93	m³	**33.23**
By hand (mechanical compaction)							
100 mm thick	2.09	0.07	6.44	0.05	2.40	m²	**8.89**
150 mm thick	3.13	0.08	8.05	0.08	3.60	m²	**11.73**
250 mm thick	5.22	0.13	12.07	0.13	6.00	m²	**18.20**
over 250 mm thick	9.50	0.20	19.32	0.13	10.93	m³	**30.38**
TYPE 3 PERMEABLE BASES							
Type 3 open aggregate granular fill base; PC £59.29/m³; incoprating optional triaxial geogrid and recomended geofabric							
By machine							
100 mm thick	5.93	0.02	2.25	0.52	10.32	m²	**13.09**
150 mm thick	8.89	0.03	3.14	1.14	13.72	m²	**18.00**
200 mm thick	11.86	0.04	4.03	1.68	17.13	m²	**22.84**
250 mm thick	14.82	0.05	4.92	1.68	20.54	m²	**27.14**
over 250 mm thick	65.22	0.19	18.38	6.80	78.50	m³	**103.68**
By hand (mechanical compaction)							
100 mm thick	5.93	0.09	8.53	0.05	13.85	m²	**22.43**
150 mm thick	8.89	0.11	10.14	0.08	17.25	m²	**27.47**
250 mm thick	28.44	0.17	16.58	0.13	36.19	m²	**52.90**
over 250 mm thick	59.29	0.51	48.78	0.13	71.68	m³	**120.59**
CONCRETE BASES							
Concrete bases; on 50 mm blinding							
Plain concrete; Site mixed concrete 1:3:6							
100 mm thick	10.12	0.03	3.02	0.37	14.77	m²	**18.16**
150 mm thick	15.18	0.03	3.37	0.37	20.58	m²	**24.32**
200 mm thick	20.24	0.04	3.83	0.37	26.40	m²	**30.60**
250 mm thick	25.30	0.05	4.47	0.37	32.22	m²	**37.06**
over 250 mm thick	101.20	0.21	19.93	0.37	119.51	m³	**139.81**
Mesh reinforced concrete A 142; Site mixed concrete 1:3:6							
100 mm thick	10.12	0.03	3.37	0.37	18.85	m²	**22.59**
150 mm thick	15.18	0.04	3.83	0.37	24.67	m²	**28.87**
200 mm thick	20.24	0.04	4.19	0.37	30.49	m²	**35.05**
250 mm thick	25.30	0.05	5.00	0.37	36.31	m²	**41.68**

35 SITE WORKS – BASES

Item – Overhead and Profit Included	PC £	Labour hours	Labour £	Plant £	Material £	Unit	Total rate £
CONCRETE BASES – cont							
Concrete bases – cont							
Mesh reinforced concrete A 142;							
Site mixed concrete 1:2:4							
100 mm thick	11.37	0.03	3.37	0.37	20.29	m²	**24.03**
150 mm thick	17.05	0.04	3.83	0.37	26.82	m²	**31.02**
200 mm thick	22.73	0.04	4.19	0.37	33.36	m²	**37.92**
250 mm thick	28.41	0.05	5.00	0.37	39.89	m²	**45.26**
Mesh reinforced concrete A 142;							
Ready mixed concrete C20							
100 mm thick	9.10	0.03	3.37	0.37	17.68	m²	**21.42**
150 mm thick	13.65	0.04	3.83	0.37	22.91	m²	**27.11**
200 mm thick	18.20	0.04	4.19	0.37	28.14	m²	**32.70**
250 mm thick	22.75	0.05	5.00	0.37	33.37	m²	**38.74**
CELLULAR CONFINED BASES							
Geoweb; Greenfix; cellular							
confinement system for load							
support or permeable bases							
below pavings							
To graded and compacted substrate							
(not included); on 50 mm sharp sand							
base filled with angular drainage							
aggregate 20–5 mm							
Geoweb 75 mm depth	7.26	0.05	4.39	1.94	16.87	m²	**23.20**
Geoweb 100 mm depth	8.63	0.05	4.83	2.20	20.38	m²	**27.41**
Geoweb 150 mm depth	10.55	0.06	5.37	2.31	26.09	m²	**33.77**
Geoweb 200 mm depth	13.06	0.08	8.05	2.51	32.48	m²	**43.04**
Geoweb 300 mm depth	19.72	0.10	9.66	2.92	47.12	m²	**59.70**
SUBBASE REPLACEMENT							
SYSTEMS							
Subbase replacement system for							
SUDS paving; Permavoid Ltd;							
geocellular, interlocking, high							
strength; high void capacity							
(95%); typically a 4:1 ratio when							
compared with alternative							
granular systems; allows for							
water storage to be confined at							
shallow depth within the subbase							
layer; inclusive of connection,							
ties, pins; does not include for							
encapsulation in geotextile/							
geomembranes							

35 SITE WORKS – BASES

Item – Overhead and Profit Included	PC £	Labour hours	Labour £	Plant £	Material £	Unit	Total rate £
708 × 354 mm							
85 mm thick	52.89	0.06	5.38	–	60.82	m²	**66.20**
150 mm thick	71.20	0.06	5.38	–	81.88	m²	**87.26**
300 mm thick	142.40	0.11	10.73	–	163.76	m²	**174.49**
Encapsulation of Permavoid; 150 mm thick systems; geomembrane geofabric or combination of both for filtration or attenuation of water; please see the manufacturer's design guide							
geomembrane, heat welded	–	0.02	1.61	–	5.19	m²	**6.80**
Permatex 300 geotextile	–	0.11	10.73	–	4.48	m²	**15.21**
SURFACE TREATMENTS TO GRANULAR BASES							
Surface treatments							
Sand blinding; to hardcore base (not included); 25 mm thick	1.05	0.03	1.07	–	1.21	m²	**2.28**
Sand blinding; to hardcore base (not included); 50 mm thick	2.10	0.05	1.61	–	2.42	m²	**4.03**
Filter fabrics; to hardcore base (not included)	0.52	0.01	0.32	–	0.60	m²	**0.92**

35 SITE WORKS – KERBS AND EDGINGS

Item – Overhead and Profit Included	PC £	Labour hours	Labour £	Plant £	Material £	Unit	Total rate £
FOUNDATIONS TO KERBS AND EDGINGS							
Foundations to kerbs and edgings							
Excavating trenches; width 300 mm;							
3 tonne excavator; disposal off site							
depth 200 mm	–	–	–	3.36	2.40	m	5.76
depth 300 mm	–	–	–	4.84	3.60	m	8.44
depth 400 mm	–	–	–	5.28	4.80	m	10.08
By hand; barrowing to spoil heap							
and disposal off site							
depth 300	–	0.06	5.80	–	39.94	m	45.74
depth 400	–	0.08	7.73	–	39.94	m	47.67
Excavating trenches; width 450 mm;							
3 tonne excavator; disposal off site							
depth 300 mm	–	–	–	6.05	5.39	m	11.44
depth 400 mm	–	–	–	6.92	7.19	m	14.11
Foundations; to kerbs, edgings or channels; in situ concrete; 21 N/mm^2–20 mm aggregate (1:2:4) site mixed); one side against earth face, other against formwork (not included); site mixed concrete							
Site mixed concrete							
150 mm wide × 100 mm deep	–	0.04	4.30	–	1.96	m	6.26
150 mm wide × 150 mm deep	–	0.06	5.36	–	2.94	m	8.30
200 mm wide × 150 mm deep	–	0.07	6.44	–	3.92	m	10.36
300 mm wide × 150 mm deep	–	0.06	5.65	–	5.88	m	11.53
600 mm wide × 200 mm deep	–	0.06	5.64	–	15.69	m	21.33
Ready mixed concrete							
150 mm wide × 100 mm deep	–	0.04	4.30	–	1.56	m	5.86
150 mm wide × 150 mm deep	–	0.06	5.36	–	2.36	m	7.72
200 mm wide × 150 mm deep	–	0.07	6.44	–	3.14	m	9.58
300 mm wide × 150 mm deep	–	0.08	7.52	–	4.70	m	12.22
600 mm wide × 200 mm deep	–	0.10	9.20	–	12.56	m	21.76
Formwork; sides of foundations (this will usually be required to one side of each kerb foundation adjacent to road subbases)							
100 mm deep	–	0.01	1.35	–	0.17	m	1.52
150 mm deep	–	0.01	1.35	–	0.25	m	1.60

35 SITE WORKS – KERBS AND EDGINGS

Item – Overhead and Profit Included	PC £	Labour hours	Labour £	Plant £	Material £	Unit	Total rate £
PRECAST CONCRETE KERBS							
Precast concrete kerbs, channels, edgings etc.; Marshalls Mono; bedding, jointing and pointing in cement: mortar (1:3); including haunching with in situ concrete; 11.50 N/mm²–40 mm aggregate one side							
Kerbs; straight							
150 × 305 mm; HB1	12.43	0.17	16.10	–	17.78	m	**33.88**
125 × 255 mm; HB2; SP	6.10	0.15	14.31	–	10.51	m	**24.82**
125 × 150 mm; BN	4.22	0.13	12.88	–	8.35	m	**21.23**
Dropper kerbs; left and right handed							
125 × 255–150 mm; DL1 or DR1	10.15	0.17	16.10	–	15.17	m	**31.27**
125 × 255–150 mm; DL2 or DR2	10.15	0.17	16.10	–	15.17	m	**31.27**
Quadrant kerbs							
305 mm radius	17.82	0.17	16.10	–	22.82	nr	**38.92**
455 mm radius	20.13	0.17	16.10	–	26.65	nr	**42.75**
Straight kerbs or channels; to radius; 125 × 255 mm							
3.00 m radius (5 units per quarter circle)	11.37	0.22	21.46	–	16.69	m	**38.15**
4.50 m radius (2 units per quarter circle)	11.37	0.20	19.34	–	16.69	m	**36.03**
6.10 m radius (11 units per quarter circle)	11.37	0.19	18.66	–	16.69	m	**35.35**
7.60 m radius (14 units per quarter circle)	11.37	0.19	18.40	–	16.69	m	**35.09**
9.15 m radius (17 units per quarter circle)	11.37	0.19	17.89	–	16.69	m	**34.58**
10.70 m radius (20 units per quarter circle)	11.37	0.19	17.89	–	16.69	m	**34.58**
12.20 m radius (22 units per quarter circle)	11.37	0.18	16.95	–	16.69	m	**33.64**
Kerbs; Conservation Kerb units; to simulate natural granite kerbs							
255 × 150 × 914 mm; laid flat	32.09	0.19	18.40	–	43.16	m	**61.56**
150 × 255 × 914 mm; laid vertical	32.09	0.19	18.40	–	42.14	m	**60.54**
145 × 255 mm; radius internal 3.25 m	41.69	0.28	26.84	–	53.18	m	**80.02**
150 × 255 mm; radius external 3.40 m	39.81	0.28	26.84	–	51.03	m	**77.87**
145 × 255 mm; radius internal 6.50 m	37.25	0.22	21.46	–	48.07	m	**69.53**
145 × 255 mm; radius external 6.70 m	37.03	0.22	21.46	–	47.83	m	**69.29**
150 × 255 mm; radius internal 9.80 m	36.83	0.22	21.46	–	47.59	m	**69.05**

35 SITE WORKS – KERBS AND EDGINGS

Item – Overhead and Profit Included	PC £	Labour hours	Labour £	Plant £	Material £	Unit	Total rate £
PRECAST CONCRETE KERBS – cont							
Precast concrete kerbs, channels, edgings etc. – cont							
Kerbs – cont							
150 × 255 mm; radius external							
10.00 m	34.92	0.22	21.46	–	45.39	m	**66.85**
305 × 305 × 255 mm; solid							
quadrants	47.04	0.19	18.40	–	59.33	nr	**77.73**
Marshalls Mono; small element precast concrete kerb system; Keykerb Large (KL) upstand of 100–125 mm; on 150 mm concrete foundation including haunching with in situ concrete (1:3:6) 1 side							
Bullnosed or half battered;							
100 × 127 × 200 mm							
laid straight	18.27	0.27	25.76	–	22.23	m	**47.99**
radial blocks laid to curve; 8							
blocks/¼ circle – 500 mm radius	14.16	0.33	32.20	–	17.51	m	**49.71**
radial blocks laid to curve; 8 radial							
blocks, alternating 8 standard							
blocks/¼ circle – 1000 mm radius	23.30	0.42	40.25	–	28.01	m	**68.26**
radial blocks, alternating							
16 standard blocks/¼ circle –							
1500 mm radius	21.41	0.50	48.30	–	25.84	m	**74.14**
internal angle 90°	7.80	0.07	6.44	–	8.97	nr	**15.41**
external angle	7.80	0.07	6.44	–	8.97	nr	**15.41**
Marshalls Mono; small element precast concrete kerb system; Keykerb Small (KS) upstand of 25–50 mm; on 150 mm concrete foundation including haunching with in situ concrete (1:3:6) 1 side							
Half battered							
laid straight	12.83	0.27	25.76	–	15.97	m	**41.73**
radial blocks laid to curve; 8							
blocks/¼ circle; 500 mm radius	10.64	0.33	32.20	–	13.47	m	**45.67**
radial blocks laid to curve; 8 radial							
blocks, alternating 8 standard							
blocks/¼ circle; 1000 mm radius	17.06	0.42	40.25	–	20.84	m	**61.09**
radial blocks, alternating							
16 standard blocks/¼ circle;							
1500 mm radius	18.29	0.50	48.30	–	22.25	m	**70.55**
internal angle 90°	7.80	0.07	6.44	–	8.97	nr	**15.41**
external angle	7.80	0.07	6.44	–	8.97	nr	**15.41**

35 SITE WORKS – KERBS AND EDGINGS

Item – Overhead and Profit Included	PC £	Labour hours	Labour £	Plant £	Material £	Unit	Total rate £
PRECAST CONCRETE EDGINGS							
Precast concrete edging units; including haunching with in situ concrete; 11.50 N/mm²–40 mm aggregate both sides							
Edgings; rectangular, bullnosed or chamfered							
50 × 150 mm	2.22	0.11	10.73	–	8.37	m	**19.10**
125 × 150 mm bullnosed	4.60	0.11	10.73	–	11.11	m	**21.84**
50 × 200 mm	3.37	0.11	10.73	–	9.69	m	**20.42**
50 × 250 mm	3.73	0.11	10.73	–	10.11	m	**20.84**
50 × 250 mm flat top	4.14	0.11	10.73	–	10.58	m	**21.31**
Conservation edgings							
150 × 915 × 63 mm	14.39	0.14	13.80	–	20.15	ea	**33.95**
Precast concrete block edgings; PC £11.76/m²; 200 × 100 × 60 mm; on prepared base (not included); haunching one side							
Edgings; butt joints							
stretcher course	1.18	0.06	5.36	–	1.60	m	**6.96**
header course	2.35	0.09	8.59	–	3.20	m	**11.79**
STONE KERBS							
Dressed natural stone kerbs – General							
Preamble: The kerbs are to be good, sound and uniform in texture and free from defects; worked straight or to radius, square and out of wind, with the top front and back edges parallel or concentric to the dimensions specified. All drill and pick holes shall be removed from dressed faces. Standard dressings shall be designated as either fine picked, single axed or nidged or rough punched.							
Dressed natural stone kerbs; CED Ltd; on concrete foundations (not included); including haunching with in situ concrete; 11.50 N/mm²–40 mm aggregate one side							
Granite kerbs; 125 × 250 mm							
special quality; straight; random lengths	25.05	0.27	25.76	–	32.90	m	**58.66**
Granite kerbs; 125 × 250 mm; curved to mean radius 3 m							
special quality; random lengths	34.52	0.30	29.30	–	43.79	m	**73.09**

35 SITE WORKS – KERBS AND EDGINGS

Item – Overhead and Profit Included	PC £	Labour hours	Labour £	Plant £	Material £	Unit	Total rate £
STONE EDGINGS							
New granite sett edgings; CED Ltd; 100 × 100 × 100 mm; bedding in cement: mortar (1:4); including haunching with in situ concrete; 11.50 N/mm²–40 mm aggregate one side; concrete foundations (not included)							
Edgings 100 × 100 × 100 mm; ref 1R 'better quality' with 20 mm pointing gaps							
100 mm wide	5.10	0.44	42.83	–	9.94	m	**52.77**
2 rows; 220 mm wide	10.19	0.74	71.48	–	16.99	m	**88.47**
3 rows; 340 mm wide	15.29	1.00	96.60	–	21.72	m	**118.32**
Edgings 100 × 100 × 200 mm; ref 1R 'better quality' with 20 mm pointing gaps							
100 mm wide	5.19	0.25	24.15	–	10.04	m	**34.19**
2 rows; 220 mm wide	10.38	0.50	48.30	–	17.20	m	**65.50**
3 rows; 340 mm wide	15.58	0.75	72.45	–	22.06	m	**94.51**
Edgings 100 × 100 × 100 mm; ref 8R 'standard quality' with 20 mm pointing gaps							
100 mm wide	4.26	0.44	42.83	–	8.97	m	**51.80**
2 rows; 220 mm wide	8.52	0.74	71.48	–	15.06	m	**86.54**
3 rows; 340 mm wide	12.78	1.00	96.60	–	18.84	m	**115.44**
Edgings 100 × 100 × 200 mm; ref 8R 'standard quality' with 20 mm pointing gaps							
100 mm wide	4.54	0.25	24.15	–	9.30	m	**33.45**
2 rows; 220 mm wide	9.09	0.50	48.30	–	15.72	m	**64.02**
3 rows; 340 mm wide	13.63	0.75	72.45	–	19.83	m	**92.28**
Reclaimed granite setts edgings; CED Ltd; 100 × 100 mm; bedding in cement: mortar (1:4); including haunching with in situ concrete; 11.50 N/mm²–40 mm aggregate one side; on concrete foundations (not included)							
Edgings; with 20 mm pointing gaps							
100 mm wide	4.12	0.44	42.83	–	8.81	m	**51.64**
2 rows; 220 mm wide	8.23	0.74	71.48	–	14.73	m	**86.21**
3 rows; 340 mm wide	12.35	1.00	96.60	–	18.34	m	**114.94**

35 SITE WORKS – KERBS AND EDGINGS

Item – Overhead and Profit Included	PC £	Labour hours	Labour £	Plant £	Material £	Unit	Total rate £
Natural Yorkstone edgings; Johnsons Wellfield Quarries; sawn 6 sides; 50 mm thick; on prepared base measured separately; bedding on 25 mm cement: sand (1:3); cement: sand (1:3) joints							
Edgings							
100 mm wide × random lengths	10.09	0.17	16.10	–	12.09	m	**28.19**
100 × 100 mm	10.09	0.17	16.10	–	16.38	m	**32.48**
100 × 200 mm	20.18	0.17	16.10	–	27.97	m	**44.07**
250 mm wide × random lengths	20.50	0.13	12.88	–	28.34	m	**41.22**
500 mm wide × random lengths	40.99	0.11	10.73	–	51.91	m	**62.64**
Yorkstone edgings; 600 mm long × 250 mm wide; cut to radius							
1.00 m to 3.00 m	56.75	0.17	16.10	–	65.73	m	**81.83**
3.00 m to 5.00 m	56.75	0.15	14.31	–	65.73	m	**80.04**
exceeding 5.00 m	56.75	0.13	12.88	–	65.73	m	**78.61**
Natural stone, slate or granite flag edgings; CED Ltd; on prepared base (not included); bedding on 25 mm cement: sand (1:3); cement: sand (1:3) joints							
Edgings; silver grey							
100 mm wide × random lengths	3.00	0.33	32.20	–	3.92	m	**36.12**
100 × 100 mm	3.43	0.50	48.30	–	4.42	m	**52.72**
100 mm long × 200 mm wide	5.57	0.20	19.32	–	6.88	m	**26.20**
250 mm wide × random lengths	5.89	0.25	24.15	–	11.55	m	**35.70**
300 mm wide × random lengths	6.42	0.25	24.15	–	12.16	m	**36.31**
BRICK OR BLOCK EDGINGS							
Brick or block stretchers; bedding in cement: mortar (1:4); on 150 mm deep concrete foundations (not included)							
Single course							
concrete paving blocks; PC £12.66/m²; 200 × 100 × 60 mm	2.51	0.10	9.90	–	3.88	m	**13.78**
engineering bricks; PC £358.00/1000; 215 × 102.5 × 65 mm	1.67	0.13	12.88	–	2.91	m	**15.79**
Two courses							
concrete paving blocks; PC £12.66/m²; 200 × 100 × 60 mm	5.02	0.13	12.88	–	7.56	m	**20.44**
engineering bricks; PC £358.00/1000; 215 × 102.5 × 65 mm	3.35	0.19	18.40	–	5.62	m	**24.02**

35 SITE WORKS – KERBS AND EDGINGS

Item – Overhead and Profit Included	PC £	Labour hours	Labour £	Plant £	Material £	Unit	Total rate £
BRICK OR BLOCK EDGINGS – cont							
Brick or block stretchers – cont							
Three courses							
concrete paving blocks; PC £12.66/m²; 200 × 100 × 60 mm	7.54	0.15	14.31	–	10.44	m	**24.75**
engineering bricks; PC £358.00/ 1000; 215 × 102.5 × 65 mm	5.02	0.22	21.46	–	7.54	m	**29.00**
Bricks on edge; bedding in cement: mortar (1:4); on 150 mm deep concrete foundations							
One brick wide							
engineering bricks; 215 × 102.5 × 65 mm	5.02	0.19	18.40	–	7.74	m	**26.14**
paving bricks; 215 × 102.5 × 65 mm	8.00	0.19	18.40	–	11.17	m	**29.57**
Two courses; stretchers laid on edge; 225 mm wide							
engineering bricks; 215 × 102.5 × 65 mm	10.03	0.40	38.64	–	15.26	m	**53.90**
paving bricks; 215 × 102.5 × 65 mm	16.00	0.40	38.64	–	22.11	m	**60.75**
Edge restraints; to brick paving; on prepared base (not included); 65 mm thick bricks; PC £600.00/ 1000							
Header course							
200 × 100 mm; butt joints	6.00	0.09	8.59	–	7.38	m	**15.97**
200 × 100 mm; mortar joints	5.59	0.17	16.10	–	9.45	m	**25.55**
200 × 100 mm × 50 mm; on edge; mortar joints	8.38	0.23	22.55	–	10.59	m	**33.14**
215 × 102.5 mm mortar joints	5.33	0.17	16.10	–	7.08	m	**23.18**
215 × 102.5 mm × 65 mm; on edge; mortar joints	8.00	0.25	24.16	–	10.14	m	**34.30**
Stretcher course; on flat or on edge							
200 × 100 mm; butt joints	3.00	0.06	5.36	–	3.69	m	**9.05**
210 × 105 mm; mortar joints	2.72	0.11	10.73	–	3.59	m	**14.32**

35 SITE WORKS – KERBS AND EDGINGS

Item – Overhead and Profit Included	PC £	Labour hours	Labour £	Plant £	Material £	Unit	Total rate £
Edge restraints; to brick paving; on prepared base (not included); Penter bricks by Wienerberger; haunching not included							
Header courses; laid on edge; butt jointed on sand bed							
Wienerberger Dutch pavers							
200 × 64 × 85 mm	9.13	0.20	12.88	–	10.84	m	**23.72**
Blue Baggeridge							
200 × 100 × 62 mm	13.05	0.21	13.56	–	15.36	m	**28.92**
Wienerberger Dutch pavers							
200 × 48 × 85 mm	9.61	0.22	14.31	–	11.40	m	**25.71**
Wienerberger Dutch tumbled							
pavers 200 × 50 × 65 mm	7.25	0.22	14.31	–	8.68	m	**22.99**
Wienerberger chamfered							
collection 200 × 100 × 62 mm	10.57	0.21	13.56	–	12.51	m	**26.07**
Header courses; laid on edge; 10 mm flowable mortar jointed on 35 mm mortar bed							
Wienerberger Dutch pavers							
200 × 48 × 85 mm	7.95	0.36	23.00	–	10.52	m	**33.52**
Wienerberger Dutch tumbled							
pavers 200 × 50 × 65 mm	8.57	0.40	25.76	–	11.24	m	**37.00**
Blue Baggeridge							
200 × 100 × 62 mm	11.24	0.33	21.46	–	14.29	m	**35.75**
Wienerberger Dutch pavers							
200 × 64 × 85 mm	7.89	0.33	21.46	–	10.44	m	**31.90**
Wienerberger chamfered							
collection 200 × 100 × 62 mm	9.11	0.33	21.25	–	11.85	m	**33.10**
Wienerberger Dutch tumbled							
pavers 150 × 150 × 65 mm	5.63	0.29	18.40	–	7.84	m	**26.24**
Stretcher courses; laid flat or on edge; butt jointed on mortar bed							
Wienerberger Dutch pavers							
200 × 64 × 85 mm	2.92	0.08	5.36	–	3.70	m	**9.06**
Blue Baggeridge							
200 × 100 × 62 mm	4.05	0.08	5.36	–	5.00	m	**10.36**
Wienerberger Dutch pavers							
200 × 48 × 85 mm	2.31	0.08	5.36	–	3.00	m	**8.36**
Wienerberger chamfered							
collection 200 × 100 × 62 mm	3.28	0.08	5.36	–	4.12	m	**9.48**
Wienerberger Dutch tumbled							
pavers 200 × 50 × 65 mm	2.36	0.08	5.36	–	3.06	m	**8.42**
Stretcher courses; laid flat or on edge; 10 mm flowable mortar jointed on 35 mm mortar bed							
Wienerberger Dutch pavers							
200 × 64 × 85 mm	2.78	0.10	6.44	–	4.58	m	**11.02**
Blue Baggeridge							
200 × 100 × 62 mm	3.85	0.10	6.44	–	5.81	m	**12.25**
Wienerberger chamfered							
collection 200 × 100 × 62 mm	3.12	0.10	6.44	–	4.97	m	**11.41**

35 SITE WORKS – KERBS AND EDGINGS

Item – Overhead and Profit Included	PC £	Labour hours	Labour £	Plant £	Material £	Unit	Total rate £
BRICK OR BLOCK EDGINGS – cont							
Edge restraints – cont							
Stretcher courses – cont							
Wienerberger Dutch pavers							
200 × 48 × 85 mm	2.19	0.10	6.44	–	3.89	m	**10.33**
Wienerberger Dutch tumbled							
pavers 200 × 50 × 65 mm	2.24	0.10	6.44	–	3.96	m	**10.40**
Header courses; laid flat; butt jointed on mortar bed							
Wienerberger Dutch pavers							
200 × 64 × 85 mm	6.87	0.17	10.74	–	8.25	m	**18.99**
Blue Baggeridge							
200 × 100 × 62 mm	8.10	0.17	10.74	–	9.66	m	**20.40**
Wienerberger Dutch pavers							
200 × 48 × 85 mm	5.42	0.17	10.74	–	6.59	m	**17.33**
Wienerberger Dutch tumbled							
pavers 200 × 50 × 65 mm	7.25	0.17	10.74	–	8.68	m	**19.42**
Wienerberger Dutch tumbled							
pavers 150 × 150 × 65 mm	6.00	0.17	10.74	–	7.26	m	**18.00**
Wienerberger chamfered							
collection 200 × 100 × 62 mm	6.56	0.17	10.74	–	7.89	m	**18.63**
Header courses; laid flat; 10 mm flowable mortar jointed on 35 mm mortar bed							
Wienerberger Dutch pavers							
200 × 64 × 85 mm	6.15	0.29	18.40	–	8.44	m	**26.84**
Blue Baggeridge							
200 × 100 × 62 mm	7.36	0.29	18.40	–	9.84	m	**28.24**
Wienerberger Dutch tumbled							
pavers 200 × 50 × 65 mm	6.29	0.31	19.81	–	8.60	m	**28.41**
Wienerberger Dutch pavers							
200 × 48 × 85 mm	4.85	0.29	18.40	–	6.96	m	**25.36**
Wienerberger Dutch tumbled							
pavers 150 × 150 × 65 mm	5.63	0.29	18.40	–	7.84	m	**26.24**
Wienerberger chamfered							
collection 200 × 100 × 62 mm	5.96	0.17	10.74	–	8.23	m	**18.97**
Haunching							
Haunching with in situ concrete; 11.50 N/mm^2–40 mm aggregate one side							
one side	3.54	0.13	8.05	–	4.07	m	**12.12**

35 SITE WORKS – KERBS AND EDGINGS

Item – Overhead and Profit Included	PC £	Labour hours	Labour £	Plant £	Material £	Unit	Total rate £
Variation in brick prices; add or subtract the following amounts for every £100/1000 difference in the PC price							
Edgings; mortar jointed							
100 mm wide	0.50	–	–	–	0.58	m	**0.58**
200 mm wide	0.95	–	–	–	1.09	m	**1.09**
102.5 mm wide	0.47	–	–	–	0.54	m	**0.54**
215 mm wide	0.93	–	–	–	1.07	m	**1.07**
CHANNELS							
Channels; bedding in cement: mortar (1:3); joints pointed flush; on concrete foundations (not included)							
Three courses stretchers; 350 mm wide; quarter bond to form dished channels							
engineering bricks; PC £358.00/ 1000; 215 × 102.5 × 65 mm	5.02	0.33	32.20	–	7.54	m	**39.74**
paving bricks; PC £600.00/1000; 215 × 102.5 × 65 mm	8.00	0.33	32.20	–	10.97	m	**43.17**
Three courses granite setts; 340 mm wide; to form dished channels							
340 mm wide	13.82	0.67	64.40	–	19.96	m	**84.36**
Precast concrete channels etc.; Marshalls Mono; bedding, jointing and pointing in cement: mortar (1:3); including haunching with in situ concrete; 11.50 N/mm²–40 mm aggregate one side							
Channels; square							
125 × 225 × 915 mm long; CS1	6.36	0.13	12.88	–	19.49	m	**32.37**
125 × 150 × 915 mm long; CS2	5.45	0.13	12.88	–	16.39	m	**29.27**
Channels; dished							
305 × 150 × 915 mm long; CD	18.20	0.13	12.88	–	34.09	m	**46.97**
150 × 100 × 915 mm	10.50	0.13	12.88	–	19.56	m	**32.44**

35 SITE WORKS – KERBS AND EDGINGS

Item – Overhead and Profit Included	PC £	Labour hours	Labour £	Plant £	Material £	Unit	Total rate £
TIMBER EDGINGS							
Timber edging boards; fixed with 50 × 50 × 750 mm timber pegs at 1000 mm centres (excavations and hardcore under edgings not included)							
Straight							
22 × 150 mm treated softwood edge boards	2.25	0.07	4.30	–	2.59	m	**6.89**
50 × 150 mm treated softwood edge boards	3.01	0.10	6.44	–	3.46	m	**9.90**
Curved							
22 × 150 mm treated softwood edge boards	2.25	0.13	8.05	–	2.59	m	**10.64**
50 × 150 mm treated softwood edge boards	3.01	0.14	9.20	–	3.46	m	**12.66**
METAL EDGINGS							
Halestem edge L-shaped galvanized steel edging; Everedge; high tensile galvanized steel L shaped edging with 6 mm rolled top edge; fixed to roadway base and edge profile with 5 anchor pins per length; laid to straight or curvilinear road edge; subsequently filled with macadam or other paving material (not included); Rigid or Flexible profiles							
Depth of macadam							
50 mm	9.84	0.07	4.30	–	11.32	m	**15.62**
75 mm	11.52	0.08	4.96	–	13.25	m	**18.21**
100 mm	13.32	0.08	5.36	–	15.32	m	**20.68**
150 mm	14.96	0.07	4.60	–	17.20	m	**21.80**
ProEdge 2.5 mm thick steel edging with 100 mm long spikes. Everedge							
Galvanized steel							
75 mm deep	3.34	0.10	6.44	–	3.84	m	**10.28**
100 mm deep	3.70	0.10	6.44	–	4.26	m	**10.70**
125 mm deep	4.09	0.10	6.44	–	4.70	m	**11.14**
150 mm deep	4.36	0.10	6.44	–	5.01	m	**11.45**
Galvanized steel; brown							
75 mm deep	3.34	0.10	6.44	–	3.84	m	**10.28**
100 mm deep	3.70	0.10	6.44	–	4.26	m	**10.70**
120 mm deep	4.09	0.10	6.44	–	4.70	m	**11.14**
150 mm deep	4.36	0.10	6.44	–	5.01	m	**11.45**

35 SITE WORKS – KERBS AND EDGINGS

Item – Overhead and Profit Included	PC £	Labour hours	Labour £	Plant £	Material £	Unit	Total rate £
Cor-Ten							
75 mm deep	3.34	0.10	6.44	–	3.84	m	**10.28**
100 mm deep	3.70	0.10	6.44	–	4.26	m	**10.70**
125 mm deep	4.09	0.10	6.44	–	4.70	m	**11.14**
Mild steel edgings; 6 mm galvanized steel; Outdoor design							
Bespoke edgings; complete with bolts for adjoining sections and mild steel pins at 1.50 m centres							
100 mm straight	18.00	0.07	4.30	–	20.70	m	**25.00**
150 mm straight	26.00	0.10	6.44	–	29.90	m	**36.34**
300 mm straight	33.00	0.13	8.05	–	37.95	m	**46.00**
Corten prefabricated retaining walls; Adezz limited							
200 mm high on prepared concrete base (not included); top edge of 50 × 25 mm; bolted together internally with 30 mm M8 bolts							
straight sections 1.00 m long	26.10	0.20	12.88	–	30.02	m	**42.90**
straight sections 1.50 m long	25.75	0.17	10.74	–	29.61	m	**40.35**
straight sections 2.00 m long	25.32	0.13	8.05	–	29.12	m	**37.17**
inside corner or outside corner 500 mm × 500 mm	25.06	0.25	16.10	–	28.82	m	**44.92**
inside or outside corner 1.00 m × 1.00 m	24.80	0.30	19.32	–	28.52	m	**47.84**
inside or outside curve 500 mm	72.04	0.20	12.88	–	82.85	m	**95.73**
inside or outside curve; 1.00 m lengths	72.04	0.25	16.10	–	82.85	m	**98.95**
inside or outside curve; 1.50 m lengths	72.03	0.17	10.74	–	82.83	m	**93.57**
inside or outside curve; 2.00 m lengths	72.03	0.14	9.20	–	82.83	m	**92.03**
400 mm high; on prepared concrete base (not included); top edge of 50 × 25 mm; bolted together internally with 30 mm M8 bolts							
straight sections 1.00 m long	46.98	0.22	14.17	–	54.03	m	**68.20**
straight sections 1.50 m long	46.29	0.18	11.81	–	53.23	m	**65.04**
straight sections 2.00 m long	45.67	0.14	8.86	–	52.52	m	**61.38**
inside corner or outside corner 500 mm × 500 mm	44.89	0.28	17.71	–	51.62	m	**69.33**
inside or outside corner 1.00 m × 1.00 m	44.63	0.33	21.25	–	51.32	m	**72.57**
inside or outside curve 500 mm	120.06	0.22	14.17	–	138.07	m	**152.24**
inside or outside curve; 1.00 m lengths	120.06	0.28	17.71	–	138.07	m	**155.78**
inside or outside curve; 1.50 m lengths	131.71	0.18	11.81	–	151.47	m	**163.28**
inside or outside curve; 2.00 m lengths	119.80	0.16	10.12	–	137.77	m	**147.89**

35 SITE WORKS – KERBS AND EDGINGS

Item – Overhead and Profit Included	PC £	Labour hours	Labour £	Plant £	Material £	Unit	Total rate £
METAL EDGINGS – cont							
Corten prefabricated retaining walls – cont							
600 mm high; on prepared concrete base (not included); top edge of 50 × 25 mm; bolted together internally with 30 mm M8 bolts							
straight sections 1.00 m long	90.31	0.24	15.58	–	103.86	m	**119.44**
straight sections 1.50 m long	88.74	0.20	13.00	–	102.05	m	**115.05**
straight sections 2.00 m long	88.22	0.15	9.74	–	101.45	m	**111.19**
inside corner or outside corner 500 mm × 500 mm	86.13	0.30	19.48	–	99.05	m	**118.53**
inside or outside corner 1.00 m × 1.00 m	85.87	0.36	23.38	–	98.75	m	**122.13**
inside or outside curve 500 mm	210.88	0.24	15.58	–	242.51	m	**258.09**
inside or outside curve; 1.00 m lengths	210.89	0.30	19.48	–	242.52	m	**262.00**
inside or outside curve; 1.50 m lengths	210.89	0.20	13.00	–	242.52	m	**255.52**
inside or outside curve; 2.00 m lengths	210.90	0.17	11.13	–	242.54	m	**253.67**

35 SITE WORKS – PAVINGS

Item – Overhead and Profit Included	PC £	Labour hours	Labour £	Plant £	Material £	Unit	Total rate £
SPECIALIZED MORTARS FOR PAVINGS							
SteinTec Ltd							
Primer for specialized mortar; SteinTec Tuffbond priming mortar immediately prior to paving material being placed on bedding mortar maximum thickness 1.5 mm							
(1.5 kg/m²)	1.27	0.02	0.64	–	1.46	m²	**2.10**
Tuffbed; 2 pack hydraulic mortar for paving applications							
30 mm thick	59.32	0.17	5.37	–	68.22	m²	**73.59**
40 mm thick	79.10	0.20	6.44	–	90.96	m²	**97.40**
50 mm thick	98.87	0.25	8.05	–	113.70	m²	**121.75**
SteinTec jointing mortar; Grey or Natural; joint size							
100 × 100 × 100 mm; 10 mm joints; 31.24 kg/m²	22.90	0.04	3.45	–	26.33	m²	**29.78**
100 × 100 × 100 mm; 20 mm joints; 31.24 kg/m²	40.31	0.05	4.83	–	46.36	m²	**51.19**
200 × 100 × 65 mm; 10 mm joints; 15.7 kg/m²	11.51	0.02	2.14	–	13.24	m²	**15.38**
215 × 112.5 × 65 mm; 10 mm joints; 14.3 kg/m²	10.48	0.02	2.14	–	12.05	m²	**14.19**
300 × 450 × 65 mm; 10 mm joints; 6.24 kg/m²	4.57	0.02	2.14	–	5.26	m²	**7.40**
300 × 450 × 65 mm; 20 mm joints; 11.98 kg/m²	8.78	0.04	3.45	–	10.10	m²	**13.55**
450 × 450 × 65 mm; 20 mm joints; 9.75 kg/m²	7.15	0.15	7.03	–	8.22	m²	**15.25**
450 × 600 × 65 mm; 20 mm joints; 8.6 kg/m²	6.30	0.02	2.14	–	7.24	m²	**9.38**
600 × 600 × 65 mm; 20 mm joints; 7.43 kg/m²	5.45	0.02	2.14	–	6.27	m²	**8.41**
SteinTec jointing mortar; Dark grey or beige; joint size							
100 × 100 × 100 mm; 10 mm joints; 31.24 kg/m²	27.44	0.04	3.45	–	31.56	m²	**35.01**
100 × 100 × 100 mm; 20 mm joints; 31.24 kg/m²	48.31	0.05	4.83	–	55.56	m²	**60.39**
200 × 100 × 65 mm; 10 mm joints; 15.7 kg/m²	13.79	0.02	2.14	–	15.86	m²	**18.00**
215 × 112.5 × 65 mm; 10 mm joints; 14.3 kg/m²	12.56	0.02	2.14	–	14.44	m²	**16.58**
300 × 450 × 65 mm; 10 mm joints; 6.24 kg/m²	5.48	0.02	2.14	–	6.30	m²	**8.44**
300 × 450 × 65 mm; 20 mm joints; 11.98 kg/m²	10.52	0.04	3.45	–	12.10	m²	**15.55**
450 × 450 × 65 mm; 20 mm joints; 9.75 kg/m²	8.56	0.15	7.03	–	9.84	m²	**16.87**

35 SITE WORKS – PAVINGS

Item – Overhead and Profit Included	PC £	Labour hours	Labour £	Plant £	Material £	Unit	Total rate £
SPECIALIZED MORTARS FOR PAVINGS – cont							
450 × 600 × 65 mm; 20 mm joints; 8.6 kg/m²	7.55	0.02	2.14	–	8.68	m²	**10.82**
600 × 600 × 65 mm; 20 mm joints; 7.43 kg/m²	6.53	0.02	2.14	–	7.51	m²	**9.65**
TREE ROOT ZONE PROTECTION							
Tree root zone protection; Greenfix Ltd Carefully hand dig around base of tree but away from trunk to required depth; Lay Treetex geofabric to subgrade; Lay Geoweb cellular root protection to required depth and backfill with angular material.							
To graded and compacted substrate (not included); on 50 mm sharp sand base filled with angular drainage aggregate 20–5 mm; not mechanical traffic on the area of the root zone permitted; Excavated material loaded outside the root protection zone to spoil heaps							
Geoweb 75 mm depth	7.26	0.07	6.44	1.16	16.87	m²	**24.47**
Geoweb 100 mm depth	8.63	0.08	7.24	1.35	20.38	m²	**28.97**
Geoweb 150 mm depth	10.55	0.10	9.18	1.46	26.09	m²	**36.73**
Geoweb 200 mm depth	13.06	0.10	10.00	2.20	32.48	m²	**44.68**
Geoweb 300 mm depth	19.72	0.13	12.72	3.51	47.12	m²	**63.35**
EXCAVATION OF PATHWAYS							
Excavation and path preparation							
Excavating; 300 mm deep; to width of path; depositing excavated material at sides of excavation							
width 1.00 m	–	–	–	3.99	–	m²	**3.99**
width 1.50 m	–	–	–	3.32	–	m²	**3.32**
width 2.00 m	–	–	–	2.85	–	m²	**2.85**
width 3.00 m	–	–	–	2.39	–	m²	**2.39**
Excavating trenches; in centre of pathways; 100 mm flexible drain pipes; filling with clean broken stone or gravel rejects							
300 × 450 mm deep	5.86	0.03	2.14	1.99	6.74	m	**10.87**
Hand trimming and compacting reduced surface of pathway; by machine							
width 1.00 m	–	0.02	1.61	0.15	–	m	**1.76**
width 1.50 m	–	0.01	1.43	0.14	–	m	**1.57**
width 2.00 m	–	0.01	1.29	0.12	–	m	**1.41**
width 3.00 m	–	0.01	1.29	0.12	–	m	**1.41**

35 SITE WORKS – PAVINGS

Item – Overhead and Profit Included	PC £	Labour hours	Labour £	Plant £	Material £	Unit	Total rate £
Permeable membranes; to trimmed and compacted surface of pathway							
Terram 1000	0.52	0.01	0.64	–	0.60	m²	**1.24**
FILLING TO MAKE UP LEVELS FOR PAVINGS AND SURFACES							
Filling to make up levels							
Obtained off site; hardcore; PC £32.40/m³							
150 mm thick	5.18	0.01	1.29	0.71	5.96	m²	**7.96**
Type 1 granular fill base; PC £24.75/ tonne (£55.45/m³ compacted)							
100 mm thick	5.45	0.01	0.90	0.54	6.27	m²	**7.71**
150 mm thick	8.17	0.01	0.81	0.83	9.40	m²	**11.04**
Surface treatments							
Sand blinding; to hardcore (not included)							
50 mm thick	2.31	0.01	1.29	–	2.66	m²	**3.95**
Filter fabric; to hardcore (not included)	0.52	–	0.32	–	0.60	m²	**0.92**
BRICK PAVINGS							
Bricks – General							
Preamble: Bricks shall be hard, well burnt, non-dusting, resistant to frost and sulphate attack and true to shape, size and sample.							
Movement of materials							
Mechanically offloading bricks; loading wheelbarrows; transporting maximum 25 m distance	–	0.07	6.44	–	–	m²	**6.44**

35 SITE WORKS – PAVINGS

Item – Overhead and Profit Included	PC £	Labour hours	Labour £	Plant £	Material £	Unit	Total rate £
BRICK PAVINGS – cont							
Variation in brick prices; add or subtract the following amounts for every £100/1000 difference in the PC price							
Edgings; mortar jointed							
100 mm wide	0.50	–	–	–	0.58	m	**0.58**
200 mm wide	0.95	–	–	–	1.09	m	**1.09**
102.5 mm wide	0.47	–	–	–	0.54	m	**0.54**
215 mm wide	0.93	–	–	–	1.07	m	**1.07**
Edge restraints; to brick paving; on prepared base (not included); haunching one side; Penter bricks by Wienerberger; haunched							
Header courses; laid on edge; butt jointed on sand bed; haunched							
Wienerberger Dutch pavers							
200 × 64 × 85 mm	9.13	0.20	12.88	–	14.33	m	**27.21**
Blue Baggeridge							
200 × 100 × 62 mm	13.05	0.21	13.56	–	18.85	m	**32.41**
Wienerberger Dutch pavers							
200 × 48 × 85 mm	9.61	0.22	14.31	–	14.89	m	**29.20**
Wienerberger Dutch tumbled pavers 200 × 50 × 65 mm	7.25	0.22	14.31	–	12.18	m	**26.49**
Wienerberger chamfered collection 200 × 100 × 62 mm	10.57	0.21	13.56	–	16.00	m	**29.56**
Header courses; laid on edge; 10 mm flowable mortar jointed on 35 mm mortar bed; haunched							
Wienerberger Dutch pavers							
200 × 48 × 85 mm	7.95	0.36	23.00	–	14.01	m	**37.01**
Wienerberger Dutch tumbled pavers 200 × 50 × 65 mm	8.57	0.40	25.76	–	14.72	m	**40.48**
Blue Baggeridge							
200 × 100 × 62 mm	11.24	0.33	21.46	–	17.79	m	**39.25**
Wienerberger Dutch pavers							
200 × 64 × 85 mm	7.89	0.33	21.46	–	13.94	m	**35.40**
Wienerberger chamfered collection 200 × 100 × 62 mm	9.11	0.33	21.25	–	15.34	m	**36.59**
Wienerberger Dutch tumbled pavers 150 × 150 × 65 mm	5.63	0.29	18.40	–	11.34	m	**29.74**
Stretcher courses; laid flat or on edge; butt jointed on mortar bed; haunched							
Wienerberger Dutch pavers							
200 × 64 × 85 mm	2.92	0.08	5.36	–	7.20	m	**12.56**
Blue Baggeridge							
200 × 100 × 62 mm	4.05	0.08	5.36	–	8.50	m	**13.86**
Wienerberger Dutch pavers							
200 × 48 × 85 mm	2.31	0.08	5.36	–	6.50	m	**11.86**

35 SITE WORKS – PAVINGS

Item – Overhead and Profit Included	PC £	Labour hours	Labour £	Plant £	Material £	Unit	Total rate £
Wienerberger chamfered collection 200 × 100 × 62 mm	3.28	0.08	5.36	–	7.61	m	**12.97**
Wienerberger Dutch tumbled pavers 200 × 50 × 65 mm	2.36	0.08	5.36	–	6.55	m	**11.91**
Stretcher courses; laid flat or on edge; 10 mm flowable mortar jointed on 35 mm mortar bed; haunched							
Wienerberger Dutch pavers 200 × 64 × 85 mm	2.78	0.10	6.44	–	8.06	m	**14.50**
Blue Baggeridge 200 × 100 × 62 mm	3.85	0.10	6.44	–	9.29	m	**15.73**
Wienerberger chamfered collection 200 × 100 × 62 mm	3.12	0.10	6.44	–	8.45	m	**14.89**
Wienerberger Dutch pavers 200 × 48 × 85 mm	2.19	0.10	6.44	–	7.38	m	**13.82**
Wienerberger Dutch tumbled pavers 200 × 50 × 65 mm	2.24	0.10	6.44	–	7.44	m	**13.88**
Header courses; laid flat; butt jointed on mortar bed; haunched							
Wienerberger Dutch pavers 200 × 64 × 85 mm	6.87	0.17	10.74	–	11.74	m	**22.48**
Blue Baggeridge 200 × 100 × 62 mm	8.10	0.17	10.74	–	13.16	m	**23.90**
Wienerberger Dutch pavers 200 × 48 × 85 mm	5.42	0.17	10.74	–	10.07	m	**20.81**
Wienerberger Dutch tumbled pavers 200 × 50 × 65 mm	7.25	0.17	10.74	–	12.18	m	**22.92**
Wienerberger Dutch tumbled pavers 150 × 150 × 65 mm	6.00	0.17	10.74	–	10.74	m	**21.48**
Wienerberger chamfered collection 200 × 100 × 62 mm	6.56	0.17	10.74	–	11.39	m	**22.13**
Header courses; laid flat; 10 mm flowable mortar jointed on 35 mm mortar bed; haunched							
Wienerberger Dutch pavers 200 × 64 × 85 mm	6.15	0.29	18.40	–	11.94	m	**30.34**
Blue Baggeridge 200 × 100 × 62 mm	7.36	0.29	18.40	–	13.33	m	**31.73**
Wienerberger Dutch tumbled pavers 200 × 50 × 65 mm	6.29	0.31	19.81	–	12.10	m	**31.91**
Wienerberger Dutch pavers 200 × 48 × 85 mm	4.85	0.29	18.40	–	10.44	m	**28.84**
Wienerberger Dutch tumbled pavers 150 × 150 × 65 mm	5.63	0.29	18.40	–	11.34	m	**29.74**
Wienerberger chamfered collection 200 × 100 × 62 mm	5.96	0.17	10.74	–	11.72	m	**22.46**

35 SITE WORKS – PAVINGS

Item – Overhead and Profit Included	PC £	Labour hours	Labour £	Plant £	Material £	Unit	Total rate £
BRICK PAVINGS – cont							
Clay brick pavings; Wienerberger; on prepared base (not included); bedding on 50 mm sharp sand; kiln dried sand joints brushed in; Herringbone bond							
Pavings; Blue Baggeridge range 200 × 100 × 62 mm wirecut chamfered paviors							
laid flat	40.49	0.44	42.37	–	50.58	m²	**92.95**
Pavings; Wienerberger chamfered collection 200 × 100 × 62 mm							
laid flat	32.80	0.44	42.37	–	41.80	m²	**84.17**
Pavings; Wienerberger Dutch pavers 200 × 48 × 85 mm; PC £450.00/1000							
laid flat	27.13	0.52	49.79	–	35.28	m²	**85.07**
laid on edge	47.97	0.92	88.62	–	59.27	m²	**147.89**
Pavings; Wienerberger Dutch pavers 200 × 64 × 85 mm; PC £570.00/1000							
laid flat	34.35	0.52	49.79	–	43.58	m²	**93.37**
laid on edge	45.65	0.68	66.16	–	56.59	m²	**122.75**
Pavings; Wienerberger Dutch tumbled pavers 200 × 50 × 65 mm; PC £460.00/1000							
laid flat	36.27	0.44	42.37	–	45.79	m²	**88.16**
laid on edge	47.15	0.88	84.73	–	58.33	m²	**143.06**
Pavings; Wienerberger Dutch tumbled pavers 150 × 150 × 65 mm; PC £460.00/1000							
laid flat	41.00	0.39	37.58	–	51.32	m²	**88.90**
Clay brick pavings; laid to running stretcher or stack bond only; on prepared base (not included); bedding on cement: sand (1:4) pointing with flowable mortar – Steintec							
Pavings; Blue Baggeridge range 200 × 100 × 62 mm wirecut chamfered paviors							
laid flat	35.05	0.44	42.55	–	75.52	m²	**118.07**
Pavings; Wienerberger chamfered collection 200 × 100 × 62 mm							
laid flat	28.40	0.37	35.78	–	67.86	m²	**103.64**
Pavings; Wienerberger Dutch pavers 200 × 48 × 85 mm; PC £450.00/1000							
laid flat	23.12	0.44	42.37	–	63.64	m²	**106.01**
laid on edge	47.97	0.92	88.62	–	143.78	m²	**232.40**

35 SITE WORKS – PAVINGS

Item – Overhead and Profit Included	PC £	Labour hours	Labour £	Plant £	Material £	Unit	Total rate £
Pavings; Wienerberger Dutch pavers 200 × 64 × 85 mm; PC £570.00/1000							
laid flat	34.35	0.52	49.79	–	74.72	m²	**124.51**
laid on edge	45.65	0.39	37.73	–	126.91	m²	**164.64**
Pavings; Wienerberger Dutch tumbled pavers 200 × 50 × 65 mm; PC £460.00/1000							
laid flat	20.41	0.38	36.73	–	58.68	m²	**95.41**
laid on edge	47.15	0.50	48.30	–	121.00	m²	**169.30**
Pavings; Wienerberger Dutch tumbled pavers 150 × 150 × 65 mm; PC £460/1000							
laid flat	35.15	0.34	33.09	–	81.16	m²	**114.25**
Cutting							
curved cutting	–	0.50	16.10	4.36	–	m	**20.46**
raking cutting	–	0.25	8.05	3.42	–	m	**11.47**
CELLULAR CONFINED GRANULAR PAVINGS							
StablePAVE; Cedar Nursery; honeycomb shaped cellular polypropylene containment grids for retaining filled decorative aggregate surfaces (not included) for driveways, cycle paths and footpaths; to graded compacted substrate (not included) and 50 mm sharp sand base; integrated geofabric base							
Foot traffic, wheelchairs, bikes and pushchairs							
StablePAVE Eco;							
1200 × 800 × 25 mm deep	10.16	0.08	4.03	–	11.68	m²	**15.71**
Driveways: vehicles, wheelchairs, bikes and pushchairs (requires compacted granular base to support vehicle loadings)							
StablePAVE Trade 30;							
2400 × 1200 × 40 mm deep	2.10	0.08	4.37	–	12.88	m²	**17.25**
Aggregate filling to StablePAVE Eco							
shingle	1.15	0.03	3.22	2.53	1.32	m²	**7.07**
decorative aggregate; PC £115.00/tonne	5.20	0.03	3.22	2.53	5.98	m²	**11.73**
Aggregate filling to StablePAVE Trade 30							
shingle	1.84	0.03	2.70	3.06	2.12	m²	**7.88**
decorative aggregate; CED Ltd; PC £115.00/tonne	8.32	0.03	2.70	3.06	9.57	m²	**15.33**

35 SITE WORKS – PAVINGS

Item – Overhead and Profit Included	PC £	Labour hours	Labour £	Plant £	Material £	Unit	Total rate £
CELLULAR CONFINED GRANULAR PAVINGS – cont							
StablePAVE HD; Cedar Nursery; honeycomb shaped polypropylene cellular containment system for retaining filled decorative aggregate surfaces for driveways, cyclepaths and footpaths; on compacted subgrade and subbase (both not included)							
Driveways: vehicles, wheelchairs, bikes and pushchairs (requires compacted granular base to support vehicle loadings)							
StablePAVE HD; 600 × 500 × 32 mm deep	2.10	0.08	4.63	–	14.81	m^2	**19.44**
Aggregate filling to StablePAVE HD							
shingle	1.47	0.03	3.02	3.69	1.69	m^2	**8.40**
decorative aggregate; PC £115.00/tonne	6.65	0.03	3.02	3.69	7.65	m^2	**14.36**
StableGRASS 38 for Grass reinforcement							
Polyethylene soil stabilizing panels; to soil surfaces brought to grade (not included); filling with excavated material							
panels; 50 cm × 50 cm × 38 mm deep	9.70	0.03	3.22	5.20	11.15	m^2	**19.57**
Extra for filling StableGRASS with topsoil; seeding with rye grass at 50 g/m^2	1.51	0.07	6.47	0.22	1.74	m^2	**8.43**
Ground Grid; Recycled plastic; Hahn Plastics Ltd; Interlocking; filling with angular granular material (5–15 mm)							
HDGG 8 cm thick × 40 cm wide × 60 cm long grid for HGV use; Strength: 18,600 kN/m^2 (1,896 tonnes/m^2); laid to prepared base (not included)							
HDGG; Units 40 cm wide × 60 cm long × 80 mm thick; 100–500 m^2 Volume of angular gravel fill is 1.8 tonnes per 25 m^2 or 7.2 kg/m^2	19.50	0.03	2.80	2.54	23.06	m^2	**28.40**

35 SITE WORKS – PAVINGS

Item – Overhead and Profit Included	PC £	Labour hours	Labour £	Plant £	Material £	Unit	Total rate £
Geocell; Greenfix; cellular confinement system for load support or permeable bases below pavings							
To graded and compacted substrate (not included); on 50 mm sharp sand base filled with angular drainage aggregate 20–5 mm							
Geocell 75 mm depth	7.26	0.05	4.39	1.94	16.87	m²	**23.20**
Geocell 100 mm depth	8.63	0.05	4.83	2.20	20.38	m²	**27.41**
Geocell 150 mm depth	10.55	0.06	5.37	2.31	26.09	m²	**33.77**
Geocell 200 mm depth	13.06	0.08	8.05	2.51	32.48	m²	**43.04**
Geocell 300 mm depth	19.72	0.10	9.66	2.92	47.12	m²	**59.70**
Cellular pavings by Grass Concrete							
Honeycomb cellular polyproylene interconnecting paviors with integral downstead anti-shear cleats including topsoil but excluding edge restraints; to granular subbase (not included)							
635 × 330 × 42 mm overall laid to a module of 622 × 311 × 32 mm	–	–	–	–	–	m²	**37.28**
GRASS CONCRETE							
Grass concrete – General							
Preamble: Grass seed should be a perennial ryegrass mixture, with the proportion depending on expected traffic. Hardwearing winter sportsground mixtures are suitable for public areas. Loose gravel, shingle or sand is liable to be kicked out of the blocks; rammed hoggin or other stabilized material should be specified.							
Grass concrete; Grass Concrete Ltd; on blinded granular Type 1 subbase (not included)							
Grasscrete in situ concrete continuously reinforced surfacing; including expansion joints at 10 m centres; soiling; seeding							
GC2; 150 mm thick; traffic up to 40.00 tonnes	–	–	–	–	–	m²	**55.18**
GC1; 100 mm thick; traffic up to 13.30 tonnes	–	–	–	–	–	m²	**42.89**
GC3; 76 mm thick; traffic up to 4.30 tonnes	–	–	–	–	–	m²	**37.35**

35 SITE WORKS – PAVINGS

Item – Overhead and Profit Included	PC £	Labour hours	Labour £	Plant £	Material £	Unit	Total rate £
GRASS CONCRETE – cont							
Embankments; Grass Concrete Ltd							
Grasscrete; in situ reinforced concrete surfacing; to 20 mm thick sand blinding layer (not included); including soiling and seeding							
GC1; 100 mm thick	–	–	–	–	47.23	m²	**47.23**
GC2; 150 mm thick	–	–	–	–	61.57	m²	**61.57**
Grassblock 103; solid matrix precast concrete blocks; to 20 mm thick sand blinding layer; excluding edge restraint; including soiling and seeding							
406 × 406 × 103 mm; fully interlocking	–	–	–	–	57.59	m²	**57.59**
Grass concrete; Grass Concrete Ltd; 406 × 406 mm blocks; on 20 mm sharp sand; on blinded MOT type 1 subbase (not included); level and to falls only; including filling with topsoil; seeding with dwarf rye grass at PC £4.50/kg							
Pavings							
GB103; 103 mm thick	24.97	0.40	12.88	0.22	31.72	m²	**44.82**
GB83; 83 mm thick	23.70	0.36	11.71	0.22	30.26	m²	**42.19**
Grass concrete; Marshalls Plc; on 25 mm sharp sand; including filling with topsoil; seeding with dwarf rye grass at PC £4.50/kg							
Concrete grass pavings							
Grassguard 130; for light duty applications (80 mm prepared base not included)	30.35	0.38	12.07	0.22	38.17	m²	**50.46**
Grassguard 160; for medium duty applications (80–150 mm prepared base not included)	34.04	0.46	14.86	0.22	42.41	m²	**57.49**
Grassguard 180; for heavy duty applications (150 mm prepared base not included)	38.11	0.60	19.32	0.22	47.09	m²	**66.63**
Full mortar bedding							
Extra over pavings for bedding on 25 mm cement: sand (1:4); in lieu of spot bedding on sharp sand	–	0.03	0.81	–	2.14	m²	**2.95**

35 SITE WORKS – PAVINGS

Item – Overhead and Profit Included	PC £	Labour hours	Labour £	Plant £	Material £	Unit	Total rate £
COBBLE PAVINGS							
Cobble pavings – General Cobbles should be embedded by hand, tight-butted, endwise to a depth of 60% of their length. A dry grout of rapid-hardening cement: sand (1:2) shall be brushed over the cobbles until the interstices are filled to the level of the adjoining paving. Surplus grout shall then be brushed off and a light, fine spray of water applied over the area.							
Cobble pavings Cobbles; to present a uniform colour in panels; or varied in colour as required							
Scottish Beach Cobbles; 50–75 mm	42.67	1.11	107.33	–	57.51	m²	**164.84**
Scottish Beach Cobbles; 75–100 mm	42.67	0.83	80.50	–	57.51	m²	**138.01**
Scottish Beach Cobbles; 100–200 mm	42.67	0.67	64.40	–	57.51	m²	**121.91**
GRANULAR PAVINGS							
Footpath gravels; porous self binding gravel CED Ltd; Cedec gravel; self-binding; laid to inert (non-limestone) base measured separately; compacting							
red, silver or gold; 50 mm thick	14.49	0.01	0.81	0.43	16.66	m²	**17.90**
Grundon Ltd; Coxwell self-binding path gravels laid and compacted to excavation or base measured separately							
50 mm thick	6.95	0.01	0.81	0.43	7.99	m²	**9.23**
Breedon Special Aggregates; Golden Gravel or equivalent; rolling wet; on hardcore base (not included); for pavements; to falls and crossfalls and to slopes not exceeding 15° from horizontal; over 300 mm wide							
50 mm thick	5.22	0.01	0.81	0.33	6.00	m²	**7.14**
75 mm thick	7.83	0.03	3.22	0.44	9.00	m²	**12.66**

35 SITE WORKS – PAVINGS

Item – Overhead and Profit Included	PC £	Labour hours	Labour £	Plant £	Material £	Unit	Total rate £
GRANULAR PAVINGS – cont							
Footpath gravels – cont							
Breedon Special Aggregates; Wayfarer specially formulated fine gravel for use on golf course pathways							
50 mm thick	5.44	0.01	0.81	0.33	6.26	m²	**7.40**
75 mm thick	8.15	0.02	1.61	0.59	9.37	m²	**11.57**
Hoggin (stabilized); PC £49.50/m³ on hardcore base (not included); to falls and crossfalls and to slopes not exceeding 15° from horizontal; over 300 mm wide							
100 mm thick	7.42	0.01	1.07	0.56	8.53	m²	**10.16**
150 mm thick	11.14	0.02	1.61	0.84	12.81	m²	**15.26**
Ballast; as dug; watering; rolling; on hardcore base (not included)							
100 mm thick	5.74	0.01	1.07	0.56	6.60	m²	**8.23**
150 mm thick	8.61	0.02	1.61	0.84	9.90	m²	**12.35**
Footpath gravels; porous loose gravels							
Breedon Special Aggregates; Breedon Buff decorative limestone chippings in Warwick Gold, Coastal cream							
50 mm thick	3.36	–	0.32	0.10	3.86	m²	**4.28**
75 mm thick	6.86	–	0.40	0.14	7.89	m²	**8.43**
Breedon Special Aggregates; Brindle, Moorland Black, Shierglas and Galloway chippings							
50 mm thick	3.36	–	0.32	0.10	3.86	m²	**4.28**
75 mm thick	5.04	–	0.40	0.14	5.80	m²	**6.34**
Breedon Special Aggregates; 40 mm slate chippings; plum/blue							
50 mm thick	3.36	–	0.32	0.10	3.86	m²	**4.28**
75 mm thick	5.04	–	0.40	0.14	5.80	m²	**6.34**
Washed shingle; on prepared base (not included)							
25–50 mm size; 25 mm thick	1.21	0.01	0.81	0.09	1.39	m²	**2.29**
25–50 mm size; 75 mm thick	3.62	0.03	2.42	0.29	4.16	m²	**6.87**
50–75 mm size; 25 mm thick	1.21	0.01	0.64	0.10	1.39	m²	**2.13**
50–75 mm size; 75 mm thick	3.62	0.02	2.14	0.36	4.16	m²	**6.66**
Pea shingle; on prepared base (not included)							
10–15 mm size; 25 mm thick	1.35	0.01	0.58	0.09	1.55	m²	**2.22**
5–10 mm size; 75 mm thick	4.05	0.02	1.69	0.29	4.66	m²	**6.64**

35 SITE WORKS – PAVINGS

Item – Overhead and Profit Included	PC £	Labour hours	Labour £	Plant £	Material £	Unit	Total rate £
PRECAST CONCRETE BLOCK PAVINGS							
Precast concrete block edgings; PC £11.76/m²; 200 × 100 × 60 mm; on prepared base (not included); haunching one side							
Edgings; butt joints							
stretcher course	1.18	0.06	5.36	–	4.62	m	**9.98**
header course	2.35	0.09	8.59	–	6.22	m	**14.81**
Precast concrete paving blocks; Keyblok Marshalls Plc; on prepared base (not included); on 50 mm compacted sharp sand bed; blocks laid in 7 mm loose sand and vibrated; joints filled with sharp sand and vibrated; level and to falls only							
Herringbone bond							
200 × 100 × 60 mm; natural grey	12.35	0.28	27.06	0.18	17.27	m²	**44.51**
200 × 100 × 60 mm; colours	13.19	0.28	27.06	0.18	18.24	m²	**45.48**
200 × 100 × 80 mm; natural grey	14.64	0.28	27.06	0.18	19.91	m²	**47.15**
200 × 100 × 80 mm; colours	15.13	0.28	27.06	0.18	20.47	m²	**47.71**
Basketweave bond							
200 × 100 × 60 mm; natural grey	12.35	0.25	24.15	0.18	17.27	m²	**41.60**
200 × 100 × 60 mm; colours	13.19	0.25	24.15	0.18	18.24	m²	**42.57**
200 × 100 × 80 mm; natural grey	14.64	0.25	24.15	0.18	19.91	m²	**44.24**
200 × 100 × 80 mm; colours	15.13	0.25	24.15	0.18	20.47	m²	**44.80**
Tegula traditional Concrete setts; on 50 mm sand; compacted; vibrated; joints filled with sand; natural or coloured; well rammed hardcore base (not included)							
Marshalls Plc; Tegula cobble paving; Mixed sizes 240 × 160 mm; 160 x160 mm; 120 × 160 mm							
60 mm thick; random sizes	28.08	0.19	18.40	0.18	35.60	m²	**54.18**
60 mm thick; single size	28.08	0.15	14.64	0.18	35.60	m²	**50.42**
80 mm thick; random sizes	31.11	0.19	18.40	0.18	39.09	m²	**57.67**
80 mm thick; single size	31.11	0.15	14.64	0.18	39.09	m²	**53.91**
cobbles; 80 × 80 × 60 mm thick;							
traditional	50.25	0.19	17.89	0.18	61.10	m²	**79.17**

35 SITE WORKS – PAVINGS

Item – Overhead and Profit Included	PC £	Labour hours	Labour £	Plant £	Material £	Unit	Total rate £
PRECAST CONCRETE BLOCK PAVINGS – cont							
Marshalls Plc Textured granite aggregate sett paving; on primary aggregate 6–2 mm; compacted; vibrated; joints filled with grit; Silver grey; Graphite; Charcoal; well rammed hardcore base (not included)							
Mistral; Mixed sizes 240 × 160 mm; 160 x160 mm; 120 × 160 mm							
60 mm thick; mixed sizes	37.43	0.25	24.15	0.18	47.79	m²	**72.12**
80 mm thick; mixed sizes	42.66	0.29	27.60	0.18	53.81	m²	**81.59**
Modal textured blocks; mixed shades and sizes; 200 × 100 mm– 400 × 200 mm; (larger sizes shown under slab paving)							
single size 80 mm thick	43.72	0.29	27.60	0.18	55.03	m²	**82.81**
mixed sizes to pattern 80 mm thick	43.72	0.33	32.20	0.18	55.03	m²	**87.41**
Modal smooth blocks; mixed shades and sizes; 200 × 100 mm– 400 × 200 mm; (larger sizes shown under slab paving)							
single size 80 mm thick	58.48	0.29	27.60	0.18	72.00	m²	**99.78**
mixed sizes to pattern 80 mm thick	58.48	0.33	32.20	0.18	72.00	m²	**104.38**
PRECAST CONCRETE SUDS PERMEABLE PAVING							
SUDS paving; Concrete Block Permeable Paving (CBPP)							
Note: Sustainable Drainage Systems (SUDS) aim to reduce flood risk by controlling the rate and volume of surface water run-off from developments. Permeable Paving is a SUDS technique. In addition to managing the quantity of surface water run-off more effectively, one of the primary benefits of CBPP is the potential enhancement of water quality. Permeable pavements improve water quality by mirroring nature in providing filtration and allowing for natural biodegradation of hydrocarbons and the dilution of other contaminants as water passes through the system.							

35 SITE WORKS – PAVINGS

Item – Overhead and Profit Included	PC £	Labour hours	Labour £	Plant £	Material £	Unit	Total rate £
Bases for Sustainable Urban Drainage System (SUDS) Type 3 open graded subbase for use below permeable pavings 40 down open graded material laid loose on prepared subgrade; within edge restraints to paving area (not included); graded to levels and falls as per paving manufacturer's instructions							
100 mm thick	5.93	0.02	2.25	0.52	10.32	m²	**13.09**
150 mm thick	8.89	0.03	3.14	1.14	13.72	m²	**18.00**
200 mm thick	11.86	0.04	4.03	1.68	17.13	m²	**22.84**
250 mm thick	14.82	0.05	4.92	1.68	20.54	m²	**27.14**
over 250 mm thick	65.22	0.19	18.38	6.80	78.50	m³	**103.68**
Pervious surfacing materials; Marshalls Mono SuDs blocks; permeable block paving system; incorporating a 5 mm spacer design that provides a 5 mm void allowing ingress of water through to the subbase storage system (priced separately); prices shown are for the highest cost product in each range Priora; Rectangular various colours; bedded on 50 mm thick 2–6.3 mm clean, angular, free draining uncompacted aggregate; joints infilled with 3 mm clean grit; various colours available; prices shown below are for the highest cost product							
200 × 100 × 60 mm	15.38	0.20	19.32	2.61	21.10	m²	**43.03**
200 × 100 × 80 mm	19.07	0.25	24.15	2.61	25.33	m²	**52.09**
Tegula Priora; various colours; bedded on 50 mm thick 2–6.3 mm clean, angular, free draining uncompacted aggregate; joints infilled with 3 mm clean grit; prices shown below are for the highest cost product							
120 × 160 × 60 mm	28.85	0.13	12.55	2.61	36.59	m²	**51.75**
240 × 160 × 80 mm	33.72	0.13	12.55	2.61	42.19	m²	**57.35**

35 SITE WORKS – PAVINGS

Item – Overhead and Profit Included	PC £	Labour hours	Labour £	Plant £	Material £	Unit	Total rate £
PRECAST CONCRETE SUDS PERMEABLE PAVING – cont							
Pervious surfacing materials – cont							
Modal Priora; Textured; various colours; bedded on 50 mm thick 2–6.3 mm clean, angular, free draining uncompacted aggregate; joints infilled with 3 mm clean grit; prices shown below are for the highest cost product; 80 mm thick							
200 × 100	46.78	0.17	16.10	2.61	57.21	m^2	**75.92**
200 × 200	46.78	0.13	12.88	2.61	57.21	m^2	**72.70**
200 × 300	46.78	0.13	12.07	2.61	57.21	m^2	**71.89**
300 × 300	46.78	0.14	13.80	2.61	57.21	m^2	**73.62**
300 × 600	46.78	0.25	24.15	2.61	57.21	m^2	**83.97**
Mixed size formats	46.78	0.29	27.60	2.61	57.21	m^2	**87.42**
Conservation priora; mix of natural granite aggregates with a choice of surface finish; bedded on 50 mm thick 2–6.3 mm clean, angular, free draining uncompacted aggregate; joints infilled with 3 mm clean grit; prices shown below are for the highest cost product; 65 mm thick							
200 × 400 mm	46.88	0.17	16.10	2.61	57.33	m^2	**76.04**
400 × 400 mm	46.88	0.14	13.80	2.61	57.33	m^2	**73.74**
600 × 400 mm	46.88	0.25	24.15	2.61	57.33	m^2	**84.09**
Mixed formats	46.88	0.29	27.60	2.61	57.33	m^2	**87.54**
PRECAST CONCRETE FLAG PAVINGS							
Note: Precast flag pavings. The prices shown below do not always reflect the full ranges of colours and sizes avaialble; Readers should consult the supplier listed via the directory at the front of this book.							
Flag paving; precast concrete pavings; Marshalls Mono; on prepared subbase (not included); bedding on 25 mm thick cement: sand mortar (1:4); butt joints; straight both ways; on 50 mm thick sharp sand base							

35 SITE WORKS – PAVINGS

Item – Overhead and Profit Included	PC £	Labour hours	Labour £	Plant £	Material £	Unit	Total rate £
Pavings; Standard; Grey							
450 × 450 × 50 mm	18.39	0.15	14.31	–	27.96	m²	**42.27**
600 × 300 × 50 mm	21.83	0.15	14.31	–	31.91	m²	**46.22**
400 × 400 × 65 mm	18.38	0.13	12.88	–	27.93	m²	**40.81**
450 × 600 × 50 mm	13.59	0.15	14.31	–	22.44	m²	**36.75**
600 × 600 × 50 mm	11.92	0.13	12.88	–	20.52	m²	**33.40**
750 × 600 × 50 mm	11.18	0.13	12.88	–	19.67	m²	**32.55**
900 × 600 × 50 mm	9.46	0.13	12.88	–	17.69	m²	**30.57**
Pavings; Buff							
400 × 400 × 65 mm	18.38	0.13	12.88	–	27.93	m²	**40.81**
600 × 600 × 50 mm	18.28	0.13	12.88	–	27.83	m²	**40.71**
900 × 600 × 50 mm	15.28	0.13	12.88	–	24.38	m²	**37.26**
Pavings; Pimple textured Buff or Natural; (Prices shown are for Buff) Natural slightly less							
450 × 450 × 50 mm	15.26	0.15	14.31	–	24.36	m²	**38.67**
600 × 300 × 50 mm	15.26	0.15	14.31	–	24.36	m²	**38.67**
400 × 400 × 65 mm	15.26	0.13	12.88	–	24.36	m²	**37.24**
450 × 600 × 50 mm	15.26	0.15	14.31	–	24.36	m²	**38.67**
600 × 600 × 50 mm	15.26	0.13	12.88	–	24.36	m²	**37.24**
750 × 600 × 50 mm	15.26	0.13	12.88	–	24.36	m²	**37.24**
900 × 600 × 50 mm	15.26	0.13	12.88	–	24.36	m²	**37.24**
Pavings; Pimple textured Buff or Natural; (Prices shown are for Buff) Natural slightly less							
450 × 450 × 50 mm	15.26	0.15	14.31	–	24.36	m²	**38.67**
600 × 300 × 50 mm	15.26	0.15	14.31	–	24.36	m²	**38.67**
400 × 400 × 65 mm	15.26	0.13	12.88	–	24.36	m²	**37.24**
450 × 600 × 50 mm	15.26	0.15	14.31	–	24.36	m²	**38.67**
600 × 600 × 50 mm	15.26	0.13	12.88	–	24.36	m²	**37.24**
750 × 600 × 50 mm	15.26	0.13	12.88	–	24.36	m²	**37.24**
900 × 600 × 50 mm	15.26	0.13	12.88	–	24.36	m²	**37.24**
Precast concrete flag pavings; Marshalls Plc; on prepared subbase measured separately; bedding on 25 mm thick cement: sand mortar (1:4); pointed straight both ways Steintec jointing mixture							
Urbex paving; Practical economical paving alternative to Pimple paving; Riven or Textured in Natural or Buff							
450 × 450 × 35 mm	16.23	0.32	31.17	–	27.76	m²	**58.93**
600 × 300 × 35 mm	16.23	0.33	32.20	–	27.91	m²	**60.11**
600 × 600 × 35 mm	16.23	0.27	25.76	–	26.06	m²	**51.82**
600 × 600 × 63 mm	52.84	0.27	25.76	–	68.16	m²	**93.92**

35 SITE WORKS – PAVINGS

Item – Overhead and Profit Included	PC £	Labour hours	Labour £	Plant £	Material £	Unit	Total rate £
PRECAST CONCRETE FLAG PAVINGS – cont							
Precast concrete flag pavings – cont							
Celestia paving; Smooth or skimmed or textured in various colour options; Luna, Nebula, Grava, Lyra with minor cost variations beteween colours and textures. Highest material cost used in the calculations below; Smooth pavings are generally higher priced							
450 × 450 × 50 mm	48.41	0.32	31.17	–	63.72	m²	**94.89**
600 × 600 × 50 mm	48.41	0.27	25.76	–	63.07	m²	**88.83**
600 × 200 × 50 mm	48.41	0.45	43.91	–	63.07	m²	**106.98**
900 × 300 × 50 mm	48.41	0.27	25.76	–	63.07	m²	**88.83**
Modal textured flags; mixed shades and sizes; (smaller sizes shown under slab paving); 80 mm thick							
400 × 400 mm	43.72	0.14	13.52	–	57.07	m²	**70.59**
450 × 450 mm	43.72	0.16	15.02	–	57.07	m²	**72.09**
600 × 200 mm	43.72	0.16	15.02	–	57.07	m²	**72.09**
600 × 300 mm	43.72	0.16	15.84	–	57.07	m²	**72.91**
600 × 600 mm	43.72	0.14	13.52	–	57.07	m²	**70.59**
900 × 600 mm	43.72	0.14	13.52	–	57.07	m²	**70.59**
mixed sizes to pattern	43.72	0.26	25.36	0.18	57.97	m²	**83.51**
Precast concrete flag pavings; Marshalls Plc; Conservation paving; Natural granite aggregate appearance on prepared subbase measured separately; bedding on 25 mm thick cement: sand mortar (1:4); pointed straight both ways Steintec jointing mixture							
Conservation paving; Smooth skimmed or textured in various colour options; Silver grey, Charcoal, Heather grey, and Buff with minor cost variations between colours and textures. Highest material cost used in the calculations below; Smooth pavings are generally higher priced							
450 × 300 × 50 mm	52.84	0.33	32.20	–	69.45	m²	**101.65**
450 × 450 × 50 mm	52.84	0.32	31.17	–	68.82	m²	**99.99**
450 × 600 × 63 mm	52.84	0.15	14.31	–	68.16	m²	**82.47**

35 SITE WORKS – PAVINGS

Item – Overhead and Profit Included	PC £	Labour hours	Labour £	Plant £	Material £	Unit	Total rate £
Precast concrete vehicular paving flags; Marshalls Plc; on prepared base (not included); on 50 mm compacted sharp sand bed; flags laid in 7 mm loose sand and vibrated; joints filled with sharp sand and vibrated; level and to falls only							
Trafica paving blocks;							
450 × 450 × 70 mm							
Perfecta finish; colour natural	44.42	0.17	16.10	0.18	53.74	m²	70.02
Perfecta finish; colour buff	48.76	0.17	16.10	0.18	58.74	m²	75.02
Saxon finish; colour natural	43.71	0.17	16.10	0.18	52.92	m²	69.20
Saxon finish; colour buff	48.36	0.17	16.10	0.18	58.27	m²	74.55
900 × 600 mm with additional							
central support per slab	84.66	0.31	20.13	–	97.36	m²	117.49
Paving flags to pedestals on roof decks or podiums; Rates for installing flags to pedestals; Prices do not include supply or movement of the slabs to the laying position							
Paving flags; laid at uniform sizes with one pedestal per corner of each paving flag; paving flag sizes; laid on adjustable pedestal 285–400 mm high above the flag							
300 × 300 mm	205.36	0.63	40.25	–	236.16	m²	276.41
400 × 400 mm	120.80	0.63	40.25	–	138.92	m²	179.17
450 × 450 mm	90.60	0.63	40.25	–	104.19	m²	144.44
600 × 450 mm with additional							
central support per flag	79.73	0.42	26.83	–	91.69	m²	118.52
900 × 600 mm with additional							
central support per flag	45.90	0.31	20.13	–	52.78	m²	72.91
Paving flags; laid at uniform sizes with one pedestal per corner of each paving flag; paving flag sizes; laid on adjustable pedestal 645–850 mm high above the flag							
300 × 300 mm	378.76	0.63	40.25	–	435.57	m²	475.82
400 × 400 mm	222.80	0.63	40.25	–	256.22	m²	296.47
450 × 450 mm	167.10	0.63	40.25	–	192.16	m²	232.41
600 × 450 mm with additional							
central support per flag	147.05	0.42	26.83	–	169.11	m²	195.94
900 × 600 mm with additional							
central support per flag	84.66	0.31	20.13	–	97.36	m²	117.49

35 SITE WORKS – PAVINGS

Item – Overhead and Profit Included	PC £	Labour hours	Labour £	Plant £	Material £	Unit	Total rate £
PRECAST CONCRETE FLAG PAVINGS FIBRE REINFORCED							
Precast concrete flag pavings; Marshalls Plc; Fibre reinforced pavings to match other flag paving families; on prepared subbase measured separately; bedding on 25 mm thick cement: sand mortar (1:4); pointed straight both ways Steintec jointing mixture							
Fibre Reinforced Paving							
Conservation 450 × 600 × 63 mm							
Textured	71.49	0.27	25.76	–	89.62	m²	**115.38**
Standard paving 600/900/							
750 × 600 × 63 mm	21.04	0.27	25.76	–	31.59	m²	**57.35**
Standard paving							
450 × 600 × 63 mm	25.34	0.27	25.76	–	36.54	m²	**62.30**
DETERRENT AND TACTILE PAVINGS							
Pedestrian deterrent or tactile pavings; Marshalls Plc; on prepared base (not included); bedding on 25 mm cement: sand (1:3); cement: sand (1:3) joints							
Blister paving							
450 × 450 × 50 mm	41.35	0.15	14.86	–	52.97	m²	**67.83**
450 × 450 × 70 mm	46.29	0.17	16.10	–	58.65	m²	**74.75**
400 × 400 × 50 mm	34.44	0.15	14.86	–	45.02	m²	**59.88**
400 × 400 × 65 mm	49.25	0.16	15.46	–	62.05	m²	**77.51**
Hazard warning flags							
400 × 400 × 50 mm	46.44	0.20	19.32	–	58.82	m²	**78.14**

35 SITE WORKS – PAVINGS

Item – Overhead and Profit Included	PC £	Labour hours	Labour £	Plant £	Material £	Unit	Total rate £
Edge restraints; to block paving; on prepared base (not included); 200 × 100 × 80 mm; PC £13.94/m²; haunching one side							
Header course							
200 × 100 mm; butt joints	2.79	0.09	8.59	–	6.72	m	**15.31**
Stretcher course							
200 × 100 mm; butt joints	1.39	0.06	5.36	–	4.88	m	**10.24**
STONE PAVINGS							
Natural stone paving – General							
Preamble: Provide paving slabs of the specified thickness in random sizes but not less than 25 slabs per 10 m² of surface area, to be laid in parallel courses with joints alternately broken and laid to falls.							
Natural stone, slate or granite flag pavings; CED Ltd; on prepared base (not included); bedding on 25 mm cement: sand (1:3); Steintec Tuff top joints							
Yorkstone; riven laid random rectangular							
new slabs; 40–60 mm thick	116.78	0.57	55.21	–	144.31	m²	**199.52**
reclaimed slabs; Cathedral grade; 50–75 mm thick	127.05	0.93	90.16	–	165.60	m²	**255.76**
Donegal quartzite slabs; standard tiles							
200 mm × random lengths × 15–25 mm	84.00	0.67	64.40	–	104.78	m²	**169.18**
250 mm × random lengths × 15–25 mm	84.00	0.67	64.40	–	104.25	m²	**168.65**
300 mm × random lengths × 15–25 mm	84.00	0.67	64.40	–	103.98	m²	**168.38**
350 mm × random lengths × 15–25 mm	84.00	0.67	64.40	–	103.72	m²	**168.12**
400 mm × random lengths × 15–25 mm	84.00	0.67	64.40	–	103.45	m²	**167.85**
450 mm × random lengths × 15–25 mm	84.00	0.67	64.40	–	103.45	m²	**167.85**

35 SITE WORKS – PAVINGS

Item – Overhead and Profit Included	PC £	Labour hours	Labour £	Plant £	Material £	Unit	Total rate £
STONE PAVINGS – cont							
Natural Yorkstone pavings; Johnsons Wellfield Quarries; sawn 6 sides; 50 mm thick; on prepared base measured separately; bedding on 25 mm cement: sand (1:3); Steintec Tuff top joints							
Paving							
laid to random rectangular pattern	95.48	0.91	87.81	–	119.58	m²	**207.39**
laid to coursed laying pattern;							
3 sizes	88.27	0.80	77.28	–	111.02	m²	**188.30**
Paving; single size							
600 × 600 mm	88.27	0.67	64.40	–	110.23	m²	**174.63**
600 × 400 mm	88.27	0.67	64.40	–	111.02	m²	**175.42**
300 × 200 mm	94.58	0.83	80.50	–	123.02	m²	**203.52**
215 × 102.5 mm	100.41	0.83	80.50	–	136.56	m²	**217.06**
Paving; cut to template off site;							
600 × 600 mm; radius							
1.00 m	210.17	1.00	96.60	–	251.46	m²	**348.06**
2.50 m	210.17	0.91	87.81	–	250.41	m²	**338.22**
5.00 m	210.17	0.87	84.01	–	250.41	m²	**334.42**
CED Ltd; Indian sandstone, riven pavings or edgings; calibrated +/−26 mm thick; on prepared base measured separately; bedding on 25 mm cement: sand (1:3); Steintec Tuff top joints							
Paving							
laid to random rectangular pattern	21.59	0.67	64.40	–	34.59	m²	**98.99**
laid to coursed laying pattern;							
3 sizes	21.59	0.63	60.38	–	34.59	m²	**94.97**
Paving; single size							
600 × 600 mm	21.59	0.67	64.40	–	31.18	m²	**95.58**
600 × 400 mm	21.59	0.67	64.40	–	31.96	m²	**96.36**
Natural stone, slate or granite flag pavings; CED Ltd; on prepared base (not included); bedding on 25 mm cement: sand (1:3); Steintec Tuff top joints							
Granite paving; sawn 6 sides;							
textured top							
new slabs; silver grey; 50 mm thick	53.62	0.67	64.40	–	71.43	m²	**135.83**
new slabs; blue grey; 50 mm thick	54.86	0.67	64.40	–	72.84	m²	**137.24**
new slabs; yellow grey; 50 mm thick	59.70	0.67	64.40	–	78.41	m²	**142.81**
new slabs; black; 50 mm thick	93.45	0.67	64.40	–	117.22	m²	**181.62**

35 SITE WORKS – PAVINGS

Item – Overhead and Profit Included	PC £	Labour hours	Labour £	Plant £	Material £	Unit	Total rate £
Edgings; silver grey; blue grey yellow or black basalt textured							
100 mm wide × 150 deep	21.95	0.33	32.20	–	25.79	m	**57.99**
Natural Portland stone pavings; Albion Stone; sawn 6 sides; 40 mm thick; on prepared base measured separately; bedding on 25 mm cement: sand (1:3); white cement: sand (1:3) joints							
Paving; 40 mm thick							
laid to coursed laying pattern;							
3 sizes	152.00	0.87	84.01	–	185.43	m²	**269.44**
Paving 40 mm thick; single size							
1200 × 750 mm	132.00	1.11	107.33	–	162.43	m²	**269.76**
300 × 200 mm	152.00	0.40	38.64	–	189.70	m²	**228.34**
600 × 600 mm	152.00	0.32	30.59	–	185.43	m²	**216.02**
600 × 400 mm	152.00	0.30	28.98	–	185.43	m²	**214.41**
Paving; 50 mm thick							
laid to coursed laying pattern;							
3 sizes	165.00	0.87	84.01	–	200.38	m²	**284.39**
Paving 50 mm thick; single size							
1200 × 750 mm	165.00	1.25	120.75	–	200.38	m²	**321.13**
300 × 200 mm	165.00	0.40	38.64	–	204.65	m²	**243.29**
600 × 600 mm	165.00	0.32	30.59	–	200.38	m²	**230.97**
600 × 400 mm	165.00	0.30	28.98	–	200.38	m²	**229.36**
PORCELAIN PAVINGS							
Porcelain pavings; London Stone Ltd; 20 mm thick; on prepared base measured separately; bedding to 35 mm mortar bed primed with slurry primer; Grouted with Ardex flowable grout 3–5 mm joints							
Paving 20 mm thick in range of colours; Higher rate paving ranges; PC £56.27/m²; less than 150 m²							
596 mm × 596 mm	56.27	0.40	38.64	–	72.59	m²	**111.23**
748 mm × 748 mm	56.27	0.36	35.12	–	72.34	m²	**107.46**
895 mm × 895 mm	56.27	0.33	32.20	–	72.17	m²	**104.37**
895 mm × 446 mm	56.27	0.40	38.64	–	72.60	m²	**111.24**
1194 mm × 596 mm	56.27	0.33	32.20	–	72.28	m²	**104.48**
1194 mm × 297 mm	56.27	0.33	32.20	–	72.91	m²	**105.11**
Paving 20 mm thick in range of colours; Lower rate paving ranges; PC £45.99/m²							
596 mm × 596 mm	45.99	0.40	38.64	–	60.77	m²	**99.41**
748 mm × 748 mm	45.99	0.36	35.12	–	60.51	m²	**95.63**
895 mm × 895 mm	45.99	0.33	32.20	–	60.35	m²	**92.55**

35 SITE WORKS – PAVINGS

Item – Overhead and Profit Included	PC £	Labour hours	Labour £	Plant £	Material £	Unit	Total rate £
PORCELAIN PAVINGS – cont							
Porcelain pavings – cont							
Paving 20 mm thick in range of colours – cont							
895 mm × 446 mm	45.99	0.40	38.64	–	60.78	m²	**99.42**
1194 mm × 596 mm	45.99	0.33	32.20	–	60.46	m²	**92.66**
1194 mm × 297 mm	45.99	0.33	32.20	–	61.09	m²	**93.29**
Paving 20 mm thick in range of colours; Higher rate paving ranges; PC £53.45/m²; above 150 m²							
596 mm × 596 mm	53.45	0.40	38.64	–	69.34	m²	**107.98**
748 mm × 748 mm	53.45	0.36	35.12	–	69.09	m²	**104.21**
895 mm × 895 mm	53.45	0.33	32.20	–	68.93	m²	**101.13**
895 mm × 446 mm	53.45	0.40	38.64	–	69.36	m²	**108.00**
1194 mm × 596 mm	53.45	0.33	32.20	–	69.03	m²	**101.23**
1194 mm × 297 mm	53.45	0.33	32.20	–	69.67	m²	**101.87**
Paving 20 mm thick in range of colours; Lower rate paving ranges; PC £43.68/m²							
596 mm × 596 mm	43.68	0.40	38.64	–	58.11	m²	**96.75**
748 mm × 748 mm	43.68	0.36	35.12	–	57.86	m²	**92.98**
895 mm × 895 mm	43.68	0.33	32.20	–	57.70	m²	**89.90**
895 mm × 446 mm	43.68	0.40	38.64	–	58.12	m²	**96.76**
1194 mm × 596 mm	43.68	0.33	32.20	–	57.80	m²	**90.00**
1194 mm × 297 mm	43.68	0.33	32.20	–	58.43	m²	**90.63**
Cutting Porcelain paving							
straight	–	0.50	16.10	1.50	–	m	**17.60**
curved	–	0.80	25.76	1.50	–	m	**27.26**
STONE SETTS							
Granite setts; CED Ltd; bedding on 25 mm cement: sand (1:3); SteinTec TuffTop							
Natural granite setts; 100 × 100 mm to 125 × 150 mm × 150–250 mm length; riven surface; silver grey; 10 mm joints							
new; high grade	52.80	0.33	32.20	–	129.17	m²	**161.37**
new; standard grade	46.22	0.33	32.20	–	121.60	m²	**153.80**
reclaimed; cleaned	41.64	0.33	32.20	–	116.33	m²	**148.53**
Natural granite setts; 100 × 100 mm to 125 × 100 mm length; riven surface; silver grey; 10 mm joints							
new; high grade	51.50	0.33	32.20	–	148.20	m²	**180.40**
new; standard grade	45.08	0.33	32.20	–	140.82	m²	**173.02**
reclaimed; cleaned	41.64	0.33	32.20	–	150.01	m²	**182.21**

35 SITE WORKS – PAVINGS

Item – Overhead and Profit Included	PC £	Labour hours	Labour £	Plant £	Material £	Unit	Total rate £
Granite setts; CED Ltd bedding on 30 mm SteinTec mortar; surfaces primed with SteinTec primer; Jointing with Stentec TuffTop							
Natural granite setts; 100 × 100 mm to 125 × 150 mm × 150–250 mm length; riven surface; silver grey							
new; standard grade	46.36	0.35	32.84	–	185.35	m²	**218.19**
new; high grade	52.78	0.35	32.84	–	221.01	m²	**253.85**
reclaimed; cleaned	42.91	0.35	32.84	–	209.67	m²	**242.51**
RECONSTITUTED STONE PAVINGS							
Reconstituted Yorkstone aggregate pavings; Marshalls Plc; Saxon on prepared subbase measured separately; bedding on 25 mm thick cement: sand mortar (1:4); on 50 mm thick sharp sand base							
Square and rectangular paving in buff; butt joints straight both ways							
300 × 300 × 35 mm	60.72	0.27	25.76	–	77.56	m²	**103.32**
600 × 300 × 35 mm	41.83	0.22	20.93	–	55.83	m²	**76.76**
450 × 450 × 50 mm	33.93	0.25	24.15	–	46.75	m²	**70.90**
600 × 600 × 35 mm	30.02	0.17	16.10	–	42.25	m²	**58.35**
600 × 600 × 50 mm	29.54	0.18	17.71	–	41.70	m²	**59.41**
Square and rectangular paving in natural; butt joints straight both ways							
300 × 300 × 35 mm	50.84	0.27	25.76	–	66.19	m²	**91.95**
450 × 450 × 50 mm	30.58	0.23	22.54	–	42.89	m²	**65.43**
600 × 300 × 35 mm	36.79	0.25	24.15	–	50.04	m²	**74.19**
600 × 600 × 35 mm	26.96	0.17	16.10	–	38.73	m²	**54.83**
600 × 600 × 50 mm	27.08	0.20	19.32	–	38.87	m²	**58.19**
RESIN BOUND SURFACING							
Resin bound paving; Sureset Ltd; fully mixed permeable decorative aggregate paving; laid to macadam binder and base course (not included); thickness dependent on loading application and aggregate size							
18 mm thick – 6 mm aggregate							
Natural gravel							
areas 100–300 m²	–	–	–	–	–	m²	**56.10**
areas 300–500 m²	–	–	–	–	–	m²	**51.70**
areas over 500 m²	–	–	–	–	–	m²	**44.00**

35 SITE WORKS – PAVINGS

Item – Overhead and Profit Included	PC £	Labour hours	Labour £	Plant £	Material £	Unit	Total rate £
RESIN BOUND SURFACING – cont							
Resin bound paving – cont							
18 mm thick – cont							
Crushed rock							
areas 100–300 m^2	–	–	–	–	–	m^2	58.30
areas 300–500 m^2	–	–	–	–	–	m^2	50.60
areas over 500 m^2	–	–	–	–	–	m^2	47.30
Marble							
areas 100–300 m^2	–	–	–	–	–	m^2	82.50
areas 300–500 m^2	–	–	–	–	–	m^2	69.30
areas over 500 m^2	–	–	–	–	–	m^2	63.80
Spectrum							
areas 100–300 m^2	–	–	–	–	–	m^2	85.80
areas 300–500 m^2	–	–	–	–	–	m^2	72.60
areas over 500 m^2	–	–	–	–	–	m^2	66.00
Bound aggregates; Addagrip Terraco Ltd; natural decorative resin bound surface dressing laid to concrete, macadam or to plywood panels priced separately							
Lucerne Silver; Pecan; Scandanavian Pearl; Blue Granite							
3 mm aggregate							
6 mm depth	–	–	–	–	–	m^2	33.00
Chocolate Terrabound; 6 mm							
16 mm depth	–	–	–	–	–	m^2	38.50
Terracotta 6 mm							
16 mm or 18 mm depth	–	–	–	–	–	m^2	44.00
Midnight Grey; 6 mm							
15 mm depth	–	–	–	–	–	m^2	44.00
Chocolate; 6 mm							
15 mm depth	–	–	–	–	–	m^2	44.00
18 mm depth	–	–	–	–	–	m^2	55.00
Acorn 3 mm							
12 mm	–	–	–	–	–	m^2	44.00
Terrabase; rustic bound porous paving							
16 mm deep	–	–	–	–	–	m^2	71.50
Resin bound macadam pavings; Bituchem; to pedestrian or vehicular hard landscape areas; laid to base course (not included)							
Natratex wearing course; clear resin bound macadam							
25 mm thick to pedestrian areas	25.00	0.04	4.03	–	28.75	m^2	32.78
30 mm thick to vehicular areas	30.00	0.01	1.07	–	34.50	m^2	35.57

35 SITE WORKS – PAVINGS

Item – Overhead and Profit Included	PC £	Labour hours	Labour £	Plant £	Material £	Unit	Total rate £
Colourtex coloured resin bound macadam							
25 mm thick	25.00	0.04	4.03	–	28.75	m^2	**32.78**
Resin bound paving to individual tree pits; SureSet Ltd; fully mixed permeable decorative aggregate paving; laid to well compacted stone Type 3 (not included)							
30–40 mm thick – average tree pit size 1.50 × 1.50 tree pit aggregate							
Natural gravel							
areas 100–300 m^2	–	–	–	–	–	nr	**121.00**
SudsTech; Langford Direct Ltd; resin bound porous paving solution; fully sustainable urban drainage Systems (SuDS) compliant, blended aggregates mixed with tensile resins, in collaboration with proprietary flexible base							
Permeable wearing top course 20 mm (resin bound aggregate) to subbase of resin bound recycled tyres and aggregate 30 mm laid to Type 3 granular subbase (not included)							
SudsTech – 2 part system overall 50 mm thick; standard colours	–	–	–	–	–	m^2	**86.90**
SudsTech – 2 part system; overall 50 mm thick; premium colours	–	–	–	–	–	m^2	**97.90**
Trailflex; SudsTech Ltd; Single flexible surface of resin bound recycled tyres and aggregate; alternative to macadam or concrete surfacing; traffic of up to 18 tonnes							
Permeable flexible wearing course; 40 mm thick; resin bound recycled tyre and stone; laid to Type 3 base (not included)							
less than 300 m^2	–	–	–	–	–	m^2	**49.50**
300 m^2–900 m^2	–	–	–	–	–	m^2	**46.20**
900 m^2–2000 m^2	–	–	–	–	–	m^2	**44.00**
Over 2000 m^2	–	–	–	–	–	m^2	**42.90**

35 SITE WORKS – PAVINGS

Item – Overhead and Profit Included	PC £	Labour hours	Labour £	Plant £	Material £	Unit	Total rate £
RESIN BOUND SURFACING – cont							
Trailflex – cont							
Permeable flexible wearing course; 35 mm thick; resin bound recycled tyre and stone; laid to Type 3 base (not included); footpaths; cycle tracks							
less than 300 m²	–	–	–	–	–	m²	47.30
300 m²–900 m²	–	–	–	–	–	m²	42.90
900 m²–2000 m²	–	–	–	–	–	m²	41.80
Over 2000 m²	–	–	–	–	–	m²	41.25
Surface to tree pits; Langford Direct Ltd; resin bound porous recycled tyre and aggregate paving solution; fully sustainable urban drainage Systems (SuDS) compliant							
Laid to surface of tree pits as permeable mulch layer; min 10 nr pits average size 1.20 m × 1.20 m; to Type 3 base (SudsTech only)							
SudsTech (2 part system) to Type 3 base (not included)	–	–	–	–	–	m²	152.90
Trailflex (1 part system)	–	–	–	–	–	m²	107.80
RESIN BONDED SURFACING							
Bonded aggregates; Addagrip Terraco Ltd; natural decorative resin bonded surface dressing laid to concrete, macadam or to plywood panels priced separately							
Primer coat to macadam or concrete base	–	–	–	–	–	m²	5.50
Brittany Bronze; 1–3 mm							
buff adhesive	–	–	–	–	–	m²	27.50
red adhesive	–	–	–	–	–	m²	27.50
green adhesive	–	–	–	–	–	m²	27.50
Brittany Bronze; 2–5 mm							
buff adhesive	–	–	–	–	–	m²	30.80
Chinese bauxite; 1–3 mm							
buff adhesive	–	–	–	–	–	m²	30.80

35 SITE WORKS – PAVINGS

Item – Overhead and Profit Included	PC £	Labour hours	Labour £	Plant £	Material £	Unit	Total rate £
MACADAM SURFACING							
Coated macadam/asphalt roads/ pavings – General							
Preamble: The prices for all in situ finishings to roads and footpaths include for work to falls, crossfalls or slopes not exceeding 15° from horizontal; for laying on prepared bases (not included) and for rolling with an appropriate roller.							
Users should note the new terminology for the surfaces described below which is to European standard descriptions. The now redundant descriptions for each course are shown in brackets.							
Macadam surfacing; surface (wearing) course; 20 mm of 6 mm dense bitumen macadam							
Machine lay; areas 1000 m² and over							
limestone aggregate	–	–	–	–	–	m²	**9.35**
granite aggregate	–	–	–	–	–	m²	**9.57**
Hand lay; areas 400 m² and over							
limestone aggregate	–	–	–	–	–	m²	**17.05**
granite aggregate	–	–	–	–	–	m²	**17.27**
Macadam surfacing; surface (wearing) course; 30 mm of 10 mm dense bitumen macadam							
Machine lay; areas 1000 m² and over							
limestone aggregate	–	–	–	–	–	m²	**13.75**
granite aggregate	–	–	–	–	–	m²	**13.53**
Hand lay; areas 400 m² and over							
limestone aggregate	–	–	–	–	–	m²	**20.68**
granite aggregate	–	–	–	–	–	m²	**20.46**
Macadam surfacing; surface (wearing) course; 40 mm of 10 mm dense bitumen macadam							
Machine lay; areas 1000 m² and over							
limestone aggregate	–	–	–	–	–	m²	**17.82**
granite aggregate	–	–	–	–	–	m²	**18.04**
Hand lay; areas 400 m² and over							
limestone aggregate	–	–	–	–	–	m²	**27.06**
granite aggregate	–	–	–	–	–	m²	**27.28**

35 SITE WORKS – PAVINGS

Item – Overhead and Profit Included	PC £	Labour hours	Labour £	Plant £	Material £	Unit	Total rate £
MACADAM SURFACING – cont							
Macadam surfacing; binder (base) course; 50 mm of 20 mm dense bitumen macadam							
Machine lay; areas 1000 m² and over							
limestone aggregate	–	–	–	–	–	m²	12.10
granite aggregate	–	–	–	–	–	m²	12.32
Hand lay; areas 400 m² and over							
limestone aggregate	–	–	–	–	–	m²	19.58
granite aggregate	–	–	–	–	–	m²	19.91
Macadam surfacing; Construction Co Ltd; binder (base) course; 60 mm of 20 mm dense bitumen macadam							
Machine lay; areas 1000 m² and over							
limestone aggregate	–	–	–	–	–	m²	14.52
granite aggregate	–	–	–	–	–	m²	14.30
Hand lay; areas 400 m² and over							
limestone aggregate	–	–	–	–	–	m²	23.43
granite aggregate	–	–	–	–	–	m²	23.87
Macadam surfacing; base (roadbase) course; 75 mm of 28 mm dense bitumen macadam							
Machine lay; areas 1000 m² and over							
limestone aggregate	–	–	–	–	–	m²	20.96
granite aggregate	–	–	–	–	–	m²	20.63
Hand lay; areas 400 m² and over							
limestone aggregate	–	–	–	–	–	m²	38.50
granite aggregate	–	–	–	–	–	m²	38.77
Macadam surfacing; base (roadbase) course; 100 mm of 28 mm dense bitumen macadam							
Machine lay; areas 1000 m² and over							
limestone aggregate	–	–	–	–	–	m²	27.50
granite aggregate	–	–	–	–	–	m²	27.94
Hand lay; areas 400 m² and over							
limestone aggregate	–	–	–	–	–	m²	51.26
granite aggregate	–	–	–	–	–	m²	51.70

35 SITE WORKS – PAVINGS

Item – Overhead and Profit Included	PC £	Labour hours	Labour £	Plant £	Material £	Unit	Total rate £
Resin bound macadam pavings; Bituchem; to pedestrian or vehicular hard landscape areas; laid to base course (not included) Natratex wearing course; clear resin bound macadam							
25 mm thick to pedestrian areas	25.00	0.04	4.03	–	28.75	m²	**32.78**
30 mm thick to vehicular areas	30.00	0.01	1.07	–	34.50	m²	**35.57**
Colourtex coloured resin bound macadam							
25 mm thick	25.00	0.04	4.03	–	28.75	m²	**32.78**
MARKING OF CAR PARKS							
Marking car parks Car parking space division strips; laid hot at 115°C; on bitumen macadam surfacing							
minimum daily rate	–	–	–	–	–	item	**550.00**
Stainless metal road studs							
100 × 100 mm	6.20	0.08	8.05	–	7.13	nr	**15.18**
SUSPENDED PAVING							
Beam and Block suspended base for external use; FP McCann Ltd; Prestressed precast concrete beam; profile 115 mm wide at base, 80 mm wide at top 152 mm deep; laid to perimeter or spanned support retaining walls (not included); block infill of 440 × 225 × 100 mm solid blocks; screed of 75 mm concrete reinforced with mesh to receive surface treatments Block centres between beams							
520 mm; blocks laid 440 mm wide	12.27	0.30	19.47	7.35	48.25	m²	**75.07**
408 mm; alternate blocks 440 mm wide and 215 mm wide between beams	15.58	0.36	23.43	9.71	45.38	m²	**78.52**
295 mm; blocks laid 215 mm wide between beams	24.54	0.36	23.43	14.71	51.58	m²	**89.72**

35 SITE WORKS – PAVINGS

Item – Overhead and Profit Included	PC £	Labour hours	Labour £	Plant £	Material £	Unit	Total rate £
FORMWORK FOR CONCRETE PAVINGS							
Formwork for in situ concrete							
Sides of foundations							
height not exceeding 250 mm	0.37	0.01	1.07	–	0.72	m	**1.79**
height 250–500 mm	0.49	0.01	1.29	–	1.29	m	**2.58**
height 500 mm–1.00 m	0.49	0.02	1.61	–	1.90	m	**3.51**
height exceeding 1.00 m	1.46	0.67	64.40	–	4.92	m²	**69.32**
Extra over formwork for curved work							
6 m radius	–	0.25	16.10	–	–	m	**16.10**
REINFORCEMENT FOR CONCRETE PAVINGS							
Reinforcement mesh; side laps 150 mm; head laps 300 mm; mesh 200 × 200 mm; in roads, footpaths or pavings							
Fabric							
A142 (2.22 kg/m²)	3.55	0.03	2.68	–	4.08	m²	**6.76**
A193 (3.02 kg/m²)	4.09	0.02	1.71	–	4.70	m²	**6.41**
A252 (3.02 kg/m²)	5.16	0.02	1.93	–	5.93	m²	**7.86**
IN SITU CONCRETE PAVINGS							
Unreinforced concrete; on prepared subbase (not included)							
Roads; 21.00 N/mm²–20 mm aggregate (1:2:4) mechanically mixed on site							
100 mm thick	11.93	0.04	4.03	–	13.72	m²	**17.75**
150 mm thick	17.90	–	0.20	–	20.58	m²	**20.78**
Reinforced in situ concrete; mechanically mixed on site; on hardcore base (not included); reinforcement (not included)							
Roads; 11.50 N/mm²–40 mm aggregate (1:3:6)							
100 mm thick	10.37	0.03	3.22	–	11.93	m²	**15.15**
150 mm thick	15.56	0.20	19.32	–	17.89	m²	**37.21**
200 mm thick	21.25	0.27	25.76	–	24.44	m²	**50.20**
250 mm thick	25.93	0.33	32.20	0.52	29.82	m²	**62.54**
300 mm thick	31.12	0.40	38.64	0.52	35.79	m²	**74.95**
Roads; 21.00 N/mm²–20 mm aggregate (1:2:4)							
100 mm thick	11.65	0.13	12.88	–	13.40	m²	**26.28**
150 mm thick	17.48	0.20	19.32	–	20.10	m²	**39.42**
200 mm thick	23.30	0.27	25.76	–	26.80	m²	**52.56**
250 mm thick	29.12	0.33	32.20	0.52	33.49	m²	**66.21**
300 mm thick	34.95	0.40	38.64	0.52	40.19	m²	**79.35**

35 SITE WORKS – PAVINGS

Item – Overhead and Profit Included	PC £	Labour hours	Labour £	Plant £	Material £	Unit	Total rate £
Roads; 25.00 N/mm²–20 mm aggregate C 25/30 ready mixed							
100 mm thick	9.56	0.13	12.88	–	10.99	m²	23.87
150 mm thick	14.34	0.20	19.32	–	16.49	m²	35.81
200 mm thick	19.12	0.27	25.76	–	21.99	m²	47.75
250 mm thick	23.89	0.33	32.20	0.52	27.47	m²	60.19
300 mm thick	28.67	0.40	38.64	0.52	32.97	m²	72.13
Reinforced in situ concrete; ready mixed; discharged directly into location from supply vehicle; normal Portland cement; on hardcore base (not included); reinforcement (not included)							
Roads; 11.50 N/mm²–40 mm aggregate (1:3:6)							
100 mm thick	8.70	0.05	5.15	–	10.00	m²	15.15
150 mm thick	13.05	0.08	7.73	–	15.01	m²	22.74
200 mm thick	17.40	0.12	11.59	–	20.01	m²	31.60
250 mm thick	21.75	0.18	17.39	0.52	25.01	m²	42.92
300 mm thick	26.10	0.22	21.25	0.52	30.02	m²	51.79
Roads; 21.00 N/mm²–20 mm aggregate (1:2:4)							
100 mm thick	9.10	2.08	201.25	–	10.46	m²	211.71
150 mm thick	13.65	0.08	7.73	–	15.70	m²	23.43
200 mm thick	18.20	0.12	11.59	–	20.93	m²	32.52
250 mm thick	22.75	0.18	17.39	0.52	26.16	m²	44.07
300 mm thick	27.30	0.22	21.25	0.52	31.39	m²	53.16
Roads; 26.00 N/mm²–20 mm aggregate (1:1.5:3)							
100 mm thick	9.15	0.05	5.15	–	10.52	m²	15.67
150 mm thick	13.72	0.08	7.73	–	15.78	m²	23.51
200 mm thick	18.20	0.12	11.59	–	20.93	m²	32.52
250 mm thick	22.88	0.18	17.39	0.52	26.31	m²	44.22
300 mm thick	27.45	0.22	21.25	0.52	31.57	m²	53.34
Roads; PAV1 concrete – 35 N/mm²; Designated mix							
100 mm thick	10.09	0.05	5.15	–	11.60	m²	16.75
150 mm thick	15.14	0.08	7.73	–	17.41	m²	25.14
200 mm thick	20.18	0.12	11.59	–	23.21	m²	34.80
250 mm thick	25.22	0.18	17.39	0.52	29.00	m²	46.91
300 mm thick	30.27	0.22	21.25	0.52	34.81	m²	56.58
Concrete sundries							
Treating surfaces of unset concrete; grading to cambers; tamping with 75 mm thick steel shod tamper or similar	–	0.13	8.58	–	–	m²	8.58

35 SITE WORKS – PAVINGS

Item – Overhead and Profit Included	PC £	Labour hours	Labour £	Plant £	Material £	Unit	Total rate £
EXPANSION JOINTS IN PAVINGS							
Expansion joints							
13 mm thick joint filler; formwork							
width or depth not exceeding							
150 mm	1.63	0.04	2.88	–	2.52	m	**5.40**
width or depth 150–300 mm	3.25	0.06	3.59	–	5.05	m	**8.64**
width or depth 300–450 mm	4.88	0.07	4.31	–	7.57	m	**11.88**
25 mm thick joint filler; formwork							
width or depth not exceeding							
150 mm	3.00	0.04	2.88	–	4.11	m	**6.99**
width or depth 150–300 mm	6.00	0.06	3.59	–	8.21	m	**11.80**
width or depth 300–450 mm	9.00	0.07	4.31	–	12.32	m	**16.63**
Sealants; sealing top 25 mm of joint with rubberized bituminous compound	1.25	0.06	3.59	–	1.44	m	**5.03**
WOODCHIP SURFACES							
Wood chip surfaces; Melcourt Industries Ltd (items labelled FSC are Forest Stewardship Council certified)							
Wood chips; to surface of pathways by machine; material delivered in 80 m³ loads; levelling and spreading by hand (excavation and preparation not included)							
Walk Chips; 100 mm thick; FSC (25 m³ loads)	4.69	0.01	1.07	0.49	5.39	m²	**6.95**
Walk Chips; 100 mm thick; FSC (80 m³ loads)	3.33	0.01	1.07	0.49	3.83	m²	**5.39**
Woodfibre; 100 mm thick; FSC (80 m³ loads)	3.56	0.01	1.07	0.49	4.09	m²	**5.65**

35 SITE WORKS – PAVINGS

Item – Overhead and Profit Included	PC £	Labour hours	Labour £	Plant £	Material £	Unit	Total rate £
STEPS							
In situ concrete steps; ready mixed concrete; C15; on 100 mm hardcore base; base concrete thickness 100 mm treads and risers cast simultaneously; inclusive of A142 mesh within the concrete; step riser 170 mm high; inclusive of all shuttering and concrete labours; floated finish; to graded ground (not included)							
Treads 250 mm wide							
one riser	5.94	1.32	127.51	–	10.35	m	**137.86**
two risers	9.51	1.98	191.27	–	18.88	m	**210.15**
three risers	13.77	2.64	255.02	–	29.84	m	**284.86**
four risers	18.50	3.96	382.54	–	37.49	m	**420.03**
six risers	23.43	5.28	510.05	–	41.69	m	**551.74**
Lay surface to concrete in situ steps above							
granite treads; pre-cut off site; 305 mm wide × 50 mm thick bullnose finish inclusive of 20 mm overhang; granite riser 20 mm thick all laid on 15 mm mortar bed	68.81	0.66	63.76	–	80.56	m	**144.32**
riven Yorkstone tread and riser 50 mm thick; cut on site	48.83	0.82	79.69	–	57.59	m	**137.28**
riven Yorkstone treads 225 mm wide × 50 mm thick cut on site; two courses brick risers laid on flat stretcher bond; bricks PC £500.00/1000	23.89	0.99	95.63	–	34.38	m	**130.01**
Marshalls Saxon paving flag 450 × 450 mm cut to 225 mm wide × 38 mm thick; cut on site; two courses brick risers laid on flat stretcher bond; bricks PC £500.00/1000	6.93	0.99	95.63	–	14.88	m	**110.51**
Treads 450 mm wide							
one riser	10.69	1.65	159.39	–	17.66	m	**177.05**
two risers	21.84	2.31	223.15	–	36.24	m	**259.39**
three risers	30.87	2.97	286.90	–	49.94	m	**336.84**
four risers	53.40	4.29	414.41	–	83.58	m	**497.99**
six risers	67.87	5.94	573.80	–	97.28	m	**671.08**

35 SITE WORKS – PAVINGS

Item – Overhead and Profit Included	PC £	Labour hours	Labour £	Plant £	Material £	Unit	Total rate £
STEPS – cont							
Lay surface to concrete in situ steps							
granite treads; pre-cut off site;							
470 mm wide × 50 mm thick							
bullnose finish inclusive of 20 mm							
overhang; granite riser 20 mm							
thick all laid on 15 mm mortar bed	68.81	1.32	127.51	–	80.93	m	208.44
riven Yorkstone tread and riser							
50 mm thick; cut on site	66.35	1.49	143.45	–	78.10	m	221.55
riven Yorkstone treads 460 mm							
wide 50 mm thick cut on site; two							
courses brick risers laid on flat							
stretcher bond; bricks PC							
£500.00/1000	48.83	1.65	159.39	–	63.35	m	222.74
Marshalls Saxon paving flag							
450 × 450 × 38 mm thick cut on							
site; two courses brick risers laid							
on flat stretcher bond; bricks PC							
£500.00/1000	13.85	0.91	87.66	–	22.84	m	110.50
Precast concrete steps; FP							
McCann Ltd; Precast units laid to							
graded blinded and levelled bases							
(not included); to receive surface							
treatments priced separately;							
waist 200 mm thick; finished sizes							
with paving surface treatments of							
40 mm thick including bedding							
5 Risers; 280 mm deep × 175 mm							
high							
1.20 m wide	355.05	3.00	289.80	43.08	408.31	nr	741.19
2.00 m wide	710.09	4.00	386.40	86.16	816.60	nr	1289.16
2.50 m wide	887.66	5.00	483.00	215.41	1020.81	nr	1719.22
3.00 m wide	1065.14	7.00	676.20	215.41	1224.91	nr	2116.52
4.00 m wide	1420.19	8.00	772.80	301.56	1633.22	nr	2707.58
10 Risers; 280 mm deep × 175 mm							
high							
1.20 m wide	852.18	4.00	386.40	86.16	980.01	nr	1452.57
2.00 m wide	1420.36	6.00	579.60	172.32	1633.41	nr	2385.33
2.50 m wide	1775.50	8.00	772.80	344.64	2041.82	nr	3159.26
3.00 m wide	2130.54	8.00	772.80	344.64	2450.12	nr	3567.56
4.00 m wide	2840.73	8.00	772.80	344.64	3266.84	nr	4384.28
5 Risers; 450 mm deep × 175 mm							
high							
1.20 m wide	653.68	3.00	289.80	43.08	751.73	nr	1084.61
2.00 m wide	1089.41	4.00	386.40	86.16	1252.82	nr	1725.38
2.50 m wide	1361.85	5.00	483.00	215.41	1566.13	nr	2264.54
3.00 m wide	1234.65	7.00	676.20	215.41	1419.85	nr	2311.46
4.00 m wide	2178.90	8.00	772.80	301.56	2505.74	nr	3580.10

35 SITE WORKS – PAVINGS

Item – Overhead and Profit Included	PC £	Labour hours	Labour £	Plant £	Material £	Unit	Total rate £
10 Risers; 450 mm deep × 175 mm high							
1.20 m wide	1307.71	3.00	289.80	43.08	1503.87	nr	**1836.75**
2.00 m wide	2179.51	4.00	386.40	86.16	2506.44	nr	**2979.00**
2.50 m wide	2724.39	5.00	483.00	215.41	3133.05	nr	**3831.46**
3.00 m wide	3267.52	7.00	676.20	215.41	3757.65	nr	**4649.26**
4.00 m wide	4353.77	8.00	772.80	301.56	5006.84	nr	**6081.20**
Extra over for integrated landing in step flights; 1.00 m deep							
1.20 m wide	–	–	–	–	241.78	nr	**241.78**
2.00 m wide	–	–	–	–	402.96	nr	**402.96**
2.50 m wide	–	–	–	–	503.70	nr	**503.70**
3.00 m wide	–	–	–	–	604.45	nr	**604.45**
4.00 m wide	–	–	–	–	805.93	nr	**805.93**
Step treads; Marshalls Plc; Precast step treads to match paving types							
Conservation							
Step Tread 450 × 450 × 50 mm – Chamfered	154.51	1.00	96.60	–	179.85	m	**276.45**
Step Tread 450 × 450 × 50 mm	147.71	1.00	96.60	–	172.02	m	**268.62**
Step Tread 400 × 400 × 65 mm	164.82	1.05	101.43	–	191.70	m	**293.13**
Step Unit 350 × 1000 × 150 mm	250.27	1.00	96.60	–	289.96	m	**386.56**
Step Unit 400 × 1000 × 150 mm	267.62	1.00	96.60	–	309.91	m	**406.51**
Saxon Step Tread 400 × 400 × 50 mm	137.38	1.05	101.43	–	160.14	m	**261.57**
Saxon Step Tread 450 × 450 × 50 mm	128.90	1.00	96.60	–	150.39	m	**246.99**
Portland stone steps; Albion Stone							
Stone 40 mm thick; Laid to preformed step surface (not included) on 35 mm bedding of white cement mortar; jointed with white cement mortar							
Treads 350 mm deep	47.00	0.50	32.20	–	59.41	m	**91.61**
Treads 500 mm deep	66.00	0.75	48.30	–	105.71	m	**154.01**
Treads 600 mm deep	79.00	0.90	57.96	–	99.36	m	**157.32**
Landings; maximum stone size 1200 × 750 mm	152.00	1.25	120.75	–	189.70	m²	**310.45**
Stone 50 mm thick; Laid to preformed step surface (not included) on 35 mm bedding of white cement mortar; jointed with white cement mortar							
Treads 350 mm deep	121.20	0.55	35.42	–	144.74	m	**180.16**
Treads 500 mm deep	144.30	0.75	48.30	–	195.75	m	**244.05**
Treads 600 mm deep	160.80	1.00	64.40	–	193.43	m	**257.83**
Landings	165.00	1.25	120.75	–	204.65	m²	**325.40**

35 SITE WORKS – PAVINGS

Item – Overhead and Profit Included	PC £	Labour hours	Labour £	Plant £	Material £	Unit	Total rate £
STEPS – cont							
Portland stone steps – cont							
Risers to steps; fixed with white cement mortar							
140 mm high × 30 mm thick	21.15	0.50	32.20	–	24.32	m	**56.52**
Stone steps; Haddonstone Ltd to step base in concrete or brick substructure; Cast stone steps in Tecstone in colours Bath, Terracotta, Portland, Slate or Sienna; continuosly bedded on full mortar bed							
Steps							
Bullnose Step 1000 × 350 × 50	56.00	0.25	24.15	–	66.75	m	**90.90**
Bullnose Step 1000 × 375 × 50	65.00	0.25	24.15	–	77.25	m	**101.40**
Risers							
Step Riser 1000 × 175 × 50	27.00	0.13	12.07	–	31.13	m	**43.20**
Step Riser 1000 × 150 × 50	24.00	0.13	12.07	–	27.68	m	**39.75**

35 SITE WORKS – DECKING

Item – Overhead and Profit Included	PC £	Labour hours	Labour £	Plant £	Material £	Unit	Total rate £
SUPPORTS FOR DECKING SYSTEMS							
Timber decking; Exterior Decking Ltd; timber decking substructure/ carcassing costs							
Roof decks exclusive of beams; timber craned (not included); joists at 400 mm centres							
47 × 47 mm	–	–	–	–	–	m²	32.65
47 × 100 mm	–	–	–	–	–	m²	36.68
47 × 150 mm	–	–	–	–	–	m²	40.48
Commercial applications; decks with double beams at 300 mm centres							
47 × 100 mm	–	–	–	–	–	m²	29.10
47 × 150 mm	–	–	–	–	–	m²	33.84
Domestic garden applications; decks with double beams at 300 mm centres							
47 × 150 mm	–	–	–	–	–	m²	51.90
Structural posts to decks; tanalized treated softwood; rebated including excavation and setting in concrete							
100 × 100 mm; single beam; single rebate	–	–	–	–	–	m	12.51
150 × 150 mm; double beam; double rebate	–	–	–	–	–	m	17.59
Adjustable pedestals to rooftops for decking and pavings							
Decking; 300 mm joists at 1 pedestal per linear metre (3.3/m²) along the length of the joists;							
17 mm high	2.81	0.08	5.31	–	3.23	m²	8.54
28 mm high	3.79	0.08	5.31	–	4.36	m²	9.67
175–285 mm high	23.03	0.08	5.43	–	26.48	m²	31.91
285–400 mm high	39.86	0.08	5.43	–	45.84	m²	51.27
355–515 mm high	49.01	0.08	5.43	–	56.36	m²	61.79
465–625 mm high	56.69	0.09	5.69	–	65.19	m²	70.88
545–740 mm high	65.83	0.09	5.85	–	75.70	m²	81.55
645–850 mm high	73.52	0.09	5.85	–	84.55	m²	90.40
720–960 mm high	82.67	0.09	5.85	–	95.07	m²	100.92
830–1070 mm high	90.35	0.09	5.85	–	103.90	m²	109.75
Decking; 400 mm joists at 1 pedestal per linear metre (2.5/m²) along the length of the joists;							
17 mm high	2.13	0.07	4.30	–	2.45	m²	6.75
28 mm high	2.88	0.07	4.30	–	3.31	m²	7.61
175–285 mm high	17.45	0.07	4.60	–	20.07	m²	24.67

35 SITE WORKS – DECKING

Item – Overhead and Profit Included	PC £	Labour hours	Labour £	Plant £	Material £	Unit	Total rate £
SUPPORTS FOR DECKING SYSTEMS – cont							
Adjustable pedestals to rooftops for decking and pavings – cont							
Decking – cont							
285–400 mm high	30.20	0.07	4.60	–	34.73	m²	**39.33**
355–515 mm high	37.13	0.07	4.60	–	42.70	m²	**47.30**
465–625 mm high	42.95	0.08	4.96	–	49.39	m²	**54.35**
545–740 mm high	49.88	0.08	4.96	–	57.36	m²	**62.32**
645–850 mm high	55.70	0.08	4.96	–	64.06	m²	**69.02**
720–960 mm high	62.63	0.08	4.96	–	72.02	m²	**76.98**
830–1070 mm high	68.45	0.08	4.96	–	78.72	m²	**83.68**
Adjustable pedestals to rooftops for decking and pavings; Note other heights available up to 1.07 m							
Paving slabs; laid at uniform sizes with one pedestal per corner of each paving slab; paving slab sizes; laid on adjustable pedestal 28 mm high above the slab							
300 × 300 mm	19.55	0.50	32.20	–	22.48	m²	**54.68**
400 × 400 mm	11.50	0.50	32.20	–	13.23	m²	**45.43**
450 × 450 mm	8.63	0.50	32.20	–	9.92	m²	**42.12**
600 × 450 mm with additional central support per slab	7.59	0.33	21.46	–	8.73	m²	**30.19**
900 × 600 mm with additional central support per slab	4.37	0.25	16.10	–	5.03	m²	**21.13**
TIMBER DECKING							
Market prices of timber decking materials; Exterior Decking Ltd							
Hardwood decking; stainless steel screw fixings included in rates shown							
Ipe; pre-drilled and countersunk	86.48	–	–	–	99.45	m²	**99.45**
Hardwood decking; Exterpark invisible fixings included in rates shown							
Exterpark Magnet Accoya® 21 × 144 mm	64.54	–	–	–	74.22	m²	**74.22**
Exterpark Magnet Iroko 21 × 120 mm	70.91	–	–	–	81.55	m²	**81.55**
Teak Rustic FSC 21 mm thick	125.00	–	–	–	143.75	m²	**143.75**
Exterpark magnet Indonesian Teak 20 × 120 mm	185.83	–	–	–	213.70	m²	**213.70**
Softwood decking							
Pine smooth; 145 × 38 mm	43.85	–	–	–	50.43	m²	**50.43**

35 SITE WORKS – DECKING

Item – Overhead and Profit Included	PC £	Labour hours	Labour £	Plant £	Material £	Unit	Total rate £
Timber decking; Exterior Decking Ltd; timber decking surfaces; fixed to substructure (not included)							
Softwood deck timbers; screw fixed pine; smooth; 145 × 38 mm fixed with 65 × 5 mm coated deck screws	48.11	0.25	16.10	–	55.33	m²	**71.43**
Hardwood deck timbers; screw fixed Ipe; S4S E4E; 145 × 21 mm; pre-drilled and countersunk; S4S e4e	86.48	0.75	48.30	–	104.35	m²	**152.65**
Hardwood deck timbers; invisibly fixed							
Ipe; 21 mm thick	52.50	0.50	16.10	–	70.65	m²	**86.75**
Teak Rustic FSC 21 mm	125.00	0.50	16.10	–	154.02	m²	**170.12**
Exterpark Ipe 21 mm thickness 21 × 120 mm	98.66	0.50	16.10	–	123.73	m²	**139.83**
Hardwood decking; Yellow Balau; grooved or smooth; 6 mm joints							
deck boards; 145 mm wide × 21 mm thick	42.76	1.00	32.20	–	51.29	m²	**83.49**
Western Red Cedar; Clear grade; 6 mm joints smooth or reeded							
90 mm wide × 25 mm thick	62.44	1.00	32.20	–	74.09	m²	**106.29**
140 mm wide × 25 mm thick	70.43	1.00	32.20	–	83.28	m²	**115.48**
Surface treatments to timber decks; Owatrol decking oil							
Seasonite	3.35	–	–	–	3.85	m²	**3.85**
Textrol	5.38	–	–	–	6.19	m²	**6.19**
Handrails and base rail; fixed to posts at 2.00 m centres							
posts; 100 × 100 × 1370 mm high	23.99	1.00	32.20	–	28.19	m	**60.39**
posts; turned; 1220 mm high	23.95	1.00	32.20	–	28.14	m	**60.34**
Handrails; balusters							
square balusters at 100 mm centres	27.80	0.50	16.10	–	32.57	m	**48.67**
square balusters at 300 mm centres	9.26	0.33	10.72	–	11.13	m	**21.85**
turned balusters at 100 mm centres	55.00	0.50	16.10	–	63.85	m	**79.95**
turned balusters at 300 mm centres	18.32	0.33	10.63	–	21.55	m	**32.18**

35 SITE WORKS – DECKING

Item – Overhead and Profit Included	PC £	Labour hours	Labour £	Plant £	Material £	Unit	Total rate £
COMPOSITE DECKING							
Composite decking; London Stone Ltd; Millboard composite decking							
Substructure for composite or timber decking; Millboard; London Stone Ltd; Pro-Plas recyled substructural elements to support timber or composite decking boards; joists and double bearers 125 × 50 mm; joists laid at 300 mm centres							
4 m² deck; 2.00 × 2.00 m	51.60	1.50	84.00	–	54.80	m²	**138.80**
10 m² deck; 3.33 × 3.33 m	53.66	1.20	67.20	–	56.86	m²	**124.06**
25 m² deck; 5.00 × 5.00 m	47.06	1.18	65.97	–	50.26	m²	**116.23**
10 m² deck; 10.00 × 5.00 m	46.65	1.00	56.00	–	49.85	m²	**105.85**
Composite deck surface; Millboard; 176 mm wide; fixed with stainless screws 4.5 × 70 mm							
Enhanced grain range	45.00	0.33	21.46	–	55.71	m²	**77.17**
Weathered Oak range	57.00	0.33	21.46	–	69.51	m²	**90.97**

35 SITE WORKS – STREET FURNITURE

Item – Overhead and Profit Included	PC £	Labour hours	Labour £	Plant £	Material £	Unit	Total rate £
BARRIERS							
Barriers – General							
Preamble: The provision of car and truck control barriers may form part of the landscape contract. Barriers range from simple manual counterweighted poles to fully automated remote-control security gates, and the exact degree of control required must be specified. Complex barriers may need special maintenance and repair.							
Barriers; Barriers Direct							
Manually operated pole barriers; counterbalance; to tubular steel supports; bolting to concrete foundation (foundation not included); aluminium boom; various finishes, including arm rest							
clear opening up to 3.00 m	671.35	6.00	193.20	–	800.28	nr	**993.48**
clear opening up to 4.00 m	671.35	6.00	193.20	–	800.28	nr	**993.48**
clear opening 5.00 m	685.10	6.00	193.20	–	816.10	nr	**1009.30**
clear opening 6.00 m	1051.05	6.00	193.20	–	1236.94	nr	**1430.14**
clear opening 7.00 m	1114.95	6.00	193.20	–	1310.43	nr	**1503.63**
Electrically operated pole barriers; enclosed fan-cooled motor; double worm reduction gear; overload clutch coupling with remote controls; aluminium boom; various finishes, including catch pole (exclusive of electrical connections by electrician)							
Barriers Direct 2.65 m boom with catchpost	2010.50	6.00	193.20	–	2340.31	nr	**2533.51**
Barriers Direct 4.5 m boom with catchpost	2010.50	6.00	193.20	–	2340.31	nr	**2533.51**
Barriers Direct 6.0 m boom with catchpost	2010.50	6.00	193.20	–	2340.31	nr	**2533.51**
BASES FOR STREET FURNITURE							
Bases for street furniture							
Excavating; filling with concrete 1:3:6; bases for street furniture							
300 × 450 × 500 mm deep	–	0.75	24.15	–	11.08	nr	**35.23**
300 × 600 × 500 mm deep	–	1.11	35.74	–	14.79	nr	**50.53**
300 × 900 × 500 mm deep	–	1.67	53.77	–	22.18	nr	**75.95**
1750 × 900 × 300 mm deep	–	2.33	75.03	–	77.64	nr	**152.67**
2000 × 900 × 300 mm deep	–	2.67	85.97	–	88.73	nr	**174.70**
2400 × 900 × 300 mm deep	–	3.20	103.04	–	106.47	nr	**209.51**
2400 × 1000 × 300 mm deep	–	3.56	114.63	–	118.30	nr	**232.93**

35 SITE WORKS – STREET FURNITURE

Item – Overhead and Profit Included	PC £	Labour hours	Labour £	Plant £	Material £	Unit	Total rate £
BASES FOR STREET FURNITURE – cont							
Bases for street furniture – cont							
Precast concrete flags; to concrete bases (not included); bedding and jointing in cement: mortar (1:4)							
450 × 600 × 50 mm	13.59	1.17	37.56	–	24.40	m²	**61.96**
Precast concrete paving blocks; to concrete bases (not included); bedding in sharp sand; butt joints							
200 × 100 × 65 mm	12.05	0.50	16.10	–	16.72	m²	**32.82**
200 × 100 × 80 mm	13.94	0.50	16.10	–	19.10	m²	**35.20**
Engineering paving bricks; to concrete bases (not included); bedding and jointing in sulphate-resisting cement: lime: sand mortar (1:1:6)							
over 300 mm wide	14.84	0.56	18.09	–	33.90	m²	**51.99**
Edge restraints to pavings; haunching in concrete (1:3:6)							
200 × 300 mm	6.07	0.10	3.22	–	6.98	m	**10.20**
Bases; Earth Anchors Ltd							
Rootfast ancillary anchors; A1; 500 mm long × 25 mm dia. and top strap F2; including bolting to site furniture (site furniture not included)	11.90	0.50	16.10	–	13.69	set	**29.79**
installation tool for above	23.50	–	–	–	27.02	nr	**27.02**
Rootfast ancillary anchors; A4; heavy duty 40 mm square fixed head anchors; including bolting to site furniture (site furniture not included)	29.00	0.33	10.73	–	33.35	set	**44.08**
Rootfast ancillary anchors; F4; vertical socket; including bolting to site furniture (site furniture not included)	12.75	0.05	1.61	–	14.66	set	**16.27**
Rootfast ancillary anchors; F3; horizontal socket; including bolting to site furniture (site furniture not included)	14.00	0.05	1.61	–	16.10	set	**17.71**
installation tools for the above	71.00	–	–	–	81.65	nr	**81.65**
Rootfast ancillary anchors; A3; anchored bases; including bolting to site furniture (site furniture not included)	48.00	0.33	10.73	–	55.20	set	**65.93**
installation tools for the above	71.00	–	–	–	81.65	nr	**81.65**

35 SITE WORKS – STREET FURNITURE

Item – Overhead and Profit Included	PC £	Labour hours	Labour £	Plant £	Material £	Unit	Total rate £
BINS: LITTER/WASTE/ DOGWASTE/SALT GRIT							
Dog waste bins							
Earth Anchors Ltd							
HG45 A; steel; 45 l; earth anchored; post mounted	107.00	0.33	10.73	–	123.05	nr	**133.78**
HG45 A; steel; 45 l; earth anchored with pedal operation	127.00	0.33	10.73	–	146.05	nr	**156.78**
Furnitubes International Ltd							
PED 701; Pedigree; post mounted cast iron dog waste bins; 1250 mm total height above ground; 400 mm square bin	544.00	0.75	24.15	–	633.96	nr	**658.11**
LUK745 P; Lucky; steel dog bin; post mounted; 47 l	364.00	1.50	48.30	–	426.96	nr	**475.26**
LUK745 W; Lucky; steel dog bin; wall mounted; 47 l	322.00	0.75	24.15	–	378.66	nr	**402.81**
Litter bins; precast concrete in textured white or exposed aggregate finish; with wire baskets and drainage holes							
Bins; Marshalls Plc							
Boulevard 700; concrete circular litter bin	566.41	0.50	16.10	–	651.37	nr	**667.47**
Bins; Neptune Outdoor Furniture Ltd							
AR16; 42 l	465.00	0.50	16.10	–	534.75	nr	**550.85**
AR14; 100 l	520.01	0.50	16.10	–	598.01	nr	**614.11**
Bins; Townscape Products Ltd							
Braunton; with basket and canopy; finished in natural round exposed aggregate	357.61	1.50	48.30	–	411.25	nr	**459.55**
Litter bins; all-steel							
Bins; Marshall Plc							
MSF Central; steel litter bin	407.11	1.00	32.20	–	468.18	nr	**500.38**
MSF Central; stainless steel litter bin	777.54	0.67	21.46	–	894.17	nr	**915.63**
Bins; Earth Anchors Ltd							
Ranger; 100 l; pedestal mounted	453.00	0.50	16.10	–	520.95	nr	**537.05**
Big Ben; 82 l; steel frame and liner; colour coated finish; earth anchored	273.00	1.00	32.20	–	313.95	nr	**346.15**
Beau; 42 l; steel frame and liner; colour coated finish; earth anchored	253.00	1.00	32.20	–	290.95	nr	**323.15**

35 SITE WORKS – STREET FURNITURE

Item – Overhead and Profit Included	PC £	Labour hours	Labour £	Plant £	Material £	Unit	Total rate £
BINS: LITTER/WASTE/ DOGWASTE/SALT GRIT – cont							
Litter bins; cast iron							
Bins; Marshalls Plc							
MSF5501 Heritage; cast iron litter bin	809.17	1.00	32.20	2.69	930.55	nr	**965.44**
Bins; Furnitubes International Ltd							
Covent Garden COV 702; side opening, 500 mm square ×							
1050 mm high; 105 l capacity	714.00	0.50	16.10	–	826.56	nr	**842.66**
Covent Garden COV 803; side opening, 500 mm dia. × 1025 mm							
high; 85 l capacity	575.00	0.50	16.10	–	666.71	nr	**682.81**
Covent Garden COV 912; open top; 500 mm A/F octagonal ×							
820 mm high; 85 l capacity	478.00	0.50	16.10	–	555.16	nr	**571.26**
Litter bins; timber faced; hardwood slatted casings with removable metal litter containers; ground or wall fixing							
Bins; Woodscape Ltd							
square; 580 × 580 × 950 mm high; with lockable lid	685.00	0.50	16.10	–	787.75	nr	**803.85**
round; 580 mm dia. × 950 mm high; with lockable lid	860.00	0.50	16.10	–	989.00	nr	**1005.10**
Plastic litter and grit bins; glassfibre reinforced polyester grit bins; yellow body; hinged lids							
Bins; Wybone Ltd; Victoriana glass fibre; cast iron effect litter bins; including lockable liner							
LBV/2; 521 × 521 × 673 mm high; open top; square shape; 0.078 m^3 capacity	156.99	1.00	32.20	–	186.00	nr	**218.20**
LVC/3; 457 mm dia. × 648 mm high; closed top; drum shape; 0.084 m^3 capacity; with lockable liner	360.99	1.00	32.20	–	420.60	nr	**452.80**
Grit bins; Furnitubes International Ltd							
Q3 Grit & Salt bin, yellow glass fibre, integral lock, 100 l	–	–	–	–	335.80	nr	**335.80**

35 SITE WORKS – STREET FURNITURE

Item – Overhead and Profit Included	PC £	Labour hours	Labour £	Plant £	Material £	Unit	Total rate £
Cigarette bins							
Furnitubes International Ltd							
ZEN 275 Zenith cigarette bin; stainless steel; wall or post mounted; 1.7 l	39.00	0.75	24.15	–	53.21	nr	**77.36**
LVR250 PC Liverpool cigarette bin; steel; wall mounted; 1.9 l	32.99	0.50	16.10	–	37.94	nr	**54.04**
SMK500F Smoke King cigarette bin; cast aluminium; bolt down; 3.5 l	219.99	0.75	24.15	–	261.35	nr	**285.50**
BOLLARDS							
Excavating; for bollards and barriers; by hand							
Holes for bollards							
400 × 400 × 400 mm; disposing off site	–	0.42	13.42	–	3.07	nr	**16.49**
600 × 600 × 600 mm; disposing off site	–	0.97	31.23	–	8.62	nr	**39.85**
Concrete bollards – General							
Preamble: Precast concrete bollards are available in a very wide range of shapes and sizes. The bollards listed here are the most commonly used sizes and shapes; manufacturer's catalogues should be consulted for the full range. Most manufacturers produce bollards to match their suites of street furniture, which may include planters, benches, litter bins and cycle stands. Most parallel sided bollards can be supplied in removable form, with a reduced shank, precast concrete socket and lifting hole to permit removal with a bar.							
Concrete bollards							
Marshalls Plc; cylinder; straight or tapered; 200–400 mm dia.; plain grey concrete; setting into firm ground (excavating and backfilling not included)							
Bridgford; 915 mm high above ground	112.52	2.00	64.40	–	143.95	nr	**208.35**

35 SITE WORKS – STREET FURNITURE

Item – Overhead and Profit Included	PC £	Labour hours	Labour £	Plant £	Material £	Unit	Total rate £
BOLLARDS – cont							
Concrete bollards – cont							
Marshalls Plc; cylinder; straight or tapered; 200–400 mm dia.; Beadalite reflective finish; setting into firm ground (excavating and backfilling not included)							
Wexham concrete bollard; exposed silver grey	227.59	2.00	64.40	–	269.55	nr	**333.95**
Wexham Major concrete bollard; exposed silver grey	354.01	2.00	64.40	–	421.66	nr	**486.06**
Marshalls Plc; Precast concrete verge markers; various shapes;							
Spherical 500; exposed silver grey	480.10	1.00	32.20	–	557.70	nr	**589.90**
Cube 600; silver grey	464.47	1.00	32.20	–	539.73	nr	**571.93**
Monoscape Woodhouse; smooth grey	142.91	1.00	32.20	–	169.94	nr	**202.14**
Townscape Products Ltd; Linby; natural grey concrete							
seats; 1800 × 600 × 736 mm high	527.81	2.00	64.40	–	606.98	nr	**671.38**
Bollards							
Service bollards; Furnitubes International Ltd; stainless steel; for housing services connection points to electricity, water etc. (services connections not included)							
Kenton KEN717; 900 × 250 mm dia.	661.00	2.00	64.40	–	767.97	nr	**832.37**
Zenith ZEN707; 900 × 250 mm dia.	620.00	2.00	64.40	–	720.82	nr	**785.22**
Other bollards							
Removable parking posts; Marshalls Plc							
RT/RD4; domestic telescopic bollard	208.61	2.00	64.40	–	247.72	nr	**312.12**
RT/R8; heavy duty telescopic bollard	297.11	2.00	64.40	–	349.50	nr	**413.90**
Plastic bollards; Marshalls Plc							
Lismore; three ring recycled plastic bollard	98.61	2.00	64.40	–	121.22	nr	**185.62**
Cast iron bollards – General							
Preamble: The following bollards are particularly suitable for conservation areas. Logos for civic crests can be incorporated to order.							

35 SITE WORKS – STREET FURNITURE

Item – Overhead and Profit Included	PC £	Labour hours	Labour £	Plant £	Material £	Unit	Total rate £
Cast iron bollards							
Bollards; Furnitubes International Ltd (excavating and backfilling not included)							
Doric Round; 920 mm high × 170 mm dia.	193.00	2.00	64.40	–	229.40	nr	**293.80**
Manchester Round; 975 mm high; 225 mm square base	197.00	2.00	64.40	–	234.00	nr	**298.40**
Kenton; heavy duty galvanized steel; 900 mm high; 350 mm dia.	283.00	2.00	64.40	–	332.90	nr	**397.30**
Bollards; Marshalls Plc (excavating and backfilling not included)							
MSF103; cast iron bollard; Small Manchester	183.33	2.00	64.40	–	218.65	nr	**283.05**
MSF102; cast iron bollard; Manchester	183.33	2.00	64.40	–	218.65	nr	**283.05**
Cast iron bollards with rails – General							
Preamble: The following cast iron bollards are suitable for conservation areas.							
Cast iron bollards with rails							
Cast iron posts with steel tubular rails; Broxap Street Furniture; setting into firm ground (excavating not included)							
Weaver Polyurethane single rail post	68.00	1.50	48.30	–	81.73	nr	**130.03**
Type C mild steel tubular rail; including connector	1.39	0.03	1.07	–	3.20	m	**4.27**
Steel bollards							
Steel bollards; Marshalls Plc (excavating and backfilling not included)							
SSB01; stainless steel bollard; 101 × 1250 mm	145.39	2.00	64.40	–	175.02	nr	**239.42**
RS001; stainless steel bollard; 114 × 1500 mm	151.71	2.00	64.40	–	182.29	nr	**246.69**
RB119 Brunel; steel bollard; 168 × 1500 mm	150.45	2.00	64.40	–	187.56	nr	**251.96**
Timber bollards							
Woodscape Ltd; durable hardwood							
RP 250/1500; 250 mm dia. × 1500 mm long	351.20	1.00	32.20	–	404.20	nr	**436.40**
SP 250/1500; 250 mm square × 1500 mm long	351.20	1.00	32.20	–	404.20	nr	**436.40**
SP 150/1200; 150 mm square × 1200 mm long	118.20	1.00	32.20	–	136.25	nr	**168.45**

35 SITE WORKS – STREET FURNITURE

Item – Overhead and Profit Included	PC £	Labour hours	Labour £	Plant £	Material £	Unit	Total rate £
BOLLARDS – cont							
Timber bollards – cont							
Woodscape Ltd – cont							
SP 125/750; 125 mm square × 750 mm long	74.10	1.00	32.20	–	85.54	nr	**117.74**
RP 125/750; 125 mm dia. × 750 mm long	74.10	1.00	32.20	–	85.54	nr	**117.74**
Deterrent bollards							
Semi-mountable vehicle deterrent and kerb protection bollards; Furnitubes International Ltd (excavating and backfilling not included)							
bell decorative	720.00	2.00	64.40	–	838.60	nr	**903.00**
half bell	762.00	2.00	64.40	–	886.90	nr	**951.30**
full bell ®	930.00	2.00	64.40	–	1080.10	nr	**1144.50**
three quarter bell	829.00	2.00	64.40	–	963.95	nr	**1028.35**
CYCLE HOLDERS/CYCLE SHELTERS							
Cycle holders							
Cycle stands; Marshalls Plc							
Sheffield; steel cycle stand; RCS1	56.89	0.50	16.10	–	65.42	nr	**81.52**
Sheffield; stainless steel cycle stand; RSCS1	152.98	0.50	16.10	–	175.93	nr	**192.03**
Cycle holders; Autopa; VELOPA; galvanized steel							
R; fixing to wall or post; making good	102.66	1.00	32.20	–	128.98	nr	**161.18**
SR(V); fixing in ground; making good	160.98	1.00	32.20	–	196.05	nr	**228.25**
Sheffield cycle stands; ragged steel	39.59	1.00	32.20	–	56.45	nr	**88.65**
Cycle holders; Townscape Products Ltd							
Guardian cycle holders; tubular steel frame; setting in concrete; 1250 × 550 × 775 mm high; making good	393.12	1.00	32.20	–	463.01	nr	**495.21**
Penny cycle stands; 600 mm dia.; exposed aggregate bollards with 8 nr cycle holders; in galvanized steel; setting in concrete; making good	867.32	1.00	32.20	–	1008.34	nr	**1040.54**

35 SITE WORKS – STREET FURNITURE

Item – Overhead and Profit Included	PC £	Labour hours	Labour £	Plant £	Material £	Unit	Total rate £
Cycle holders; Broxap Street Furniture							
Neath cycle rack; BX/MW/AG; for 6 nr cycles; semi-vertical; galvanized and polyester powder coated; 1.32 m wide × 2.542 m long × 1.80 m high	470.00	10.00	322.00	–	562.35	nr	884.35
Wardale cycle shelter for 10 nr cycles; galvanized; 2.13 m wide × 3.05 m long × 2.15 m high	1160.00	10.00	322.00	–	1355.85	nr	1677.85
Toast Rack double sided free-standing cycle rack; BX/MW/GH; for 10 nr cycles; galvanized and polyester coated; 3.25 m long	260.00	4.00	128.80	–	313.12	nr	441.92
Cycle shelters							
Furnitubes International Ltd; Academy free-standing bolt down shelter, galvanized steel frame and roof; 3450 mm overall width; 2150 mm maximum height; 2670 mm depth							
corrugated roof	1882.00	6.00	193.20	–	2181.03	nr	2374.23
clear polycarbonate roof	1861.00	6.00	193.20	–	2156.88	nr	2350.08
Concrete cycle blocks; **Townscape Products Ltd; on** **100 mm concrete (1:2:4); on** **150 mm hardcore; bedding in** **cement: sand (1:4)**							
Cycle blocks							
Cycle Bloc left hand; in natural round exposed aggregate	90.25	0.25	8.05	–	104.89	nr	112.94
Cycle Bloc right hand; in exposed aggregate concrete	90.25	0.25	8.05	–	104.89	nr	112.94
Flagpoles							
Flagpoles; ground mounted; in glass fibre; smooth white finish; including hinged baseplate, nylon halyard system and revolving gold onion finial; setting in concrete; to manufacturer's recommendations (excavating not included)							
6 m high; external halyard system	218.00	4.00	128.80	–	278.93	nr	407.73
6 m high; internal halyard system	293.00	4.00	128.80	–	365.18	nr	493.98
8 m high; external halyard system	258.00	6.00	193.20	–	324.93	nr	518.13
8 m high; internal halyard system	333.00	6.00	193.20	–	411.18	nr	604.38
10 m high; external halyard system	388.00	5.00	161.00	–	474.43	nr	635.43
10 m high; internal halyard system	463.00	5.00	161.00	–	560.68	nr	721.68

35 SITE WORKS – STREET FURNITURE

Item – Overhead and Profit Included	PC £	Labour hours	Labour £	Plant £	Material £	Unit	Total rate £
Flagpoles – cont							
Flagpoles; wall mounted; vertical or angled poles; in glass fibre; smooth white finish; including base, top bracket, external nylon halyard rope, cleat and revolving gold onion finial							
3 m pole	185.00	2.00	64.40	–	212.75	nr	**277.15**
Banner flagpoles; 90 mm dia. aluminium pole; suitable for 1 × 2 m banner							
6 m high	332.39	4.00	128.80	–	410.48	nr	**539.28**
GENERALLY							
Note: The costs of delivery from the supplier are generally not included in these items. Readers should check with the supplier to verify the delivery costs.							
Furniture/equipment – General Preamble: The following items include fixing to manufacturer's instructions; holding down bolts or other fittings and making good (excavating, backfilling and tarmac, concrete or paving bases not included).							
PERGOLAS							
Pergolas; construct timber pergola; posts 150 × 150 mm × 2.40 m finished height in 600 mm deep minimum concrete pits; beams of 200 × 50 mm × 2.00 m wide; fixed to posts with dowels; rafters 200 × 38 mm notched to beams; inclusive of all mechanical excavation and disposal off site Pergola 2.00 m wide in green oak; posts at 2.00 m centres; dowel fixed							
rafters at 600 mm centres	170.76	8.40	302.68	4.50	292.58	m	**599.76**
rafters at 400 mm centres	189.13	10.80	389.16	4.50	322.25	m	**715.91**
Pergola 3.00 m wide in green oak; posts at 1.50 m centres							
rafters at 600 mm centres	122.99	7.20	273.24	5.39	253.57	m	**532.20**
rafters at 400 mm centres	150.49	8.00	303.60	5.39	297.98	m	**606.97**

35 SITE WORKS – STREET FURNITURE

Item – Overhead and Profit Included	PC £	Labour hours	Labour £	Plant £	Material £	Unit	Total rate £
Pergola 2.00 m wide in prepared softwood; beams fixed with bolts; posts at 2.00 m centres							
rafters at 600 mm centres	72.51	5.00	189.75	4.50	180.42	m	**374.67**
rafters at 400 mm centres	92.55	6.00	227.70	4.50	212.01	m	**444.21**
Pergola 3.00 m wide in prepared softwood; posts at 1.50 m centres							
rafters at 600 mm centres	67.04	7.20	273.24	5.39	189.22	m	**467.85**
rafters at 400 mm centres	81.23	8.00	303.60	5.39	218.33	m	**527.32**
PICNIC TABLES/SUNDRY FURNITURE							
Picnic benches – General Preamble: The following items include for fixing to ground to manufacturer's instructions or concreting in.							
Picnic tables and benches Picnic tables and benches; Broxap Street Furniture							
Springfield timber picnic unit 1800 mm long softwood	269.00	2.00	64.40	–	320.27	nr	**384.67**
Picnic tables and benches; Woodscape Ltd							
table and benches built in; 2 m long	2595.00	2.00	64.40	–	2995.18	nr	**3059.58**
STREET FURNITURE RANGES							
Lifebuoy stations Lifebuoy stations; Earth Anchors Ltd							
Rootfast lifebuoy station complete with post AP44; SOLAS approved lifebuoys 590 mm and lifeline	165.00	2.00	64.40	–	202.83	nr	**267.23**
installation tool for above	–	–	–	–	98.90	nr	**98.90**
PLANT CONTAINERS							
Market prices of containers							
Plant containers; terracotta Capital Garden Products Ltd							
Large Pot LP63; weathered terracotta; 1170 × 1600 mm dia.	–	–	–	–	991.30	nr	**991.30**
Large Pot LP38; weathered terracotta; 610 × 970 mm dia.	–	–	–	–	422.05	nr	**422.05**
Large Pot LP23; weathered terracotta; 480 × 580 mm dia.	–	–	–	–	240.35	nr	**240.35**

35 SITE WORKS – STREET FURNITURE

Item – Overhead and Profit Included	PC £	Labour hours	Labour £	Plant £	Material £	Unit	Total rate £
PLANT CONTAINERS – cont							
Plant containers – cont							
Capital Garden Products Ltd – cont							
Indian style Shimmer Pot 2322;							
585 × 560 mm dia.	–	–	–	–	211.60	nr	**211.60**
Indian style Shimmer Pot 1717;							
430 × 430 mm dia.	–	–	–	–	126.50	nr	**126.50**
Indian style Shimmer Pot 1314;							
330 × 355 mm dia.	–	–	–	–	119.60	nr	**119.60**
Plant containers; faux lead							
Capital Garden Products Ltd							
Trough 2508 Tudor Rose;							
620 × 220 × 230 mm high	–	–	–	–	64.40	nr	**64.40**
Tub 2004 Elizabethan; 510 mm square	–	–	–	–	112.70	nr	**112.70**
Tub 1513 Elizabethan; 380 mm square	–	–	–	–	63.25	nr	**63.25**
Tub 1601 Tudor Rose;							
420 × 400 mm dia.	–	–	–	–	79.35	nr	**79.35**
Plant containers; window boxes							
Capital Garden Products Ltd							
Adam 5401; faux lead;							
1370 × 270 × 210 mm high	–	–	–	–	129.95	nr	**129.95**
Oakleaf OAK24; terracotta;							
610 × 230 × 240 mm high	–	–	–	–	97.75	nr	**97.75**
Swag 2402; faux lead;							
610 × 200 × 210 mm high	–	–	–	–	50.60	nr	**50.60**
Plant containers; timber							
Plant containers; hardwood;							
Neptune Outdoor Furniture Ltd							
Woodland T38–4D;							
1500 × 1500 × 900 mm high	–	–	–	–	1414.50	nr	**1414.50**
Woodland T38–3C;							
1000 × 1500 × 700 mm high	–	–	–	–	1023.50	nr	**1023.50**
Woodland T38–2 A;							
1000 × 500 × 500 mm high	–	–	–	–	517.50	nr	**517.50**
Hudson T42–4D;							
1500 × 1500 × 900 mm high	–	–	–	–	1460.50	nr	**1460.50**
Hudson T42–3C;							
1000 × 1500 × 700 mm high	–	–	–	–	1069.50	nr	**1069.50**
Hudson T42–2 A;							
1000 × 500 × 500 mm high	–	–	–	–	563.50	nr	**563.50**
Plant containers; hardwood;							
Woodscape Ltd							
square; 900 × 900 × 420 mm high	–	–	–	–	569.94	nr	**569.94**

35 SITE WORKS – STREET FURNITURE

Item – Overhead and Profit Included	PC £	Labour hours	Labour £	Plant £	Material £	Unit	Total rate £
Plant containers; precast concrete							
Plant containers; Marshalls Plc							
Boulevard 700; circular base and ring	788.94	2.00	64.40	29.88	907.28	nr	**1001.56**
Boulevard 1200; circular base and ring	996.30	1.00	32.20	29.88	1145.74	nr	**1207.82**
Bespoke Planters; Outdoor Design; planters to required size and shape; with seamless polished corners; preformed drainage holes and lip detail; inclusive of placement; drainage layer 50 mm thick and potting soil							
Galvanized 1.5 mm steel; powder coated							
500 mm × 500 mm × 500 mm	325.00	0.50	32.20	–	382.33	nr	**414.53**
1500 mm × 500 mm × 500 mm	403.00	0.80	51.52	–	487.95	nr	**539.47**
1800 mm × 750 mm × 600 mm	513.00	1.60	103.04	21.54	644.08	nr	**768.66**
2500 mm × 1000 mm × 600 mm	629.00	1.50	96.60	21.54	829.09	nr	**947.23**
Corten 2.0 mm							
500 mm × 500 mm × 500 mm	336.00	0.50	32.20	–	394.98	nr	**427.18**
1500 mm × 500 mm × 500 mm	427.00	0.80	51.52	–	515.55	nr	**567.07**
1800 mm × 750 mm × 600 mm	551.00	1.60	103.04	21.54	687.78	nr	**812.36**
2500 mm × 1000 mm × 600 mm	687.00	1.50	96.60	21.54	895.79	nr	**1013.93**
Stainless steel 1.5 mm; Grained finish							
500 mm × 500 mm × 500 mm	416.00	0.50	32.20	–	486.98	nr	**519.18**
1500 mm × 500 mm × 500 mm	602.00	0.80	51.52	–	716.79	nr	**768.31**
1800 mm × 750 mm × 600 mm	828.00	1.60	103.04	21.54	1006.33	nr	**1130.91**
2500 mm × 1000 mm × 600 mm	1106.00	1.50	96.60	21.54	1377.64	nr	**1495.78**

SEATING

Outdoor seats

Outdoor seats; concrete framed – General

Preamble: Prices for the following concrete framed seats and benches with hardwood slats include for fixing (where necessary) by bolting into existing paving or concrete bases (bases not included) or building into walls or concrete foundations (walls and foundations not included). Delivery generally not included.

35 SITE WORKS – STREET FURNITURE

Item – Overhead and Profit Included	PC £	Labour hours	Labour £	Plant £	Material £	Unit	Total rate £
SEATING – cont							
Outdoor seats; concrete framed							
Outdoor seats; Townscape Products Ltd							
Belgrave seat; exposed aggregate 1800 × 645 × 730 mm	479.83	2.00	128.80	–	568.54	nr	**697.34**
Maidstone seats;							
1800 × 610 × 785 mm	668.18	2.00	128.80	–	785.14	nr	**913.94**
Outdoor seats; concrete							
Outdoor seats; Marshalls Ltd							
Boulevard 2000 concrete seat	922.96	2.00	128.80	–	1082.00	nr	**1210.80**
Outdoor seats; Furnitubes International Ltd							
Amesbury concrete seat, 2.2 m long	772.00	2.00	128.80	–	908.40	nr	**1037.20**
Marlborough concrete seat, 1.8 m long	714.00	2.00	128.80	–	841.70	nr	**970.50**
Outdoor seats; metal framed – General							
Preamble: Metal framed seats with hardwood backs to various designs can be bolted to ground anchors.							
Outdoor seats; metal framed							
Outdoor seats; Furnitubes International Ltd							
NS6 Newstead; steel standards with iroko slats; 1.80 m long	439.00	2.00	128.80	–	525.45	nr	**654.25**
NEB6 New Forest Single Bench; cast iron standards with iroko slats; 1.83 m long	420.00	2.00	128.80	–	503.60	nr	**632.40**
EA6 Eastgate; cast iron standards with iroko slats; 1.86 m long	562.00	2.00	128.80	–	666.90	nr	**795.70**
NE6 New Forest Seat; cast iron standards with iroko slats; 1.83 m long	541.00	2.00	128.80	–	642.75	nr	**771.55**
Zenith long bench, satin stainless steel, iroko slats, 1.8 m long	1312.00	2.00	128.80	–	1529.40	nr	**1658.20**
Ashburton; steel with recycled plastic slats; 1.8 m long	587.00	2.00	128.80	–	695.65	nr	**824.45**
Outdoor seats; Neptune Outdoor Furniture Ltd; Kara; treated hardwood slats to heavy duty steel frame							
1.50 m long	530.00	2.00	128.80	–	620.42	nr	**749.22**
1.80 m long	555.00	2.00	128.80	–	649.17	nr	**777.97**
2.00 m long	580.00	2.00	128.80	–	677.92	nr	**806.72**

35 SITE WORKS – STREET FURNITURE

Item – Overhead and Profit Included	PC £	Labour hours	Labour £	Plant £	Material £	Unit	Total rate £
Outdoor seats; Neptune Outdoor Furniture Ltd; Beaufort, powder coated steel frame with treated hardwood slats							
1.50 m long	495.00	2.00	128.80	–	580.17	nr	**708.97**
1.80 m long	535.00	2.00	128.80	–	626.17	nr	**754.97**
2.00 m long	555.00	2.00	128.80	–	649.17	nr	**777.97**
Outdoor seats; Neptune Outdoor Furniture Ltd; Avenue – treated hardwood slats to cast iron seat and bench ends							
1.80 m long	560.00	2.00	128.80	–	654.92	nr	**783.72**
2.10 m long	580.00	2.00	128.80	–	677.92	nr	**806.72**
Outdoor seats; Broxap Street Furniture; metal and timber							
Eastgate	398.00	2.00	128.80	–	468.63	nr	**597.43**
Outdoor seats; Earth Anchors Ltd; Forest-Saver; steel frame and recycled slats							
bench; 1.8 m	209.00	1.00	64.40	–	240.35	nr	**304.75**
seat; 1.8 m	375.00	1.00	64.40	–	431.25	nr	**495.65**
Outdoor seats; Earth Anchors Ltd; Evergreen; cast iron frame and recycled slats							
bench; 1.8 m	456.00	1.00	64.40	–	524.40	nr	**588.80**
seat; 1.8 m	586.00	1.00	64.40	–	673.90	nr	**738.30**
Outdoor seats; all steel							
Outdoor seats; Marshalls Plc							
MSF Central; stainless steel seat	1536.15	0.50	32.20	–	1766.57	nr	**1798.77**
MSF Central; steel seat	783.88	2.00	128.80	–	901.46	nr	**1030.26**
MSF 502; cast iron Heritage seat	676.41	2.00	128.80	–	777.87	nr	**906.67**
Outdoor seats; Earth Anchors Ltd; Ranger							
bench; 1.8 m	290.00	1.00	64.40	–	333.50	nr	**397.90**
seat; 1.8 m	439.00	1.00	64.40	–	504.85	nr	**569.25**
Outdoor seats; all timber							
Outdoor seats; Woodcraft UK; Mendip; teak							
1.524 m long	674.00	1.00	64.40	–	800.92	nr	**865.32**
1.829 m long	718.00	1.00	64.40	–	851.52	nr	**915.92**
2.438 m long (including centre leg)	946.00	1.00	64.40	–	1113.72	nr	**1178.12**
Outdoor seats; Woodcraft UK; Wykeham; hardwood							
1.5 m	496.00	1.00	64.40	–	596.22	nr	**660.62**

35 SITE WORKS – STREET FURNITURE

Item – Overhead and Profit Included	PC £	Labour hours	Labour £	Plant £	Material £	Unit	Total rate £
SEATING – cont							
Outdoor seats – cont							
Outdoor seats; Woodscape Ltd; solid hardwood							
seat type 3 with back; 2.00 m long; free-standing	1025.00	2.00	128.80	–	1204.57	nr	1333.37
seat type 3 with back; 2.00 m long; building in	1110.00	4.00	257.60	–	1302.32	nr	1559.92
seat type 4; 2.00 m long; fixing to wall	490.00	2.00	128.80	–	589.32	nr	718.12
seat type 4 with back; 2.00 m long; fixing to wall	600.00	3.00	193.20	–	695.80	nr	889.00
seat type 4 with back; 2.50 m long; fixing to wall	665.00	3.00	193.20	–	770.55	nr	963.75
seat type 5 with back; 2.00 m long; free-standing	1055.00	2.00	128.80	–	1239.07	nr	1367.87
seat type 5 with back; 2.50 m long; free-standing	1125.00	2.00	128.80	–	1319.57	nr	1448.37
seat type 5 with back; 2.00 m long; building in	1140.00	2.00	128.80	–	1336.82	nr	1465.62
seat type 5 with back; 2.50 m long; building in	1210.00	3.00	193.20	–	1417.32	nr	1610.52
bench type 1; 2.00 m long; free-standing	840.00	2.00	128.80	–	991.82	nr	1120.62
bench type 1; 2.00 m long; building in	920.00	4.00	257.60	–	1083.82	nr	1341.42
bench type 2; 2.00 m long; free-standing	795.00	2.00	128.80	–	940.07	nr	1068.87
bench type 2; 2.50 m long; free-standing	840.00	2.00	128.80	–	991.82	nr	1120.62
bench type 2; 2.00 m long; building in	875.00	4.00	257.60	–	1032.07	nr	1289.67
bench type 2; 2.50 m long; building in	920.00	4.00	257.60	–	1083.82	nr	1341.42
bench type 2; 2.00 m long overall; curved to 5 m radius; building in	1040.00	4.00	257.60	–	1221.82	nr	1479.42
Outdoor seats; Neptune Outdoor Furniture Ltd; Southampton – hardwood slats to exposed aggregate concrete base							
1.50 m long	420.00	2.00	128.80	–	508.82	nr	637.62
1.80 m long	430.00	2.00	128.80	–	520.32	nr	649.12
2.410 m long	450.00	2.00	128.80	–	543.32	nr	672.12
Outdoor seats; tree benches/seats							
Tree bench; Neptune Outdoor Furniture Ltd; Beaufort; hexagonal; timber							
SF34–15 A; 1500 mm dia.	960.00	0.50	32.20	–	1104.00	nr	1136.20

35 SITE WORKS – STREET FURNITURE

Item – Overhead and Profit Included	PC £	Labour hours	Labour £	Plant £	Material £	Unit	Total rate £
Tree seat; Neptune Outdoor Furniture Ltd; Beaufort; hexagonal; timber; with back							
SF32–10 A; 720 mm dia.	1230.00	0.50	32.20	–	1414.50	nr	**1446.70**
SF32–20 A; 1720 mm dia.	1540.00	0.50	32.20	–	1771.00	nr	**1803.20**
Anti-skate board devices							
Stainless steel; Furnitubes International Ltd							
for fitting to seats	20.00	0.20	6.44	–	23.00	nr	**29.44**
for fitting to benches	20.00	0.20	6.44	–	23.00	nr	**29.44**
SIGNAGE							
Directional signage – Finger posts; cast aluminium							
Signage; Furnitubes International Ltd							
FFL1 Lancer; cast aluminium finials	53.00	0.07	2.14	–	60.95	nr	**63.09**
FAAIS; arrow end type cast aluminium directional arms; single line; 90 mm wide	152.00	2.00	64.40	–	174.80	nr	**239.20**
FAAID; arrow end type cast aluminium directional arms; double line; 145 mm wide	187.00	0.13	4.30	–	215.05	nr	**219.35**
FAAIS; arrow end type cast aluminium directional arms; treble line; 200 mm wide	230.00	0.20	6.44	–	264.50	nr	**270.94**
FCK1211G Kingston; composite standard root columns	690.00	2.00	64.40	–	808.35	nr	**872.75**
Park signage							
Entrance signs and map boards; vitreous enamel							
entrance map board; 1250 × 1000 mm high with two support posts; all associated works to post bases	–	–	–	–	–	nr	**100.84**
information board with two locking cabinets; 1250 × 1000 mm high with two support posts; all associated works to post bases	–	–	–	–	–	nr	**307.04**

35 SITE WORKS – STREET FURNITURE

Item – Overhead and Profit Included	PC £	Labour hours	Labour £	Plant £	Material £	Unit	Total rate £
SIGNAGE – cont							
Miscellaneous park signage; vitreous enamel							
'No dogs' sign; 100 × 100 mm; fixing to fencing or gates	–	–	–	–	–	nr	**17.08**
'Dog exercise area' sign; 300 × 400 mm; fixing to fencing or gates	–	–	–	–	–	nr	**30.55**
'Nature conservation area' sign; 900 × 400 mm high with two support posts; all associated works to post bases	–	–	–	–	–	nr	**46.86**
Finger Posts; Oak FSC; Landmark Timber Products							
Fresh-sawn oak post 100 × 100 mm; Air-dried oak finger-arms; 550 × 150 × 25 mm; mortized and fixed with hardwood dowels; Engraved and in-filled with durable paint; in concrete footing; County range							
Post 2.80 m overall with single finger	165.00	1.00	64.40	22.57	256.63	nr	**343.60**
Post 3.00 m overall with 2 fingers	259.00	1.00	64.40	22.57	364.73	nr	**451.70**
Post 3.20 m overall with 3 fingers	353.00	1.20	77.28	22.57	472.83	nr	**572.68**
Post 3.20 m overall with 4 fingers	356.00	1.30	83.72	22.57	575.18	nr	**681.47**
RECYCLED PLASTIC STREET FURNITURE							
Recycled plastic Street furniture; Hahn Plastics Ltd; inclusive of excavations concrete bases and anchors							
Sets							
Calero Table & Bench set							
150 cm × 180 cm × 75 cm	724.00	1.00	64.40	–	858.42	nr	**922.82**
Alton Table & Bench set							
200 cm × 205 cm × 80 cm	1520.00	1.00	64.40	–	1773.82	nr	**1838.22**
Serengeti picnic set							
200 cm × 174 cm × 76 cm	741.00	1.00	64.40	–	877.97	nr	**942.37**
Benches							
Wall Street 955 bench							
200 × 56 cm × 45 cm	899.00	0.75	48.30	–	1059.67	nr	**1107.97**
Hyde Park bench							
165 cm × 46 cm × 48 cm	206.00	0.75	48.30	–	262.72	nr	**311.02**
High line bench							
2.00 m × 480 mm × 450 mm	411.00	0.75	48.30	–	498.47	nr	**546.77**

35 SITE WORKS – STREET FURNITURE

Item – Overhead and Profit Included	PC £	Labour hours	Labour £	Plant £	Material £	Unit	Total rate £
Seats							
Caserata Wall mounted seat							
150 cm × 70 cm × 65 cm	383.00	0.75	48.30	–	466.27	nr	**514.57**
Eclipse seat							
150 cm × 60 cm × 89 cm	247.00	0.75	48.30	–	309.87	nr	**358.17**
Sapo Seniors' seat							
200 cm × 61 cm × 97 cm	696.00	0.75	48.30	–	826.22	nr	**874.52**
Turon Round tree seat – octagonal							
330 cm × 305 cm × 80 cm	4303.00	3.00	193.20	–	5044.02	nr	**5237.22**
Planters; inclusive of placement and filling with planting medium and drainage; excludes planting							
Vinca 2 Planter							
108 cm × 62 cm × 44 cm	474.00	1.00	64.40	–	582.52	nr	**646.92**
Iberis 2 Planter							
110 cm × 110 cm × 47 cm	424.00	1.00	64.40	–	529.79	nr	**594.19**
Litter Bin							
Collec Litter Bin 48 cm × 66 cm; ground mounted with wire basket insert; circular; 70 litre	259.46	1.00	64.40	–	311.29	nr	**375.69**
Resco Litter Bin 45 cm × 45 cm × 66 cm high 50 litre; square; brown	254.38	1.00	64.40	–	305.45	nr	**369.85**
SMOKING SHELTERS							
Smoking shelters							
Furnitubes International Ltd; Ashby free-standing bolt down; aluminium frame; PET glazing; 2050 mm length × 2330 mm high							
3290 mm depth; maximum 3 people	1957.23	4.00	128.80	–	2267.55	nr	**2396.35**
6050 mm depth; maximum 6 people	2389.86	5.00	161.00	–	2765.07	nr	**2926.07**
STONES AND BOULDERS							
Standing stones; CED Ltd; erect standing stones; vertical height above ground; in concrete base; including excavation setting in concrete to ⅓ depth and crane offload into position							
Purple schist							
1.00 m high	226.36	1.50	48.30	22.18	286.11	nr	**356.59**
1.25 m high	409.50	1.50	48.30	22.18	491.98	nr	**562.46**
1.50 m high	409.50	2.00	64.40	26.61	493.46	nr	**584.47**
2.00 m high	409.50	3.00	96.60	44.36	536.38	nr	**677.34**
2.50 m high	409.50	1.50	48.30	88.71	597.12	nr	**734.13**

35 SITE WORKS – STREET FURNITURE

Item – Overhead and Profit Included	PC £	Labour hours	Labour £	Plant £	Material £	Unit	Total rate £
STONES AND BOULDERS – cont							
Rockery stone – General							
Preamble: Rockery stone prices vary considerably with source, carriage, distance and load. Typical PC prices are in the range of £60–80 per tonne collected.							
Rockery stone; CED Ltd							
Boulders; maximum distance 25 m; by machine							
750 mm dia.	226.36	0.60	19.32	32.60	260.31	nr	**312.23**
1 m dia.	226.36	2.00	64.40	27.16	260.31	nr	**351.87**
1.5 m dia.	226.36	2.00	64.40	74.75	260.31	nr	**399.46**
2 m dia.	226.36	2.00	64.40	74.75	260.31	nr	**399.46**
Boulders; maximum distance 25 m; by hand							
750 mm dia.	226.36	0.75	24.15	–	260.31	nr	**284.46**
1 m dia.	226.36	1.89	60.71	–	260.31	nr	**321.02**
TREE GRILLES/TREE PROTECTION							
Tree grilles; cast iron							
Cast iron tree grilles; Furnitubes International Ltd							
GS 1070 Greenwich; two part; 1000 mm square × 700 mm dia. tree hole	195.00	2.00	64.40	–	224.25	nr	**288.65**
GC 1270 Greenwich; two part, 1200 mm dia. × 700 mm dia. tree hole	168.00	2.00	64.40	–	193.20	nr	**257.60**
Cast iron tree grilles; Marshalls Plc							
Heritage; cast iron grille plus frame; 1 × 1 m	499.73	3.00	96.60	–	591.50	nr	**688.10**
Cast iron tree grilles; Townscape Products Ltd							
Baltimore; 1200 mm square Ductile(SG) finished in Black RAL9004	557.18	2.00	64.40	–	640.76	nr	**705.16**

35 SITE WORKS – SPORTS, PLAY EQUIPMENT, SURFACES

Item – Overhead and Profit Included	PC £	Labour hours	Labour £	Plant £	Material £	Unit	Total rate £
MARKET PRICES OF SPORTS AND PLAY SURFACING MATERIALS							
Market prices of surfacing materials							
Surfacings; Melcourt Industries Ltd (items labelled FSC are Forest Stewardship Council certified)							
Playbark® 10/50; per 25 m³ load	–	–	–	–	66.35	m³	**66.35**
Playbark® 10/50; per 80 m³ load	–	–	–	–	53.35	m³	**53.35**
Playbark® 8/25; per 25 m³ load	–	–	–	–	64.85	m³	**64.85**
Playbark® 8/25; per 80 m³ load	–	–	–	–	51.85	m³	**51.85**
Playchips®; per 25 m³ load; FSC	–	–	–	–	46.75	m³	**46.75**
Playchips®; per 80 m³ load; FSC	–	–	–	–	33.75	m³	**33.75**
Kushyfall; per 25 m³ load; FSC	–	–	–	–	49.20	m³	**49.20**
Kushyfall; per 80 m³ load; FSC	–	–	–	–	36.20	m³	**36.20**
Softfall; per 25 m³ load	–	–	–	–	33.90	m³	**33.90**
Softfall; per 80 m³ load	–	–	–	–	20.90	m³	**20.90**
Playsand; per 10 t load	–	–	–	–	116.01	m³	**116.01**
Playsand; per 20 t load	–	–	–	–	103.77	m³	**103.77**
Walk Chips; per 25 m³ load; FSC	–	–	–	–	44.70	m³	**44.70**
Walk Chips; per 80 m³ load; FSC	–	–	–	–	31.70	m³	**31.70**
Woodfibre; per 25 m³ load; FSC	–	–	–	–	46.95	m³	**46.95**
Woodfibre; per 80 m³ load; FSC	–	–	–	–	33.95	m³	**33.95**
SPORTS EQUIPMENT							
Sports equipment							
Tennis posts; steel; suitable for hard or grass tennis courts; including winder, sockets and dust cap							
round	144.00	1.00	32.20	–	165.60	set	**197.80**
square	171.00	1.00	32.20	–	196.65	set	**228.85**
Tennis nets; not including posts or fixings							
Tournament	115.00	3.00	96.60	–	141.66	set	**238.26**
Match	90.00	3.00	96.60	–	112.91	set	**209.51**
Club	75.00	3.00	96.60	–	95.66	set	**192.26**
Football goals; full size; socketed; including international net supports and nets							
aluminium	1069.00	4.00	128.80	–	1238.76	set	**1367.56**
steel; heavyweight	915.00	5.00	161.00	–	1313.67	set	**1474.67**
Football goals; full size; free-standing; including nets							
aluminium	2657.00	4.00	128.80	–	3055.55	set	**3184.35**
steel	1584.00	4.00	128.80	–	1821.60	set	**1950.40**

35 SITE WORKS – SPORTS, PLAY EQUIPMENT, SURFACES

Item – Overhead and Profit Included	PC £	Labour hours	Labour £	Plant £	Material £	Unit	Total rate £
SPORTS EQUIPMENT – cont							
Sports equipment – cont							
Mini-football goals; free-standing; including nets							
aluminium	1210.00	4.00	128.80	–	1391.50	set	**1520.30**
steel	442.00	4.00	128.80	–	508.30	set	**637.10**
Rugby posts; socketed							
aluminium; 10 m high	1425.00	6.00	193.20	–	1663.84	set	**1857.04**
aluminium; 12 m high	1530.00	6.00	193.20	–	1784.59	set	**1977.79**
steel; 12 m high	1334.00	6.00	193.20	–	1559.19	set	**1752.39**
Hockey goals; steel; including backboards and nets							
socketed	791.00	1.00	32.20	–	909.65	set	**941.85**
free-standing	1024.00	1.00	32.20	–	1177.60	set	**1209.80**
Cricket cages; steel; including netting							
free-standing	1416.00	12.00	386.40	–	1628.40	set	**2014.80**
wheelaway	1008.00	12.00	386.40	–	1159.20	set	**1545.60**
PLAYGROUND EQUIPMENT							
Playground equipment – General							
Preamble: The range of equipment manufactured or available in the UK is so great that comprehensive coverage would be impossible, especially as designs, specifications and prices change fairly frequently. The following information should be sufficient to give guidance to anyone designing or equipping a playground. In comparing prices note that only outline specification details are given here and that other refinements which are not mentioned may be the reason for some difference in price between two apparently identical elements. The fact that a particular manufacturer does not appear under one item heading does not necessarily imply that they do not make it. Landscape designers are advised to check that equipment complies with ever more stringent safety standards before specifying.							

35 SITE WORKS – SPORTS, PLAY EQUIPMENT, SURFACES

Item – Overhead and Profit Included	PC £	Labour hours	Labour £	Plant £	Material £	Unit	Total rate £
Playground equipment – Installation The rates below include for installation of the specified equipment by the manufacturers. Most manufacturers will offer an option to install the equipment they have supplied.							
Sports and social areas; Freegame multi-use games areas; Kompan Ltd; enclosed sports areas; complete with surfacing boundary and goals and targets; galvanized steel framework with high density polyethylene panels, galvanized steel goals and equipment; surfacing priced separately							
FRE1211 Classic Multigoal; 7 m Pitch; complete; suitable for multiple ball sports; suitable for use with natural, artificial or hard landscape surfaces; fully enclosed including two end sports walls	–	–	–	–	–	nr	5065.28
FRE2110 Cosmos; 12 × 20 m	–	–	–	–	–	nr	28412.36
FRE2116 Cosmos; 19 × 36 m	–	–	–	–	–	nr	41113.11
FRE3000; Meeting Point; shelter or social area	–	–	–	–	–	nr	4994.21
Swings – General Preamble: Prices for the following vary considerably. Those given represent the middle of the range and include multiple swings and timber, rubber or tyre seats; ground fixing and priming only.							
Swings Swings							
Flat seat swing	–	–	–	–	–	nr	2090.00
Birds nest swing	–	–	–	–	–	nr	3685.00
Swing Q10144 (Estimated)	–	–	–	–	–	nr	2365.00
Swing with 2 × Cradle seats 137414M	–	–	–	–	–	nr	2365.00
Swings; Wicksteed Ltd							
Flexi Swing; 1 bay 1 flat seat	–	–	–	–	–	nr	2439.84
Flexi Swing; 1 bay 2 flat seats	–	–	–	–	–	nr	2995.36
Flexi Swing; 1 bay 2 cradle seats	–	–	–	–	–	nr	3176.87

35 SITE WORKS – SPORTS, PLAY EQUIPMENT, SURFACES

Item – Overhead and Profit Included	PC £	Labour hours	Labour £	Plant £	Material £	Unit	Total rate £
PLAYGROUND EQUIPMENT – cont							
Swings – cont							
Swings – cont							
Flexi Swing; 1 bay 1 flat seat & 1 cradle seat	–	–	–	–	–	nr	3085.57
Hurricane Swing; 2 button seats	–	–	–	–	–	nr	7285.46
Pendulum Swing; 1 basket seat	–	–	–	–	–	nr	9800.12
1.8 m Timber Swing; 2 cradle seats	–	–	–	–	–	nr	1842.53
2.4 m Timber Swing; 2 cradle seats	–	–	–	–	–	nr	2169.24
2.4 m Timber Swing; 2 flat seats	–	–	–	–	–	nr	3454.07
2.4 m Birds Nest Timber Swing; 1 basket seat	–	–	–	–	–	nr	1861.24
Swings; Kompan Ltd							
M951P Sunflower swing; 1–3 years	–	–	–	–	–	nr	1602.72
Slides							
Slides; Wicksteed Ltd							
Pedestal slides; 3.40 m	–	–	–	–	–	nr	4337.39
Pedestal slides; 4.40 m	–	–	–	–	–	nr	5647.53
Pedestal slides; 5.80 m	–	–	–	–	–	nr	6570.45
Embankment slides; 3.40 m	–	–	–	–	–	nr	3322.07
Embankment slides; 4.40 m	–	–	–	–	–	nr	4319.79
Embankment slides; 5.80 m	–	–	–	–	–	nr	5497.91
Embankment slides; 7.30 m	–	–	–	–	–	nr	6986.27
Embankment slides; 9.10 m	–	–	–	–	–	nr	8750.70
Embankment slides; 11.00 m	–	–	–	–	–	nr	10465.62
Slides; Kompan Ltd							
M351P slide	–	–	–	–	–	nr	2392.32
M326P Aladdin's Cave slide	–	–	–	–	–	nr	2709.01
Moving equipment – General							
Preamble: The following standard items of playground equipment vary considerably in quality and price; the following prices are middle of the range.							
Moving equipment							
Roundabouts; Wicksteed Ltd							
Dual axis roundabout; Dizzy	–	–	–	–	–	nr	4221.88
Dual axis roundabout; Whizzy	–	–	–	–	–	nr	2875.46
Dual axis roundabout; Solar Spinner	–	–	–	–	–	nr	3567.37

35 SITE WORKS – SPORTS, PLAY EQUIPMENT, SURFACES

Item – Overhead and Profit Included	PC £	Labour hours	Labour £	Plant £	Material £	Unit	Total rate £
Roundabouts; Kompan Ltd							
Supernova GXY 916; multifunctional spinning and balancing disc; capacity approximately 15 children	–	–	–	–	–	nr	5242.24
Lappset Ltd							
Spinning machine	–	–	–	–	–	nr	12223.20
Waltz	–	–	–	–	–	nr	5736.50
Polka	–	–	–	–	–	nr	4435.20
Inclusive roundabout	–	–	–	–	–	nr	7200.00
Seesaws							
Seesaws; Lappset UK Ltd							
137237M Finno seesaw	–	–	–	–	–	nr	2871.00
Octopus seesaw Q10279	–	–	–	–	–	nr	1320.00
Seesaws; Wicksteed Ltd							
Buddy Board	–	–	–	–	–	nr	2610.34
Glow Worm	–	–	–	–	–	nr	2231.94
Cobra	–	–	–	–	–	nr	2603.74
Sidewinder	–	–	–	–	–	nr	2388.14

Play sculptures – General
Preamble: Many variants on the shapes of playground equipment are available, simulating spacecraft, trains, cars, houses etc., and these designs are frequently changed. The basic principles remain constant but manufacturers' catalogues should be checked for the latest styles.

Climbing equipment and play structures – General
Preamble: Climbing equipment generally consists of individually designed modules. Play structures generally consist of interlinked and modular pieces of equipment and sculptures. These may consist of climbing, play and skill based modules, nets and various other activities. Both are set into either safety surfacing or defined sand pit areas. The equipment below outlines a range from various manufacturers. Individual catalogues should be consulted in each instance. Safety areas should be allowed round all equipment.

35 SITE WORKS – SPORTS, PLAY EQUIPMENT, SURFACES

Item – Overhead and Profit Included	PC £	Labour hours	Labour £	Plant £	Material £	Unit	Total rate £
PLAYGROUND EQUIPMENT – cont							
Climbing equipment and play structures							
Climbing equipment; Wicksteed Ltd							
Funrun Fitness Trail; Under Starter's Orders; set of 12 units	–	–	–	–	–	nr	22302.98
Climbing equipment; Lappset UK Ltd							
137072M Finno Climbing Frame	–	–	–	–	–	nr	6248.00
230015 Halo Activity Tower	–	–	–	–	–	nr	14564.00
Climbing equipment and play structures							
Finno Activity Tower	–	–	–	–	–	nr	14564.00
Finno 3D Cube Climber	–	–	–	–	–	nr	6248.00
Finno Hades	–	–	–	–	–	nr	13860.00
Castle Finno	–	–	–	–	–	nr	30635.00
Finno Fishing Ship	–	–	–	–	–	nr	34050.50
Santa Maria	–	–	–	–	–	nr	127600.00
Halo Climber	–	–	–	–	–	nr	35695.00
Halo Cubic	–	–	–	–	–	nr	99000.00
Flora Goblin's climbing	–	–	–	–	–	nr	1265.00
Flora Troll's shelter	–	–	–	–	–	nr	7448.10
Flora Goblin's track	–	–	–	–	–	nr	2349.60
Play tower Q10945	–	–	–	–	–	nr	7364.50
ABC Play tower Q10519	–	–	–	–	–	nr	11110.00
Fire truck Q10259	–	–	–	–	–	nr	4180.00
Hags-SMP Playgrounds Ltd							
Nexus – The Core; multi-play structure	–	–	–	–	–	nr	12617.00
Spring equipment							
Spring based equipment for 1–8 year olds; Kompan Ltd							
M101P Crazy Hen	–	–	–	–	–	nr	605.89
M128P Crazy Daisy	–	–	–	–	–	nr	1003.33
M141P Spring Seesaw	–	–	–	–	–	nr	1902.49
M155P Quartet Seesaw	–	–	–	–	–	nr	1487.44
Spring based equipment for under 12s; Lappset UK Ltd							
010501 Horse Springer	–	–	–	–	–	nr	861.30
Tractor rocker 010515	–	–	–	–	–	nr	1958.00
Pig rocker 01516	–	–	–	–	–	nr	1023.00
Froggy 22034	–	–	–	–	–	nr	1320.00
Sandpits							
Market prices							
play pit sand; Boughton Loam Ltd	185.71	–	–	–	213.57	m^3	213.57

35 SITE WORKS – SPORTS, PLAY EQUIPMENT, SURFACES

Item – Overhead and Profit Included	PC £	Labour hours	Labour £	Plant £	Material £	Unit	Total rate £
Kompan Ltd							
Basic550; 276 × 154 × 31 cm deep	–	–	–	–	–	nr	2322.14
Play area Furniture							
Picnic table Accessibility adapted NF 2694	–	–	–	–	–	nr	660.00
NATURAL SPORTS PITCHES							
Natural sports pitches; Agripower Ltd; sports pitches; gravel raft construction. Prices shown are for sports pitches ordered as part of a landscape subcontract. If sports pitches are ordered directly from the sports pitch contractor these rates would reduce by approximately 10%.							
Professional standard							
cricket wicket; county standard	–	–	–	–	–	nr	40880.81
Playing fields; Sport England Compliant							
football pitch; 6500 m^2	–	–	–	–	–	nr	86401.47
cricket wicket; playing field standard	–	–	–	–	–	nr	20942.01
track and infield	–	–	–	–	–	nr	122291.30
Existing area to 1:100 existing grade; 6500 m^2	1515.72	16.00	515.20	1031.78	2240.07	nr	3787.05
SPORTS FIELD CONSTRUCTION							
Note: The playing fields in this section are constructed to a lower standard than the specialist sports fields elsewhere in this book and would comply with standards for general amenity in parks, recreation grounds or school playing fields.							
Surface preparations to existing playing area – Prices shown are per 1000 m^2							
Spray turf; Glyphosate single application by tractor;	–	–	–	75.24	3.67	1000 m^2	78.91
Strip existing turf 25 mm thick	–	–	–	257.02	–	1000 m^2	257.02
Disposal green waste off site	–	–	–	62.97	825.00	1000 m^2	887.97

35 SITE WORKS – SPORTS, PLAY EQUIPMENT, SURFACES

Item – Overhead and Profit Included	PC £	Labour hours	Labour £	Plant £	Material £	Unit	Total rate £
NATURAL SPORTS PITCHES – cont							
Cultivation and soil preparation – Prices shown are per 1000 m²							
Light to medium soils							
Decompaction operations by tractor; ripping and breaking up ground	–	–	–	361.47	–	1000 m²	**361.47**
Cultivation and stone burying harrow and power raking in single operation	–	–	–	276.00	–	1000 m²	**276.00**
Clay soils							
Decompaction operations by tractor; ripping and breaking up ground	–	–	–	525.78	–	1000 m²	**525.78**
Strip and remove 100 mm thick surface	–	–	–	251.88	–	1000 m²	**251.88**
Disposal on site	–	–	–	336.61	–	1000 m²	**336.61**
Disposal off site 100 mm thick arisings	–	2.78	89.52	–	3993.74	1000 m²	**4083.26**
Cultivation, stone burying, harrow and power raking in single operation	–	–	–	480.70	–	1000 m²	**480.70**
Imported new topsoil root zone 150 mm deep	–	–	–	1620.84	5117.04	1000 m²	**6737.88**
Surface operations to prepared ground							
fertilizing 35 g/m²	42.00	3.00	96.60	–	48.30	nr	**144.90**
seeding operations by mechanical seeder; sportsfield seed 35 g/m² by mechanical seeder	233.19	–	–	54.34	268.17	nr	**322.51**
PLAY SURFACES							
Playgrounds; Rubaflex in situ playground surfacing, porous; on prepared stone/granular base Type 1 (not included) and macadam base course (not included)							
Black							
15 mm thick (0.50 m critical fall height)	–	–	–	–	–	m²	**38.50**
35 mm thick (1.00 m critical fall height)	–	–	–	–	–	m²	**45.10**
60 mm thick (1.50 m critical fall height)	–	–	–	–	–	m²	**52.80**

35 SITE WORKS – SPORTS, PLAY EQUIPMENT, SURFACES

Item – Overhead and Profit Included	PC £	Labour hours	Labour £	Plant £	Material £	Unit	Total rate £
Playgrounds; Wetpour safety surfacing in situ playground surfacing, porous; on prepared stone/granular base Type 1 (not included) and macadam base course (not included)							
Coloured							
15 mm thick (0.50 m critical fall height)	–	–	–	–	–	m²	**44.00**
35 mm thick (1.00 m critical fall height)	–	–	–	–	–	m²	**48.40**
60 mm thick (1.50 m critical fall height)	–	–	–	–	–	m²	**62.70**
Playgrounds; Meon UK; porous safety surfaces; on prepared Type 1 base (not included)							
Critical fall height 500 mm							
natural colours	–	–	–	–	–	m²	**72.77**
bright colours	–	–	–	–	–	m²	**68.64**
Critical fall height 1.70 m							
natural colours	–	–	–	–	–	m²	**101.72**
bright colours	–	–	–	–	–	m²	**86.85**
Playgrounds Meon UK; Rubber crumb resin surfacing							
Mulch resin surfacing; Black or colours							
10–50 m²	–	–	–	–	–	m²	**96.10**
over 50 m²	–	–	–	–	–	m²	**85.12**
Playgrounds; Melcourt Industries Ltd; specifiers and users should contact the supplier for performance specifications of the materials below							
Play surfaces; on drainage layer (not included); minimum 300 mm settled depth; prices for 80 m³ loads unless specified							
Playbark® 8/25; 8–25 mm particles; red/brown; 25 m³ loads	19.45	0.35	11.29	–	22.37	m²	**33.66**
Playbark® 8/25; 8–25 mm particles; red/brown	17.11	0.35	11.29	–	19.68	m²	**30.97**
Playbark® 10/50; 10–50 mm particles; red/brown	19.56	0.35	11.29	–	22.49	m²	**33.78**
Playchips®; FSC graded woodchips	11.25	0.35	11.29	–	12.94	m²	**24.23**
Kushyfall; 'fiberised' woodchips	12.07	0.35	11.29	–	13.88	m²	**25.17**
Softfall; conifer shavings	6.97	0.35	11.29	–	8.02	m²	**19.31**

35 SITE WORKS – SPORTS, PLAY EQUIPMENT, SURFACES

Item – Overhead and Profit Included	PC £	Labour hours	Labour £	Plant £	Material £	Unit	Total rate £
TENNIS COURTS							
Tennis courts; En Tout Cas Ltd Hard playing surfaces to SAPCA Code of Practice minimum requirements; laid on 40 mm thick macadam base on 200 mm thick stone foundation; to include lines, nets, posts, brick edging, 2.75 m high fence and perimeter drainage; based on court size 573 m^2							
Playdek; all weather porous macadam and acrylic colour coating hardcourt	–	–	–	–	–	nr	40700.00
TennisTurf Advantage Pro; artificial grass	–	–	–	–	–	nr	55000.00
Sportsturf; short pile sand dressed artificial grass	–	–	–	–	–	nr	55000.00
Matchplay; impervious acrylic hardcourt	–	–	–	–	–	nr	52800.00
Savanna; half sand filled synthetic turf	–	–	–	–	–	nr	57200.00
Matchplay; acrylic tournament surface	–	–	–	–	–	nr	60500.00
Omniclay; synthetic clay system	–	–	–	–	–	nr	58300.00
SPORTS SURFACES							
Artificial surfaces and finishes – General Preamble: Advice should also be sought from the Technical Unit for Sport, the appropriate regional office of the Sports Council or the National Playing Fields Association. Some of the following prices include base work whereas others are for a specialist surface only on to a base prepared and costed separately.							
Artificial sports pitches; Agripower Ltd; third generation rubber crumb Sports pitches to respective National governing body standards; inclusive of all excavation, drainage, lighting, fencing, etc.							
football pitch; 6500 m^2	–	–	–	–	–	nr	640800.41
rugby union multi-use pitch; 120 × 75 m	–	–	–	–	–	nr	890348.90
hockey pitch; water-based; 101.4 × 63 m	–	–	–	–	–	nr	747391.47

35 SITE WORKS – SPORTS, PLAY EQUIPMENT, SURFACES

Item – Overhead and Profit Included	PC £	Labour hours	Labour £	Plant £	Material £	Unit	Total rate £
athletic track; 8 lane; International amateur athletic federation	–	–	–	–	–	nr	752407.52
Multi-sport pitches; inclusive of all excavation, drainage, lighting, fencing, etc.							
sand dressed; 101.4 × 63 m	–	–	–	–	–	nr	535463.36
sand filled; 101.4 × 63 m	–	–	–	–	–	nr	485302.85
polymeric (synthetic bound rubber)	–	–	–	–	–	m²	178.07
macadam	–	–	–	–	–	m²	112.23
Sports areas; Agripower Ltd							
Sports tracks; polyurethane rubber surfacing; on bitumen-macadam (not included); prices for 5500 m² minimum							
International	–	–	–	–	–	m²	59.57
Club Grade	–	–	–	–	–	m²	50.31
Sports areas; multi-component polyurethane rubber surfacing; on bitumen-macadam; finished with polyurethane or acrylic coat; green or red							
Permaprene	–	–	–	–	–	m²	61.15

35 SITE WORKS – WATER AND WATER FEATURES

Item – Overhead and Profit Included	PC £	Labour hours	Labour £	Plant £	Material £	Unit	Total rate £
CLARIFICATION NOTES ON LABOUR COSTS IN THIS SECTION							
General groundworks team Generally a three man team is used in this section; The column Labour hours reports team hours. The column Labour £ reports the total cost of the team for the unit of work shown							
3 man team	–	1.00	96.60	–	–	hr	96.60
2 man team	–	1.00	64.40	–	–	hr	64.40
BOREHOLES							
Borehole drilling; Agripower Ltd Drilling operations to average 100 m depth; lining with 150 mm dia. sieve							
drilling to 100 m	–	–	–	–	–	nr	15951.05
rate per m over 100 m	–	–	–	–	–	m	159.51
pump, rising main, cable and kiosk	–	–	–	–	–	nr	6646.27
COLD WATER							
Blue MDPE polythene pipes; type 50; for cold water services; with compression fittings; bedding on 100 mm DOT type 1 granular fill material Pipes							
20 mm dia.	0.49	0.08	2.58	–	6.20	m	8.78
25 mm dia.	1.13	0.08	2.58	–	6.93	m	9.51
32 mm dia.	1.38	0.08	2.58	–	7.22	m	9.80
50 mm dia.	2.27	0.10	3.22	–	8.25	m	11.47
60 mm dia.	3.64	0.10	3.22	–	9.82	m	13.04
Hose union bib taps; including fixing to wall; making good surfaces							
15 mm	16.79	0.75	24.15	–	20.69	nr	44.84
22 mm	43.89	0.75	24.15	–	52.54	nr	76.69
Stopcocks; including fixing to wall; making good surfaces							
15 mm	5.09	0.75	24.15	–	7.23	nr	31.38
22 mm	8.90	0.75	24.15	–	11.61	nr	35.76
Standpipes; to existing 25 mm water mains							
1.00 m high	–	–	–	–	–	nr	247.50
Hose junction bib taps; to standpipes							
19 mm	–	–	–	–	–	nr	88.00

35 SITE WORKS – WATER AND WATER FEATURES

Item – Overhead and Profit Included	PC £	Labour hours	Labour £	Plant £	Material £	Unit	Total rate £
RAINWATER HARVESTING							
Rainwater harvesting tanks							
Notes: Rainwater harvesting tanks must be installed on granular or concrete bases. If installed below ground the tanks may require construction of drained underground chambers to support them. Please see the appropriate sections in this book for excavation, disposal, bases, retaining walls and drainage.							
Excavation for underground tanks; excavation inclusive of earthwork retention for self supporting tanks							
8 tonne tracked excavator (bucket volume 0.28 m³)							
maximum depth not exceeding 1.00 m	–	0.06	2.01	3.39	–	m³	**5.40**
maximum depth not exceeding 2.00 m	–	0.07	2.30	3.88	–	m³	**6.18**
maximum depth not exceeding 3.00 m	–	0.09	2.93	4.93	–	m³	**7.86**
Disposal							
Excavated material; off site; to tip; mechanically loaded (JCB)							
inert	–	–	–	–	–	m³	**39.94**
Type 1 granular fill base; PC £24.75/tonne (£54.45/m³ compacted)							
By machine							
100 mm thick	5.45	0.03	0.90	0.54	6.27	m²	**7.71**
150 mm thick	8.17	0.03	0.81	0.83	9.40	m²	**11.04**
Backfilling to surround of rainwater tank; carefully compacting as work proceeds							
Arising from the excavations							
average thickness exceeding 0.25 m; depositing in layers 150 mm maximum thickness	–	0.03	1.07	7.76	–	m³	**8.83**

35 SITE WORKS – WATER AND WATER FEATURES

Item – Overhead and Profit Included	PC £	Labour hours	Labour £	Plant £	Material £	Unit	Total rate £
RAINWATER HARVESTING – cont							
Rainwater harvesting tanks; Rainwater Harvesting Ltd; self supporting underground or above ground tanks							
Rainwater tank with max 1.00 m cover in pedestrian areas; complete with tank dome and pedestrian lid, submersible automatic pump system, supra filtration system, 25 m black and green rainwater pipe, rainwater labelling kit, 125 mm fine filter (excavation, backfilling base, trenching and inlet pipework all not included)							
3600 litre; 1.78 m high × 1.8 m dia.	475.00	3.00	193.20	–	766.25	nr	**959.45**
4500 litre; 1.96 m high × 2.07 m long × 2.01 m dia.	990.00	3.50	225.40	–	1358.50	nr	**1583.90**
10000 litre; 2.3 m long × 2.3 m wide × 2.9 m high	1933.23	7.00	450.80	–	2443.21	nr	**2894.01**
Lighter weight tank; complete with tank dome and pedestrian lid, submersible automatic pump system, supra filtration system, 25 m black and green rainwater pipe, rainwater labelling kit, 125 mm fine filter (excavation, backfilling base, trenching and inlet pipework all not included)							
1650 litre; 2.10 × 1.05 × 1.22 m	1110.00	5.00	322.00	–	1496.50	nr	**1818.50**
2650 litre; 2.10 × 1.30 × 1.50 m	1555.00	6.00	386.40	–	2008.25	nr	**2394.65**
Low profile tanks for reduced excavation; complete with rainwater filters and pressure pump							
1500 litre; 2.10 × 1.25 × 1.02 m	974.01	4.00	257.60	–	1340.11	nr	**1597.71**
3000 litre; 2.46 × 2.1 × 1.05 m	1404.00	6.00	386.40	–	1834.60	nr	**2221.00**
5000 litre; 2.89 × 2.30 × 1.26 m	1890.90	6.50	418.60	–	2394.53	nr	**2813.13**
10000 litre; 2.89 × 4.60 × 1.26 m	2263.77	7.00	450.80	–	2823.34	nr	**3274.14**
Rainwater Harvesting Ltd; accessories for rainwater harvesting tanks							
Downpipe filter for connection to roof downpipes							
for roof areas up to 50 m^2	93.89	0.25	8.05	–	107.97	nr	**116.02**
for roof areas up to 80 m^2	40.50	0.25	8.05	–	46.58	nr	**54.63**
for roof areas up to 100 m^2	92.59	0.25	8.05	–	106.48	nr	**114.53**

35 SITE WORKS – WATER AND WATER FEATURES

Item – Overhead and Profit Included	PC £	Labour hours	Labour £	Plant £	Material £	Unit	Total rate £
Pumps; inclusive of electrical installation and connections							
HydroForce Series 3 pump	288.25	0.75	28.75	–	331.49	nr	**360.24**
Filters for rainwater tanks; complete with overflow siphons and rodent guard	359.62	–	–	–	413.56	nr	**413.56**
Extra over to cover rainwater tanks in geofabric prior to backfilling; 10 m² of terram per tank; average cost	–	0.25	16.10	–	5.65	nr	**21.75**
IRRIGATION							
Irrigation Infrastructure and pipework; main or ring main supply							
Excavate and lay mains supply pipe to supply irrigated area							
PE 80 20 mm	–	–	–	–	–	m	**5.90**
PE 80 25 mm	–	–	–	–	–	m	**6.13**
Pro-Flow PEHD 63 mm PE	–	–	–	–	–	m	**10.51**
Pro-Flow PEHD 50 mm PE	–	–	–	–	–	m	**8.58**
Pro-Flow PEHD 40 mm PE	–	–	–	–	–	m	**7.39**
Pro-Flow PEHD 32 mm PE	–	–	–	–	–	m	**6.64**
Irrigation systems; head control to sprinkler stations							
Valves installed at supply point on irrigation supply manifold; includes for pressure control and filtration							
2 valve unit	–	–	–	–	–	nr	**159.75**
4 valve unit	–	–	–	–	–	nr	**249.39**
6 valve unit	–	–	–	–	–	nr	**352.70**
12 valve unit	–	–	–	–	–	nr	**676.30**
Solenoid valve; 25 mm with chamber; extra over for each active station	–	–	–	–	–	nr	**128.46**
Cable to solenoid valves (alternative to valves on manifold)							
4 core	–	–	–	–	–	m	**1.90**
6 core	–	–	–	–	–	m	**9.23**
12 core	–	–	–	–	–	m	**3.86**

35 SITE WORKS – WATER AND WATER FEATURES

Item – Overhead and Profit Included	PC £	Labour hours	Labour £	Plant £	Material £	Unit	Total rate £
IRRIGATION – cont							
Irrigation; water supply and control							
Header tank and submersible pump and pressure stat							
1000 litre (500 gallon 25 mm/10 days to 1500 m²) 2 m³/hr pump	–	–	–	–	–	nr	592.50
4540 litre (1000 gallon 25 mm/10 days to 3500 m²)	–	–	–	–	–	nr	1200.21
9080 litre (2000 gallon 25 mm/10 days to 7000 m²)	–	–	–	–	–	nr	1987.19
Electric multi-station controllers 240 V							
WeatherMatic Smartline; 4 station; standard	–	–	–	–	–	nr	180.54
WeatherMatic Smartline; 6 station; standard	–	–	–	–	–	nr	202.82
WeatherMatic Smartline; 12 station; standard	–	–	–	–	–	nr	343.86
WeatherMatic Smartline; 12 station; radio controlled	–	–	–	–	–	nr	706.42
WeatherMatic Smartline; 12 station; with moisture sensor	–	–	–	–	–	nr	1187.55
WeatherMatic Smartline; 12 station; radio controlled; with moisture sensor	–	–	–	–	–	nr	1426.15
Irrigation sprinklers; station consisting of multiple sprinklers; inclusive of all trenching, wiring and connections to ring main or main supply							
Sprayheads; WeatherMatic LX; including nozzle; 21 m² coverage							
100 mm (4')	–	–	–	–	–	nr	18.78
150 mm (6')	–	–	–	–	–	nr	26.52
300 mm (12')	–	–	–	–	–	nr	30.59
Matched precipitation sprinklers; WeatherMatic LX							
MP 1000; 16 m² coverage	–	–	–	–	–	nr	43.04
MP 2000; 30 m² coverage	–	–	–	–	–	nr	43.04
MP 3000; 81 m² coverage	–	–	–	–	–	nr	43.04
Gear drive sprinklers; WeatherMatic; placed to provide head to head (100%) overlap; average							
T3; 121 m² coverage	–	–	–	–	–	nr	56.53
CT70; 255 m² coverage	–	–	–	–	–	nr	99.66
CT70SS; 361 m² coverage	–	–	–	–	–	nr	110.67

35 SITE WORKS – WATER AND WATER FEATURES

Item – Overhead and Profit Included	PC £	Labour hours	Labour £	Plant £	Material £	Unit	Total rate £
Drip irrigation; driplines; inclusive of connections and draindown systems							
Metzerplas TechLand 30 cm drip emitter							
fixed on soil surface	–	–	–	–	–	m	1.30
subsurface	–	–	–	–	–	m	1.90
Metzerplas TechLand 50 cm drip emitter							
fixed on soil surface	–	–	–	–	–	m	1.09
subsurface	–	–	–	–	–	m	1.71
Commissioning and testing of irrigation system							
per station	–	–	–	–	–	nr	47.87
Annual maintenance costs of irrigation system							
Call out charge per visit	–	–	–	–	–	nr	398.83
Extra over per station	–	–	–	–	–	nr	15.96
Moisture Control moisture sensing control system							
Large garden comprising 7000 m^2 and 24 stations with four moisture sensors							
turf only	–	–	–	–	–	nr	16133.68
turf/shrub beds; 70/30	–	–	–	–	–	nr	9411.34
Medium garden comprising 3500 m^2 and 12 stations with two moisture sensors							
turf only	–	–	–	–	–	nr	10755.80
turf/shrub beds; 70/30	–	–	–	–	–	nr	13982.56
Smaller garden comprising 1000 m^2 and 6 stations with one moisture sensor							
turf only	–	–	–	–	–	nr	6587.90
turf/shrub beds; 70/30	–	–	–	–	–	nr	7260.18
turf/shrub beds; 50/50	–	–	–	–	–	nr	7797.92
Leaky Pipe Systems Ltd; Leaky Pipe; moisture leaking pipe irrigation system							
Main supply pipe inclusive of machine excavation; exclusive of connectors							
20 mm LDPE polytubing	0.96	0.05	1.61	0.54	1.10	m	3.25
16 mm LDPE polytubing	0.67	0.05	1.61	0.54	0.77	m	2.92

35 SITE WORKS – WATER AND WATER FEATURES

Item – Overhead and Profit Included	PC £	Labour hours	Labour £	Plant £	Material £	Unit	Total rate £
IRRIGATION – cont							
Leaky Pipe Systems Ltd – cont							
Water filters and cartridges							
No 10; 20 mm	–	–	–	–	71.88	nr	**71.88**
Big Blue and RR30 cartridge; 25 mm	–	–	–	–	109.25	nr	**109.25**
Water filters and pressure regulator sets; complete assemblies							
No 10; flow rate 3.1–82 litres per minute	–	–	–	–	102.35	nr	**102.35**
Leaky pipe hose; placed 150 mm subsurface for turf irrigation; distance between laterals 350 mm; excavation and backfilling priced separately							
LP12L low leak	3.51	0.04	1.29	–	4.04	m²	**5.33**
LP12H high leak	3.56	0.04	1.29	–	4.09	m²	**5.38**
Leaky pipe hose; laid to surface for landscape irrigation; distance between laterals 600 mm							
LP12L low leak	2.09	0.03	0.81	–	2.40	m²	**3.21**
LP12H high leak	2.13	0.03	0.81	–	2.45	m²	**3.26**
Leaky pipe hose; laid to surface for landscape irrigation; distance between laterals 900 mm							
LP12L low leak	1.40	0.02	0.54	–	1.61	m²	**2.15**
LP12H high leak	1.42	0.02	0.54	–	1.63	m²	**2.17**
Leaky pipe hose; laid to surface for tree irrigation laid around circumference of tree pit							
LP12L low leak	2.07	0.13	4.03	–	2.38	nr	**6.41**
LP12H high leak	2.10	0.13	4.03	–	2.42	nr	**6.45**
Accessories							
automatic multi-station controller stations; inclusive of connections	300.00	2.00	82.80	–	345.00	nr	**427.80**
solenoid valves; inclusive of wiring and connections to a multi-station controller; nominal distance from controller 25 m	35.50	0.50	16.10	–	589.95	nr	**606.05**

35 SITE WORKS – WATER AND WATER FEATURES

Item – Overhead and Profit Included	PC £	Labour hours	Labour £	Plant £	Material £	Unit	Total rate £
FOUNTAINS AND WATER FEATURES							
Waterfall construction; Fairwater Water Garden Design Consultants Ltd							
Stone placed on top of butyl liner; securing with concrete and dressing to form natural rock pools and edgings							
Boulders; m³ rate	–	–	–	–	–	m³	917.53
Boulders; tonne rate	–	–	–	–	–	tonne	508.06
Balancing tank; blockwork construction; inclusive of recirculation pump and pond level control; 110 mm balancing pipe to pond; waterproofed with preformed polypropylene membrane; pipework mains water top-up and overflow							
450 × 600 × 1000 mm	–	–	–	–	–	nr	1916.20
Extra for pump							
2000 gallons per hour; submersible	–	–	–	–	–	nr	274.66
Water features; Fairwater Water Garden Design Consultants Ltd							
Natural stream; butyl lined level changes of 1.00 m; water pumped from lower pond (not included) via balancing tank; level changes via Purbeck stone water falls							
20 m long stream	–	–	–	–	–	nr	24388.33
Bespoke free-standing water wall; steel or glass panel; self contained recirculation system and reservoir							
water wall; 2.00 m high × 850 mm wide	–	–	–	–	–	nr	11613.48

35 SITE WORKS – WATER AND WATER FEATURES

Item – Overhead and Profit Included	PC £	Labour hours	Labour £	Plant £	Material £	Unit	Total rate £
INTERNAL WATER FEATURES							
Internal Water features; Fairwater Water Garden Design Consultants Ltd							
Stainless steel rill; powder coated black 3000 mm × 250 mm; 2 nr letterbox weirs to maintain water level in rill; WRAS compliant header tank; modified balancing tank with chemical dosing unit; circulation pump and pipework							
3.00 m × 250 mm wide	–	–	–	–	–	nr	5226.07
Brimming bowl with catchment tray; Fairwater Water Garden Design Consultants Ltd							
Marble brimming bowl. recessed Polypro catchment tray 1.50 × 1.50 × 120 mm; pipework; WRAS compliant header tank balancing tank with Chemical dosing unit; circulation pump							
1.00 m dia.	–	–	–	–	–	nr	4355.07
Design mock-up and commissioning on the above	–	–	–	–	–	nr	4064.72
Maintenance 12 months; monthly	–	–	–	–	–	nr	1742.02
WATER AND RIVER BANK REMEDIATION/ECOLOGY							
River and lake bed dredging; Blackdown Environmental Limited; Specialist water reclamation engineers; Operations include for removal of excavated material away from the water body bank and spreading locally in layers not exceeding 300 mm deep							
By long reach excavator							
Mobilization for Desilting	–	–	–	–	–	nr	3300.07
Dewatering and mechanical excavation with long reach excavator and tracked excavator to stockpile on bank	–	–	–	–	–	m³	13.20
Mobilization £12,000 craning in barges, pontoon and tug	–	–	–	–	–	nr	13200.28

35 SITE WORKS – WATER AND WATER FEATURES

Item – Overhead and Profit Included	PC £	Labour hours	Labour £	Plant £	Material £	Unit	Total rate £
Excavation by floating excavator, transport to bank by tug and barge	–	–	–	–	–	m³	27.28
Pumping of silt to geotubes for containment and drying	–	–	–	–	–	m³	24.75
By Pump dredger							
By long reach excavator using dewatering techniques							
Mobilization Truxor type pump dredger up to 1000 m³	–	–	–	–	–	nr	660.02
Mobilization pontoon type pump dredger over 1000 m³	–	–	–	–	–	nr	5500.12
Lake size 1.5 ha; 15000 m³	–	–	–	–	–	item	6600.14
Lake size 3.0 ha; 30000 m³	–	–	–	–	–	item	15400.32
Bioengineering techniques for reclamation or establishment of waterways; Blackdown Environmental Limited							
Erosion control							
Coir roll + Faggot	–	–	–	–	–	m²	41.47
Coir roll and rock roll	–	–	–	–	–	m²	82.50
Brushwood mattress revetments to encourage deposition of sediments							
200 mm deep	33.25	0.07	6.44	–	38.24	m²	44.68
500 mm deep	56.00	0.10	9.66	–	64.40	m²	74.06
Habitat creation to waterways and water bodies; Blackdown Environmental Limited; Specialist works to SSSIs and SAC sites							
Grading and enablement of river bank to allow access; temporary trackways or haul road and removal of same	–	8.00	772.80	503.76	1240.52	nr	2517.08
Creation of riffle bars and point bars using gravel beds	–	–	–	–	–	m²	221.10
Wetland establishment; Blackdown Environmental Limited;							
Restoration of riverbanks; to create habitats for wildlife							
greenfield sites	–	–	–	–	–	m²	16.50
brownfield sites	–	–	–	–	–	m²	27.50

35 SITE WORKS – WATER AND WATER FEATURES

Item – Overhead and Profit Included	PC £	Labour hours	Labour £	Plant £	Material £	Unit	Total rate £
WATER AND RIVER BANK REMEDIATION/ECOLOGY – cont							
Wetland and revetment establishment and erosion control; Blackdown Environmental Limited							
Preparation works to receive treatments below							
Grade out bank receive treatments	–	0.05	3.22	2.22	–	m²	**5.44**
Remove silt and accumulated deposits to maximum 300 mm deep within water line to reinstate freeflow of water along shoreline; Spread on river bank	–	0.05	3.22	2.71	–	m²	**5.93**
Placement of pre-grown coir rolls 200 mm dia. to base of revetments or within the water body to aid establishment of vegetation and habitat at the base of retained revetments (above)							
on tidal banks or wet river banks	26.90	0.11	10.73	–	30.94	m	**41.67**
on shorelines max 300 mm deep	43.25	0.11	10.97	–	49.74	m	**60.71**
on shorelines max 600 mm deep	66.00	0.13	12.07	–	75.90	m	**87.97**
on shorelines max 1.00 m deep	91.50	0.13	12.88	–	105.22	m	**118.10**
on shorelines max 1.50 m deep	97.50	0.17	16.10	–	112.13	m	**128.23**
Placement of pre-grown coir rolls 300 mm dia. to base of revetments or within the water body to aid establishment of vegetation and habitat at the base of retained revetments (above)							
on tidal banks or wet river banks	29.40	0.11	10.73	–	33.81	m	**44.54**
on shorelines max 300 mm deep	45.75	0.11	10.97	–	52.61	m	**63.58**
on shorelines max 600 mm deep	66.00	0.13	12.07	–	75.90	m	**87.97**
on shorelines max 1.00 m deep	94.00	0.13	12.88	–	108.10	m	**120.98**
on shorelines max 1.50 m deep	100.00	0.17	16.10	–	115.00	m	**131.10**
Stronger tidal demands– placement of pre-grown coir rolls 400 mm dia. to base of revetments or within the water body to aid establishment of vegetation and habitat at the base of retained revetments (above)							
on tidal banks or wet river banks	57.00	0.13	12.07	–	65.55	m	**77.62**
on shorelines max 300 mm deep	61.00	0.13	12.07	–	70.15	m	**82.22**
on shorelines max 600 mm deep	84.50	0.17	16.10	–	119.03	m	**135.13**
on shorelines max 1.00 m deep	110.00	0.20	19.32	–	126.50	m	**145.82**
on shorelines max 1.50 m deep	116.00	0.20	19.32	–	133.40	m	**152.72**

35 SITE WORKS – WATER AND WATER FEATURES

Item – Overhead and Profit Included	PC £	Labour hours	Labour £	Plant £	Material £	Unit	Total rate £
Placement of pre-planted coir pallets to establish reedbeds to wetlands; habitat creation projects where mature vegetation needs to be established rapidly							
on tidal banks or wet river banks	34.25	0.20	19.32	–	39.39	m	**58.71**
on shorelines max 300 mm deep	37.25	0.25	24.15	–	42.84	m	**66.99**
unplanted	23.00	0.25	24.15	–	74.75	m²	**98.90**
Brushwood faggots fascine in Hazel or Willow; 300 mm dia. to form sediment traps or build up inter-tidal sediment							
300 m dia.	–	0.10	9.66	–	27.02	m	**36.68**
400 mm dia.	–	0.10	9.66	–	29.33	m	**38.99**
Rock rolls; permanent revetment for use around reservoirs, shorelines, lake edges, streams and river banks							
300 mm dia.	–	0.07	6.44	4.20	114.89	m²	**125.53**
400 mm dia.	–	0.10	9.66	6.30	115.00	m²	**130.96**
rock mattresses to retain and in conjuction with rock rolls on riverbank	–	0.08	8.05	5.24	54.05	m²	**67.34**

35 SITE WORKS – WATER AND WATER FEATURES

Item – Overhead and Profit Included	PC £	Labour hours	Labour £	Plant £	Material £	Unit	Total rate £
LAKES AND PONDS							
Lakes and ponds – General							
Preamble: The pressure of water against a retaining wall or dam is considerable, and where water retaining structures form part of the design of water features, the landscape architect is advised to consult a civil engineer. Artificially contained areas of water in raised reservoirs over 25,000 m^3 have to be registered with the local authority and their dams will have to be covered by a civil engineer's certificate of safety.							
Typical linings – General							
Preamble: In addition to the traditional methods of forming the linings of lakes and ponds in puddled clay or concrete, there are a number of lining materials available. They are mainly used for reservoirs but can also help to form comparatively economic water features especially in soil which is not naturally water retentive. Information on the construction of traditional clay puddle ponds can be obtained from the British Trust for Conservation Volunteers, 36 St. Mary's Street, Wallingford, Oxfordshire, OX10 0EU. Tel: (01491) 39766. The cost of puddled clay ponds depends on the availability of suitable clay, the type of hand or machine labour that can be used and the use to which the pond is to be put.							

35 SITE WORKS – WATER AND WATER FEATURES

Item – Overhead and Profit Included	PC £	Labour hours	Labour £	Plant £	Material £	Unit	Total rate £
Lake liners; Fairwater Water Garden Design Consultants Ltd; to evenly graded surface of excavations (excavating not included); all stones over 75 mm; removing debris; including all welding and jointing of liner sheets							
Geotextile underlay; inclusive of spot welding to prevent dragging							
to water features	–	–	–	–	–	m²	**2.04**
to lakes or large features	–	–	–	–	–	1000 m²	**2044.02**
Butyl rubber liners; Varnamo; inclusive of site vulcanizing laid to geotextile above							
0.75 mm thick	–	–	–	–	–	m²	**9.14**
0.75 mm thick	–	–	–	–	–	1000 m²	**9139.79**
1.00 mm thick	–	–	–	–	–	m²	**14.34**
1.00 mm thick	–	–	–	–	–	1000 m²	**14342.68**
Operations over surfaces of lake liners							
Dug ballast; evenly spread over excavation already brought to grade							
150 mm thick	757.35	2.00	64.40	119.53	870.95	100 m²	**1054.88**
200 mm thick	1009.80	3.00	96.60	179.29	1161.27	100 m²	**1437.16**
300 mm thick	1514.70	3.50	112.70	209.17	1741.91	100 m²	**2063.78**
Imported topsoil; evenly spread over excavation							
100 mm thick	296.64	1.50	48.30	89.64	341.14	100 m²	**479.08**
150 mm thick	444.96	2.00	64.40	119.53	511.70	100 m²	**695.63**
200 mm thick	593.28	3.00	96.60	179.29	682.27	100 m²	**958.16**
Blinding existing subsoil with 50 mm sand	231.38	1.00	32.20	119.53	266.09	100 m²	**417.82**
Topsoil from excavation; evenly spread over excavation							
100 mm thick	–	–	–	89.64	–	100 m²	**89.64**
200 mm thick	–	–	–	119.53	–	100 m²	**119.53**
300 mm thick	–	–	–	179.29	–	100 m²	**179.29**
Extra over for screening topsoil using a Powergrid screener; removing debris	–	–	–	8.83	2.00	m³	**10.83**

35 SITE WORKS – WATER AND WATER FEATURES

Item – Overhead and Profit Included	PC £	Labour hours	Labour £	Plant £	Material £	Unit	Total rate £
LAKES AND PONDS – cont							
Lake construction; Fairwater Water Garden Design Consultants Ltd; lake construction at ground level; excavation; forming of lake; commissioning; excavated material spread on site							
Lined with existing site clay; natural edging with vegetation meeting the water							
500 m²	–	–	–	–	–	nr	11613.48
1000 m²	–	–	–	–	–	nr	19162.28
1500 m²	–	–	–	–	–	nr	26711.04
2000 m²	–	–	–	–	–	nr	32517.77
Lined with 0.75 mm butyl on geotextile underlay; natural edging with vegetation meeting the water							
500 m²	–	–	–	–	–	nr	18000.92
1000 m²	–	–	–	–	–	nr	33679.14
1500 m²	–	–	–	–	–	nr	44131.27
2000 m²	–	–	–	–	–	nr	52841.40
Extra over to the above for hard block edging to secure and protect liner; measured at lake perimeter	–	–	–	–	–	m	55.12
Lined with imported puddling clay; natural edging with vegetation meeting the water							
500 m²	–	–	–	–	–	nr	28453.06
1000 m²	–	–	–	–	–	nr	48776.67
1500 m²	–	–	–	–	–	nr	66777.60
2000 m²	–	–	–	–	–	nr	83617.15

35 SITE WORKS – WATER AND WATER FEATURES

Item – Overhead and Profit Included	PC £	Labour hours	Labour £	Plant £	Material £	Unit	Total rate £
Ornamental pools – General							
Preamble: Small pools may be lined with one of the materials mentioned under lakes and ponds, or may be in rendered brickwork, puddled clay or, for the smaller sizes, fibreglass. Most of these tend to be cheaper than waterproof concrete. Basic prices for various sizes of concrete pools are given in the Approximate Estimates section. Prices for excavation, grading, mass concrete, and precast concrete retaining walls are given in the relevant sections. The manufacturers should be consulted before specifying the type and thickness of pool liner, as this depends on the size, shape and proposed use of the pool. The manufacturer's recommendation on foundations and construction should be followed.							
Ornamental pools; Fairwater Ltd							
Pool liners; to 50 mm sand blinding to excavation (excavating not included); all stones over 50 mm; removing debris from surfaces of excavation; including underlay, all welding and jointing of liner sheets							
black polythene; 1000 gauge	–	–	–	–	–	m^2	7.00
blue polythene; 1000 gauge	–	–	–	–	–	m^2	9.24
coloured PVC; 1500 gauge	–	–	–	–	–	m^2	10.59
black PVC; 1500 gauge	–	–	–	–	–	m^2	8.00
black butyl; 0.75 mm thick	–	–	–	–	–	m^2	10.43
black butyl; 1.00 mm thick	–	–	–	–	–	m^2	11.65
black butyl; 1.50 mm thick	–	–	–	–	–	m^2	38.55
Fine gravel; 100 mm; evenly spread over area of pool; by hand	4.21	0.13	4.03	–	4.84	m^2	8.87
Selected topsoil from excavation; 100 mm; evenly spread over area of pool; by hand	–	0.13	4.03	–	–	m^2	4.03
Extra over selected topsoil for spreading imported topsoil over area of pool; by hand	29.66	–	–	–	34.11	m^3	34.11

35 SITE WORKS – WATER AND WATER FEATURES

Item – Overhead and Profit Included	PC £	Labour hours	Labour £	Plant £	Material £	Unit	Total rate £
LAKES AND PONDS – cont							
Pool surrounds and ornament; **Haddonstone Ltd; Portland Bath or** **terracotta cast stone**							
Pool surrounds; installed to pools or water feature construction priced separately; surrounds and copings to 112.5 mm internal brickwork							
C4HSKVP half small pool surround; internal dia. 1780 mm; kerb features continuous moulding enriched with ovolvo and palmette designs; inclusive of plinth and integral conch shell vases flanked by dolphins	1073.01	16.00	515.20	–	1261.75	nr	1776.95
C4SKVP small pool surround as above but with full circular construction; internal dia. 1780 mm	2824.03	48.00	1545.60	–	3280.80	nr	4826.40
C4MKVP medium pool surround; internal dia. 2705 mm; inclusive of plinth and integral vases	4236.04	48.00	1545.60	–	4940.12	nr	6485.72
C4XLKVP extra large pool surround; internal dia. 5450 mm	5648.06	140.00	4508.00	–	6573.76	nr	11081.76
Pool centre pieces and fountains; inclusive of plumbing and pumps							
C251 Gothic Fountain and Gothic Upper Base A350; free-standing fountain	1103.01	4.00	128.80	–	1272.85	nr	1401.65
HC521 Romanesque Fountain; free-standing bowl with self-circulating fountain; filled with cobbles; 815 mm dia. × 348 mm high	489.00	2.00	64.40	–	684.51	nr	748.91
C300 Lion Fountain; 610 mm high on fountain base C305; 280 mm high	248.00	2.00	64.40	–	289.58	nr	353.98
Wall fountains, watertanks and **fountains; inclusive of installation** **drainage, automatic top up, pump** **and balancing tank**							
Capital Garden Products Ltd							
Lion Wall Fountain F010; 970 × 940 × 520 mm	268.00	4.00	128.80	–	2224.40	nr	2353.20
Dolphin Wall Fountain F001; 740 × 510 mm	86.00	4.00	128.80	–	2015.10	nr	2143.90
Dutch Master Wall Fountain F012; 740 × 410 mm	81.00	4.00	128.80	–	2009.35	nr	2138.15

35 SITE WORKS – WATER AND WATER FEATURES

Item – Overhead and Profit Included	PC £	Labour hours	Labour £	Plant £	Material £	Unit	Total rate £
James II Watertank 2801; 730 × 730 × 760 mm high; 405 litres	297.00	4.00	128.80	–	2257.75	nr	**2386.55**
James II Fountain 4901 bp; 710 × 1300 × 1440 mm high; 655 litres	674.00	4.00	128.80	–	2691.30	nr	**2820.10**
Fountain surrounds; Architectural Heritage Ltd; to pools constructed separately inclusive of installation, drainage, automatic top up, pump and balancing tank							
Decorative Circular Pool Surround; age patinated artificial stone; reproduction; Overall dia. 4.26 m, overall height 220 mm	4000.00	4.00	128.80	–	6516.20	nr	**6645.00**
Great Westwood pool surround; age-patinated artificial stone pool; reproduction; Overall dia. 6 m; height 350 mm	6000.00	4.00	128.80	–	8816.20	nr	**8945.00**
Fountain kits; typical prices of submersible units comprising fountain pumps, fountain nozzles, underwater spotlights, nozzle extension armatures, underwater terminal boxes and electrical control panels							
Single aerated white foamy water columns; ascending jet 70 mm dia.; descending water up to four times larger; jet height adjustable between 1.00–1.70 m	4135.95	–	–	–	5415.24	nr	**5415.24**
Single aerated white foamy water columns; ascending jet 110 mm dia.; descending water up to four times larger; jet height adjustable between 1.50–3.00 m	6893.25	–	–	–	8944.52	nr	**8944.52**
SWIMMING POOLS							
Natural swimming pools; Fairwater Water Garden Design Consultants Ltd							
Natural swimming pool of 50 m² area; shingle regeneration zone 50 m² planted with marginal and aquatic plants	–	–	–	–	–	nr	**75487.71**

35 SITE WORKS – WATER AND WATER FEATURES

Item – Overhead and Profit Included	PC £	Labour hours	Labour £	Plant £	Material £	Unit	Total rate £
SWIMMING POOLS – cont							
Guncast Swimming Pools Ltd; excavation and blinding of swimming pools; works by machine inclusive of allowances for working space and wall thicknesses; excavation of all trenches for pool pipeworks and electrical services							
Regular shaped pools							
100 m² pool	–	–	–	–	–	nr	11141.02
72 m² pool	–	–	–	–	–	nr	8703.93
60 m² pool	–	–	–	–	–	nr	8703.93
Irregular shaped pools							
100 m² pool	–	–	–	–	–	nr	11698.07
72 m² pool	–	–	–	–	–	nr	9139.12
60 m² pool	–	–	–	–	–	nr	9139.12
Guncast Swimming Pools Ltd; construction of a rectangular swimming pool in situ concrete corner steps, graduated shell depths of 1.1–2.5 m, water depth of 1.0–2.4 m as standard skimmer pool, tiled corner steps; excludes pump house construction							
Inclusive of drainage layer, formwork reinforcement and all Gunite construction, rendered internally and bespoke mosaic surface finish; bullnosed Indian sandstone copings							
pool size 17.0 × 6.0 m; 172,500 litre	–	–	–	–	–	nr	172917.95
pool size 12.0 × 6.0 m; 125,750 litre	–	–	–	–	–	nr	129978.60
pool size 12.0 × 5.0 m; 102,150 litre	–	–	–	–	–	nr	114891.79
Guncast Swimming Pools Ltd; options and accessories for swimming pools							
Underwater lighting							
8 nr underwater floodlights 300 W 12 V; stainless steel	–	–	–	–	–	nr	6382.88
2 nr underwater floodlights 300 W 12 V; stainless steel	–	–	–	–	–	nr	1595.71
Automatic pool cleaner	–	–	–	–	–	nr	1798.82

35 SITE WORKS – WATER AND WATER FEATURES

Item – Overhead and Profit Included	PC £	Labour hours	Labour £	Plant £	Material £	Unit	Total rate £
Automatic pool covers							
insulated automatic slatted cover; 12 × 6 m inclusive of additional surface skimmer and balance pipe in pit; housed in decking (not included)	–	–	–	–	–	nr	24254.93
solar cover; 17 × 6 m Ocea; housed in wall cave pit; logic motor; concealed with stainless steel frame and removable PVC panels; tiled or rendered to match pool	–	–	–	–	–	nr	41546.74
Driglide pool cover; 17 × 6 m selected colours on trackway	–	–	–	–	–	nr	45887.09
Ozone disinfection system	–	–	–	–	–	nr	6527.95
Guncast Swimming Pools Ltd; refurbishment of an existing swimming pool; with Roman bay end steps; no leaks present in existing pool; existing pipework and existing filtration systems and pump							
Empty pool, retain existing blockwork structure; retile pool; replace copings as Indian sandstone bullnose; prepare shell; fix mosaic; pressure test							
pool size 8.50 × 4.25 m; 59,000 litre	–	–	–	–	–	nr	53964.33
Guncast Swimming Pools Ltd; options and accessories for swimming pools							
Pool heating options							
propane boiler 250,000 BTU	–	–	–	–	–	nr	4792.95
pool heating via propane boiler 400,000 BTU	–	–	–	–	–	nr	6081.15
heat pump 100 × 1000 mm concrete base	–	–	–	–	–	nr	7427.35
pool heating via heat pump	–	–	–	–	–	nr	7427.35
Diving board	–	–	–	–	–	nr	2738.82
Pool maintenance kit							
telescopic pole, wall/floor brush; skimmer net; flexible vacuum hose; underwater vacuum cleaner; pool water test kit; thermometer; pool treatment chemicals	–	–	–	–	–	nr	1067.68

36 FENCING

Item – Overhead and Profit Included	PC £	Labour hours	Labour £	Plant £	Material £	Unit	Total rate £
CLARIFICATION NOTES ON LABOUR COSTS IN THIS SECTION							
Fencing team Generally a two man fencing team is used in this section; The column 'Labour hours' reports team hours. The column 'Labour £' reports the total cost of the team for the unit of fencing shown							
2 man team	–	1.00	64.40	–	–	hr	**64.40**
TEMPORARY FENCING							
Temporary security fence; HSS Hire; mesh framed unclimbable fencing; including precast concrete supports and couplings Weekly hire; 2.85 × 2.00 m high							
weekly hire rate	–	–	–	1.35	–	m	**1.35**
erection of fencing; labour only	–	0.07	4.30	–	–	m	**4.30**
removal of fencing loading to collection vehicle	–	0.03	2.15	–	–	m	**2.15**
delivery charge	–	–	–	0.99	–	m	**0.99**
return haulage charge	–	–	–	0.74	–	m	**0.74**
Protective fencing; Chestnut pale fencing Cleft chestnut rolled fencing; fixing to 100 mm dia. chestnut posts; driving into firm ground at 3 m centres							
900 mm high	4.68	0.05	3.43	–	12.02	m	**15.45**
1200 mm high	6.85	0.08	5.15	–	14.50	m	**19.65**
Enclosures; Earth Anchors							
Rootfast anchored galvanized steel enclosures post ADP 20–1000; 1000 mm high × 20 mm dia. with AA25–750 socket and padlocking ring	35.87	0.05	3.22	–	44.39	nr	**47.61**
steel cable, orange plastic coated	0.44	–	0.07	–	0.51	m	**0.58**

36 FENCING

Item – Overhead and Profit Included	PC £	Labour hours	Labour £	Plant £	Material £	Unit	Total rate £
TIMBER FENCING							
Timber fencing; tanalized softwood fencing; all timber posts are kiln dried redwood with 15 year guarantee							
Timber lap panels; pressure treated; fixed to timber posts 75 × 75 mm in 1:3:6 concrete; at 1.90 m centres							
900 mm high	13.96	0.33	21.46	–	26.58	m	**48.04**
1200 mm high	15.42	0.38	24.15	–	28.05	m	**52.20**
1500 mm high	15.42	0.40	25.76	–	30.68	m	**56.44**
1800 mm high	15.06	0.45	28.98	–	31.27	m	**60.25**
Timber lap panels; fixed to slotted concrete posts 100 × 100 mm 1:3:6 concrete at 1.88 m centres							
900 mm high	13.96	0.38	24.15	–	37.28	m	**61.43**
1200 mm high	14.48	0.40	25.76	–	37.88	m	**63.64**
1500 mm high	20.61	0.42	27.37	–	47.27	m	**74.64**
1800 mm high	20.25	0.50	32.20	–	48.71	m	**80.91**
extra for corner posts	48.77	0.45	28.98	–	65.40	nr	**94.38**
extra over panel fencing for 300 mm high trellis tops; slats at 100 mm centres; including							
additional length of posts	9.95	0.50	32.20	–	11.44	m	**43.64**
Closeboarded fencing; to concrete posts 100 × 100 mm; 2 nr softwood arris rails; 100 × 22 mm softwood pales lapped 13 mm; including excavating and backfilling into firm ground at 3.00 m centres; setting in concrete 1:3:6; concrete gravel board 150 × 150 mm							
1050 mm high	24.46	0.55	35.42	–	31.63	m	**67.05**
1500 mm high	24.46	0.57	37.03	–	31.63	m	**68.66**
Closeboarded fencing; to concrete posts 100 × 100 mm; 3 nr softwood arris rails; 100 × 22 mm softwood pales lapped 13 mm; including excavating and backfilling into firm ground at 3.00 m centres; setting in concrete 1:3:6							
1650 mm high	32.60	0.63	40.25	–	40.99	m	**81.24**
1800 mm high	30.74	0.65	41.86	–	38.85	m	**80.71**

36 FENCING

Item – Overhead and Profit Included	PC £	Labour hours	Labour £	Plant £	Material £	Unit	Total rate £
TIMBER FENCING – cont							
Closeboarded fencing; to timber redwood posts 100 × 100 mm; 2 nr softwood arris rails; 100 × 22 mm softwood pales lapped 13 mm; including excavating and backfilling into firm ground at 3.00 m centres; setting in concrete 1:3:6							
1350 mm high	24.09	0.50	32.20	–	31.20	m	**63.40**
1650 mm high	25.29	0.63	40.25	–	32.58	m	**72.83**
1800 mm high	22.60	0.65	41.86	–	29.47	m	**71.33**
Closeboarded fencing; to timber posts 100 × 100 mm; 3 nr softwood arris rails; 100 × 22 mm softwood pales lapped 13 mm; including excavating and backfilling into firm ground at 3.00 m centres; setting in concrete 1:3:6							
1350 mm high	29.12	0.48	30.67	–	36.98	m	**67.65**
1650 mm high	30.32	0.65	41.86	–	38.36	m	**80.22**
1800 mm high	27.63	0.68	43.47	–	35.26	m	**78.73**
extra over for post 125 × 100 mm; for 1800 mm high fencing	18.84	–	–	–	21.67	m	**21.67**
extra over to the above for counter rail	1.03	–	–	–	1.18	m	**1.18**
extra over to the above for capping rail	1.03	0.05	3.22	–	1.18	m	**4.40**
Feather edge board fencing; to timber posts 100 × 100 mm; 2 nr softwood cant rails; 125 × 22 mm feather edge boards lapped 13 mm; including excavating and backfilling into firm ground at 3.00 m centres; setting in concrete 1:3:6							
1350 mm high	22.84	0.28	18.04	–	27.84	m	**45.88**
1650 mm high	24.84	0.28	18.04	–	30.13	m	**48.17**
1800 mm high	23.58	0.28	18.04	–	28.68	m	**46.72**
Feather edge board fencing; to timber posts 100 × 100 mm; 3 nr softwood cant rails; 125 × 22 mm feather edge boards lapped 13 mm; including excavating and backfilling into firm ground at 3.00 m centres; setting in concrete 1:3:6							
1350 mm high	28.36	0.29	18.40	–	34.19	m	**52.59**
1650 mm high	30.36	0.29	18.40	–	36.48	m	**54.88**
1800 mm high	27.18	0.29	18.40	–	35.03	m	**53.43**
extra over for post 125 × 100 mm; for 1800 mm high fencing	18.84	–	–	–	21.67	nr	**21.67**
extra over to the above for counter rail and capping	6.48	0.05	3.22	–	7.45	m	**10.67**

36 FENCING

Item – Overhead and Profit Included	PC £	Labour hours	Labour £	Plant £	Material £	Unit	Total rate £
Palisade fencing; 22 × 75 mm softwood vertical palings with pointed tops; nailing to two 50 × 100 mm horizontal softwood arris rails; morticed to 100 × 100 mm softwood posts with weathered tops at 3.00 m centres; setting in concrete							
900 mm high	15.13	0.45	28.98	–	20.90	m	49.88
1200 mm high	21.93	0.47	30.59	–	28.70	m	59.29
extra over for rounded tops	1.19	–	–	–	1.37	m	1.37
Post-and-rail fencing; 90 × 38 mm softwood horizontal rails; fixing with galvanized nails to 150 × 75 mm softwood posts; including excavating and backfilling into firm ground at 1.80 m centres; all treated timber							
1200 mm high; 3 horizontal rails	11.79	0.18	11.28	–	13.74	m	25.02
1200 mm high; 4 horizontal rails	13.58	0.18	11.28	–	15.62	m	26.90
Cleft rail fencing; chestnut adze tapered rails 2.80 m long; morticed into joints; to 125 × 100 mm softwood posts 1.95 m long; including excavating and backfilling into firm ground at 2.80 m centres							
two rails	11.21	0.13	8.05	–	12.89	m	20.94
three rails	14.73	0.14	9.03	–	16.94	m	25.97
four rails	18.24	0.18	11.28	–	20.98	m	32.26
Hit and Miss horizontal rail fencing; 87 × 38 mm top and bottom rails; 100 × 22 mm vertical boards arranged alternately on opposite side of rails; to 100 × 100 mm posts; including excavating and backfilling into firm ground; setting in concrete at 1.8 m centres							
treated softwood; 1600 mm high	30.06	0.60	38.64	–	37.71	m	76.35
treated softwood; 1800 mm high	32.60	0.67	42.92	–	40.95	m	83.87
treated softwood; 2000 mm high	35.16	0.70	45.08	–	43.88	m	88.96
Trellis tops to fencing							
Extra over screen fencing for 300 mm high trellis tops; slats at 100 mm centres; including additional length of posts	9.95	0.05	3.22	–	11.44	m	14.66

36 FENCING

Item – Overhead and Profit Included	PC £	Labour hours	Labour £	Plant £	Material £	Unit	Total rate £
TRELLIS							
Traditional trellis panels; The Garden Trellis Company; bespoke trellis panels for decorative, screening or security applications; timber planed all round; height of trellis 1800 mm Free-standing panels; posts 70 × 70 mm set in concrete; timber frames mitred and grooved 45 × 34 mm; heights and widths to suit; slats 32 × 10 mm joinery quality tanalized timber at 100 mm ccs; elements fixed by galvanized staples; capping rail 70 × 34 mm							
HV68 horizontal and vertical slats; softwood	145.80	1.00	32.20	–	176.89	m	**209.09**
D68 diagonal slats; softwood	162.00	1.00	32.20	–	195.52	m	**227.72**
HV68 horizontal and vertical slats; hardwood iroko	271.80	1.00	32.20	–	321.79	m	**353.99**
D68 diagonal slats; hardwood iroko or Western Red Cedar	252.00	1.00	32.20	–	299.02	m	**331.22**
Trellis panels fixed to face of existing wall or railings; timber frames mitred and grooved 45 × 34 mm; heights and widths to suit; slats 32 × 10 mm joinery quality tanalized timber at 100 mm ccs; elements fixed by galvanized staples; capping rail 70 × 34 mm							
HV68 horizontal and vertical slats; softwood	117.00	0.50	16.10	–	136.48	m	**152.58**
D68 diagonal slats; softwood	135.00	0.50	16.10	–	157.18	m	**173.28**
HV68 horizontal and vertical slats; iroko or Western Red Cedar	237.60	0.50	16.10	–	275.17	m	**291.27**
D68 diagonal slats; iroko or Western Red Cedar	255.60	0.50	16.10	–	295.87	m	**311.97**

36 FENCING

Item – Overhead and Profit Included	PC £	Labour hours	Labour £	Plant £	Material £	Unit	Total rate £
Contemporary style trellis panels; The Garden Trellis Company; bespoke trellis panels for decorative, screening or security applications; timber planed all round							
Free-standing panels 30/15; posts 70 × 70 mm set in concrete with 90 × 30 mm top capping; slats 30 × 14 mm with 15 mm gaps; vertical support at 450 mm centres							
joinery treated softwood	189.00	1.00	32.20	–	226.57	m	**258.77**
hardwood iroko or Western Red Cedar	248.40	1.00	32.20	–	294.88	m	**327.08**
Panels 30/15 face fixed to existing wall or fence; posts 70 × 70 mm set in concrete with 90 × 30 mm top capping; slats 30 × 14 mm with 15 mm gaps; vertical support at 450 mm centres							
joinery treated softwood	171.00	0.50	16.10	–	198.58	m	**214.68**
hardwood iroko or Western Red Cedar	273.60	0.50	16.10	–	316.57	m	**332.67**
Integral arches to ornamental trellis panels; The Garden Trellis Company							
Arches to trellis panels in 45 × 34 mm grooved timbers to match framing; fixed to posts							
R450 ¼ circle; joinery treated softwood	61.00	1.50	48.30	–	79.37	nr	**127.67**
R450 ¼ circle; hardwood iroko or Western Red Cedar	69.00	1.50	48.30	–	88.57	nr	**136.87**
Arches to span 1800 mm wide							
joinery treated softwood	228.00	1.50	48.30	–	271.42	nr	**319.72**
hardwood iroko or Western Red Cedar	182.00	1.50	48.30	–	218.52	nr	**266.82**
Painting or staining of trellis panels; high quality coatings							
microporous opaque paint or spirit based stain	–	–	–	–	34.50	m²	**34.50**

36 FENCING

Item – Overhead and Profit Included	PC £	Labour hours	Labour £	Plant £	Material £	Unit	Total rate £
STRAINED WIRE FENCING/FIELD FENCING							
Posts for strained wire and wire mesh							
Strained wire fencing; concrete inter posts only at 2750 mm centres; 610 mm below ground; excavating holes; filling with concrete; replacing topsoil; disposing surplus soil off site							
900 mm high	4.50	0.28	17.89	1.15	6.93	m	**25.97**
1200 mm high	5.50	0.28	17.89	1.15	8.08	m	**27.12**
1800 mm high	6.67	0.39	25.05	2.86	9.43	m	**37.34**
Extra over strained wire fencing for concrete straining posts with one strut; posts and struts 610 mm below ground; struts, cleats, stretchers, winders, bolts and eye bolts; excavating holes; filling to within 150 mm of ground level with concrete (1:12) – 40 mm aggregate; replacing topsoil; disposing surplus soil off site							
900 mm high	46.10	0.34	21.57	2.58	92.45	nr	**116.60**
1200 mm high	51.43	0.33	21.47	2.59	100.41	nr	**124.47**
1800 mm high	56.34	0.33	21.47	2.59	110.08	nr	**134.14**
Extra over strained wire fencing for concrete straining posts with two struts; posts and struts 610 mm below ground; excavating holes; filling to within 150 mm of ground level with concrete (1:12) – 40 mm aggregate; replacing topsoil; disposing surplus soil off site							
900 mm high	64.70	0.46	29.30	3.31	133.50	nr	**166.11**
1200 mm high	70.86	0.42	27.37	2.59	145.08	nr	**175.04**
1800 mm high	78.18	0.42	27.37	2.59	165.45	nr	**195.41**
Strained wire fencing; galvanized steel angle posts only at 2750 mm centres; 610 mm below ground; driving in							
900 mm high; 40 × 40 × 5 mm	3.27	0.02	0.98	–	3.76	m	**4.74**
1200 mm high; 40 × 40 × 5 mm	3.64	0.02	1.08	–	4.19	m	**5.27**
1500 mm high; 40 × 40 × 5 mm	4.57	0.03	1.96	–	5.26	m	**7.22**
1800 mm high; 40 × 40 × 5 mm	6.09	0.04	2.35	–	7.00	m	**9.35**
Galvanized steel straining posts with two struts for strained wire fencing; setting in concrete							
900 mm high; 50 × 50 × 6 mm	61.62	0.50	32.20	–	74.75	nr	**106.95**
1200 mm high; 50 × 50 × 6 mm	34.00	0.50	32.20	–	42.99	nr	**75.19**
1500 mm high; 50 × 50 × 6 mm	80.62	0.50	32.20	–	96.61	nr	**128.81**
1800 mm high; 50 × 50 × 6 mm	97.98	0.50	32.20	–	116.58	nr	**148.78**

36 FENCING

Item – Overhead and Profit Included	PC £	Labour hours	Labour £	Plant £	Material £	Unit	Total rate £
Fixing of fencing to posts							
Strained wire; to posts (posts not included); 3 mm galvanized wire; fixing with galvanized stirrups							
900 mm high; 2 wire	0.24	0.02	1.07	–	0.39	m	**1.46**
1200 mm high; 3 wire	0.36	0.02	1.50	–	0.53	m	**2.03**
1400 mm high; 3 wire	0.36	0.02	1.50	–	0.53	m	**2.03**
1800 mm high; 3 wire	0.36	0.02	1.50	–	0.53	m	**2.03**
Barbed wire; to posts (posts not included); 3 mm galvanized wire; fixing with galvanized stirrups							
900 mm high; 2 wire	0.52	0.03	2.14	–	0.71	m	**2.85**
1200 mm high; 3 wire	0.78	0.05	3.00	–	1.01	m	**4.01**
1400 mm high; 3 wire	0.78	0.05	3.00	–	1.01	m	**4.01**
1800 mm high; 3 wire	0.78	0.05	3.00	–	1.01	m	**4.01**
Fencing only to posts and strained wire							
Chain link fencing; to strained wire and posts priced separately; 3 mm galvanized wire; 51 mm mesh; galvanized steel components; fixing to line wires threaded through posts and strained with eye-bolts; posts (not included)							
1800 mm high	3.35	0.05	3.22	–	3.97	m	**7.19**
Chain link fencing; to strained wire and posts priced separately; 3.15 mm plastic coated galvanized wire (wire only 2.5 mm); 51 mm mesh; galvanized steel components; fencing with line wires threaded through posts and strained with eye-bolts; posts (not included) (Note: plastic coated fencing can be cheaper than galvanized finish as wire of a smaller cross-sectional area can be used)							
900 mm high	4.58	0.03	2.15	–	5.50	m	**7.65**
1200 mm high	6.25	0.03	2.15	–	7.42	m	**9.57**
1800 mm high	9.00	0.06	4.03	–	11.27	m	**15.30**
Extra over strained wire fencing for cranked arms and galvanized barbed wire							
1 row	2.41	0.01	0.54	–	2.77	m	**3.31**
2 rows	2.67	0.03	1.61	–	3.07	m	**4.68**
3 rows	2.93	0.03	1.61	–	3.37	m	**4.98**

36 FENCING

Item – Overhead and Profit Included	PC £	Labour hours	Labour £	Plant £	Material £	Unit	Total rate £
STRAINED WIRE FENCING/FIELD FENCING – cont							
Fencing only to posts and strained wire – cont							
Field fencing; woven wire mesh; fixed to posts and straining wires measured separately							
cattle fence; 1200 m high; 114 × 300 mm at bottom to 230 × 300 mm at top	1.64	0.05	3.22	–	2.18	m	**5.40**
sheep fence; 800 mm high; 140 × 300 mm at bottom to 230 × 300 mm at top	1.34	0.05	3.22	–	1.84	m	**5.06**
deer fence; 2.0 m high; 89 × 150 mm at bottom to 267 × 300 mm at top	2.45	0.06	4.03	–	3.12	m	**7.15**
Extra for concreting in posts	–	0.25	16.10	–	5.59	nr	**21.69**
Extra for straining post	29.66	0.38	24.15	–	34.11	nr	**58.26**
Boundary fencing; strained wire and wire mesh; Jacksons Fencing							
Tubular chain link fencing; galvanized; plastic coated; 60.3 mm dia. posts at 3.0 m centres; setting 700 mm into ground; choice of 10 mesh colours; including excavating holes, backfilling and removing surplus soil; with top rail only							
900 mm high	–	–	–	–	–	m	**46.16**
1200 mm high	–	–	–	–	–	m	**49.79**
1800 mm high	–	–	–	–	–	m	**54.65**
2000 mm high	–	–	–	–	–	m	**57.72**
Tubular chain link fencing; galvanized; plastic coated; 60.3 mm dia. posts at 3.0 m centres; cranked arms and three lines barbed wire; setting 700 mm into ground; including excavating holes; backfilling and removing surplus soil; with top rail only							
2000 mm high	–	–	–	–	–	m	**58.51**
1800 mm high	–	–	–	–	–	m	**56.86**

36 FENCING

Item – Overhead and Profit Included	PC £	Labour hours	Labour £	Plant £	Material £	Unit	Total rate £
RABBIT NETTING							
Rabbit netting							
Timber stakes; peeled kiln dried pressure treated; pointed; 1.8 m posts driven 750 mm into ground at 3 m centres (line wires and netting priced separately)							
75–100 mm stakes	5.30	0.13	8.05	–	6.09	m	14.14
Corner posts or straining posts 150 mm dia. × 2.4 m high set in concrete; centres to suit local conditions or changes of direction							
1 strut	32.85	0.50	32.20	7.36	46.14	nr	85.70
2 struts	36.68	0.50	32.20	7.36	50.54	nr	90.10
Strained wire; to posts (posts not included); 3 mm galvanized wire; fixing with galvanized stirrups							
900 mm high; 2 wire	0.24	0.02	1.07	–	0.39	m	1.46
1200 mm high; 3 wire	0.36	0.02	1.50	–	0.53	m	2.03
Rabbit netting; 31 mm 19 gauge 1050 mm high netting fixed to posts (line wires and straining posts or corner posts all priced separately)							
900 mm high; turned in	0.97	0.02	1.29	–	1.12	m	2.41
900 mm high; buried 150 mm in trench	0.97	0.04	2.69	–	1.12	m	3.81
TUBULAR FENCING							
Boundary fencing; Steelway-Fensecure Ltd							
Classic two rail tubular fencing; top and bottom with stretcher bars and straining wires in between; comprising 60.3 mm tubular posts at 3.00 m centres; setting in concrete; 35 mm top rail tied with aluminium and steel fittings; 50 × 50 × 355/ 2.5 mm PVC coated chain link; all components galvanized and coated in green nylon							
964 mm high	–	–	–	–	–	m	58.18
1269 mm high	–	–	–	–	–	m	62.33
1574 mm high	–	–	–	–	–	m	64.22
1878 mm high	–	–	–	–	–	m	65.21
2188 mm high	–	–	–	–	–	m	66.46
2458 mm high	–	–	–	–	–	m	74.26
2948 mm high	–	–	–	–	–	m	93.39
3562 mm high	–	–	–	–	–	m	97.28

36 FENCING

Item – Overhead and Profit Included	PC £	Labour hours	Labour £	Plant £	Material £	Unit	Total rate £
TUBULAR FENCING – cont							
Boundary fencing – cont							
End posts; Classic range; 60.3 mm dia.; setting in concrete							
964 mm high	–	–	–	–	–	nr	77.09
1269 mm high	–	–	–	–	–	nr	83.72
1574 mm high	–	–	–	–	–	nr	94.24
1878 mm high	–	–	–	–	–	nr	102.85
2188 mm high	–	–	–	–	–	nr	112.38
2458 mm high	–	–	–	–	–	nr	127.46
2948 mm high	–	–	–	–	–	nr	147.62
3562 mm high	–	–	–	–	–	nr	168.50
Corner posts; 60.3 mm dia.; setting in concrete							
964 mm high	–	–	–	–	–	nr	83.94
1269 mm high	–	–	–	–	–	nr	62.33
1574 mm high	–	–	–	–	–	nr	89.98
1878 mm high	–	–	–	–	–	nr	90.97
2188 mm high	–	–	–	–	–	nr	98.66
2458 mm high	–	–	–	–	–	nr	106.46
2948 mm high	–	–	–	–	–	nr	132.03
3562 mm high	–	–	–	–	–	nr	142.37
WELDED MESH FENCING							
Boundary fencing							
Paladin welded mesh colour coated green fencing; fixing to metal posts at 2.975 m centres with manufacturer's fixings; setting 600 mm deep in firm ground; including excavating holes; backfilling and removing surplus excavated material; includes 10 year product guarantee							
1800 mm high	26.28	0.33	21.25	–	30.22	m	51.47
2000 mm high	20.75	0.33	21.25	–	33.47	m	54.72
2400 mm high	24.50	0.33	21.25	–	39.35	m	60.60
Extra over welded galvanized plastic coated mesh fencing for concreting in posts	9.72	0.06	3.58	–	11.18	m	14.76
Boundary fencing; Lang & Fulton							
Orsogril rectangular steel bar mesh fence panels; pleione pattern; bolting to 60 × 8 mm uprights at 2 m centres; mesh 62 × 66 mm; setting in concrete							
930 mm high panels	49.00	0.08	8.05	–	75.22	m	83.27
1326 mm high panels	67.00	0.10	9.66	–	95.92	m	105.58
1722 mm high panels	83.00	0.17	16.09	–	114.32	m	130.41

36 FENCING

Item – Overhead and Profit Included	PC £	Labour hours	Labour £	Plant £	Material £	Unit	Total rate £
SECURITY FENCING							
Security fencing; Jacksons Fencing							
Barbican galvanized steel paling fencing; on 60 × 60 mm posts at 3 m centres; setting in concrete							
1250 mm high	–	–	–	–	–	m	69.07
1500 mm high	–	–	–	–	–	m	73.80
2000 mm high	–	–	–	–	–	m	83.78
2500 mm high	–	–	–	–	–	m	100.22
Gates; to match Barbican galvanized steel paling fencing							
width 1 m	–	–	–	–	–	nr	1292.76
width 2 m	–	–	–	–	–	nr	1318.82
width 3 m	–	–	–	–	–	nr	1364.14
width 4 m	–	–	–	–	–	nr	1407.20
width 8 m	–	–	–	–	–	pair	1943.12
width 9 m	–	–	–	–	–	pair	2390.65
width 10 m	–	–	–	–	–	pair	2573.07
Security fencing; Jacksons Fencing; intruder guards							
Viper Spike Intruder Guards; to existing structures; including fixing bolts							
Viper 1; 40 × 5 mm × 1.1 m long; with base plate	39.90	1.00	64.40	–	45.88	nr	110.28
Viper 3; 160 × 190 mm wide; U shape; to prevent intruders climbing pipes	43.00	0.50	32.20	–	49.45	nr	81.65
Security fencing; Jacksons Fencing							
Razor Barb Concertina; spiral wire security barriers; fixing to 600 mm steel ground stakes							
Ref 3275; 450 mm dia. roll; medium barb; galvanized	3.25	0.05	3.22	–	4.01	m	7.23
Ref 3276; 730 mm dia. roll; medium barb; galvanized	4.96	0.05	3.22	–	5.97	m	9.19
Ref 3277; 950 mm dia. roll; medium barb; galvanized	6.05	0.05	3.22	–	7.22	m	10.44
Three lines Barbed Tape; medium barb; on 50 × 50 mm mild steel angle posts; setting in concrete							
Ref 3283; barbed tape; medium barb; galvanized	1.25	0.21	13.56	–	14.66	m	28.22

36 FENCING

Item – Overhead and Profit Included	PC £	Labour hours	Labour £	Plant £	Material £	Unit	Total rate £
SECURITY FENCING – cont							
Security fencing – cont							
Five lines Barbed Tape; medium barb; on 50 × 50 mm mild steel angle posts; setting in concrete							
Ref 3283; barbed tape; medium barb; galvanized	2.08	0.17	10.74	–	15.62	m	**26.36**
TRIP RAILS							
Trip rails; Birdsmouth; Jacksons Fencing							
Diamond rail fencing; Birdsmouth; posts 100 × 100 mm softwood planed at 1.35 m centres set in 1:3:6 concrete; rail 75 × 75 mm nominal secured with galvanized straps nailed to posts							
posts 900 mm (600 mm above ground); at 1.35 m centres	12.15	0.25	16.10	0.49	16.30	m	**32.89**
posts 1.20 m (900 mm above ground); at 1.35 m centres	14.07	0.25	16.10	0.49	18.52	m	**35.11**
posts 900 mm (600 mm above ground); at 1.80 m centres	10.36	0.19	12.11	0.37	13.69	m	**26.17**
posts 1.20 m (900 mm above ground); at 1.80 m centres	11.81	0.19	12.11	0.37	15.32	m	**27.80**
WINDBREAK FENCING							
Windbreak fencing							
Fencing; English Woodlands; Shade and Shelter Netting windbreak fencing; green; to 100 mm dia. treated softwood posts; setting 450 mm into ground; fixing with 50 × 25 mm treated softwood battens nailed to posts; including excavating and backfilling into firm ground; setting in concrete at 3 m centres							
1200 mm high	1.30	0.08	5.01	–	8.96	m	**13.97**
1800 mm high	1.90	0.08	5.01	–	9.65	m	**14.66**

36 FENCING

Item – Overhead and Profit Included	PC £	Labour hours	Labour £	Plant £	Material £	Unit	Total rate £
BALL STOP FENCING							
Ball stop fencing;							
Steelway-Fensecure Ltd							
Ball stop net; 30 × 30 mm netting							
fixed to 60.3 mm dia. 12 mm solid							
bar lattice; galvanized; dual posts;							
top, middle and bottom rails							
4.5 m high	–	–	–	–	–	m	142.56
5.0 m high	–	–	–	–	–	m	176.63
6.0 m high	–	–	–	–	–	m	197.74
7.0 m high	–	–	–	–	–	m	216.29
8.0 m high	–	–	–	–	–	m	235.84
9.0 m high	–	–	–	–	–	m	256.32
10.0 m high	–	–	–	–	–	m	277.62
Ball stop net; corner posts							
4.5 m high	–	–	–	–	–	m	126.75
5.0 m high	–	–	–	–	–	m	138.75
6.0 m high	–	–	–	–	–	m	277.99
7.0 m high	–	–	–	–	–	m	191.99
8.0 m high	–	–	–	–	–	m	218.66
9.0 m high	–	–	–	–	–	m	245.28
10 m high	–	–	–	–	–	m	271.89
Ball stop net; end posts							
4.5 m high	–	–	–	–	–	m	105.48
5.0 m high	–	–	–	–	–	m	117.46
6.0 m high	–	–	–	–	–	m	144.09
7.0 m high	–	–	–	–	–	m	170.72
8.0 m high	–	–	–	–	–	m	197.33
9.0 m high	–	–	–	–	–	m	223.95
10 m high	–	–	–	–	–	m	253.24
RAILINGS							
Railings; Steelway-Fensecure Ltd							
Mild steel bar railings of balusters at							
112 mm centres welded to flat rail							
top and bottom; bays 2.72 m long;							
bolting to 50 × 50 mm hollow square							
section posts; setting in concrete							
galvanized; 900 mm high	–	–	–	–	–	m	97.52
galvanized; 1200 mm high	–	–	–	–	–	m	118.11
galvanized; 1500 mm high	–	–	–	–	–	m	131.26
galvanized; 1800 mm high	–	–	–	–	–	m	157.48
galvanized & powder coated; 900 mm high	–	–	–	–	–	m	91.91
galvanized & powder coated; 1200 mm high	–	–	–	–	–	m	111.58
galvanized & powder coated; 1500 mm high	–	–	–	–	–	m	123.35
galvanized & powder coated; 1800 mm high	–	–	–	–	–	m	147.03

36 FENCING

Item – Overhead and Profit Included	PC £	Labour hours	Labour £	Plant £	Material £	Unit	Total rate £
RAILINGS – cont							
Railings – cont							
Mild steel blunt top railings of 20 mm balusters at 130 mm centres welded to bottom rail; passing through holes in top rail and welded; top and bottom rails 50 × 10 mm; bolting to 70 × 70 mm hollow square section posts; setting in concrete							
galvanized; 800 mm high	–	–	–	–	–	m	104.99
galvanized; 1000 mm high	–	–	–	–	–	m	110.26
galvanized; 1300 mm high	–	–	–	–	–	m	131.25
galvanized; 1500 mm high	–	–	–	–	–	m	141.76
galvanized & powder coated; 800 mm high	–	–	–	–	–	m	119.22
galvanized & powder coated; 1000 mm high	–	–	–	–	–	m	126.85
galvanized & powder coated; 1300 mm high	–	–	–	–	–	m	150.23
galvanized & powder coated; 1500 mm high	–	–	–	–	–	m	163.09
Railings; traditional pattern; 16 mm dia. verticals at 120 mm intervals with horizontal bars near top and bottom; balusters with spiked tops; 50 × 50 mm standards; including setting 520 mm into concrete at 2.72 m centres							
glavanized & powder coated; 1200 mm high	–	–	–	–	–	m	119.97
galvanized & powder coated; 1500 mm high	–	–	–	–	–	m	150.50
galvanized & powder coated; 1800 mm high	–	–	–	–	–	m	179.93
Interlaced bow-top mild steel railings; traditional park type; 16 mm dia. verticals at 80 mm intervals; welded at bottom to 40 × 10 mm flat and slotted through 38 × 8 mm top rail to form hooped top profile; 50 × 50 mm standards; setting 560 mm into concrete at 2.75 m centres							
galvanized; 900 mm high	–	–	–	–	–	m	86.47
galvanized; 1200 mm high	–	–	–	–	–	m	97.79
galvanized; 1500 mm high	–	–	–	–	–	m	134.08

36 FENCING

Item – Overhead and Profit Included	PC £	Labour hours	Labour £	Plant £	Material £	Unit	Total rate £
galvanized; 1800 mm high	–	–	–	–	–	m	182.23
galvanized & powder coated; 900 mm high	–	–	–	–	–	m	65.84
galvanized & powder coated; 1200 mm high	–	–	–	–	–	m	87.28
galvanized & powder coated; 1500 mm high	–	–	–	–	–	m	109.10
galvanized & powder coated; 1800 mm high	–	–	–	–	–	m	130.92
ESTATE FENCING/HORIZONTAL BAR FENCING							
Metal estate and parkland fencing; Britiannia Metalwork Services Ltd; continously welded, mild steel, 40 × 10 mm uprights manually driven into normal soil to a minimum depth of 600 mm at 1.00 m centres; 40 × 10 mm flat steel supports welded to uprights beneath topsoil at typically every third upright; horizontal rails threaded through holes within uprights; 20 mm solid steel top rail; 4 nr 25 × 8 mm × 16 mm dia. solid steel lower rails; painted fencing shot-blasted, primed and finished with two coats of paint							
Bare metal; 1200 mm high							
10–50 m	–	–	–	–	–	m	71.86
51–100 m	–	–	–	–	–	m	60.69
101–500 m	–	–	–	–	–	m	56.66
above 500 m	–	–	–	–	–	m	54.45
Primed and painted							
10–50 m	–	–	–	–	–	m	85.84
51–100 m	–	–	–	–	–	m	72.69
101–500 m	–	–	–	–	–	m	67.95
above 500 m	–	–	–	–	–	m	65.35
Galvanized and painted							
10–50 m	–	–	–	–	–	m	115.89
51–100 m	–	–	–	–	–	m	98.13
101–500 m	–	–	–	–	–	m	91.73
above 500 m	–	–	–	–	–	m	88.22
End posts and corner posts; 40 × 40 mm; standard hollow section; end or corner posts with capped tops set into concrete							
bare metal	–	–	–	–	–	nr	71.50
primed and painted	–	–	–	–	–	nr	93.50
galvanized and painted	–	–	–	–	–	nr	110.00

36 FENCING

Item – Overhead and Profit Included	PC £	Labour hours	Labour £	Plant £	Material £	Unit	Total rate £
ESTATE FENCING/HORIZONTAL BAR FENCING – cont							
Gates for metal estate fencing; Britannia Metalwork Services Ltd; fully welded gates to match fence; scroll at hinge end; turnover at latch end; posts topped with finials and set into concrete							
Pedestrian gates; 1.0–1.5 m wide							
bare metal	–	–	–	–	–	nr	544.50
galvanized and painted	–	–	–	–	–	nr	764.52
Field gates; 1.50–4.0 m wide							
bare metal	–	–	–	–	–	nr	764.52
galvanized and painted	–	–	–	–	–	nr	984.52
Kissing gates							
bare metal	–	–	–	–	–	nr	605.01
galvanized and painted	–	–	–	–	–	nr	825.02
Quadrant; 4 nr uprights with rails welded to fence							
bare metal	–	–	–	–	–	nr	308.00
primed and painted	–	–	–	–	–	nr	385.00
Step stile; 2 treads							
bare metal	–	–	–	–	–	nr	165.00
primed and painted	–	–	–	–	–	nr	242.00
RECYCLED PLASTIC FENCING							
Recycled plastic; Fencing and Barriers; Hahn Plastics Ltd; Recycled and recyclable; components							
Posts; inclusive of mechanical excavation; 400 × 400 × 400 mm deep; backfilling with site mixed concrete and disposal of arisings							
Hanit Ultra – square post							
10 cm × 10 cm × 1.75 m	27.08	0.25	16.10	7.74	46.59	nr	70.43
Square post – reinforced							
9 cm × 9 cm × 2 m	25.40	0.25	16.10	7.74	44.65	nr	68.49
Round post – reinforced 8 cm dia. × 2 m	18.42	0.25	16.10	7.74	36.63	nr	60.47
Square post 7 cm × 7 cm × 2 m	10.52	0.25	16.10	7.74	27.54	nr	51.38
Round post 8 cm dia. × 2 m; in concrete inclusive of mechanical excavation and disposal	10.21	0.25	16.10	7.74	27.18	nr	51.02
Cross profile stake (wire fence) 7 cm × 7 cm × 2 m; driven in	5.02	–	–	–	–	nr	–

36 FENCING

Item – Overhead and Profit Included	PC £	Labour hours	Labour £	Plant £	Material £	Unit	Total rate £
Recycled Plastic fencing; Hahn Plastics Ltd; Hanit fencing							
Fencing							
Woven Privacy Screen – Ribbon 1.50 m wide panels × 1.80 m high inclusive of recycled 9 cm × 9 cm posts set in concrete at 1.545 m centres	154.40	0.29	18.40	5.16	204.50	m	**228.06**
Picket fencing; Hahn Plastics Ltd; curved top profile; 100% recycled self-coloured hanit® material, non-rot post and panels with 20 year warranty; pre-assembled panels in grey, brown or green with either aluminium or plastic cross-bar with matching posts; hand excavation and disposal off site							
850 mm high with plastic cross-bar	117.00	0.20	12.88	–	161.58	m	**174.46**
Pale & Post fencing (palisade): Hahn Plastics Ltd; 100% recycled self-coloured hanit® Ultra material, enhanced wood grain finish, non-rot post and pales with 20 year warranty							
In black or brown woodgrain for fences; two crossrails and recycled 10 cm × 10 cm posts included; Posts at 1.80 m centres							
1.80 m high	86.70	0.40	25.76	–	107.42	m	**133.18**
2.40 m high	113.26	0.40	25.76	–	137.97	m	**163.73**
Privacy fencing: Hahn Plastics Ltd; 100% recycled self-coloured hanit®, non-rot post and tongue & grooved panels with 20 year warranty; dedicated brown plastic reinforced fence posts supplied with matching end caps with individual grey, brown or green tongue & groove panels inserted into built in recesses with a central support							
In Grey Brown or Green; Prices shown for the average priced colour; Posts at 2.00 m centres							
1.95 m high;	161.90	0.20	12.88	–	193.75	m	**206.63**
Corner post or end post	117.00	0.50	32.20	–	148.64	nr	**180.84**

36 FENCING

Item – Overhead and Profit Included	PC £	Labour hours	Labour £	Plant £	Material £	Unit	Total rate £
RECYCLED PLASTIC FENCING – cont							
Pasture fencing: Hahn Plastics Ltd; 100% recycled self-coloured hanit®, non-rot post and tongue & grooved panels with 20 year warranty; equestrian and agricultural use							
Pointed reinforced posts driven in							
1.50 m high 3 rails (inc posts)	66.42	0.15	9.75	–	76.38	m	**86.13**
1.50 m high 5 rails (inc posts)	73.48	0.15	9.75	–	84.50	m	**94.25**
GATES, KISSING GATES AND STILES							
Gates – General							
Preamble: Gates in fences; see specification for fencing as gates in traditional or proprietary fencing systems are usually constructed of the same materials and finished as the fencing itself.							
Gates; hardwood							
Hardwood field gate; five bar diamond braced; planed iroko; fixed to 150 × 150 mm softwood posts; inclusive of hinges and furniture							
Ref 1100 100; 0.9 m wide	207.99	2.00	128.80	–	327.78	nr	**456.58**
Ref 1100 101; 1.2 m wide	247.98	2.00	128.80	–	373.77	nr	**502.57**
Ref 1100 102; 1.5 m wide	275.98	2.00	128.80	–	405.97	nr	**534.77**
Ref 1100 103; 1.8 m wide	399.99	2.00	128.80	–	548.58	nr	**677.38**
Ref 1100 104; 2.1 m wide	429.99	2.00	128.80	–	583.08	nr	**711.88**
Ref 1100 105; 2.4 m wide	459.99	2.00	128.80	–	617.58	nr	**746.38**
Ref 1100 106; 2.7 m wide	489.99	2.00	128.80	–	652.08	nr	**780.88**
Ref 1100 108; 3.3 m wide	549.99	2.00	128.80	–	721.08	nr	**849.88**
Gates; treated softwood; Jacksons Fencing							
Timber field gates; including wrought iron ironmongery; five bar type; diamond braced; 1.80 m high; to 200 × 200 mm posts; setting 750 mm into firm ground							
width 2400 mm	111.80	3.00	193.20	–	308.50	nr	**501.70**
width 2700 mm	120.55	3.00	193.20	–	318.56	nr	**511.76**
width 3000 mm	129.25	3.50	225.40	–	328.57	nr	**553.97**
width 3300 mm	138.55	3.75	241.50	–	339.26	nr	**580.76**

36 FENCING

Item – Overhead and Profit Included	PC £	Labour hours	Labour £	Plant £	Material £	Unit	Total rate £
Featherboard garden gates; including ironmongery; to 100 × 120 mm posts; one diagonal brace							
1.0 × 1.2 m high	38.90	1.50	96.60	–	202.53	nr	**299.13**
1.0 × 1.5 m high	51.20	1.50	96.60	–	228.17	nr	**324.77**
1.0 × 1.8 m high	152.00	1.50	96.60	–	251.29	nr	**347.89**
Picket garden gates; including ironmongery; to match picket fence; width 1000 mm; to 100 × 120 mm posts; one diagonal brace							
950 mm high	87.60	1.50	96.60	–	156.07	nr	**252.67**
1200 mm high	92.30	1.50	96.60	–	161.47	nr	**258.07**
1800 mm high	108.30	1.50	96.60	–	201.03	nr	**297.63**
Gates; tubular mild steel; Jacksons Fencing Field gates; galvanized; including ironmongery; diamond braced; 1.80 m high; to tubular steel posts; setting in concrete							
width 3000 mm	132.75	2.50	161.00	–	275.25	nr	**436.25**
width 3300 mm	140.00	2.50	161.00	–	283.59	nr	**444.59**
width 3600 mm	147.38	2.50	161.00	–	292.08	nr	**453.08**
width 4200 mm	162.22	2.50	161.00	–	309.15	nr	**470.15**
Gates; sliding; Jacksons Fencing Sliding Gate; including all galvanized rails and vertical rail infill panels; special guide and shutting frame posts (Note: foundations installed by suppliers)							
access width 4.00 m; 1.5 m high gates	–	–	–	–	–	nr	**4638.55**
access width 4.00 m; 2.0 m high gates	–	–	–	–	–	nr	**4774.51**
access width 4.00 m; 2.5 m high gates	–	–	–	–	–	nr	**4910.47**
access width 6.00 m; 1.5 m high gates	–	–	–	–	–	nr	**5514.37**
access width 6.00 m; 2.0 m high gates	–	–	–	–	–	nr	**5689.98**
access width 6.00 m; 2.5 m high gates	–	–	–	–	–	nr	**5866.73**
access width 8.00 m; 1.5 m high gates	–	–	–	–	–	nr	**6331.27**
access width 8.00 m; 2.0 m high gates	–	–	–	–	–	nr	**6549.94**
access width 8.00 m; 2.5 m high gates	–	–	–	–	–	nr	**6767.48**
access width 10.00 m; 2.0 m high gates	–	–	–	–	–	nr	**8071.57**
access width 10.00 m; 2.5 m high gates	–	–	–	–	–	nr	**8341.22**

36 FENCING

Item – Overhead and Profit Included	PC £	Labour hours	Labour £	Plant £	Material £	Unit	Total rate £
GATES, KISSING GATES AND STILES – cont							
Kissing gates							
Kissing gates; Jacksons Fencing; in galvanized metal bar; fixing to fencing posts (posts not included); 1.65 × 1.30 × 1.00 m high	368.00	2.50	161.00	–	423.20	nr	**584.20**
Stiles							
Stiles; two posts; setting into firm ground; three rails; two treads	94.00	1.50	96.60	–	122.99	nr	**219.59**
GUARD RAILS							
Pedestrian guard rails and barriers							
Mild steel pedestrian guard rails; Broxap Street Furniture; 1.00 m high with 150 mm toe space; to posts at 2.00 m centres; galvanized finish							
vertical staggered in-line bar infill panel	44.00	1.00	64.40	–	58.96	m	**123.36**
vertical staggered bar infill with 230 mm visibility gap at top	46.00	1.00	64.40	–	61.26	m	**125.66**

37 SOFT LANDSCAPING

Item – Overhead and Profit Included	PC £	Labour hours	Labour £	Plant £	Material £	Unit	Total rate £
GENERALLY							
Seeding/turfing – General Preamble: The following market prices generally reflect the manufacturer's recommended retail prices. Trade and bulk discounts are often available on the prices shown. The manufacturers of these products generally recommend application rates. Note: the following rates reflect the average rate for each product.							
CLARIFICATION NOTES ON LABOUR COSTS IN THIS SECTION							
General landscape team Generally a three man team is used in this section; The column Labour hours reports team hours. The column Labour £ reports the total cost of the team for the unit of work shown							
3 man team	–	1.00	96.60	–	–	hr	**96.60**
MARKET PRICES OF SEEDING MATERIALS							
Market prices of pre-seeding materials Rigby Taylor Ltd							
turf fertilizer; Mascot Outfield 16 +6+6	–	–	–	–	2.42	100 m²	**2.42**
Boughton Loam							
screened topsoil; 100 mm	–	–	–	–	44.50	m³	**44.50**
screened Kettering loam; 3 mm	–	–	–	–	183.60	m³	**183.60**
screened Kettering loam; sterilized; 3 mm	–	–	–	–	189.90	m³	**189.90**
top dressing; sand soil mixtures; 90/10 to 50/50	–	–	–	–	99.00	m³	**99.00**

37 SOFT LANDSCAPING

Item – Overhead and Profit Included	PC £	Labour hours	Labour £	Plant £	Material £	Unit	Total rate £
MARKET PRICES OF SEEDING MATERIALS – cont							
Market prices of turf fertilizers; recommended average application rates							
Everris; turf fertilizers							
grass fertilizer; slow release; Sierraform GT Preseeder 18–22–05	–	–	–	–	4.88	100 m²	**4.88**
grass fertilizer; slow release; Sierraform GT All Season 18–06–18	–	–	–	–	4.07	100 m²	**4.07**
grass fertilizer; slow release; Sierraform GT Anti-Stress 15-00–26	–	–	–	–	4.88	100 m²	**4.88**
grass fertilizer; slow release; Sierraform GT Momentum 22–05–11	–	–	–	–	3.26	100 m²	**3.26**
grass fertilizer; slow release; Sierraform GT Spring Start 16–00–16	–	–	–	–	4.07	100 m²	**4.07**
grass fertilizer; controlled release; Sierrablen 27–05–05 (8–9 months)	–	–	–	–	5.65	100 m²	**5.65**
grass fertilizer; controlled release; Sierrablen 28–05–05 (5–6 months)	–	–	–	–	5.69	100 m²	**5.69**
grass fertilizer; Sierrablen Fine 38–00–00 (4–5 months)	–	–	–	–	2.14	100 m²	**2.14**
grass fertilizer; Sierrablen Mini 38–00–00 (2–3 months)	–	–	–	–	2.29	100 m²	**2.29**
grass fertilizer; Greenmaster Pro-Lite Turf Tonic 08–00–00	–	–	–	–	2.45	100 m²	**2.45**
grass fertilizer; Greenmaster Pro-Lite Spring & Summer 14–05–10	–	–	–	–	3.49	100 m²	**3.49**
grass fertilizer; Greenmaster Pro-Lite Zero Phosphate 14–00–10	–	–	–	–	3.47	100 m²	**3.47**
grass fertilizer; Greenmaster Pro-Lite Mosskiller 14–00–00	–	–	–	–	3.76	100 m²	**3.76**
grass fertilizer; Greenmaster Pro-Lite Invigorator 04–00–08	–	–	–	–	2.65	100 m²	**2.65**
grass fertilizer; Greenmaster Pro-Lite Autumn 06–05–10	–	–	–	–	3.76	100 m²	**3.76**

37 SOFT LANDSCAPING

Item – Overhead and Profit Included	PC £	Labour hours	Labour £	Plant £	Material £	Unit	Total rate £
grass fertilizer; Greenmaster Pro-Lite Double K 07–00–14	–	–	–	–	3.76	100 m^2	**3.76**
grass fertilizer; Greenmaster Pro-Lite NK 12–00–12	–	–	–	–	3.76	100 m^2	**3.76**
outfield turf fertilizer; Sportsmaster Standard Spring & Summer 09–07–07	–	–	–	–	3.02	100 m^2	**3.02**
outfield turf fertilizer; Sportsmaster Standard Autumn 04–12–12	–	–	–	–	3.28	100 m^2	**3.28**
outfield turf fertilizer; Sportsmaster Standard Zero Phosphate 12–00–09	–	–	–	–	3.44	100 m^2	**3.44**
TPMC; tree planting and mulching compost	–	–	–	–	2.60	75 l	**2.60**
Rigby Taylor Ltd; 35 g/m^2							
grass fertilizer; Mascot Microfine 12+0+10 + 2% Mg + 2% Fe	–	–	–	–	5.99	100 m^2	**5.99**
grass fertilizer; Mascot Microfine 8+0+6 + 2% Mg + 4% Fe	–	–	–	–	5.04	100 m^2	**5.04**
grass fertilizer; Mascot Microfine Organic OC1 8+0+0 + 2% Fe	–	–	–	–	5.21	100 m^2	**5.21**
grass fertilizer; Mascot Microfine Organic OC2 5+2+10	–	–	–	–	5.64	100 m^2	**5.64**
grass fertilizer; Mascot Delta Sport 12+4+8 + 0.5% Fe	–	–	–	–	4.64	100 m^2	**4.64**
grass fertilizer; Mascot Fine Turf 12+0+9 + 1% Mg + 1% Fe + seaweed	–	–	–	–	5.46	100 m^2	**5.46**
outfield fertilizer; Mascot Outfield 16+6+6	–	–	–	–	3.38	100 m^2	**3.38**
outfield fertilizer; Mascot Outfield 4+10+10	–	–	–	–	3.46	100 m^2	**3.46**
outfield fertilizer; Mascot Outfield 9+5+5	–	–	–	–	3.27	100 m^2	**3.27**
outfield fertilizer; Mascot Outfield 12+4+4	–	–	–	–	3.32	100 m^2	**3.32**
liquid fertilizer; Mascot Microflow-C 26+0+0 + trace elements; 200–1400 ml/100 m^2	–	–	–	–	6.67	100 m^2	**6.67**
liquid fertilizer; Mascot Microflow-CX 6+2+18 + trace elements; 200–1400 ml/100 m^2	–	–	–	–	6.42	100 m^2	**6.42**
liquid fertilizer; Mascot Microflow-CX 14+0+7 + trace elements; 200–1400 ml/100 m^2	–	–	–	–	5.66	100 m^2	**5.66**
liquid fertilizer; Mascot Microflow-CX 17+2+5 + trace elements; 200–1400 ml/100 m^2	–	–	–	–	6.37	100 m^2	**6.37**

37 SOFT LANDSCAPING

Item – Overhead and Profit Included	PC £	Labour hours	Labour £	Plant £	Material £	Unit	Total rate £
MARKET PRICES OF SEEDING MATERIALS – cont							
Market prices of grass seed							
Preamble: The prices shown are for supply only at one number 20 kg or 25 kg bag purchase price unless otherwise stated. Rates shown are based on the manufacturer's maximum recommendation for each seed type. Trade and bulk discounts are often available on the prices shown for quantities of more than one bag.							
Bowling greens; fine lawns; ornamental turf; croquet lawns							
Germinal GB; A1 Green; 35 g/m²	–	–	–	–	31.06	100 m²	**31.06**
DLF Seeds Ltd; J Green; 34–50 g/m²	–	–	–	–	56.85	100 m²	**56.85**
DLF Seeds Ltd; J Premier Green; 34–50 g/m²	–	–	–	–	98.20	100 m²	**98.20**
DLF Seeds Ltd; Promaster 20; 35–50 g/m²	–	–	–	–	26.85	100 m²	**26.85**
DLF Seeds Ltd; Promaster 10; 35–50 g/m²	–	–	–	–	52.20	100 m²	**52.20**
Tennis courts; cricket squares							
Germinal GB; A2 Lawns; 35 g/m²	–	–	–	–	21.96	100 m²	**21.96**
Germinal GB; A5 Cricket; 35 g/m²	–	–	–	–	27.50	100 m²	**27.50**
DLF Seeds Ltd; J Premier Green; 34–50 g/m²	–	–	–	–	98.20	100 m²	**98.20**
DLF Seeds Ltd; J Intense; 18–25 g/m²	–	–	–	–	13.90	100 m²	**13.90**
DLF Seeds Ltd; Promaster 35; 35 g/m²	–	–	–	–	20.20	100 m²	**20.20**
Amenity grassed areas; general purpose lawns							
Germinal GB; A3 Banks; 25–50 g/m²	–	–	–	–	27.50	100 m²	**27.50**
DLF Seeds Ltd; J Rye Fairway; 18–25 g/m²	–	–	–	–	14.43	100 m²	**14.43**
DLF Seeds Ltd; J Intense; 18–25 g/m²	–	–	–	–	13.90	100 m²	**13.90**
DLF Seeds Ltd; Promaster 50; 25–35 g/m²	–	–	–	–	18.69	100 m²	**18.69**
DLF Seeds Ltd; Promaster 120; 25–35 g/m²	–	–	–	–	21.49	100 m²	**21.49**

37 SOFT LANDSCAPING

Item – Overhead and Profit Included	PC £	Labour hours	Labour £	Plant £	Material £	Unit	Total rate £
Conservation; country parks; slopes and banks							
Germinal GB; A4 Parkland; 17–35 g/m^2	–	–	–	–	20.56	100 m^2	**20.56**
Germinal GB; A16 Country Park; 8–19 g/m^2	–	–	–	–	13.15	100 m^2	**13.15**
Germinal GB; A17 Legume; 2 g/m^2	–	–	–	–	22.75	100 m^2	**22.75**
Shaded areas							
Germinal GB; A6 Shade; 50 g/m^2	–	–	–	–	42.63	100 m^2	**42.63**
Germinal GB; J Green; 34–50 g/m^2	–	–	–	–	56.85	100 m^2	**56.85**
DLF Seeds Ltd; Promaster 60; 35–50 g/m^2	–	–	–	–	30.95	100 m^2	**30.95**
Sports pitches; rugby; football pitches							
Germinal GB; A7 Sportsground; 20 g/m^2	–	–	–	–	13.32	100 m^2	**13.32**
DLF Seeds Ltd; J Pitch; 18–30 g/m^2	–	–	–	–	16.41	100 m^2	**16.41**
DLF Seeds Ltd; Promaster 70; 15–35 g/m^2	–	–	–	–	17.15	100 m^2	**17.15**
DLF Seeds Ltd; Promaster 75; 15–35 g/m^2	–	–	–	–	21.00	100 m^2	**21.00**
DLF Seeds Ltd; Promaster 80; 17–35 g/m^2	–	–	–	–	17.75	100 m^2	**17.75**
Rigby Taylor; Mascot R11 Football & Rugby; 35 g/m^2	–	–	–	–	23.45	100 m^2	**23.45**
Rigby Taylor; Mascot R13 General Playing Fields; 35 g/m^2	–	–	–	–	23.45	100 m^2	**23.45**
Outfields							
Germinal GB; A7 Sportsground; 20 g/m^2	–	–	–	–	13.32	100 m^2	**13.32**
Germinal GB; A9 Outfield; 17–35 g/m^2	–	–	–	–	20.56	100 m^2	**20.56**
DLF Seeds Ltd; J Fairway; 12–25 g/m^2	–	–	–	–	14.25	100 m^2	**14.25**
DLF Seeds Ltd; Promaster 40; 35 g/m^2	–	–	–	–	20.89	100 m^2	**20.89**
DLF Seeds Ltd; Promaster 70; 15–35 g/m^2	–	–	–	–	17.15	100 m^2	**17.15**
Rigby Taylor; R4 Species Rich Fairways & Outfields; 35 g/m^2	–	–	–	–	31.15	100 m^2	**31.15**
Hockey pitches							
DLF Seeds Ltd; J Fairway; 12–25 g/m^2	–	–	–	–	14.25	100 m^2	**14.25**
DLF Seeds Ltd; Promaster 70; 15–35 g/m^2	–	–	–	–	17.15	100 m^2	**17.15**

37 SOFT LANDSCAPING

Item – Overhead and Profit Included	PC £	Labour hours	Labour £	Plant £	Material £	Unit	Total rate £
MARKET PRICES OF SEEDING MATERIALS – cont							
Market prices of grass seed – cont							
Parks							
Germinal GB; A7 Sportsground 20 g/m²	–	–	–	–	13.32	100 m²	**13.32**
Germinal GB; A9 Outfield; 17–35 g/m²	–	–	–	–	20.56	100 m²	**20.56**
DLF Seeds Ltd Promaster 120; 25–35 g/m²	–	–	–	–	21.49	100 m²	**21.49**
Informal playing fields							
DLF Seeds Ltd; J Rye Fairway; 18–25 g/m²	–	–	–	–	14.43	100 m²	**14.43**
DLF Seeds Ltd; Promaster 45; 35 g/m²	–	–	–	–	20.20	100 m²	**20.20**
Caravan sites							
Germinal GB; A9 Outfield; 17–35 g/m²	–	–	–	–	20.56	100 m²	**20.56**
Sports pitch re-seeding and repair							
Germinal GB; A8 Pitch Renovator; 20–35 g/m²	–	–	–	–	19.51	100 m²	**19.51**
Germinal GB; A20 Ryesport; 20–35 g/m²	–	–	–	–	20.56	100 m²	**20.56**
DLF Seeds Ltd; J Pitch; 18–30 g/m²	–	–	–	–	16.41	100 m²	**16.41**
DLF Seeds Ltd; Promaster 80; 17–35 g/m²	–	–	–	–	17.75	100 m²	**17.75**
DLF Seeds Ltd; Promaster 81; 17–35 g/m²	–	–	–	–	19.21	100 m²	**19.21**
Rigby Taylor; R14 Sports Pitches 100% premium perennial ryegrass; 35 g/m²	–	–	–	–	25.20	100 m²	**25.20**
Racecourses; gallops; polo grounds; horse rides							
Germinal GB; Racecourse; 25–30 g/m²	–	–	–	–	18.30	100 m²	**18.30**
DLF Seeds Ltd; J Intense; 18–25 g/m²	–	–	–	–	13.90	100 m²	**13.90**
DLF Seeds Ltd; Promaster 65; 17–35 g/m²	–	–	–	–	20.61	100 m²	**20.61**
Motorway and road verges							
Germinal GB; A18 Road Verge; 6–15 g/m²	–	–	–	–	9.94	100 m²	**9.94**
DLF Seeds Ltd; Promaster 85; 10 g/m²	–	–	–	–	5.63	100 m²	**5.63**
DLF Seeds Ltd; Promaster 120; 25–35 g/m²	–	–	–	–	21.49	100 m²	**21.49**

37 SOFT LANDSCAPING

Item – Overhead and Profit Included	PC £	Labour hours	Labour £	Plant £	Material £	Unit	Total rate £
Golf courses; tees							
Germinal GB; A10 Tees, 35–50 g/m²	–	–	–	–	28.75	100 m²	**28.75**
DLF Seeds Ltd; J Premier Fairway; 18–30 g/m²	–	–	–	–	20.31	100 m²	**20.31**
DLF Seeds Ltd; J Intense; 18–25 g/m²	–	–	–	–	13.90	100 m²	**13.90**
DLF Seeds Ltd; Promaster 40; 35 g/m²	–	–	–	–	20.89	100 m²	**20.89**
DLF Seeds Ltd; Promaster 45; 35 g/m²	–	–	–	–	20.20	100 m²	**20.20**
Golf courses; greens							
Germinal GB; A11 Golf Greens; 35 g/m²	–	–	–	–	27.56	100 m²	**27.56**
Germinal GB; A13 Roughs; 8 g/m²	–	–	–	–	3.80	100 m²	**3.80**
DLF Seeds Ltd; J All Bent; 8 g/m²	–	–	–	–	23.27	100 m²	**23.27**
DLF Seeds Ltd; J Green; 34–50 g/m²	–	–	–	–	56.85	100 m²	**56.85**
DLF Seeds Ltd; Promaster 5; 35–50 g/m²	–	–	–	–	36.80	100 m²	**36.80**
Rigby Taylor; R1 Golf & Bowls Fescue/Bent; 35 g/m²	–	–	–	–	52.67	100 m²	**52.67**
Golf courses; fairways							
Germinal GB; A12 Fairways; 15–25 g/m²	–	–	–	–	12.81	100 m²	**12.81**
DLF Seeds Ltd; J Rye Fairway; 18–30 g/m²	–	–	–	–	17.31	100 m²	**17.31**
DLF Seeds Ltd; J Premier Fairway; 18–30 g/m²	–	–	–	–	20.31	100 m²	**20.31**
DLF Seeds Ltd; Promaster 40; 35 g/m²	–	–	–	–	20.89	100 m²	**20.89**
DLF Seeds Ltd; Promaster 45; 35 g/m²	–	–	–	–	20.20	100 m²	**20.20**
Rigby Taylor; Mascot R116 Fescue Rye Fairway; 35 g/m²	–	–	–	–	69.30	100 m²	**69.30**
Golf courses; roughs							
DLF Seeds Ltd; Promaster 25; 17–35 g/m²	–	–	–	–	21.84	100 m²	**21.84**
Rigby Taylor; Mascot R5 Golf Links & Rough; 35 g/m²	–	–	–	–	6.65	100 m²	**6.65**
Waste land; spoil heaps; quarries							
Germinal GB; A15 Reclamation; 15–20 g/m²	–	–	–	–	13.15	100 m²	**13.15**
DLF Seeds Ltd; Promaster 95; 12–35 g/m²	–	–	–	–	26.04	100 m²	**26.04**

37 SOFT LANDSCAPING

Item – Overhead and Profit Included	PC £	Labour hours	Labour £	Plant £	Material £	Unit	Total rate £
MARKET PRICES OF SEEDING MATERIALS – cont							
Market prices of grass seed – cont							
Low maintenance; housing estates; amenity grassed areas							
Germinal GB; A19 Housing; 25–35 g/m²	–	–	–	–	19.25	100 m²	**19.25**
Germinal GB; A22 Low Maintenance; 25–35 g/m²	–	–	–	–	22.75	100 m²	**22.75**
DLF Seeds Ltd; Promaster 120; 25–35 g/m²	–	–	–	–	21.49	100 m²	**21.49**
Saline coastal; roadside areas							
Germinal GB; A21 Saline; 15–20 g/m²	–	–	–	–	12.70	100 m²	**12.70**
DLF Seeds Ltd; Promaster 90; 15–35 g/m²	–	–	–	–	22.43	100 m²	**22.43**
Turf production							
Germinal GB; A25 Meadow Ley; 160 kg/ha	–	–	–	–	1220.00	ha	**1220.00**
Germinal GB; A24 Wear & Tear; 185 kg/ha	–	–	–	–	1061.44	ha	**1061.44**
Market prices of wild flora seed mixtures							
Acid soils							
Germinal GB; WF1 (Annual Flowering); 1–2 g/m²	–	–	–	–	22.13	100 m²	**22.13**
Neutral soils							
Germinal GB; WF3 (Neutral Soils); 0.5–1 g/m²	–	–	–	–	11.06	100 m²	**11.06**
Market prices of wild flora and grass seed mixtures							
General purpose							
DLF Seeds Ltd; Pro Flora 8 Old English Country Meadow Mix; 5 g/m²	–	–	–	–	14.50	100 m²	**14.50**
DLF Seeds Ltd; Pro Flora 9 General purpose; 5 g/m²	–	–	–	–	11.50	100 m²	**11.50**
Acid soils							
Germinal GB; WFG2 (Annual Meadow); 5 g/m²	–	–	–	–	27.50	100 m²	**27.50**
DLF Seeds Ltd; Pro Flora 2 Acidic soils; 5 g/m²	–	–	–	–	12.25	100 m²	**12.25**
Neutral soils							
Germinal GB; WFG4 (Neutral Meadow); 5 g/m²	–	–	–	–	29.70	100 m²	**29.70**
DLF Seeds Ltd; Pro Flora 3 Damp loamy soils; 5 g/m²	–	–	–	–	17.15	100 m²	**17.15**

37 SOFT LANDSCAPING

Item – Overhead and Profit Included	PC £	Labour hours	Labour £	Plant £	Material £	Unit	Total rate £
Calcareous soils							
Germinal GB; WFG5 (Calcareous Soils); 5 g/m^2	–	–	–	–	28.10	100 m^2	**28.10**
DLF Seeds Ltd; Pro Flora 4 Calcareous soils; 5 g/m^2	–	–	–	–	14.40	100 m^2	**14.40**
Heavy clay soils							
Germinal GB; WFG6 (Clay Soils); 5 g/m^2	–	–	–	–	32.75	100 m^2	**32.75**
DLF Seeds Ltd; Pro Flora 5 Wet loamy soils; 5 g/m^2	–	–	–	–	27.50	100 m^2	**27.50**
Sandy soils							
Germinal GB; WFG7 (Free Draining Soils); 5 g/m^2	–	–	–	–	37.50	100 m^2	**37.50**
Germinal GB; WFG11 (Ireland); 5 g/m^2	–	–	–	–	28.00	100 m^2	**28.00**
DLF Seeds Ltd; Pro Flora 6 Dry free draining loamy soils; 5 g/m^2	–	–	–	–	30.55	100 m^2	**30.55**
Shaded areas							
Germinal GB; WFG8 (Woodland and Hedgerow); 5 g/m^2	–	–	–	–	29.25	100 m^2	**29.25**
DLF Seeds Ltd; Pro Flora 7 Hedgerow and light shade; 5 g/m^2	–	–	–	–	34.50	100 m^2	**34.50**
Educational							
Wetlands							
Germinal GB; WFG9 (Wetlands and Ponds); 5 g/m^2	–	–	–	–	35.00	100 m^2	**35.00**
DLF Seeds Ltd; Pro Flora 5 Wet loamy soils; 5 g/m^2	–	–	–	–	27.50	100 m^2	**27.50**
Scrub and moorland							
Germinal GB; WFG10 (Cornfield Annuals); 5 g/m^2	–	–	–	–	38.05	100 m^2	**38.05**
Hedgerow							
DLF Seeds Ltd; Pro Flora 7 Hedgerow and light shade; 5 g/m^2	–	–	–	–	34.50	100 m^2	**34.50**
Vacant sites							
DLF Seeds Ltd; Pro Flora 1 Cornfield annuals; 5 g/m^2	–	–	–	–	30.00	100 m^2	**30.00**
Regional Environmental mixes							
Germinal GB; RE1 (Traditional Hay); 5 g/m^2	–	–	–	–	29.95	100 m^2	**29.95**
Germinal GB; RE2 (Lowland Meadow); 5 g/m^2	–	–	–	–	37.75	100 m^2	**37.75**
Germinal GB; RE3 (Riverflood Plain/Water Meadow); 5 g/m^2	–	–	–	–	37.25	100 m^2	**37.25**

37 SOFT LANDSCAPING

Item – Overhead and Profit Included	PC £	Labour hours	Labour £	Plant £	Material £	Unit	Total rate £
MARKET PRICES OF SEEDING MATERIALS – cont							
Market prices of wild flora and grass seed mixtures – cont							
Regional Environmental mixes – cont							
Germinal GB; RE4 (Lowland Limestone); 5 g/m²	–	–	–	–	36.75	100 m²	**36.75**
Germinal GB; RE5 (Calcareous Sub-mountain Restoration); 5 g/m²	–	–	–	–	41.10	100 m²	**41.10**
Germinal GB; RE6 (Upland Limestone); 5 g/m²	–	–	–	–	51.45	100 m²	**51.45**
Germinal GB; RE7 (Acid Sub-mountain Restoration); 5 g/m²	–	–	–	–	36.25	100 m²	**36.25**
Germinal GB; RE8 (Coastal Reclamation); 5 g/m²	–	–	–	–	45.45	100 m²	**45.45**
Germinal GB; RE9 (Farmland Mixture); 5 g/m²	–	–	–	–	32.85	100 m²	**32.85**
Germinal GB; RE10 (Marginal Land); 5 g/m²	–	–	–	–	46.75	100 m²	**46.75**
Germinal GB; RE11 (Heath Scrubland); 5 g/m²	–	–	–	–	17.80	100 m²	**17.80**
Germinal GB; RE12 (Drought Land); 5 g/m²	–	–	–	–	–	100 m²	**-**
CULTIVATION AND SOIL PREPARATION							
Cultivation by tractor; Agripower Ltd							
Ripping up subsoil; using approved subsoiling machine; minimum depth 250 mm below topsoil; at 1.20 m centres; in							
gravel or sandy clay	–	–	–	9.41	–	100 m²	**9.41**
soil compacted by machines	–	–	–	10.44	–	100 m²	**10.44**
clay	–	–	–	13.43	–	100 m²	**13.43**
chalk or other soft rock	–	–	–	15.67	–	100 m²	**15.67**
Extra for subsoiling at 1 m centres	–	–	–	2.69	–	100 m²	**2.69**
Breaking up existing ground; using tractor drawn Blec ground preparation equipment in one operation							
Single pass							
100 mm deep	–	–	–	13.03	–	100 m²	**13.03**
150 mm deep	–	–	–	16.28	–	100 m²	**16.28**
200 mm deep	–	–	–	18.61	–	100 m²	**18.61**

37 SOFT LANDSCAPING

Item – Overhead and Profit Included	PC £	Labour hours	Labour £	Plant £	Material £	Unit	Total rate £
Cultivating ploughed ground; using disc, drag or chain harrow							
4 passes	–	–	–	14.80	–	100 m²	**14.80**
Rolling cultivated ground lightly; using self-propelled agricultural roller	–	0.07	2.15	0.85	–	100 m²	**3.00**
Cultivation by pedestrian operated rotavator							
Breaking up existing ground; tine cultivator or rotavator							
100 mm deep	–	0.28	8.86	1.76	–	100 m²	**10.62**
150 mm deep	–	0.31	10.12	2.01	–	100 m²	**12.13**
200 mm deep	–	0.40	12.88	2.56	–	100 m²	**15.44**
Breaking up existing ground; tine cultivator or rotavator but in heavy clay or wet soils							
100 mm deep	–	0.44	14.17	2.82	–	100 m²	**16.99**
150 mm deep	–	0.66	21.25	4.23	–	100 m²	**25.48**
200 mm deep	–	0.82	26.57	5.29	–	100 m²	**31.86**
Importing and storing selected and approved topsoil; inclusive of settlement							
small quantities (less than 15 m³)	80.34	–	–	–	92.39	m³	**92.39**
over 15 m³	29.66	–	–	–	34.11	m³	**34.11**
Spreading and lightly consolidating approved topsoil (imported or from spoil heaps); in layers not exceeding 150 mm; travel distance from spoil heaps not exceeding 100 m; by machine (imported topsoil not included)							
minimum depth 100 mm	–	1.55	49.91	64.54	–	100 m²	**114.45**
minimum depth 150 mm	–	2.33	75.14	97.13	–	100 m²	**172.27**
minimum depth 300 mm	–	4.67	150.26	194.25	–	100 m²	**344.51**
minimum depth 450 mm	–	6.99	225.08	291.05	–	100 m²	**516.13**
Spreading and lightly consolidating approved topsoil (imported or from spoil heaps); in layers not exceeding 150 mm; travel distance from spoil heaps not exceeding 100 m; by hand (imported topsoil not included)							
minimum depth 100 mm	–	20.00	644.13	–	–	100 m²	**644.13**
minimum depth 150 mm	–	30.01	966.20	–	–	100 m²	**966.20**
minimum depth 300 mm	–	60.01	1932.39	–	–	100 m²	**1932.39**
minimum depth 450 mm	–	90.02	2898.57	–	–	100 m²	**2898.57**
Extra over for spreading topsoil to slopes 15–30°; by machine or hand	–	–	–	–	–	10%	–
Extra over for spreading topsoil to slopes over 30°; by machine or hand	–	–	–	–	–	25%	–

Prices for Measured Works – Major Works

37 SOFT LANDSCAPING

Item – Overhead and Profit Included	PC £	Labour hours	Labour £	Plant £	Material £	Unit	Total rate £
CULTIVATION AND SOIL PREPARATION – cont							
Cultivation by pedestrian operated rotavator – cont							
Extra over for spreading topsoil from spoil heaps; travel exceeding 100 m; by machine							
100–150 m	–	0.01	0.39	0.07	–	m³	**0.46**
150–200 m	–	0.02	0.60	0.10	–	m³	**0.70**
200–300 m	–	0.03	0.90	0.15	–	m³	**1.05**
Extra over spreading topsoil for travel exceeding 100 m; by hand							
100 m	–	1.20	38.64	–	–	m³	**38.64**
200 m	–	1.67	53.67	–	–	m³	**53.67**
300 m	–	2.50	80.50	–	–	m³	**80.50**
Evenly grading; to general surfaces to bring to finished levels							
by machine (tractor mounted rotavator)	–	–	–	0.06	–	m²	**0.06**
by pedestrian operated rotavator	–	–	0.13	0.03	–	m²	**0.16**
by hand	–	0.01	0.32	–	–	m²	**0.32**
Extra over grading for slopes 15–30°; by machine or hand	–	–	–	–	–	10%	**–**
Extra over grading for slopes over 30°; by machine or hand	–	–	–	–	–	25%	**–**
Apply screened topdressing to grass surfaces; spread using Tru-Lute							
sand soil mixes 90/10 to 50/50	–	–	0.07	0.10	0.17	m²	**0.34**
Spread only existing cultivated soil to final levels using Tru-Lute							
cultivated soil	–	–	0.07	0.10	–	m²	**0.17**
Clearing stones; disposing off site							
by hand; stones not exceeding 50 mm in any direction; loading to skip 4.6 m³	–	0.01	0.32	0.05	–	m²	**0.37**
by mechanical stone rake; stones not exceeding 50 mm in any direction; loading to 15 m³ truck by mechanical loader	–	–	0.07	0.14	0.02	m²	**0.23**
Lightly cultivating; weeding; to fallow areas; disposing debris off site							
by hand	–	0.01	0.46	–	0.20	m²	**0.66**

37 SOFT LANDSCAPING

Item – Overhead and Profit Included	PC £	Labour hours	Labour £	Plant £	Material £	Unit	Total rate £
SOIL ADDITIVES AND IMPROVEMENTS							
Surface applications; soil additives; pre-seeding; material delivered to a maximum of 25 m from area of application; applied; by machine							
Soil conditioners; to cultivated ground; mushroom compost; delivered in 25 m³ loads; including turning in							
1 m³ per 40 m² = 25 mm thick	0.66	0.02	0.53	0.22	0.76	m²	**1.51**
1 m³ per 20 m² = 50 mm thick	1.32	0.03	0.99	0.31	1.52	m²	**2.82**
1 m³ per 13.33 m² = 75 mm thick	1.98	0.04	1.29	0.41	2.28	m²	**3.98**
1 m³ per 10 m² = 100 mm thick	2.65	0.05	1.61	0.52	3.05	m²	**5.18**
Soil conditioners; to cultivated ground; mushroom compost; delivered in 35 m³ loads; including turning in							
1 m³ per 40 m² = 25 mm thick	0.66	0.02	0.53	0.22	0.76	m²	**1.51**
1 m³ per 20 m² = 50 mm thick	1.32	0.03	0.99	0.31	1.52	m²	**2.82**
1 m³ per 13.33 m² = 75 mm thick	1.98	0.04	1.29	0.41	2.28	m²	**3.98**
1 m³ per 10 m² = 100 mm thick	2.65	0.05	1.61	0.52	3.05	m²	**5.18**
Surface applications and soil additives; pre-seeding; material delivered to a maximum of 25 m from area of application; applied; by hand							
Soil conditioners; to cultivated ground; mushroom compost; delivered in 25 m³ loads; including turning in							
1 m³ per 40 m² = 25 mm thick	0.66	0.02	0.71	–	0.76	m²	**1.47**
1 m³ per 20 m² = 50 mm thick	1.32	0.04	1.43	–	1.52	m²	**2.95**
1 m³ per 13.33 m² = 75 mm thick	1.98	0.07	2.14	–	2.28	m²	**4.42**
1 m³ per 10 m² = 100 mm thick	2.65	0.08	2.58	–	3.05	m²	**5.63**
Soil conditioners; to cultivated ground; mushroom compost; delivered in 60 m³ loads; including turning in							
1 m³ per 40 m² = 25 mm thick	0.37	0.02	0.71	–	0.43	m²	**1.14**
1 m³ per 20 m² = 50 mm thick	0.65	0.04	1.43	–	0.75	m²	**2.18**
1 m³ per 13.33 m² = 75 mm thick	1.11	0.07	2.14	–	1.28	m²	**3.42**
1 m³ per 10 m² = 100 mm thick	1.48	0.08	2.58	–	1.70	m²	**4.28**

37 SOFT LANDSCAPING

Item – Overhead and Profit Included	PC £	Labour hours	Labour £	Plant £	Material £	Unit	Total rate £
PREPARATION OF SEED BEDS							
Preparation of seedbeds – General							
Preamble: For preliminary operations see 'Cultivation' section.							
Preparation of seedbeds; soil preparation							
Lifting selected and approved topsoil from spoil heaps; passing through 6 mm screen; removing debris	–	0.08	2.68	5.69	0.05	m³	8.42
Topsoil; supply only; PC £24.72/m³; Calculations allow for 20% settlement							
25 mm	–	–	–	–	0.85	m²	0.85
50 mm	–	–	–	–	1.70	m²	1.70
100 mm	–	–	–	–	3.42	m²	3.42
150 mm	–	–	–	–	5.12	m²	5.12
200 mm	–	–	–	–	6.82	m²	6.82
250 mm	–	–	–	–	8.53	m²	8.53
300 mm	–	–	–	–	10.23	m²	10.23
400 mm	–	–	–	–	13.65	m²	13.65
450 mm	–	–	–	–	15.35	m²	15.35
Spreading topsoil to form seedbeds (topsoil not included); by machine							
25 mm deep	–	–	0.08	0.16	–	m²	0.24
50 mm deep	–	–	0.10	0.22	–	m²	0.32
75 mm deep	–	–	0.13	0.25	–	m²	0.38
100 mm deep	–	0.01	0.16	0.32	–	m²	0.48
150 mm deep	–	0.01	0.24	0.49	–	m²	0.73
Spreading only topsoil to form seedbeds (topsoil not included); by hand							
25 mm deep	–	0.03	0.81	–	–	m²	0.81
50 mm deep	–	0.03	1.08	–	–	m²	1.08
75 mm deep	–	0.04	1.38	–	–	m²	1.38
100 mm deep	–	0.05	1.61	–	–	m²	1.61
150 mm deep	–	0.08	2.42	–	–	m²	2.42
Bringing existing topsoil to a fine tilth for seeding; by raking or harrowing; stones not to exceed 6 mm; by machine	–	–	–	0.08	–	m²	0.08
Bringing existing topsoil to a fine tilth for seeding; by raking; stones not to exceed 6 mm; by hand	–	0.01	0.46	–	–	m²	0.46

37 SOFT LANDSCAPING

Item – Overhead and Profit Included	PC £	Labour hours	Labour £	Plant £	Material £	Unit	Total rate £
Preparation of seedbeds; soil treatments							
For the following operations add or subtract the following amounts for every £0.10 difference in the material cost price							
35 g/m²	–	–	–	–	0.40	100 m²	0.40
50 g/m²	–	–	–	–	0.58	100 m²	0.58
70 g/m²	–	–	–	–	0.81	100 m²	0.81
100 g/m²	–	–	–	–	1.15	100 m²	1.15
125 kg/ha	–	–	–	–	14.38	ha	14.38
150 kg/ha	–	–	–	–	17.25	ha	17.25
175 kg/ha	–	–	–	–	20.13	ha	20.13
200 kg/ha	–	–	–	–	23.00	ha	23.00
225 kg/ha	–	–	–	–	25.88	ha	25.88
250 kg/ha	–	–	–	–	28.75	ha	28.75
300 kg/ha	–	–	–	–	34.50	ha	34.50
350 kg/ha	–	–	–	–	40.25	ha	40.25
400 kg/ha	–	–	–	–	46.00	ha	46.00
500 kg/ha	–	–	–	–	57.50	ha	57.50
700 kg/ha	–	–	–	–	80.50	ha	80.50
1000 kg/ha	–	–	–	–	115.00	ha	115.00
1250 kg/ha	–	–	–	–	143.75	ha	143.75
Pre-seeding fertilizers (12+00+09); PC £0.65/kg; to seedbeds; by machine							
35 g/m²	2.29	–	–	0.79	2.63	100 m²	3.42
50 g/m²	3.27	–	–	0.79	3.76	100 m²	4.55
70 g/m²	4.58	–	–	0.79	5.27	100 m²	6.06
100 g/m²	6.54	–	–	0.79	7.52	100 m²	8.31
Pre-seeding fertilizers (12+00+09); PC £0.65/kg; to seedbeds; by hand							
35 g/m²	2.29	0.17	5.37	–	2.63	100 m²	8.00
50 g/m²	3.27	0.17	5.37	–	3.76	100 m²	9.13
70 g/m²	4.58	0.17	5.37	–	5.27	100 m²	10.64
100 g/m²	6.54	0.20	6.44	–	7.52	100 m²	13.96
SEEDING							
Seeding labours only in two operations; by machine (for seed prices see above)							
35 g/m²	–	–	–	1.35	–	100 m²	1.35
Grass seed; spreading in two operations; PC £4.50/kg (for changes in material prices please refer to table above); by machine							
35 g/m²	–	–	–	1.35	18.11	100 m²	19.46
50 g/m²	–	–	–	1.35	25.88	100 m²	27.23
70 g/m²	–	–	–	1.35	36.22	100 m²	37.57

37 SOFT LANDSCAPING

Item – Overhead and Profit Included	PC £	Labour hours	Labour £	Plant £	Material £	Unit	Total rate £
SEEDING – cont							
Seeding labours only in two operations – cont							
Grass seed – cont							
100 g/m²	–	–	–	1.35	51.75	100 m²	**53.10**
125 kg/ha	–	–	–	134.23	646.88	ha	**781.11**
150 kg/ha	–	–	–	134.23	776.25	ha	**910.48**
200 kg/ha	–	–	–	134.23	1035.00	ha	**1169.23**
250 kg/ha	–	–	–	134.23	1293.75	ha	**1427.98**
300 kg/ha	–	–	–	134.23	1552.50	ha	**1686.73**
350 kg/ha	–	–	–	134.23	1811.25	ha	**1945.48**
400 kg/ha	–	–	–	134.23	2070.00	ha	**2204.23**
500 kg/ha	–	–	–	134.23	2587.50	ha	**2721.73**
700 kg/ha	–	–	–	134.23	3622.50	ha	**3756.73**
Extra over seeding by machine for slopes over 30° (allowing for the actual area but measured in plan)							
35 g/m²	–	–	–	0.20	2.71	100 m²	**2.91**
50 g/m²	–	–	–	0.20	3.89	100 m²	**4.09**
70 g/m²	–	–	–	0.20	5.43	100 m²	**5.63**
100 g/m²	–	–	–	0.20	7.76	100 m²	**7.96**
125 kg/ha	–	–	–	20.14	97.04	ha	**117.18**
150 kg/ha	–	–	–	20.14	116.44	ha	**136.58**
200 kg/ha	–	–	–	20.14	155.25	ha	**175.39**
250 kg/ha	–	–	–	20.14	194.06	ha	**214.20**
300 kg/ha	–	–	–	20.14	232.88	ha	**253.02**
350 kg/ha	–	–	–	20.14	271.69	ha	**291.83**
400 kg/ha	–	–	–	20.14	310.50	ha	**330.64**
500 kg/ha	–	–	–	20.14	388.13	ha	**408.27**
700 kg/ha	–	–	–	20.14	543.38	ha	**563.52**
Seeding labours only in two operations; by hand (for seed prices see above)							
35 g/m²	–	0.17	5.37	–	–	100 m²	**5.37**
Grass seed; spreading in two operations; PC £4.50/kg (for changes in material prices please refer to table above); by hand							
35 g/m²	–	0.17	5.37	–	18.11	100 m²	**23.48**
50 g/m²	–	0.17	5.37	–	25.88	100 m²	**31.25**
70 g/m²	–	0.17	5.37	–	36.22	100 m²	**41.59**
100 g/m²	–	0.20	6.44	–	51.75	100 m²	**58.19**
125 g/m²	–	0.20	6.44	–	64.69	100 m²	**71.13**
Extra over seeding by hand for slopes over 30° (allowing for the actual area but measured in plan)							
35 g/m²	2.34	–	0.12	–	2.69	100 m²	**2.81**
50 g/m²	3.38	–	0.12	–	3.89	100 m²	**4.01**
70 g/m²	4.72	–	0.12	–	5.43	100 m²	**5.55**

37 SOFT LANDSCAPING

Item – Overhead and Profit Included	PC £	Labour hours	Labour £	Plant £	Material £	Unit	Total rate £
100 g/m²	6.75	–	0.14	–	7.76	100 m²	**7.90**
125 g/m²	8.41	–	0.14	–	9.67	100 m²	**9.81**
Harrowing seeded areas; light chain harrow	–	–	–	0.41	–	100 m²	**0.41**
Raking over seeded areas							
by mechanical stone rake	–	–	–	5.75	–	100 m²	**5.75**
by hand	–	0.80	25.76	–	–	100 m²	**25.76**
Rolling seeded areas; light roller							
by tractor drawn roller	–	–	–	2.53	–	100 m²	**2.53**
by pedestrian operated							
mechanical roller	–	0.08	2.68	0.51	–	100 m²	**3.19**
by hand drawn roller	–	0.17	5.37	–	–	100 m²	**5.37**
Extra over harrowing, raking or rolling seeded areas for slopes over 30°; by machine or hand	–	–	–	–	–	25%	**–**
Turf edging; to seeded areas; 300 mm wide	–	0.05	1.53	–	4.77	m²	**6.30**
PREPARATION OF TURF BEDS							
Preparation of turf beds							
Rolling turf to be lifted; lifting by hand or mechanical turf stripper; stacks to be not more than 1 m high							
cutting only preparing to lift; pedestrian turf cutter	–	0.75	24.15	16.10	–	100 m²	**40.25**
lifting and stacking; by hand	–	8.33	268.33	–	–	100 m²	**268.33**
Rolling up; moving to stacks							
distance not exceeding 100 m	–	2.50	80.50	–	–	100 m²	**80.50**
extra over rolling and moving turf to stacks to transport per additional 100 m	–	0.83	26.83	–	–	100 m²	**26.83**
Lifting selected and approved topsoil from spoil heaps							
passing through 6 mm screen; removing debris	–	0.17	5.37	11.39	–	m³	**16.76**
Extra over lifting topsoil and passing through screen for imported topsoil; plus 20% allowance for settlement	–	–	–	–	34.11	m³	**34.11**
Topsoil; PC £24.72/m³; plus 20% allowance for settlement							
25 mm deep	–	–	–	–	0.85	m²	**0.85**
50 mm deep	–	–	–	–	1.70	m²	**1.70**
100 mm deep	–	–	–	–	3.42	m²	**3.42**
150 mm deep	–	–	–	–	5.12	m²	**5.12**
200 mm deep	–	–	–	–	6.82	m²	**6.82**
250 mm deep	–	–	–	–	8.53	m²	**8.53**
300 mm deep	–	–	–	–	10.23	m²	**10.23**
400 mm deep	–	–	–	–	13.65	m²	**13.65**
450 mm deep	–	–	–	–	15.35	m²	**15.35**

37 SOFT LANDSCAPING

Item – Overhead and Profit Included	PC £	Labour hours	Labour £	Plant £	Material £	Unit	Total rate £
PREPARATION OF TURF BEDS – cont							
Preparation of turf beds – cont							
Spreading topsoil to form turf beds (topsoil not included); by machine							
25 mm deep	–	–	–	0.28	–	m²	**0.28**
50 mm deep	–	–	–	0.37	–	m²	**0.37**
75 mm deep	–	–	–	0.43	–	m²	**0.43**
100 mm deep	–	–	–	0.55	–	m²	**0.55**
150 mm deep	–	–	–	0.84	–	m²	**0.84**
Spreading topsoil to form turf beds (topsoil not included); by hand							
25 mm deep	–	0.03	0.81	–	–	m²	**0.81**
50 mm deep	–	0.03	1.08	–	–	m²	**1.08**
75 mm deep	–	0.04	1.38	–	–	m²	**1.38**
100 mm deep	–	0.05	1.61	–	–	m²	**1.61**
150 mm deep	–	0.08	2.42	–	–	m²	**2.42**
Cultivation by pedestrian operated rotavator							
Breaking up existing ground; tine cultivator or rotavator							
100 mm deep	–	0.28	8.86	1.76	–	100 m²	**10.62**
150 mm deep	–	0.31	10.12	2.01	–	100 m²	**12.13**
200 mm deep	–	0.40	12.88	2.56	–	100 m²	**15.44**
Breaking up existing ground; tine cultivator or rotavator in heavy clay or wet soils							
100 mm deep	–	0.44	14.17	2.82	–	100 m²	**16.99**
150 mm deep	–	0.66	21.25	4.23	–	100 m²	**25.48**
200 mm deep	–	0.82	26.57	5.29	–	100 m²	**31.86**
Bringing existing topsoil to a fine tilth for turfing by raking or harrowing; stones not to exceed 6 mm; by machine	–	–	0.13	0.08	–	m²	**0.21**
Bringing existing topsoil to a fine tilth for turfing by raking; stones not to exceed 6 mm; by hand							
general commercial or amenity turf	–	0.01	0.40	–	–	m²	**0.40**
higher grade fine culivation	–	0.02	0.64	–	–	m²	**0.64**

37 SOFT LANDSCAPING

Item – Overhead and Profit Included	PC £	Labour hours	Labour £	Plant £	Material £	Unit	Total rate £
TURFING							
Turfing (see also Preparation of Turf beds – above)							
Hand load and transport and unload turf from delivery area to laying area							
by small site dumper up to 50 m	–	0.01	0.64	0.04	–	m²	**0.68**
by hand 25 m	–	0.03	0.81	–	–	m²	**0.81**
by hand 50 m	–	0.03	0.92	–	–	m²	**0.92**
Turfing; laying only; to stretcher bond; butt joints; including providing and working from barrow plank runs where necessary to surfaces not exceeding 30° from horizontal							
specially selected lawn turves from previously lifted stockpile	–	0.08	2.42	–	–	m²	**2.42**
cultivated lawn turves; to large open areas	–	0.06	1.87	–	–	m²	**1.87**
cultivated lawn turves; to domestic or garden areas	–	0.08	2.51	–	–	m²	**2.51**
road verge quality turf	–	0.04	1.29	–	–	m²	**1.29**
Industrially grown turf; PC prices listed represent the general range of industrial turf prices for sportsfields and amenity purposes; prices will vary with quantity and site location							
Rolawn							
RB Medallion; sports fields, domestic lawns, general landscape; full loads 1720 m²	3.94	0.08	2.68	–	4.53	m²	**7.21**
RB Medallion; sports fields, domestic lawns, general landscape; part loads	4.15	0.08	2.68	–	4.77	m²	**7.45**
Tensar Ltd							
Tensar Mat 400 reinforced turf for embankments; laid to embankments 3.3 m² turves	4.22	0.11	3.54	–	4.85	m²	**8.39**
Inturf							
Inturf Masters; formal lawns, golf greens, bowling greens and low maintenance areas	6.42	0.10	3.07	–	7.38	m²	**10.45**
Inturf Classic; golf tees, surrounds, lawns, parks, general purpose landscaping areas and winter sports	5.42	0.05	1.69	–	6.23	m²	**7.92**
Turf laid in Big roll; Inturf							
Inturf Classic; golf tees, surrounds, lawns, parks, general purpose landscaping areas and winter sports	4.50	0.01	0.77	–	6.61	m²	**7.38**

37 SOFT LANDSCAPING

Item – Overhead and Profit Included	PC £	Labour hours	Labour £	Plant £	Material £	Unit	Total rate £
TURFING – cont							
Specialized turf products; Lindum Turf Ltd; Laid to prepared and cultivated soil (not included)							
Wildflower Meadow Turf; British wildflowers and grasses 50:50; P							
Flowering height 40–60 cm	9.74	0.08	2.68	–	11.93	m²	**14.61**
Instant use turf; LT6 Lindum turf; grown in an imported sand/soil root zone, harvested up to 40 mm thickness and installed using specialist machinery.							
Compatibility with existing root zones;							
for sports pitches or high wear areas where instant use is required inclusive of removal of existing root zone;							
minimum area 150 m²	10.50	–	–	–	31.20	m²	**31.20**
Reinforced turf; ABG ATS Advanced Turf; blended mesh elements incorporated into root zone; root zone spread and levelled over cultivated, prepared and reduced and levelled ground (not included); compacted with vibratory roller							
ABG ATS 400/B with selected turf and fertilizer 100 mm thick							
100–500 m²	28.00	0.03	1.07	2.83	32.20	m²	**36.10**
over 500 m²	26.00	0.03	1.07	2.83	29.90	m²	**33.80**
ABG ATS Turf 400/B with selected turf and fertilizer 150 mm thick							
100–500 m²	32.00	0.04	1.29	3.24	36.80	m²	**41.33**
over 500 m²	30.00	0.04	1.29	3.24	34.50	m²	**39.03**
ABG ATS Turf 400/B with selected turf and fertilizer 200 mm thick							
100–500 m²	38.00	0.05	1.61	3.85	43.70	m²	**49.16**
over 500 m²	36.00	0.05	1.61	3.85	41.40	m²	**46.86**
Firming turves with wooden beater	–	0.01	0.32	–	–	m²	**0.32**
Rolling turfed areas; light roller							
by tractor with turf tyres and roller	–	–	–	0.63	–	100 m²	**0.63**
by pedestrian operated mechanical roller	–	0.08	2.68	0.51	–	100 m²	**3.19**
by hand drawn roller	–	0.17	5.37	–	–	100 m²	**5.37**
Dressing with finely sifted topsoil; brushing into joints	0.05	0.05	1.61	–	0.06	m²	**1.67**

37 SOFT LANDSCAPING

Item – Overhead and Profit Included	PC £	Labour hours	Labour £	Plant £	Material £	Unit	Total rate £
Turfing; laying only							
to slopes over 30°; to diagonal bond (measured as plan area – add 15% to these rates for the incline area of 30° slopes)	–	0.12	3.86	–	–	m²	**3.86**
Extra over laying turfing for pegging down turves							
wooden or galvanized wire pegs; 200 mm long; 2 pegs per 0.50 m²	1.70	0.01	0.43	–	1.96	m²	**2.39**
ARTIFICIAL TURF							
Artificial grass; Easigrass Limited							
Base for area to receive artificial lawn							
Reduce area by 80 mm; lay 50 mm Type 1 and 50 mm sharp sand; remove excavated material from site	–	0.05	3.22	–	7.53	m²	**10.75**
Artificial lawns laid to cleared and levelled area; laid to 50 mm Type 1 drainage layer on 25 mm sharp sand bed and geofabric							
Lawn replacement areas							
Easi-Mayfair; 50 mm; deep pile;	–	–	–	–	–	m²	**38.49**
Easi-Belgravia; 40 mm; deep pile; realistic lawn	–	–	–	–	–	m²	**36.29**
Easi-Chelsea 40 mm	–	–	–	–	–	m²	**32.99**
Multi-use							
Easi-Kensington 35 mm deep pile	–	–	–	–	–	m²	**29.68**
Schools and Leisure							
Easi-Holland Park; 30 mm deep pile	–	–	–	–	–	m²	**28.91**
Patios, balconies terraces and golf; lighter loading solutions							
Easi-Wentworth; 15 mm pile height; 3.15 kg/m²	–	–	–	–	–	m²	**35.18**
Easi-Knightsbridge; 30 mm deep pile; 2.94 kg /m²	–	–	–	–	–	m²	**28.58**

37 SOFT LANDSCAPING

Item – Overhead and Profit Included	PC £	Labour hours	Labour £	Plant £	Material £	Unit	Total rate £
WILD FLOWER MEADOWS							
Note for this section; Prices exclude for delivery and collection of specialized machinery where used to the site.							
Topsoil stripping for wild flower planting							
Stripping to subsoil layer 300 mm deep							
moving to stockpile	–	–	–	2.46	–	m²	**2.46**
spreading locally	–	–	–	3.37	–	m²	**3.37**
Soil testing for wildflower meadows							
Collection of samples and submission to soil laboratory for analysis and recommendation for wildflower meadow establishment							
3 nr random samples	–	–	–	–	–	nr	**880.00**
Cultivation and preparation of seedbeds for wildflower meadows							
Cultivate stripped, filled or existing surface prior to application of herbicides and leave fallow for seedbank eradication period;							
by tractor drawn implements including rough grading to medium tilth	–	–	–	1.30	–	100 m²	**1.30**
by hand drawn rotavator including hand raking, and hand grading, to medium tilth	–	0.72	23.18	1.41	–	100 m²	**24.59**
Seedbank eradication for wildflower meadows							
Treat stripped intended meadow area to eradicate inherent seedbank; allow optimum 2 years of repeat eradication by herbicide returning regularly to reapply herbicides							
Existing or reclaimed brownfield area by tractor drawn application; 4 applications	–	0.04	1.29	2.02	0.31	100 m²	**3.62**
Existing or reclaimed brownfield area by tractor drawn application; 3 applications	–	0.03	0.97	1.52	0.23	100 m²	**2.72**
Existing paddocks or agricultural area; knapsack; 4 applications	–	2.12	68.26	–	0.31	100 m²	**68.57**

37 SOFT LANDSCAPING

Item – Overhead and Profit Included	PC £	Labour hours	Labour £	Plant £	Material £	Unit	Total rate £
Existing or reclaimed brownfield area by backpack spray application; 3 applications	–	1.59	51.20	–	0.23	100 m²	**51.43**
Subsoil filling to wildflower meadows							
Average thickness 300 mm; filling of poor nutrient value soil to form seedbed for wildflower area; grading to level and preparing to medium tilth							
from stockpile on site	–	–	–	26.65	–	100 m²	**26.65**
imported subsoil fill	–	–	–	26.65	0.85	100 m²	**27.50**
Seeding to wildflower meadows; Spreading of selected wildflower seed to manufacturer's specifications; by tractor drawn seeder							
General areas							
DLF Seeds Ltd; Pro Flora 8 Old English Country Meadow Mix; 5 g/m²	–	–	–	4.05	16.67	100 m²	**20.72**
Acid soils							
Germinal GB; WFG2 (Annual Meadow); 5 g/m²	–	–	–	4.05	31.63	100 m²	**35.68**
Neutral soils							
Germinal GB; WFG4 (Neutral Meadow); 5 g/m²	–	–	–	4.05	34.16	100 m²	**38.21**
DLF Seeds Ltd; Pro Flora 3 Damp loamy soils; 5 g/m²	–	–	–	4.05	19.72	100 m²	**23.77**
Heavy clay soils							
Germinal GB; WFG6 (Clay Soils); 5 g/m²	–	–	–	4.05	37.66	100 m²	**41.71**
Seeding to wildflower meadows; Spreading of selected wildflower seed to manufacturer's specifications; by pedestrian drawn seeder							
General areas							
DLF Seeds Ltd; Pro Flora 8 Old English Country Meadow Mix; 5 g/m²	–	0.25	8.05	–	16.67	100 m²	**24.72**
Acid soils							
Germinal GB; WFG2 (Annual Meadow); 5 g/m²	–	0.25	8.05	–	31.63	100 m²	**39.68**
Neutral soils							
Germinal GB; WFG4 (Neutral Meadow); 5 g/m²	–	0.25	8.05	–	34.16	100 m²	**42.21**
DLF Seeds Ltd; Pro Flora 3 Damp loamy soils; 5 g/m²	–	0.25	8.05	–	19.72	100 m²	**27.77**

37 SOFT LANDSCAPING

Item – Overhead and Profit Included	PC £	Labour hours	Labour £	Plant £	Material £	Unit	Total rate £
WILD FLOWER MEADOWS – cont							
Seeding to wildflower meadows – cont							
Heavy clay soils							
Germinal GB; WFG6 (Clay Soils); 5 g/m²	–	0.25	8.05	–	37.66	100 m²	**45.71**
Wildflower Turf; Lindum Turf							
Wildflower Meadow Turf; British wildflowers and grasses 50:50; P							
Flowering height 40–60 cm	9.74	0.08	2.68	–	11.93	m²	**14.61**
MAINTENANCE OF SEEDED/ TURFED AREAS							
Maintenance operations (Note: the following rates apply to aftercare maintenance executed as part of a landscaping contract only)							
Initial cutting; to turfed areas							
20 mm high; using pedestrian guided power driven cylinder mower; including boxing off cuttings (stone picking and rolling not included)	–	0.18	5.80	0.43	–	100 m²	**6.23**
Repairing damaged grass areas; scraping out; removing slurry; from ruts and holes; average							
100 mm deep	–	0.13	4.29	–	–	m²	**4.29**
100 mm topsoil	–	0.13	4.29	–	3.42	m²	**7.71**
Repairing damaged grass areas; sowing grass seed to match existing or as specified; to individually prepared worn patches							
35 g/m²	0.18	0.01	0.32	–	0.21	m²	**0.53**
50 g/m²	0.26	0.01	0.32	–	0.30	m²	**0.62**
Sweeping leaves; disposing off site; motorized vacuum sweeper or rotary brush sweeper							
areas of maximum 2500 m² with occasional large tree and established boundary planting; 4.6 m³ (1 skip of material to be removed)	–	0.40	12.88	4.48	–	100 m²	**17.36**

37 SOFT LANDSCAPING

Item – Overhead and Profit Included	PC £	Labour hours	Labour £	Plant £	Material £	Unit	Total rate £
Leaf clearance; clearing grassed area of leaves and other extraneous debris							
Using equipment towed by tractor							
large grassed areas with perimeters of mature trees such as sports fields and amenity areas	–	0.01	0.40	0.05	–	100 m²	0.45
large grassed areas containing ornamental trees and shrub beds	–	0.03	0.81	0.06	–	100 m²	0.87
Using pedestrian operated mechanical equipment and blowers							
grassed areas with perimeters of mature trees such as sports fields and amenity areas	–	0.04	1.29	0.04	–	100 m²	1.33
grassed areas containing ornamental trees and shrub beds	–	0.10	3.22	0.08	–	100 m²	3.30
verges	–	0.07	2.15	0.05	–	100 m²	2.20
By hand							
grassed areas with perimeters of mature trees such as sports fields and amenity areas	–	0.05	1.61	0.20	–	100 m²	1.81
grassed areas containing ornamental trees and shrub beds	–	0.08	2.68	0.33	–	100 m²	3.01
verges	–	1.00	32.20	4.03	–	100 m²	36.23
Removal of arisings							
areas with perimeters of mature trees	–	0.01	0.17	0.12	0.11	100 m²	0.40
areas containing ornamental trees and shrub beds	–	0.02	0.53	0.45	0.29	100 m²	1.27
Cutting grass to specified height; per cut							
multi-unit gang mower	–	0.59	18.94	22.25	–	ha	41.19
ride-on triple cylinder mower	–	0.01	0.45	0.17	–	100 m²	0.62
ride-on triple rotary mower	–	0.01	0.45	–	–	100 m²	0.45
pedestrian mower	–	0.18	5.80	1.01	–	100 m²	6.81
Cutting grass to banks; per cut							
side arm cutter bar mower	–	0.02	0.75	0.41	–	100 m²	1.16
Cutting rough grass; per cut							
power flail or scythe cutter	–	0.04	1.13	–	–	100 m²	1.13
Extra over cutting grass for slopes not exceeding 30°	–	–	–	–	–	10%	-
Extra over cutting grass for slopes exceeding 30°	–	–	–	–	–	40%	-
Cutting fine sward							
pedestrian operated seven-blade cylinder lawn mower	–	0.14	4.51	0.32	–	100 m²	4.83
Extra over cutting fine sward for boxing off cuttings							
pedestrian mower	–	0.03	0.90	0.07	–	100 m²	0.97

37 SOFT LANDSCAPING

Item – Overhead and Profit Included	PC £	Labour hours	Labour £	Plant £	Material £	Unit	Total rate £
MAINTENANCE OF SEEDED/ TURFED AREAS – cont							
Maintenance operations (Note: the following rates apply to aftercare maintenance executed as part of a landscaping contract only) – cont							
Cutting areas of rough grass							
scythe	–	1.00	32.20	–	–	100 m²	**32.20**
sickle	–	2.00	64.40	–	–	100 m²	**64.40**
petrol operated strimmer	–	0.30	9.67	0.69	–	100 m²	**10.36**
Cutting areas of rough grass which contain trees or whips							
petrol operated strimmer	–	0.40	12.88	0.92	–	100 m²	**13.80**
Extra over cutting rough grass for on site raking up and dumping	–	0.33	10.73	–	–	100 m²	**10.73**
Trimming edge of grass areas; edging tool							
with petrol powered strimmer	–	0.13	4.29	0.31	–	100 m	**4.60**
by hand	–	0.67	21.47	–	–	100 m	**21.47**
Marking out pitches using approved line marking compound; including initial setting out and marking							
discus, hammer, javelin or shot putt area	8.25	2.00	64.40	–	9.49	nr	**73.89**
cricket square	5.50	2.00	64.40	–	6.32	nr	**70.72**
cricket boundary	19.25	8.00	257.60	–	22.14	nr	**279.74**
grass tennis court	8.25	4.00	128.80	–	9.49	nr	**138.29**
hockey pitch	27.50	8.00	257.60	–	31.63	nr	**289.23**
football pitch	27.50	8.00	257.60	–	31.63	nr	**289.23**
rugby pitch	27.50	8.00	257.60	–	31.63	nr	**289.23**
eight lane running track; 400 m	55.00	16.00	515.20	–	63.25	nr	**578.45**
Re-marking out pitches using approved line marking compound							
discus, hammer, javelin or shot putt area	5.50	0.50	16.10	–	6.32	nr	**22.42**
cricket square	5.50	0.50	16.10	–	6.32	nr	**22.42**
grass tennis court	19.25	1.00	32.20	–	22.14	nr	**54.34**
hockey pitch	19.25	1.00	32.20	–	22.14	nr	**54.34**
football pitch	19.25	1.00	32.20	–	22.14	nr	**54.34**
rugby pitch	19.25	1.00	32.20	–	22.14	nr	**54.34**
eight lane running track; 400 m	55.00	2.50	80.50	–	63.25	nr	**143.75**
Rolling grass areas; light roller							
by tractor drawn roller	–	–	–	0.63	–	100 m²	**0.63**
by pedestrian operated mechanical roller	–	0.08	2.68	0.51	–	100 m²	**3.19**
by hand drawn roller	–	0.17	5.37	–	–	100 m²	**5.37**

37 SOFT LANDSCAPING

Item – Overhead and Profit Included	PC £	Labour hours	Labour £	Plant £	Material £	Unit	Total rate £
Aerating grass areas; to a depth of 100 mm							
using tractor-drawn aerator	–	0.06	1.87	5.80	–	100 m²	**7.67**
using pedestrian-guided motor powered solid or slitting tine turf aerator	–	0.18	5.64	2.14	–	100 m²	**7.78**
using hollow tine aerator; including sweeping up and dumping corings	–	0.50	16.10	4.29	–	100 m²	**20.39**
using hand aerator or fork	–	1.67	53.67	–	–	100 m²	**53.67**
Extra over aerating grass areas for on site sweeping up and dumping corings	–	0.17	5.37	–	–	100 m²	**5.37**
Switching off dew; from fine turf areas	–	0.20	6.44	–	–	100 m²	**6.44**
Scarifying grass areas to break up thatch; removing dead grass							
using tractor-drawn scarifier	–	0.07	2.25	2.18	–	100 m²	**4.43**
using self-propelled scarifier; including removing and disposing of grass on site	–	0.33	10.73	0.77	–	100 m²	**11.50**
Harrowing grass areas							
using drag mat	–	0.03	0.90	0.36	–	100 m²	**1.26**
using chain harrow	–	0.04	1.13	0.45	–	100 m²	**1.58**
using drag mat	–	2.80	90.17	35.49	–	ha	**125.66**
using chain harrow	–	3.50	112.71	44.37	–	ha	**157.08**
Extra for scarifying and harrowing grass areas for disposing excavated material off site; to tip; loading by machine							
slightly contaminated	–	–	–	3.46	53.16	m³	**56.62**
rubbish	–	–	–	3.46	53.16	m³	**56.62**
inert material	–	–	–	2.30	39.94	m³	**42.24**
For the following topsoil improvement and seeding operations add or subtract the following amounts for every £0.10 difference in the material cost price							
35 g/m²	–	–	–	–	0.40	100 m²	**0.40**
50 g/m²	–	–	–	–	0.58	100 m²	**0.58**
70 g/m²	–	–	–	–	0.81	100 m²	**0.81**
100 g/m²	–	–	–	–	1.15	100 m²	**1.15**
125 kg/ha	–	–	–	–	14.38	ha	**14.38**
150 kg/ha	–	–	–	–	17.25	ha	**17.25**
175 kg/ha	–	–	–	–	20.13	ha	**20.13**
200 kg/ha	–	–	–	–	23.00	ha	**23.00**
225 kg/ha	–	–	–	–	25.88	ha	**25.88**
250 kg/ha	–	–	–	–	28.75	ha	**28.75**
300 kg/ha	–	–	–	–	34.50	ha	**34.50**

37 SOFT LANDSCAPING

Item – Overhead and Profit Included	PC £	Labour hours	Labour £	Plant £	Material £	Unit	Total rate £
MAINTENANCE OF SEEDED/ TURFED AREAS – cont							
Maintenance operations (Note: the following rates apply to aftercare maintenance executed as part of a landscaping contract only) – cont							
For the following topsoil improvement – cont							
350 kg/ ha	–	–	–	–	40.25	ha	**40.25**
400 kg/ha	–	–	–	–	46.00	ha	**46.00**
500 kg/ha	–	–	–	–	57.50	ha	**57.50**
700 kg/ha	–	–	–	–	80.50	ha	**80.50**
1000 kg/ha	–	–	–	–	115.00	ha	**115.00**
1250 kg/ha	–	–	–	–	143.75	ha	**143.75**
Top dressing fertilizers (7+7+7); PC £1.12/kg; to seedbeds; by machine							
35 g/m^2	3.91	–	–	0.63	4.50	100 m^2	**5.13**
50 g/m^2	5.58	–	–	0.63	6.42	100 m^2	**7.05**
300 kg/ha	334.80	–	–	63.25	385.02	ha	**448.27**
350 kg/ha	390.60	–	–	63.25	449.19	ha	**512.44**
400 kg/ha	446.40	–	–	63.25	513.36	ha	**576.61**
500 kg/ha	558.00	–	–	101.20	641.70	ha	**742.90**
Top dressing fertilizers (7+7+7); PC £1.12/kg; to seedbeds; by hand							
35 g/m^2	3.91	0.17	5.37	–	4.50	100 m^2	**9.87**
50 g/m^2	5.58	0.17	5.37	–	6.42	100 m^2	**11.79**
70 g/m^2	7.81	0.17	5.37	–	8.98	100 m^2	**14.35**
Watering turf; evenly; at a rate of 5 l/m^2							
using movable spray lines powering 3 nr sprinkler heads with a radius of 15 m and allowing for 60% overlap (irrigation machinery costs not included)	–	0.02	0.51	–	–	100 m^2	**0.51**
using sprinkler equipment and with sufficient water pressure to run 1 nr 15 m radius sprinkler	–	0.02	0.64	–	–	100 m^2	**0.64**
using hand-held watering equipment	–	0.25	8.05	–	–	100 m^2	**8.05**

37 SOFT LANDSCAPING

Item – Overhead and Profit Included	PC £	Labour hours	Labour £	Plant £	Material £	Unit	Total rate £
PLANTING GENERAL							
Planting – General							
Preamble: Prices for all planting work are deemed to include carrying out planting in accordance with all good horticultural practice.							
MARKET PRICES OF MATERIALS							
Market prices of planting materials (Note: the rates shown generally reflect the manufacturer's recommended retail prices; trade and bulk discounts are often available on the prices shown)							
Imported topsoil; market prices; British Sugar Topsoil; sustainably sourced graded topsoil to British standards and with independent certification; rates shown allow for 20% settlement							
Landscape 20; BS3882:2007; sandy loam							
8 wheel delivery; 20 tonne loads (approximately 13.33 m³)	–	–	–	–	56.10	m³	**56.10**
articulated delivery	–	–	–	–	39.95	m³	**39.95**
Topsoil prices; Average locally sourced prices							
Topsoil							
general purpose	–	–	–	–	24.72	m³	**24.72**
Market prices of mulching materials							
Melcourt Industries Ltd; 25 m³ loads (items labelled FSC are Forest Stewardship Council certified, items marked FT are certified fire tested)							
Ornamental Bark Mulch FT	–	–	–	–	61.40	m³	**61.40**
Bark Nuggets® FT	–	–	–	–	59.20	m³	**59.20**
Amenity Bark Mulch FSC FT	–	–	–	–	43.40	m³	**43.40**
Contract Bark Mulch FSC	–	–	–	–	38.90	m³	**38.90**
Spruce Ornamental FSC FT	–	–	–	–	44.65	m³	**44.65**
Rustic Biomulch®	–	–	–	–	47.80	m³	**47.80**
Forest BioMulch®	–	–	–	–	40.80	m³	**40.80**

37 SOFT LANDSCAPING

Item – Overhead and Profit Included	PC £	Labour hours	Labour £	Plant £	Material £	Unit	Total rate £
MARKET PRICES OF MATERIALS – cont							
Market prices of mulching materials – cont							
Melcourt Industries Ltd; 70 m³ loads							
Contract Bark Mulch FSC	–	–	–	–	26.50	m³	26.50
Melcourt Industries Ltd; 80 m³ loads							
Ornamental Bark Mulch FT	–	–	–	–	48.40	m³	48.40
Bark Nuggets® FT	–	–	–	–	46.20	m³	46.20
Amenity Bark Mulch FSC	–	–	–	–	30.40	m³	30.40
Spruce Ornamental FSC FT	–	–	–	–	31.65	m³	31.65
Rustic Biomulch®	–	–	–	–	34.80	m³	34.80
Forest BioMulch®	–	–	–	–	27.80	m³	27.80
Mulch; Melcourt Industries Ltd; 25 m³ loads							
Composted Fine Bark FSC	–	–	–	–	42.65	m³	42.65
Humus 2000	–	–	–	–	32.10	m³	32.10
Spent Mushroom Compost	–	–	–	–	26.45	m³	26.45
Topgrow	–	–	–	–	45.80	m³	45.80
Mulch; Melcourt Industries Ltd; 50 m³ loads							
Humus 2000	–	–	–	–	21.60	m³	21.60
Mulch; Melcourt Industries Ltd; 60 m³ loads							
Spent Mushroom Compost	–	–	–	–	14.75	m³	14.75
Mulch; Melcourt Industries Ltd; 65 m³ loads							
Composted Fine Bark	–	–	–	–	30.65	m³	30.65
Topgrow	–	–	–	–	35.30	m³	35.30
Market prices of fertilizers							
Fertilizers; Germinal GB							
G1; 6+9+6	–	–	–	–	31.20	kg	31.20
G4; 11+5+5	–	–	–	–	33.50	kg	33.50
G6; 20+10+10	–	–	–	–	32.50	kg	32.50
G9; 12+2+9+Fe	–	–	–	–	34.75	kg	34.75
Fertilizers; Everris							
Enmag; controlled release fertilizer (8–9 months); 11+22+09; 70 g/m²	–	–	–	–	16.79	100 m²	16.79
Fertilizers; Everris; granular Osmocote Exact; 15+09+11; controlled release fertilizer; costs for recommended application rates							
transplant	–	–	–	–	0.18	nr	0.18
whip	–	–	–	–	0.27	nr	0.27
feathered	–	–	–	–	0.36	nr	0.36
light standard	–	–	–	–	0.36	nr	0.36
standard	–	–	–	–	0.45	nr	0.45
selected standard	–	–	–	–	0.63	nr	0.63
heavy standard	–	–	–	–	0.72	nr	0.72

37 SOFT LANDSCAPING

Item – Overhead and Profit Included	PC £	Labour hours	Labour £	Plant £	Material £	Unit	Total rate £
extra heavy standard	–	–	–	–	0.90	nr	**0.90**
16–18 cm girth	–	–	–	–	0.99	nr	**0.99**
18–20 cm girth	–	–	–	–	1.08	nr	**1.08**
20–22 cm girth	–	–	–	–	1.26	nr	**1.26**
22–24 cm girth	–	–	–	–	1.35	nr	**1.35**
24–26 cm girth	–	–	–	–	1.44	nr	**1.44**
Fertilizers; Everris							
TPMC tree planting and mulching compost	–	–	–	–	2.60	bag	**2.60**
Fertilizers; Rigby Taylor Ltd; fertilizer application rates 35 g/m^2 unless otherwise shown							
liquid fertilizer; Mascot Microflow-C; 25+0+0 + trace elements; 200–1400 ml/100 m^2	–	–	–	–	6.67	100 m^2	**6.67**
liquid fertilizer; Mascot Microflow-CX; 4+2+18 + trace elements; 200–1400 ml/100 m^2	–	–	–	–	6.42	100 m^2	**6.42**
liquid fertilizer; Mascot Microflow-CX; 12+0+8 + trace elements; 200–1400 ml/100 m^2	–	–	–	–	5.66	100 m^2	**5.66**
liquid fertilizer; Mascot Microflow-CX; 17+2+5 + trace elements; 200–1400 ml/100 m^2	–	–	–	–	6.37	100 m^2	**6.37**
Wetting agents; Rigby Taylor Ltd							
wetting agent; Breaker Advance Granules; 20 kg	–	–	–	–	16.00	100 m^2	**16.00**
PLANTING PREPARATION AND CULTIVATION							
Site protection; temporary protective fencing							
Cleft chestnut rolled fencing; to 100 mm dia. chestnut posts; driving into firm ground at 3 m centres; pales at 50 mm centres							
900 mm high	4.68	0.11	3.43	–	12.02	m	**15.45**
1100 mm high	6.52	0.11	3.43	–	14.13	m	**17.56**
Extra over temporary protective fencing for removing and making good (no allowance for reuse of material)	–	0.07	2.14	0.26	–	m	**2.40**

37 SOFT LANDSCAPING

Item – Overhead and Profit Included	PC £	Labour hours	Labour £	Plant £	Material £	Unit	Total rate £
PLANTING PREPARATION AND CULTIVATION – cont							
Cultivation by tractor; Agripower Ltd; Specialist subcontract prepared turf and sports area preparation							
Ripping up subsoil; using approved subsoiling machine; minimum depth 250 mm below topsoil; at 1.20 m centres; in							
gravel or sandy clay	–	–	–	7.43	–	100 m²	**7.43**
soil compacted by machines	–	–	–	8.25	–	100 m²	**8.25**
clay	–	–	–	10.61	–	100 m²	**10.61**
chalk or other soft rock	–	–	–	12.39	–	100 m²	**12.39**
Extra for subsoiling at 1 m centres	–	–	–	2.13	–	100 m²	**2.13**
Breaking up existing ground; using tractor drawn tine cultivator or rotavator with stone burier							
Single pass							
100 mm deep	–	–	–	1.30	–	100 m²	**1.30**
150 mm deep	–	–	–	1.63	–	100 m²	**1.63**
200 mm deep	–	–	–	2.16	–	100 m²	**2.16**
Cultivating ploughed ground; using disc, drag or chain harrow							
4 passes	–	–	–	–	–	100 m²	**–**
Cultivation; pedestrian operated rotavator							
Breaking up existing ground; tine cultivator or rotavator							
100 mm deep	–	0.22	7.08	1.41	–	100 m²	**8.49**
150 mm deep	–	0.28	8.86	1.76	–	100 m²	**10.62**
200 mm deep	–	0.37	11.80	2.35	–	100 m²	**14.15**
Breaking up existing ground; tine cultivator or rotavator in heavy clay or wet soils							
100 mm deep	–	0.44	14.17	2.82	–	100 m²	**16.99**
150 mm deep	–	0.66	21.25	4.23	–	100 m²	**25.48**
200 mm deep	–	0.82	26.57	5.29	–	100 m²	**31.86**
Clearing stones; disposing off site by hand; stones not exceeding 50 mm in any direction; loading to skip 4.6 m³	–	0.01	0.32	0.05	–	m²	**0.37**
by mechanical stone rake; stones not exceeding 50 mm in any direction; loading to 15 m³ truck by mechanical loader	–	–	0.07	0.14	0.02	m²	**0.23**

37 SOFT LANDSCAPING

Item – Overhead and Profit Included	PC £	Labour hours	Labour £	Plant £	Material £	Unit	Total rate £
Lightly cultivating; weeding; to fallow areas; disposing debris off site							
by hand	–	0.01	0.46	–	0.20	m²	**0.66**
Cultivation by hand							
Dig over surface of area to be planted; turn soil and level for palnting to a depth of							
100 mm	–	0.04	1.29	–	–	m²	**1.29**
200 mm	–	0.07	2.15	–	–	m²	**2.15**
300 mm	–	0.13	4.03	–	–	m²	**4.03**
500 mm	–	0.33	10.73	–	–	m²	**10.73**
TOPSOIL FILLING							
Topsoil filling; Friedland Horticulture							
Market prices of topsoil; price shown includes a 20% settlement factor							
topsoil to BS3882	29.66	–	–	–	34.11	m³	**34.11**
small quantities (less than 15 m³)	80.34	–	–	–	92.39	m³	**92.39**
Spreading and lightly consolidating approved topsoil (imported or from spoil heaps); in layers not exceeding 150 mm; travel distance from spoil heaps not exceeding 100 m; by machine (imported topsoil not included)							
minimum depth 100 mm	–	1.55	49.91	64.54	–	100 m²	**114.45**
minimum depth 150 mm	–	2.33	75.14	97.13	–	100 m²	**172.27**
minimum depth 300 mm	–	4.67	150.26	194.25	–	100 m²	**344.51**
minimum depth 450 mm	–	6.99	225.08	291.05	–	100 m²	**516.13**
Spreading and lightly consolidating approved topsoil (imported or from spoil heaps); in layers not exceeding 150 mm; travel distance from spoil heaps not exceeding 100 m; by hand (imported topsoil not included)							
minimum depth 100 mm	–	20.00	644.13	–	–	100 m²	**644.13**
minimum depth 150 mm	–	30.01	966.20	–	–	100 m²	**966.20**
minimum depth 300 mm	–	60.01	1932.39	–	–	100 m²	**1932.39**
minimum depth 450 mm	–	90.02	2898.57	–	–	100 m²	**2898.57**
Spreading topsoil; from dump not exceeding 100 m distance; by machine (topsoil not included)							
at 1 m³ per 13 m²; 75 mm thick	–	0.56	17.89	47.83	–	100 m²	**65.72**
at 1 m³ per 10 m²; 100 mm thick	–	0.89	28.62	48.27	–	100 m²	**76.89**
at 1 m³ per 6.50 m²; 150 mm thick	–	1.11	35.74	85.58	–	100 m²	**121.32**
at 1 m³ per 5 m²; 200 mm thick	–	1.48	47.66	113.76	–	100 m²	**161.42**

37 SOFT LANDSCAPING

Item – Overhead and Profit Included	PC £	Labour hours	Labour £	Plant £	Material £	Unit	Total rate £
TOPSOIL FILLING – cont							
Topsoil filling – cont							
Spreading topsoil; from dump not exceeding 100 m distance; by hand (topsoil not included)							
at 1 m³ per 13 m²; 75 mm thick	–	15.00	483.09	–	–	100 m²	483.09
at 1 m³ per 10 m²; 100 mm thick	–	20.00	644.13	–	–	100 m²	644.13
at 1 m³ per 6.50 m²; 150 mm thick	–	30.01	966.20	–	–	100 m²	966.20
at 1 m³ per 5 m²; 200 mm thick	–	36.67	1180.90	–	–	100 m²	1180.90
Imported topsoil; tipped 100 m from area of application; by machine							
at 1 m³ per 13 m²; 75 mm thick	222.48	0.56	17.89	47.83	255.85	100 m²	321.57
at 1 m³ per 10 m²; 100 mm thick	296.64	0.89	28.62	48.27	341.14	100 m²	418.03
at 1 m³ per 6.50 m²; 150 mm thick	444.96	1.11	35.74	85.58	511.70	100 m²	633.02
at 1 m³ per 5 m²; 200 mm thick	593.28	1.48	47.66	113.76	682.27	100 m²	843.69
Imported topsoil; tipped 100 m from area of application; by hand							
at 1 m³ per 13 m²; 75 mm thick	222.48	15.00	483.09	–	255.85	100 m²	738.94
at 1 m³ per 10 m²; 100 mm thick	296.64	20.00	644.13	–	341.14	100 m²	985.27
at 1 m³ per 6.50 m²; 150 mm thick	444.96	30.01	966.20	–	511.70	100 m²	1477.90
at 1 m³ per 5 m²; 200 mm thick	593.28	36.67	1180.90	–	682.27	100 m²	1863.17
Extra over for spreading topsoil to slopes 15–30°; by machine or hand	–	–	–	–	–	10%	–
Extra over for spreading topsoil to slopes over 30°; by machine or hand	–	–	–	–	–	25%	–
Extra over for spreading topsoil from spoil heaps; travel exceeding 100 m; by machine							
100–150 m	–	0.01	0.39	0.07	–	m³	0.46
150–200 m	–	0.02	0.60	0.10	–	m³	0.70
200–300 m	–	0.03	0.90	0.15	–	m³	1.05
Extra over spreading topsoil for travel exceeding 100 m; by hand							
100 m	–	0.83	26.84	–	–	m³	26.84
200 m	–	1.67	53.67	–	–	m³	53.67
300 m	–	2.50	80.50	–	–	m³	80.50
Evenly grading; to general surfaces to bring to finished levels							
by machine (tractor mounted rotavator)	–	–	–	0.07	–	m²	0.07
by pedestrian operated rotavator	–	–	0.13	0.03	–	m²	0.16
by hand	–	0.01	0.32	–	–	m²	0.32
Extra over grading for slopes 15–30°; by machine or hand	–	–	–	–	–	10%	–
Extra over grading for slopes over 30°; by machine or hand	–	–	–	–	–	25%	–
Apply screened topdressing to grass surfaces; spread using Tru-Lute							
sand soil mixes 90/10 to 50/50	–	–	0.07	0.10	0.17	m²	0.34

37 SOFT LANDSCAPING

Item – Overhead and Profit Included	PC £	Labour hours	Labour £	Plant £	Material £	Unit	Total rate £
Spread only existing cultivated soil to final levels using Tru-Lute cultivated soil	–	–	0.07	0.10	–	m²	**0.17**
SOIL IMPROVEMENTS							
Preparation of planting operations For the following topsoil improvement and planting operations add or subtract the following amounts for every £0.10 difference in the material cost price							
35 g/m²	–	–	–	–	0.40	100 m²	**0.40**
50 g/m²	–	–	–	–	0.58	100 m²	**0.58**
70 g/m²	–	–	–	–	0.81	100 m²	**0.81**
100 g/m²	–	–	–	–	1.15	100 m²	**1.15**
150 kg/ha	–	–	–	–	17.25	ha	**17.25**
200 kg/ha	–	–	–	–	23.00	ha	**23.00**
250 kg/ha	–	–	–	–	28.75	ha	**28.75**
300 kg/ha	–	–	–	–	34.50	ha	**34.50**
400 kg/ha	–	–	–	–	46.00	ha	**46.00**
500 kg/ha	–	–	–	–	57.50	ha	**57.50**
700 kg/ha	–	–	–	–	80.50	ha	**80.50**
1000 kg/ha	–	–	–	–	115.00	ha	**115.00**
1250 kg/ha	–	–	–	–	143.75	ha	**143.75**
Fertilizers; in top 150 mm of topsoil; at 35 g/m²							
Mascot Microfine; controlled release turf fertilizer; 8+0+6 + 2% Mg + 4% Fe	0.05	–	–	–	–	m²	**–**
Enmag CRF 11–21–9+6MgO 8–9 Months Planting Fertilizer 25 kg	0.09	–	0.05	–	0.10	m²	**0.15**
Mascot Outfield; turf fertilizer; 4+10+10	0.03	–	0.05	–	0.04	m²	**0.09**
Mascot Outfield; turf fertilizer; 9+5+5	0.03	–	0.05	–	0.04	m²	**0.09**
Fertilizers; in top 150 mm of topsoil at 70 g/m²							
Mascot Microfine; controlled release turf fertilizer; 8+0+6 + 2% Mg + 4% Fe	0.10	–	0.05	–	0.12	m²	**0.17**
Enmag; controlled release fertilizer (8–9 months); 11+22+09	0.18	–	0.05	–	0.21	m²	**0.26**
Mascot Outfield; turf fertilizer; 4+10+10	0.07	–	0.05	–	0.08	m²	**0.13**
Mascot Outfield; turf fertilizer; 9+5+5	0.07	–	0.05	–	0.08	m²	**0.13**

37 SOFT LANDSCAPING

Item – Overhead and Profit Included	PC £	Labour hours	Labour £	Plant £	Material £	Unit	Total rate £
SOIL IMPROVEMENTS – cont							
Preparation of planting operations – cont							
Topgrow; Melcourt; peat free tree and shrub compost (25 m³ loads); placing on beds by mechanical loader; spreading and rotavating into topsoil by tractor drawn machine							
50 mm thick	229.00	–	–	16.40	263.35	100 m²	**279.75**
100 mm thick	480.90	–	–	24.09	553.03	100 m²	**577.12**
150 mm thick	721.35	–	–	32.10	829.55	100 m²	**861.65**
200 mm thick	961.80	–	–	39.94	1106.07	100 m²	**1146.01**
Topgrow; Melcourt; peat free tree and shrub compost (65 m³ loads); placing on beds by mechanical loader; spreading and rotavating into topsoil by tractor drawn machine							
50 mm thick	176.50	–	–	16.40	202.98	100 m²	**219.38**
100 mm thick	370.65	–	–	24.09	426.25	100 m²	**450.34**
150 mm thick	555.98	–	–	32.10	639.38	100 m²	**671.48**
200 mm thick	741.30	–	–	39.94	852.49	100 m²	**892.43**
Mushroom compost; Melcourt; (25 m³ loads); from not further than 25 m from location; cultivating into topsoil by pedestrian drawn machine							
50 mm thick	132.25	2.86	92.00	1.17	152.09	100 m²	**245.26**
100 mm thick	264.50	6.05	194.78	1.17	304.17	100 m²	**500.12**
150 mm thick	396.75	8.90	286.73	1.17	456.26	100 m²	**744.16**
200 mm thick	529.00	12.90	415.53	1.17	608.35	100 m²	**1025.05**
Mushroom compost; Melcourt (25 m³ loads); placing on beds by mechanical loader; spreading and rotavating into topsoil by tractor drawn cultivator							
50 mm thick	132.25	–	–	16.40	152.09	100 m²	**168.49**
100 mm thick	264.50	–	–	24.09	304.17	100 m²	**328.26**
150 mm thick	396.75	–	–	32.10	456.26	100 m²	**488.36**
200 mm thick	529.00	–	–	39.94	608.35	100 m²	**648.29**
Manure (60 m³ loads); from not further than 25 m from location; cultivating into topsoil by pedestrian drawn machine							
50 mm thick	100.00	2.86	92.00	1.17	115.00	100 m²	**208.17**
100 mm thick	210.00	6.05	194.78	1.17	241.50	100 m²	**437.45**
150 mm thick	315.00	8.90	286.73	1.17	362.25	100 m²	**650.15**
200 mm thick	420.00	12.90	415.53	1.17	483.00	100 m²	**899.70**

37 SOFT LANDSCAPING

Item – Overhead and Profit Included	PC £	Labour hours	Labour £	Plant £	Material £	Unit	Total rate £
Surface applications and soil additives; pre-planting; from not further than 25 m from location; by machine							
Medium bark soil conditioner; Melcourt Ltd; including turning in to cultivated ground; delivered in 25 m³ loads							
1 m³ per 40 m² = 25 mm thick	1.07	–	–	0.22	1.23	m²	**1.45**
1 m³ per 20 m² = 50 mm thick	2.13	–	–	0.31	2.45	m²	**2.76**
1 m³ per 13.33 m² = 75 mm thick	3.20	–	–	0.41	3.68	m²	**4.09**
1 m³ per 10 m² = 100 mm thick	4.26	–	–	0.52	4.90	m²	**5.42**
Works by hand; surface applications and soil additives; pre-planting; from not further than 25 m from location; by hand							
Mushroom compost soil conditioner; Melcourt Industries Ltd; including turning in to cultivated ground; delivered in 25 m³ loads							
1 m³ per 40 m² = 25 mm thick	0.66	0.02	0.71	–	0.76	m²	**1.47**
1 m³ per 20 m² = 50 mm thick	1.32	0.04	1.43	–	1.52	m²	**2.95**
1 m³ per 13.33 m² = 75 mm thick	1.98	0.07	2.14	–	2.28	m²	**4.42**
1 m³ per 10 m² = 100 mm thick	2.65	0.08	2.58	–	3.05	m²	**5.63**
Medium bark soil conditioner; Melcourt Industries Ltd; including turning in to cultivated ground; delivered in 25 m³ loads							
1 m³ per 40 m² = 25 mm thick	1.07	0.02	0.71	–	1.23	m²	**1.94**
1 m³ per 20 m² = 50 mm thick	2.13	0.04	1.43	–	2.45	m²	**3.88**
1 m³ per 13.33 m² = 75 mm thick	3.20	0.07	2.14	–	3.68	m²	**5.82**
1 m³ per 10 m² = 100 mm thick	4.26	0.08	2.58	–	4.90	m²	**7.48**
Manure; 20 m³ loads; delivered not further than 25 m from location; cultivating into topsoil by pedestrian operated machine							
50 mm thick	110.00	2.86	92.00	1.17	126.50	100 m²	**219.67**
100 mm thick	210.00	6.05	194.78	1.17	241.50	100 m²	**437.45**
150 mm thick	315.00	8.90	286.73	1.17	362.25	100 m²	**650.15**
200 mm thick	420.00	12.90	415.53	1.17	483.00	100 m²	**899.70**
Fertilizer (7+7+7); PC £1.12/kg; to beds; by hand							
35 g/m²	3.91	0.17	5.37	–	4.50	100 m²	**9.87**
50 g/m²	5.58	0.17	5.37	–	6.42	100 m²	**11.79**
70 g/m²	7.81	0.17	5.37	–	8.98	100 m²	**14.35**
Fertilizers; Enmag; PC £2.40/kg; controlled release fertilizer; to beds; by hand							
35 g/m²	8.39	0.17	5.37	–	9.65	100 m²	**15.02**
50 g/m²	11.99	0.17	5.37	–	13.79	100 m²	**19.16**
70 g/m²	17.99	0.17	5.37	–	20.69	100 m²	**26.06**

37 SOFT LANDSCAPING

Item – Overhead and Profit Included	PC £	Labour hours	Labour £	Plant £	Material £	Unit	Total rate £
SOIL IMPROVEMENTS – cont							
Works by hand – cont							
Note: For machine incorporation of fertilizers and soil conditioners see 'Cultivation'.							
TREE PLANTING PREPARATIONS							
Tree planting; pre-planting operations							
Excavating tree pits; depositing soil alongside pits; by machine							
600 × 600 × 600 mm deep	–	0.15	4.74	0.79	–	nr	**5.53**
900 × 900 × 600 mm deep	–	0.33	10.63	1.78	–	nr	**12.41**
1.00 × 1.00 m × 600 mm deep	–	0.61	19.76	2.21	–	nr	**21.97**
1.25 × 1.25 m × 600 mm deep	–	0.96	30.94	3.45	–	nr	**34.39**
1.00 × 1.00 × 1.00 m deep	–	1.02	32.94	3.68	–	nr	**36.62**
1.50 × 1.50 m × 750 mm deep	–	1.73	55.58	6.20	–	nr	**61.78**
1.50 × 1.50 × 1.00 m deep	–	2.30	73.91	8.25	–	nr	**82.16**
1.75 × 1.75 × 1.00 m deep	–	3.13	100.86	11.26	–	nr	**112.12**
2.00 × 2.00 × 1.00 m deep	–	4.09	131.72	14.70	–	nr	**146.42**
Excavating tree pits; depositing soil alongside pits; by hand							
600 × 600 × 600 mm deep	–	0.44	14.17	–	–	nr	**14.17**
900 × 900 × 600 mm deep	–	1.00	32.20	–	–	nr	**32.20**
1.00 × 1.00 m × 600 mm deep	–	1.13	36.22	–	–	nr	**36.22**
1.25 × 1.25 m × 600 mm deep	–	1.93	62.15	–	–	nr	**62.15**
1.00 × 1.00 × 1.00 m deep	–	2.06	66.33	–	–	nr	**66.33**
1.50 × 1.50 × 750 mm deep	–	3.47	111.73	–	–	nr	**111.73**
1.75 × 1.50 × 750 mm deep	–	4.05	130.41	–	–	nr	**130.41**
1.50 × 1.50 × 1.00 m deep	–	4.63	149.09	–	–	nr	**149.09**
2.00 × 2.00 m × 750 mm deep	–	6.17	198.67	–	–	nr	**198.67**
2.00 × 2.00 × 1.00 m deep	–	8.23	265.02	–	–	nr	**265.02**
Breaking up subsoil in tree pits; to a depth of 200 mm	–	0.03	1.07	–	–	m^2	**1.07**
Spreading and lightly consolidating approved topsoil (imported or from spoil heaps); in layers not exceeding 150 mm; distance from spoil heaps not exceeding 100 m (imported topsoil not included); by machine							
minimum depth 100 mm	–	1.55	49.91	64.54	–	100 m^2	**114.45**
minimum depth 150 mm	–	2.33	75.14	97.13	–	100 m^2	**172.27**
minimum depth 300 mm	–	4.67	150.26	194.25	–	100 m^2	**344.51**
minimum depth 450 mm	–	6.99	225.08	291.05	–	100 m^2	**516.13**

37 SOFT LANDSCAPING

Item – Overhead and Profit Included	PC £	Labour hours	Labour £	Plant £	Material £	Unit	Total rate £
Spreading and lightly consolidating approved topsoil (imported or from spoil heaps); in layers not exceeding 150 mm; distance from spoil heaps not exceeding 100 m (imported topsoil not included); by hand							
minimum depth 100 mm	–	20.00	644.13	–	–	100 m²	**644.13**
minimum depth 150 mm	–	30.01	966.20	–	–	100 m²	**966.20**
minimum depth 300 mm	–	60.01	1932.39	–	–	100 m²	**1932.39**
minimum depth 450 mm	–	90.02	2898.57	–	–	100 m²	**2898.57**
Extra for filling tree pits with imported topsoil; PC £24.72/m³; calculations allow for 20% settlement							
depth 100 mm	–	–	–	–	4.09	m²	**4.09**
depth 150 mm	–	–	–	–	6.14	m²	**6.14**
depth 200 mm	–	–	–	–	8.19	m²	**8.19**
depth 300 mm	–	–	–	–	12.28	m²	**12.28**
depth 400 mm	–	–	–	–	16.38	m²	**16.38**
depth 450 mm	–	–	–	–	18.42	m²	**18.42**
depth 500 mm	–	–	–	–	20.47	m²	**20.47**
depth 600 mm	–	–	–	–	24.56	m²	**24.56**
depth 750 mm	–	–	–	–	25.59	m²	**25.59**
depth 1.00 m	–	–	–	–	34.11	m²	**34.11**
Backfilling of tree pits with imported topsoil; compacting in layers							
depth 500 mm	–	–	–	6.07	17.05	m²	**23.12**
depth 600 mm	–	–	–	6.93	20.47	m²	**27.40**
depth 750 mm	–	–	–	8.10	25.59	m²	**33.69**
depth 1.00 m	–	–	–	12.13	34.11	m²	**46.24**
Add or deduct the following amounts for every £1.00 change in the material price of topsoil							
depth 100 mm	–	–	–	–	0.14	m²	**0.14**
depth 150 mm	–	–	–	–	0.22	m²	**0.22**
depth 200 mm	–	–	–	–	0.29	m²	**0.29**
depth 300 mm	–	–	–	–	0.43	m²	**0.43**
depth 400 mm	–	–	–	–	0.56	m²	**0.56**
depth 450 mm	–	–	–	–	0.64	m²	**0.64**
depth 500 mm	–	–	–	–	0.71	m²	**0.71**
depth 600 mm	–	–	–	–	0.85	m²	**0.85**

37 SOFT LANDSCAPING

Item – Overhead and Profit Included	PC £	Labour hours	Labour £	Plant £	Material £	Unit	Total rate £
TREE PLANTING **PREPARATIONS – cont**							
Structural soils (Amsterdam tree sand); Heicom; non-compressive soil mixture for tree planting in areas to receive compressive surface treatments							
Backfilling and lightly compacting in layers; excavation, disposal, moving of material from delivery position and surface treatments not included; by machine							
individual tree pits (29 t loads)	57.75	0.50	16.10	2.69	66.41	m³	**85.20**
individual tree pits (20 t loads)	66.00	0.50	16.10	2.69	75.90	m³	**94.69**
in trenches (29 t loads)	57.75	0.42	13.42	2.24	66.41	m³	**82.07**
Backfilling and lightly compacting in layers; excavation, disposal, moving of material from delivery position and surface treatments not included; by hand							
individual tree pits (29 t loads)	57.75	1.60	51.52	–	66.41	m³	**117.93**
individual tree pits (20 t loads)	66.00	1.60	51.52	–	75.90	m³	**127.42**
in trenches	57.75	1.33	42.83	–	66.41	m³	**109.24**
TREE PLANTING							
Trees; planting labours only							
Bare root trees; including backfilling with previously excavated material (all other operations and materials not included)							
light standard; 6–8 cm girth	–	0.35	11.27	–	–	nr	**11.27**
standard; 8–10 cm girth	–	0.40	12.88	–	–	nr	**12.88**
selected standard; 10–12 cm girth	–	0.58	18.68	–	–	nr	**18.68**
heavy standard; 12–14 cm girth	–	0.83	26.83	–	–	nr	**26.83**
extra heavy standard; 14–16 cm girth	–	1.00	32.20	–	–	nr	**32.20**
Rootballed trees; including backfilling with previously excavated material (all other operations and materials not included)							
standard; 8–10 cm girth	–	0.50	16.10	–	–	nr	**16.10**
selected standard; 10–12 cm girth	–	0.60	19.32	–	–	nr	**19.32**
heavy standard; 12–14 cm girth	–	0.80	25.76	–	–	nr	**25.76**
extra heavy standard; 14–16 cm girth	–	1.50	48.30	–	–	nr	**48.30**
16–18 cm girth	–	1.30	41.73	38.73	–	nr	**80.46**
18–20 cm girth	–	1.60	51.52	47.81	–	nr	**99.33**
20–25 cm girth	–	4.50	144.90	148.82	–	nr	**293.72**
25–30 cm girth	–	6.00	193.20	196.52	–	nr	**389.72**
30–35 cm girth	–	11.00	354.20	350.24	–	nr	**704.44**

37 SOFT LANDSCAPING

Item – Overhead and Profit Included	PC £	Labour hours	Labour £	Plant £	Material £	Unit	Total rate £
Tree pits for urban planting							
Excavating tree pits; depositing soil							
alongside pits; by machine							
1.75 × 1.75 × 1.00 m deep	–	2.09	67.24	50.68	–	nr	**117.92**
2.00 × 2.00 × 1.00 m deep	–	2.73	87.81	66.01	–	nr	**153.82**
2.50 × 2.50 × 1.50 m deep	–	3.28	105.62	159.19	–	nr	**264.81**
Earthwork retention to tree pits							
bracing of deep treepits by plywood							
boards braced by timber stays							
Tree pit 1.00 m × 1.00 m × 1.00 m	–	0.75	48.30	–	15.63	nr	**63.93**
Tree pit 1.50 m × 1.50 m × 1.00 m	–	0.90	57.96	–	21.47	nr	**79.43**
Tree pit 2.00 m × 2.00 m × 1.50 m	–	1.00	64.40	–	27.31	nr	**91.71**
Excavate tree break-out zone for							
tree roots to develop in restricted							
and compacted urban environments;							
depositing soil alongside pits by							
machine;							
Trenches 1.50 m wide × 1.50 m							
deep average	–	0.17	5.37	8.10	–	m^3	**13.47**
Bases to tree pits; levelling base							
breaking up base by hand and							
placing of granular drainage material							
200 mm thick							
Tree pit 1.00 m × 1.00 m	–	0.25	8.05	12.13	10.56	nr	**30.74**
Tree pit 1.50 m × 1.50 m	–	0.45	14.49	21.84	23.76	nr	**60.09**
Tree pit 2.00 m × 2.00 m	–	0.75	24.15	36.40	42.23	nr	**102.78**
Tree pit 2.50 m × 2.50 m	–	0.75	24.15	36.40	65.99	nr	**126.54**
Tree planting; containerized trees;							
nursery stock; James Coles &							
Sons (Nurseries) Ltd							
Acer platanoides 'Emerald Queen';							
including backfillling with excavated							
material (other operations not							
included)							
standard; 8–10 cm girth	49.00	0.48	15.46	–	56.35	nr	**71.81**
selected standard; 10–12 cm girth	91.00	0.56	18.05	–	104.65	nr	**122.70**
heavy standard; 12–14 cm girth	105.00	0.76	24.61	–	120.75	nr	**145.36**
extra heavy standard; 14–16 cm							
girth	133.00	1.20	38.64	–	152.95	nr	**191.59**
Carpinus betulus; including							
backfillling with excavated material							
(other operations not included)							
standard; 8–10 cm girth	25.00	0.48	15.46	–	28.75	nr	**44.21**
selected standard; 10–12 cm girth	91.00	0.56	18.05	–	104.65	nr	**122.70**
heavy standard; 12–14 cm girth	119.00	0.76	24.61	–	136.85	nr	**161.46**
extra heavy standard; 14–16 cm							
girth	119.00	1.20	38.64	–	136.85	nr	**175.49**

37 SOFT LANDSCAPING

Item – Overhead and Profit Included	PC £	Labour hours	Labour £	Plant £	Material £	Unit	Total rate £
TREE PLANTING – cont							
Tree planting – cont							
Fagus sylvatica; including backfillling with excavated material (other operations not included)							
standard; 8–10 cm girth	49.00	0.48	15.46	–	56.35	nr	**71.81**
selected standard; 10–12 cm girth	105.00	0.56	18.05	–	120.75	nr	**138.80**
heavy standard; 12–14 cm girth	119.00	0.76	24.61	–	136.85	nr	**161.46**
extra heavy standard; 14–16 cm girth	133.00	1.20	38.64	–	152.95	nr	**191.59**
Prunus avium; including backfillling with excavated material (other operations not included)							
standard; 8–10 cm girth	49.00	0.40	12.88	–	56.35	nr	**69.23**
selected standard; 10–12 cm girth	77.00	0.56	18.05	–	88.55	nr	**106.60**
heavy standard; 12–14 cm girth	91.00	0.76	24.61	–	104.65	nr	**129.26**
extra heavy standard; 14–16 cm girth	105.00	1.20	38.64	–	120.75	nr	**159.39**
Quercus robur; including backfillling with excavated material (other operations not included)							
standard; 8–10 cm girth	49.00	0.48	15.46	–	56.35	nr	**71.81**
selected standard; 10–12 cm girth	84.00	0.56	18.05	–	96.60	nr	**114.65**
heavy standard; 12–14 cm girth	112.00	0.76	24.61	–	128.80	nr	**153.41**
extra heavy standard; 14–16 cm girth	119.00	1.20	38.64	–	136.85	nr	**175.49**
Betula utilis jaquemontii; multi-stemmed; including backfillling with excavated material (other operations not included)							
175/200 mm high	84.00	0.48	15.46	–	96.60	nr	**112.06**
200/250 mm high	105.00	0.56	18.05	–	120.75	nr	**138.80**
250/300 mm high	154.00	0.76	24.61	–	177.10	nr	**201.71**
300/350 mm high	265.00	1.20	38.64	–	304.75	nr	**343.39**
Tree planting; rootballed trees; advanced nursery stock and semi-mature – General							
Preamble: The cost of planting semi-mature trees will depend on the size and species and on the access to the site for tree handling machines. Prices should be obtained for individual trees and planting.							

37 SOFT LANDSCAPING

Item – Overhead and Profit Included	PC £	Labour hours	Labour £	Plant £	Material £	Unit	Total rate £
Tree planting; bare root trees; nursery stock; James Coles & Sons (Nurseries) Ltd							
Acer platanoides; including backfillling with excavated material (other operations not included)							
light standard; 6–8 cm girth	10.50	0.35	11.27	–	12.07	nr	**23.34**
standard; 8–10 cm girth	18.25	0.40	12.88	–	20.99	nr	**33.87**
selected standard; 10–12 cm girth	24.50	0.58	18.68	–	28.17	nr	**46.85**
heavy standard; 12–14 cm girth	42.00	0.83	26.83	–	48.30	nr	**75.13**
extra heavy standard; 14–16 cm girth	56.00	1.00	32.20	–	64.40	nr	**96.60**
Carpinus betulus; including backfillling with excavated material (other operations not included)							
light standard; 6–8 cm girth	14.70	0.35	11.27	–	16.91	nr	**28.18**
standard; 8–10 cm girth	25.00	0.40	12.88	–	28.75	nr	**41.63**
selected standard; 10–12 cm girth	31.00	0.58	18.68	–	35.65	nr	**54.33**
heavy standard; 12–14 cm girth	70.00	0.83	26.84	–	80.50	nr	**107.34**
Fagus sylvatica; including backfillling with excavated material (other operations not included)							
light standard; 6–8 cm girth	22.75	0.35	11.27	–	26.16	nr	**37.43**
standard; 8–10 cm girth	28.00	0.40	12.88	–	32.20	nr	**45.08**
selected standard; 10–12 cm girth	49.00	0.58	18.68	–	56.35	nr	**75.03**
heavy standard; 12–14 cm girth	70.00	0.83	26.73	–	80.50	nr	**107.23**
extra heavy standard; 14–16 cm girth	80.50	1.00	32.20	–	92.57	nr	**124.77**
Prunus avium; including backfillling with excavated material (other operations not included)							
light standard; 6–8 cm girth	22.75	0.36	11.72	–	26.16	nr	**37.88**
standard; 8–10 cm girth	28.00	0.40	12.88	–	32.20	nr	**45.08**
selected standard; 10–12 cm girth	25.00	0.58	18.68	–	28.75	nr	**47.43**
heavy standard; 12–14 cm girth	39.25	0.83	26.83	–	45.14	nr	**71.97**
extra heavy standard; 14–16 cm girth	49.00	1.00	32.20	–	56.35	nr	**88.55**
Quercus robur; including backfillling with excavated material (other operations not included)							
light standard; 6–8 cm girth	18.25	0.35	11.27	–	20.99	nr	**32.26**
standard; 8–10 cm girth	30.75	0.40	12.88	–	35.36	nr	**48.24**
selected standard; 10–12 cm girth	42.00	0.58	18.68	–	48.30	nr	**66.98**
heavy standard; 12–14 cm girth	67.25	0.83	26.83	–	77.34	nr	**104.17**
Robinia pseudoacacia Frisia; including backfillling with excavated material (other operations not included)							
light standard; 6–8 cm girth	21.00	0.35	11.27	–	24.15	nr	**35.42**
standard; 8–10 cm girth	28.00	0.40	12.88	–	32.20	nr	**45.08**
selected standard; 10–12 cm girth	44.75	0.58	18.68	–	51.46	nr	**70.14**

37 SOFT LANDSCAPING

Item – Overhead and Profit Included	PC £	Labour hours	Labour £	Plant £	Material £	Unit	Total rate £
TREE PLANTING – cont							
Tree planting; rootballed trees; nursery stock; James Coles & Sons (Nurseries) Ltd							
Acer platanoides; including backfillling with excavated material (other operations not included)							
standard; 8–10 cm girth	27.52	0.48	15.46	–	31.65	nr	**47.11**
selected standard; 10–12 cm girth	42.00	0.56	18.05	–	48.30	nr	**66.35**
heavy standard; 12–14 cm girth	63.00	0.76	24.61	–	72.45	nr	**97.06**
extra heavy standard; 14–16 cm girth	80.50	1.20	38.64	–	92.57	nr	**131.21**
Carpinus betulus; including backfillling with excavated material (other operations not included)							
standard; 8–10 cm girth	49.77	0.48	15.46	–	57.24	nr	**72.70**
selected standard; 10–12 cm girth	70.88	0.56	18.05	–	81.51	nr	**99.56**
heavy standard; 12–14 cm girth	85.00	0.76	24.61	–	97.75	nr	**122.36**
extra heavy standard; 14–16 cm girth	101.50	1.20	38.64	–	116.72	nr	**155.36**
Fagus sylvatica; including backfillling with excavated material (other operations not included)							
standard; 8–10 cm girth	37.27	0.48	15.46	–	42.86	nr	**58.32**
selected standard; 10–12 cm girth	61.88	0.56	18.05	–	71.16	nr	**89.21**
heavy standard; 12–14 cm girth	85.96	0.76	24.61	–	98.85	nr	**123.46**
extra heavy standard; 14–16 cm girth	98.16	1.20	38.64	–	112.88	nr	**151.52**
Prunus avium; including backfillling with excavated material (other operations not included)							
standard; 8–10 cm girth	9.27	0.40	12.88	–	10.66	nr	**23.54**
selected standard; 10–12 cm girth	37.88	0.56	18.05	–	43.56	nr	**61.61**
heavy standard; 12–14 cm girth	55.21	0.76	24.61	–	63.49	nr	**88.10**
extra heavy standard; 14–16 cm girth	17.66	1.20	38.64	–	76.66	nr	**115.30**
Quercus robur; including backfillling with excavated material (other operations not included)							
standard; 8–10 cm girth	42.75	0.48	15.46	–	49.16	nr	**64.62**
selected standard; 10–12 cm girth	59.50	0.56	18.05	–	68.43	nr	**86.48**
heavy standard; 12–14 cm girth	88.25	0.76	24.61	–	101.49	nr	**126.10**
extra heavy standard; 14–16 cm girth	119.00	1.20	38.64	–	136.85	nr	**175.49**
Robinia pseudoacacia 'Frisia'; including backfillling with excavated material (other operations not included)							
standard; 8–10 cm girth	40.00	0.48	15.46	–	46.00	nr	**61.46**
selected standard; 10–12 cm girth	62.25	0.56	18.05	–	71.59	nr	**89.64**
heavy standard; 12–14 cm girth	98.00	0.76	24.61	–	112.70	nr	**137.31**

37 SOFT LANDSCAPING

Item – Overhead and Profit Included	PC £	Labour hours	Labour £	Plant £	Material £	Unit	Total rate £
Tree planting; Airpot container grown trees; advanced nursery stock and semi-mature; Hillier Nurseries							
Acer platanoides 'Emerald Queen'; including backfilling with excavated material (other operations not included)							
16–18 cm girth	115.00	1.98	63.76	58.27	132.25	nr	**254.28**
18–20 cm girth	147.00	2.18	70.13	62.66	169.05	nr	**301.84**
20–25 cm girth	210.00	2.38	76.51	69.93	241.50	nr	**387.94**
25–30 cm girth	275.00	2.97	95.63	107.41	316.25	nr	**519.29**
30–35 cm girth	295.00	3.96	127.51	116.55	339.25	nr	**583.31**
Prunus avium 'Flora Plena'; including backfilling with excavated material (other operations not included)							
16–18 cm girth	115.00	1.98	63.76	58.27	132.25	nr	**254.28**
18–20 cm girth	147.00	1.60	51.52	62.66	169.05	nr	**283.23**
20–25 cm girth	210.00	2.38	76.51	69.93	241.50	nr	**387.94**
25–30 cm girth	275.00	2.97	95.63	107.41	316.25	nr	**519.29**
30–35 cm girth	405.00	3.96	127.51	116.55	465.75	nr	**709.81**
Quercus palustris 'Pin Oak'; including backfilling with excavated material (other operations not included)							
16–18 cm girth	130.00	1.98	63.76	58.27	149.50	nr	**271.53**
18–20 cm girth	189.00	1.60	51.52	62.66	217.35	nr	**331.53**
20–25 cm girth	283.00	2.38	76.51	69.93	325.45	nr	**471.89**
25–30 cm girth	375.00	2.97	95.63	107.41	431.25	nr	**634.29**
30–35 cm girth	615.00	3.96	127.51	116.55	707.25	nr	**951.31**
Tree planting; Multi-Stem trees; Hillier Nurseries; Airpot container grown trees; semi-mature							
Betula Pendula; including backfilling with excavated material (other operations not included)							
300–400 cm	164.00	3.00	289.80	107.41	188.60	nr	**585.81**
400–500 cm	303.00	3.25	313.95	107.41	348.45	nr	**769.81**
500–600 cm	454.00	3.50	338.10	113.68	522.10	nr	**973.88**
600–700 cm	1129.01	3.75	362.25	117.99	1298.36	nr	**1778.60**
700–800 cm	1750.02	4.00	386.40	120.14	2012.52	nr	**2519.06**
800–900 cm	2195.02	4.50	434.70	132.11	2524.27	nr	**3091.08**
Betula jacquemontii; including backfilling with excavated material (other operations not included)							
300–400 cm	425.00	3.0	289.80	107.41	488.75	nr	**885.96**
400–500 cm	598.00	3.25	313.95	107.41	687.70	nr	**1109.06**
500–600 cm	998.00	3.50	338.10	113.68	1147.70	nr	**1599.48**
600–700 cm	1650.00	4.00	386.40	120.14	1897.50	nr	**2404.04**
700–800 cm	2425.00	4.50	434.70	132.11	2788.75	nr	**3355.56**

37 SOFT LANDSCAPING

Item – Overhead and Profit Included	PC £	Labour hours	Labour £	Plant £	Material £	Unit	Total rate £
TREE PLANTING – cont							
Tree planting; Airpot container grown trees; semi-mature and mature trees; Hillier Nurseries; planting and back filling; planted by telehandler or by crane; delivery included; all other operations priced separately							
Semi-mature trees; indicative prices							
40–45 cm girth	635.00	4.00	128.80	75.98	730.25	nr	**935.03**
45–50 cm girth	870.00	4.00	128.80	75.98	1000.50	nr	**1205.28**
55–60 cm girth	1560.00	6.00	193.20	75.98	1794.00	nr	**2063.18**
60–70 cm girth	2750.00	7.00	225.40	102.89	3162.50	nr	**3490.79**
70–80 cm girth	2850.00	7.50	241.50	129.79	3277.50	nr	**3648.79**
80–90 cm girth	4950.00	8.00	257.60	151.96	5692.50	nr	**6102.06**
Tree planting; rootballed trees; advanced nursery stock and semi-mature; Lorenz von Ehren							
Acer platanoides 'Emerald Queen'; including backfilling with excavated material (other operations not included)							
16–18 cm girth	110.00	1.30	41.73	4.50	126.50	nr	**172.73**
18–20 cm girth	150.00	1.60	51.52	4.50	172.50	nr	**228.52**
20–25 cm girth	180.00	4.50	144.90	21.10	207.00	nr	**373.00**
25–30 cm girth	250.00	6.00	193.20	26.22	287.50	nr	**506.92**
30–35 cm girth	380.00	11.00	354.20	35.02	437.00	nr	**826.22**
Aesculus carnea 'Briotti'; including backfilling with excavated material (other operations not included)							
16–18 cm girth	220.00	1.30	41.73	4.50	253.00	nr	**299.23**
18–20 cm girth	280.00	1.60	51.52	4.50	322.00	nr	**378.02**
20–25 cm girth	340.00	4.50	144.90	21.10	391.00	nr	**557.00**
25–30 cm girth	440.00	6.00	193.20	26.22	506.00	nr	**725.42**
30–35 cm girth	580.00	11.00	354.20	35.02	667.00	nr	**1056.22**
Prunus avium 'Plena'; including backfilling with excavated material (other operations not included)							
16–18 cm girth	140.00	1.30	41.73	4.50	161.00	nr	**207.23**
18–20 cm girth	160.00	1.60	51.52	4.50	184.00	nr	**240.02**
20–25 cm girth	180.00	4.50	144.90	21.10	207.00	nr	**373.00**
25–30 cm girth	240.00	6.00	193.20	26.22	276.00	nr	**495.42**
30–35 cm girth	380.00	11.00	354.20	30.70	437.00	nr	**821.90**
Quercus palustris 'Pin Oak'; including backfilling with excavated material (other operations not included)							
16–18 cm girth	170.00	1.30	41.73	4.50	195.50	nr	**241.73**
18–20 cm girth	210.00	1.60	51.52	4.50	241.50	nr	**297.52**
20–25 cm girth	260.00	4.50	144.90	21.10	299.00	nr	**465.00**

37 SOFT LANDSCAPING

Item – Overhead and Profit Included	PC £	Labour hours	Labour £	Plant £	Material £	Unit	Total rate £
25–30 cm girth	360.00	6.00	193.20	26.22	414.00	nr	633.42
30–35 cm girth	480.00	11.00	354.20	35.02	552.00	nr	941.22
Tilia cordata 'Greenspire'; including backfilling with excavated material (other operations not included)							
16–18 cm girth	120.00	1.30	41.73	4.50	138.00	nr	184.23
18–20 cm girth	140.00	1.60	51.52	4.50	161.00	nr	217.02
20–25 cm girth	170.00	4.50	144.90	21.10	195.50	nr	361.50
25–30 cm girth; 5 × transplanted; 4.0–5.0 m tall	210.00	6.00	193.20	26.22	241.50	nr	460.92
30–35 cm girth; 5 × transplanted; 5.0–7.0 m tall	440.00	11.00	354.20	35.02	506.00	nr	895.22
Betula pendula (3 stems); including backfilling with excavated material (other operations not included)							
3.0–3.5 m high	150.00	1.98	63.76	58.27	172.50	nr	294.53
3.5–4.0 m high	200.00	1.60	51.52	62.66	230.00	nr	344.18
4.0–4.5 m high	300.00	2.38	76.51	69.93	345.00	nr	491.44
4.5–5.0 m high	400.00	2.97	95.63	107.41	460.00	nr	663.04
5.0–6.0 m high	500.00	3.96	127.51	116.55	575.00	nr	819.06
6.0–7.0 m high	600.00	4.50	144.90	136.23	690.00	nr	971.13
Pinus sylvestris; including backfilling with excavated material (other operations not included)							
3.0–3.5 m high	860.00	1.98	63.76	58.27	989.00	nr	1111.03
3.5–4.0 m high	620.00	1.60	51.52	62.66	713.00	nr	827.18
4.0–4.5 m high	840.00	2.38	76.51	69.93	966.00	nr	1112.44
4.5–5.0 m high	1160.00	2.97	95.63	107.41	1334.00	nr	1537.04
5.0–6.0 m high	1830.00	3.96	127.51	116.55	2104.50	nr	2348.56
6.0–7.0 m high	3050.00	4.50	144.90	140.67	3507.50	nr	3793.07
Pleached trees; Carpinus betulus; specimen trees; Lorenz Von Ehren; supply and planting only; excavation, support and tree pit additives not included							
Box shaped trees to provide 'floating hedge' or screen effect; planting distance 1 tree per m run; trunk and box alignment; wire rootballs							
4 × transplanted; 20–25 cm; 1.00 m centres	400.00	5.00	161.00	21.10	460.00	m	642.10
5 × transplanted; 25–30 cm; 1.00 m centres	560.00	6.50	209.30	26.22	644.00	m	879.52
5 × transplanted; 30–35 cm; 1.20 m centres	984.00	11.00	354.20	35.02	1131.60	m	1520.82

37 SOFT LANDSCAPING

Item – Overhead and Profit Included	PC £	Labour hours	Labour £	Plant £	Material £	Unit	Total rate £
TREE PLANTING – cont							
Pleached trees; Tilia europaea 'Pallida'; specimen trees; Lorenz Von Ehren; supply and planting only; excavation, support and tree pit additives not included							
Box pleached trees to provide 'floating hedge' or screen effect; wire rootballs							
4 × transplanted; 20–25 cm; 1.00 m centres	390.00	5.00	161.00	21.10	448.50	m	**630.60**
4 × transplanted; 25–30 cm; 1.00 m centres	480.00	5.00	161.00	21.10	552.00	m	**734.10**
5 × transplanted; 30–35 cm; 1.20 m centres	483.31	5.00	161.00	21.10	555.81	m	**737.91**
5 × transplanted; 35–40 cm; 1.50 m centres	513.36	5.00	161.00	21.10	590.36	m	**772.46**
6 × transplanted; 40–45 cm; 1.80 m centres	550.04	5.00	161.00	21.10	632.55	m	**814.65**
6 × transplanted; 45–50 cm; 1.80 m centres	700.06	5.00	161.00	21.10	805.07	m	**987.17**
6 × transplanted; 50–60 cm; 2.00 m centres	840.00	5.00	161.00	21.10	966.00	m	**1148.10**
Espalier pleached trees; specimen trees; Lorenz Von Ehren; supply and planting only; excavation, support and tree pit additives not included							
Frame pleached trees; to provide floating screen effects; wire rootballs							
Carpinus betulus; 20–25 cm; 1.00 m centres	540.00	5.00	161.00	21.10	621.00	m	**803.10**
Carpinus betulus; 25–30 cm; 1.00 m centres	720.00	5.00	161.00	21.10	828.00	m	**1010.10**
Tilia europaea 'Pallida'; 20–25 cm; 1.00 m centres	480.00	5.00	161.00	21.10	552.00	m	**734.10**
Tilia europaea 'Pallida'; 25–30 cm; 1.00 m centres	830.00	5.00	161.00	21.10	954.50	m	**1136.60**
Tilia europaea 'Pallida'; 30–35 cm; 1.20 m centres	833.30	5.00	161.00	21.10	958.29	m	**1140.39**

37 SOFT LANDSCAPING

Item – Overhead and Profit Included	PC £	Labour hours	Labour £	Plant £	Material £	Unit	Total rate £
Umbrella shaped pleaches							
Umbrella or roof pleached trees to							
provide umbrella effect; wire							
rootballs							
Platanus acerifolia; clear stem;							
4 × transplanted; 20–25 cm	570.00	5.00	161.00	21.10	655.50	nr	**837.60**
Platanus acerifolia; specimen; 5 ×							
transplanted; 30–35 cm	940.00	5.00	161.00	21.10	1081.00	nr	**1263.10**
Platanus acerifolia; specimen; 5 ×							
transplanted; 25–30 cm	760.00	5.00	161.00	21.10	874.00	nr	**1056.10**
Tree planting; containerized trees;							
Lorenz von Ehren; to the tree							
prices above add for Airpot							
containerization only							
Tree size							
20–25 cm	–	–	–	–	115.00	nr	**115.00**
25–30 cm	–	–	–	–	143.75	nr	**143.75**
30–35 cm	–	–	–	–	172.50	nr	**172.50**
35–40 cm	–	–	–	–	230.00	nr	**230.00**
40–45 cm	–	–	–	–	345.00	nr	**345.00**
45–50 cm	–	–	–	–	460.00	nr	**460.00**
50–60 cm	–	–	–	–	575.00	nr	**575.00**
60–70 cm	–	–	–	–	805.00	nr	**805.00**
70–80 cm	–	–	–	–	1035.00	nr	**1035.00**
80–90 cm	–	–	–	–	1265.00	nr	**1265.00**
Tree planting; rootballed trees;							
semi-mature and mature trees;							
Lorenz von Ehren; planting and							
back filling; planted by							
telehandler or by crane; delivery							
included; all other operations							
priced separately							
Semi-mature trees							
40–45 cm girth	1000.00	8.00	257.60	81.33	1150.00	nr	**1488.93**
45–50 cm girth	2000.00	8.00	257.60	104.06	2300.00	nr	**2661.66**
50–60 cm girth	3000.00	10.00	322.00	127.37	3450.00	nr	**3899.37**
60–70 cm girth	4000.00	15.00	483.00	328.78	4600.00	nr	**5411.78**
70–80 cm girth	5000.00	18.00	579.60	309.28	5750.00	nr	**6638.88**
80–90 cm girth	7500.00	18.00	579.60	309.28	8625.00	nr	**9513.88**
90–100 cm girth	9500.00	18.00	579.60	309.28	10925.00	nr	**11813.88**

37 SOFT LANDSCAPING

Item – Overhead and Profit Included	PC £	Labour hours	Labour £	Plant £	Material £	Unit	Total rate £
TREE PLANTING ACCESSORIES							
Gt ArborRaft System; Greentech Ltd; Strong geocellular system replaces traditional subbase layer in urban tree planting enabling shallow tree planting in a nutrient rich uncompacted environment that better reflects natural environment encouraging tree establishment and growth							
70 t Loading capacity; single layer installation to fully graded and levelled partially backfilled tree pit							
ArborRaft 150 mm unit							
(708 mm × 354 mm × 150 mm)	71.95	0.17	10.74	–	82.74	m²	**93.48**
ArborRaft 85 mm unit							
(708 mm × 354 mm × 85 mm)	67.41	0.14	9.20	–	77.52	m²	**86.72**
tree planting soil with nutrients compaction resistance and water retention qualities; structure remains open for water, oxygen and nutrients to reach developing root system of the tree							
ArborRaft tree pit soil 29 t loads; by machine	41.85	–	–	15.47	48.13	m³	**63.60**
ArborRaft tree pit soil 20 t loads; by machine	57.94	–	–	15.47	66.63	m³	**82.10**
ArborRaft tree pit soil bagged; by machine	82.40	–	–	25.78	94.76	m³	**120.54**
ArborRaft tree pit soil bagged; by hand	82.40	0.50	32.20	–	94.76	m³	**126.96**
StrataCell; GreenBlue Urban; Tree planting accessories for Urban tree planting; Load bearing tree planting modules for load support in tree environments; Placed to excavated tree pit;							
single layer;							
500 mm × 500 mm × 250 mm deep; interlocking load bearing cells; backfilling with imported topsoil							
7.5 tonne load limit	156.10	0.25	8.05	2.99	193.74	m²	**204.78**
40 tonne load limit	885.20	0.25	8.05	2.99	1032.22	m²	**1043.26**
double layer;							
500 mm × 500 mm × 500 mm deep; interlocking load bearing cells; backfilling with imported topsoil							
7.5 tonne load limit	1248.80	0.50	16.10	5.99	1464.58	m²	**1486.67**
40 tonne load limit	1770.40	0.50	16.10	5.99	2064.42	m²	**2086.51**

37 SOFT LANDSCAPING

Item – Overhead and Profit Included	PC £	Labour hours	Labour £	Plant £	Material £	Unit	Total rate £
Root director; GreenBlue Urban; preformed root barrier system with integral root deflecting ribs Placed over tree rootball; under pavements; overall size							
510 × 595 × 310 mm	54.07	0.25	16.10	–	62.18	nr	**78.28**
650 × 855 × 455 mm	89.19	0.33	21.25	–	102.57	nr	**123.82**
975 × 1370 × 545 mm	137.34	0.45	28.98	–	157.94	nr	**186.92**
1300 × 1805 × 500 mm	230.20	0.60	38.64	–	264.73	nr	**303.37**
Tree staking J Toms Ltd; extra over trees for tree stake(s); driving 500 mm into firm ground; trimming to approved height; including two tree ties to approved pattern							
one stake; 1.50 m long × 32 mm × 32 mm	1.64	0.20	6.44	–	1.89	nr	**8.33**
two stakes; 900 mm long × 25 mm × 25 mm	1.60	0.30	9.66	–	1.84	nr	**11.50**
two stakes; 1.50 m long × 32 mm × 32 mm	2.38	0.30	9.66	–	2.74	nr	**12.40**
three stakes; 1.50 m long × 32 mm × 32 mm	3.57	0.36	11.59	–	4.11	nr	**15.70**
Tree anchors Platipus Anchors Ltd; extra over trees for tree anchors							
RF1P rootball kit; for 75–220 mm girth × 2–4.5 m high; inclusive of Plati-Mat PM1	31.00	1.00	32.20	–	35.65	nr	**67.85**
RF2P; rootball kit; for 220–450 mm girth × 4.5–7.5 m high; inclusive of Plati-Mat PM2	51.00	1.33	42.83	–	58.65	nr	**101.48**
RF3P; rootball kit; for 450–750 mm girth × 7.5–12 m high; inclusive of Plati-Mat PM3	111.00	1.50	48.30	–	127.65	nr	**175.95**
installation tools; drive rod for RF1 kits	–	–	–	–	73.60	nr	**73.60**
installation tools; drive rod for RF2 kits	–	–	–	–	110.40	nr	**110.40**
Extra over trees for land drain to tree pits; 100 mm dia. perforated flexible agricultural drain; including excavating drain trench; laying pipe; backfilling	1.29	1.00	32.20	–	1.48	m	**33.68**

37 SOFT LANDSCAPING

Item – Overhead and Profit Included	PC £	Labour hours	Labour £	Plant £	Material £	Unit	Total rate £
TREE PLANTING ACCESSORIES – cont							
Platipus anchors; deadman anchoring							
Deadman kits; anchoring to concrete kerbs placed in base of tree pit							
trees 12–25 cm girth; 2.5–4.0 m high	28.00	0.75	24.15	–	58.35	nr	**82.50**
trees 25–45 cm girth; 4.0–7.5 m high	47.50	1.50	48.30	–	80.78	nr	**129.08**
trees 45–75 cm girth; 7.5–12.0 m high	104.00	2.00	64.40	–	145.75	nr	**210.15**
Tree planting; tree pit additives							
Melcourt Industries Ltd; Topgrow; incorporating into topsoil at 1 part Topgrow to 3 parts excavated topsoil; supplied in 75 l bags; pit size							
600 × 600 × 600 mm	3.00	0.02	0.64	–	3.45	nr	**4.09**
900 × 900 × 900 mm	10.11	0.06	1.93	–	11.63	nr	**13.56**
1.00 × 1.00 × 1.00 m	13.85	0.24	7.73	–	15.93	nr	**23.66**
1.25 × 1.25 × 1.25 m	27.08	0.40	12.88	–	31.14	nr	**44.02**
1.50 × 1.50 × 1.50 m	46.80	0.90	28.98	–	53.82	nr	**82.80**
Melcourt Industries Ltd; Topgrow; incorporating into topsoil at 1 part Topgrow to 3 parts excavated topsoil; supplied in 60 m³ loose loads; pit size							
600 × 600 × 600 mm	1.91	0.02	0.54	–	2.20	nr	**2.74**
900 × 900 × 900 mm	6.44	0.05	1.61	–	7.41	nr	**9.02**
1.00 × 1.00 × 1.00 m	8.82	0.20	6.44	–	10.14	nr	**16.58**
1.25 × 1.25 × 1.25 m	17.24	0.33	10.73	–	19.83	nr	**30.56**
1.50 × 1.50 × 1.50 m	29.79	0.75	24.15	–	34.26	nr	**58.41**
Tree planting; root barriers							
GreenBlue Urban; linear root deflection barriers; installed to trench measured separately							
Re-Root 2000; 2.0 mm thick × 2 m wide	20.36	0.05	1.61	–	23.41	m	**25.02**
Re-Root 2000; 1.0 mm thick × 1 m wide	7.11	0.05	1.61	–	8.18	m	**9.79**
Re-Root 600; 1.0 mm thick	10.34	0.05	1.61	–	11.89	m	**13.50**

37 SOFT LANDSCAPING

Item – Overhead and Profit Included	PC £	Labour hours	Labour £	Plant £	Material £	Unit	Total rate £
GreenBlue Urban; irrigation systems; Root Rain tree pit irrigation systems							
Metro; small; 35 mm pipe dia. × 1.25 m long; for specimen shrubs and standard trees							
plastic	5.14	0.25	8.05	–	5.91	nr	**13.96**
plastic; with chain	9.60	0.25	8.05	–	11.04	nr	**19.09**
Metro; medium; 35 mm pipe dia. × 1.75 m long; for standard and selected standard trees							
plastic	6.05	0.29	9.20	–	6.96	nr	**16.16**
plastic; with chain	10.68	0.29	9.20	–	12.28	nr	**21.48**
Metro; large; 35 mm pipe dia. × 2.50 m long; for selected standards and extra heavy standards							
plastic	7.43	0.33	10.73	–	8.54	nr	**19.27**
plastic; with chain	12.05	0.33	10.73	–	13.86	nr	**24.59**
Urban; for large capacity general purpose irrigation to parkland and street verge planting							
RRUrb1; 3.0 m pipe	–	0.29	9.20	–	10.98	nr	**20.18**
RRUrb2; 5.0 m pipe	12.06	0.33	10.73	–	13.87	nr	**24.60**
RRUrb3; 8.0 m pipe	15.81	0.40	12.88	–	18.18	nr	**31.06**
Civic; heavy cast aluminium inlet; for heavily trafficked locations							
5.0 m pipe	44.14	0.33	10.73	–	50.76	nr	**61.49**
8.0 m pipe	49.57	0.40	12.88	–	57.01	nr	**69.89**
Tree irrigation kits; Platipus anchors; direct water delivery to the rootball area of trees							
Piddler tree irrigation system; permeable membrane delivering targeted irrigation to rootball surround							
rootballs up to 550 mm dia.; Irrigation Kit PID0	6.50	0.25	8.05	–	7.47	nr	**15.52**
rootballs up to 900 mm dia.; Irrigation Kit PID1	12.00	0.30	9.66	–	13.80	nr	**23.46**
rootballs up to 1.55 m dia.; Irrigation Kit PID2	15.50	0.40	12.88	–	17.83	nr	**30.71**
rootballs up to 2.40 m dia.; Irrigation Kit PID3	19.00	0.45	14.49	–	21.85	nr	**36.34**
rootballs up to 3.10 m dia.; Irrigation Kit PID4	26.50	0.50	16.10	–	30.48	nr	**46.58**

37 SOFT LANDSCAPING

Item – Overhead and Profit Included	PC £	Labour hours	Labour £	Plant £	Material £	Unit	Total rate £
TREE PLANTING ACCESSORIES – cont							
GreenBlue Urban; Arboresin porous tree pit surfacing resin to galvanized tree pit frames							
Two part support frame system							
support frame 1.20 m × 1.20 m	206.15	1.20	77.28	–	1053.74	nr	**1131.02**
support frame 1.50 m × 1.50 m	322.11	1.40	90.16	–	1506.95	nr	**1597.11**
support frame 1.20 m dia.	161.77	1.10	70.84	–	1092.78	nr	**1163.62**
TREE PLANTING OPERATIONS AFTER PLANTING							
Mulching of tree pits; Melcourt Industries Ltd; FSC (Forest Stewardship Council) and FT (fire tested) certified							
Spreading mulch; to individual trees; maximum distance 25 m (mulch not included)							
50 mm thick	–	0.05	1.56	–	–	m²	**1.56**
75 mm thick	–	0.07	2.35	–	–	m²	**2.35**
100 mm thick	–	0.10	3.13	–	–	m²	**3.13**
Mulch; Bark Nuggets®; to individual trees; delivered in 80 m³ loads; maximum distance 25 m							
50 mm thick	2.43	0.05	1.56	–	2.79	m²	**4.35**
75 mm thick	3.64	0.07	2.35	–	4.19	m²	**6.54**
100 mm thick	4.85	0.10	3.13	–	5.58	m²	**8.71**
Mulch; Bark Nuggets®; to individual trees; delivered in 25 m³ loads; maximum distance 25 m							
50 mm thick	3.11	0.05	1.56	–	3.58	m²	**5.14**
75 mm thick	4.67	0.05	1.61	–	5.37	m²	**6.98**
100 mm thick	6.22	0.07	2.15	–	7.15	m²	**9.30**
Mulch; Amenity Bark Mulch; to individual trees; delivered in 80 m³ loads; maximum distance 25 m							
50 mm thick	1.60	0.05	1.56	–	1.84	m²	**3.40**
75 mm thick	2.40	0.07	2.35	–	2.76	m²	**5.11**
100 mm thick	3.19	0.10	3.13	–	3.67	m²	**6.80**
Mulch; Amenity Bark Mulch; to individual trees; delivered in 25 m³ loads; maximum distance 25 m							
50 mm thick	2.28	0.05	1.56	–	2.62	m²	**4.18**
75 mm thick	3.42	0.07	2.35	–	3.93	m²	**6.28**
100 mm thick	4.56	0.07	2.15	–	5.24	m²	**7.39**

37 SOFT LANDSCAPING

Item – Overhead and Profit Included	PC £	Labour hours	Labour £	Plant £	Material £	Unit	Total rate £
Resin bound mulch or surface to tree pits; Dekorgrip Ltd Resin bound system with a typical makeup of 10 mm aggregate to create voids to allow and promote the free flow of air and water around tree beds to surface of tree pit							
Dekortree permeable resin tree pits 50 mm uV stabilized 10–25 m²	–	0.75	48.30	–	109.83	m²	**158.13**
Dekortree permeable resin tree pits 50 mm uV stabilized over 25 m²	–	0.75	48.30	–	96.10	m²	**144.40**
Dekortree 50 mm; Standard resin 10–25 m²	–	0.75	48.30	–	85.12	m²	**133.42**
Dekortree 50 mm uV Standard over 25 m²	–	0.75	48.30	–	72.35	m²	**120.65**
SHRUB PLANTING							
Shrub planting – General Preamble: For preparation of planting areas see 'Cultivation' at the beginning of the section on planting.							
Market prices of Shrubs The prices for shrubs are so varied that it is not possible to provide a list in this publication. For general budgeting purposes however the following average prices may be used with deciduous shrubs being slightly less and evergreeens slighty more							
1 litre shrub	–	–	–	–	1.73	nr	**1.73**
2 litre shrub	–	–	–	–	3.22	nr	**3.22**
3 litre shrub	–	–	–	–	3.91	nr	**3.91**
5 litre shrub	–	–	–	–	7.82	nr	**7.82**
10 litre specimen shrub	–	–	–	–	14.49	nr	**14.49**
25 litre specimen shrub	–	–	–	–	48.30	nr	**48.30**
Market prices of high impact specimen shrubs; Griffin Nurseries; Prices are varied and the prices below reflect the price ranges for plants as described; sizes;							
25 litre container	44.00	–	–	–	50.60	nr	**50.60**
25 litre container; specialist feature plant specimen	62.00	–	–	–	71.30	nr	**71.30**
40 litre container	73.50	–	–	–	84.53	nr	**84.53**

37 SOFT LANDSCAPING

Item – Overhead and Profit Included	PC £	Labour hours	Labour £	Plant £	Material £	Unit	Total rate £
SHRUB PLANTING – cont							
Market prices of Shrubs – cont							
Market prices of high impact							
specimen shrubs – cont							
40 litre container; specialist							
feature plant specimen	112.00	–	–	–	128.80	nr	**128.80**
80 litre container	99.00	–	–	–	113.85	nr	**113.85**
80 litre container; specialist							
feature plant specimen	225.00	–	–	–	258.75	nr	**258.75**
130 litre container; specialist							
feature plant specimen	325.00	–	–	–	373.75	nr	**373.75**
Shrub planting							
Setting out; selecting planting from							
holding area; loading to							
wheelbarrows; planting as plan or as							
directed; distance from holding area							
maximum 50 m; plants 2–3 litre							
containers							
single plants not grouped	–	0.04	1.29	–	–	nr	**1.29**
plants in groups of 3–5 nr	–	0.03	0.81	–	–	nr	**0.81**
plants in groups of 10–100 nr	–	0.02	0.53	–	–	nr	**0.53**
plants in groups of 100 nr							
minimum	–	0.01	0.37	–	–	nr	**0.37**
Forming planting holes; in cultivated							
ground (cultivating not included); by							
mechanical auger; trimming holes by							
hand; depositing excavated material							
alongside holes							
250 mm dia.	–	0.03	1.07	0.08	–	nr	**1.15**
250 × 250 mm	–	0.04	1.29	0.15	–	nr	**1.44**
300 × 300 mm	–	0.08	2.42	0.18	–	nr	**2.60**
Planting shrubs; hand excavation;							
forming planting holes; in cultivated							
ground (cultivating not included);							
depositing excavated material							
alongside holes; Placing plants							
previously set out alongside and							
backfilling (plants not included)							
100 × 100 × 100 mm deep; with							
mattock or hoe	–	0.01	0.36	–	–	nr	**0.36**
250 × 250 × 300 mm deep	–	0.01	0.43	–	–	nr	**0.43**
300 × 300 × 300 mm deep	–	0.02	0.54	–	–	nr	**0.54**
400 × 400 × 400 mm deep	–	0.08	2.68	–	–	nr	**2.68**
500 × 500 × 500 mm deep	–	0.17	5.37	–	–	nr	**5.37**
600 × 600 × 600 mm deep	–	0.25	8.05	–	–	nr	**8.05**
900 × 900 × 600 mm deep	–	0.50	16.10	–	–	nr	**16.10**
1.00 × 1.00 m x 600 mm deep	–	1.00	32.20	–	–	nr	**32.20**
1.25 × 1.25 m x 600 mm deep	–	1.33	42.93	–	–	nr	**42.93**

37 SOFT LANDSCAPING

Item – Overhead and Profit Included	PC £	Labour hours	Labour £	Plant £	Material £	Unit	Total rate £
Hand excavation; forming planting holes; in uncultivated ground; depositing excavated material alongside holes							
100 × 100 × 100 mm deep; with mattock or hoe	–	0.01	0.43	–	–	nr	**0.43**
250 × 250 × 250 mm deep	–	0.03	0.81	–	–	nr	**0.81**
300 × 300 × 300 mm deep	–	0.06	2.01	–	–	nr	**2.01**
400 × 400 × 400 mm deep	–	0.25	8.05	–	–	nr	**8.05**
500 × 500 × 500 mm deep	–	0.33	10.48	–	–	nr	**10.48**
600 × 600 × 600 mm deep	–	0.55	17.71	–	–	nr	**17.71**
900 × 900 × 600 mm deep	–	1.25	40.25	–	–	nr	**40.25**
1.00 × 1.00 m x 600 mm deep	–	1.54	49.51	–	–	nr	**49.51**
1.25 × 1.25 m x 600 mm deep	–	2.41	77.68	–	–	nr	**77.68**
Bare root planting; to planting holes (forming holes not included); including backfilling with excavated material (bare root plants not included)							
bare root 1+1; 30–90 mm high	–	0.02	0.54	–	–	nr	**0.54**
bare root 1+2; 90–120 mm high	–	0.02	0.54	–	–	nr	**0.54**
Shrub planting where plants are set out at equal offset centres where the spacings are always the specified distance between the plant centres. This introduces more uniform but slightly higher distances than rectangular centres							
2 litre containerized plants; in cultivated ground (cultivating not included); PC £2.80/nr							
1.00 m centres; 1.43 plants/m²	–	0.02	0.61	–	4.60	m²	**5.21**
900 mm centres; 1.69 plants/m²	–	0.02	0.72	–	5.44	m²	**6.16**
750 mm centres; 2.6 plants/m²	–	0.04	1.15	–	8.37	m²	**9.52**
600 mm centres; 3.04 plants/m²	–	0.04	1.30	–	9.79	m²	**11.09**
500 mm centres; 4.5 plants/m²	–	0.06	1.93	–	14.49	m²	**16.42**
450 mm centres; 6.75 plants/m²	–	0.09	2.90	–	21.73	m²	**24.63**
400 mm centres; 7.50 plants/m²	–	0.17	5.42	–	32.27	m²	**37.69**
300 mm centres; 12.7 plants/m²	–	0.17	5.44	–	41.12	m²	**46.56**
Shrub planting; 3 litre containerized plants; in cultivated ground (cultivating not included); PC £3.40/nr							
1.00 m centres; 1.43 plants/m²	–	0.02	0.61	–	5.59	m²	**6.20**
900 mm centres; 1.69 plants/m²	–	0.02	0.72	–	6.61	m²	**7.33**
750 mm centres; 2.6 plants/m²	–	0.04	1.15	–	10.17	m²	**11.32**
600 mm centres; 3.04 plants/m²	–	0.04	1.30	–	11.89	m²	**13.19**
500 mm centres; 4.5 plants/m²	–	0.06	1.93	–	17.59	m²	**19.52**
450 mm centres; 6.75 plants/m²	–	0.09	2.90	–	26.39	m²	**29.29**
400 mm centres; 7.50 plants/m²	–	0.17	5.42	–	39.18	m²	**44.60**
300 mm centres; 12.7 plants/m²	–	0.17	5.44	–	49.93	m²	**55.37**

37 SOFT LANDSCAPING

Item – Overhead and Profit Included	PC £	Labour hours	Labour £	Plant £	Material £	Unit	Total rate £
HEDGE PLANTING							
Hedges							
Excavating trench for hedges; depositing soil alongside trench; by machine							
300 mm deep × 300 mm wide	–	0.03	1.96	1.36	–	m	3.32
300 mm deep × 600 mm wide	–	0.05	2.93	2.06	–	m	4.99
300 mm deep × 450 mm wide	–	0.05	2.90	2.04	–	m	4.94
500 mm deep × 500 mm wide	–	0.06	4.04	2.83	–	m	6.87
500 mm deep × 700 mm wide	–	0.09	5.65	3.96	–	m	9.61
600 mm deep × 750 mm wide	–	0.11	7.26	5.08	–	m	12.34
750 mm deep × 900 mm wide	–	0.17	10.90	7.64	–	m	18.54
Excavating trench for hedges; depositing soil alongside trench; by hand							
300 mm deep × 300 mm wide	–	0.08	8.05	–	–	m	8.05
300 mm deep × 450 mm wide	–	0.13	12.07	–	–	m	12.07
300 mm deep × 600 mm wide	–	0.17	16.10	–	–	m	16.10
500 mm deep × 500 mm wide	–	0.23	21.84	–	–	m	21.84
500 mm deep × 700 mm wide	–	0.32	30.64	–	–	m	30.64
600 mm deep × 750 mm wide	–	0.41	39.38	–	–	m	39.38
750 mm deep × 900 mm wide	–	0.61	58.96	–	–	m	58.96
Setting out; notching out; excavating trench; breaking up subsoil to minimum depth 300 mm							
minimum 400 mm deep	–	0.25	8.05	–	–	m	8.05
Disposal of excavated soil to stockpile							
Up to 25 m distant							
By hand	–	2.40	77.28	–	–	m³	77.28
By machine	–	–	–	3.14	–	m³	3.14
Instant Hedge planting; Readyhedge Ltd; continuous strips of instant hedges; delivered in 1.00 m lengths; planting and backfilling with excavated material of hedging plants to trenches (not included); including minor trimming and shaping after planting							
Beech (Fagus) hedges							
height 0.80–1.00 m × 200 mm deep	51.00	0.20	12.88	8.61	58.65	m	80.14
height 1.40–1.60 m × 400 mm deep	126.00	0.20	12.88	8.61	144.90	m	166.39
height 1.60 m × 400 mm deep	146.00	0.25	16.10	10.78	167.90	m	194.78

37 SOFT LANDSCAPING

Item – Overhead and Profit Included	PC £	Labour hours	Labour £	Plant £	Material £	Unit	Total rate £
Purple Beech (Fagus) hedges							
height 0.80–1.00 m × 200 mm deep	51.00	–	–	–	58.65	m	**58.65**
height 1.20–1.40 m × 350 mm deep	153.00	0.25	16.10	10.78	175.95	m	**202.83**
Yew (Taxus) hedges							
700 mm high × 250 mm wide;	126.00	0.50	48.30	3.91	144.90	m	**197.11**
1.60–1.80 m high × 300 mm wide; 500 mm centres	144.00	0.57	55.20	3.91	165.60	m	**224.71**
Prunus lauruscerasus hedges							
height 1.40–1.60 m × 40 cm deep	133.00	0.25	16.10	10.78	152.95	m	**179.83**
Prunus lucitanica (Portuguese laurel) hedges							
height 70–90 cm × 20 cm deep	60.00	0.20	12.88	8.61	69.00	m	**90.49**
height 1.30–1.50 m × 35 cm deep	146.00	0.25	16.10	10.78	167.90	m	**194.78**
Carpinus (Hornbeam) hedges							
height 80–100 cm × 20 cm deep	51.00	0.20	12.88	8.61	58.65	m	**80.14**
height 1.60–1.80 m × 40 cm deep	133.00	0.25	16.10	10.78	152.95	m	**179.83**
Box (Buxus) hedges							
height 30–35 cm × 20 cm deep	53.00	0.10	6.44	4.31	60.95	m	**71.70**
height 40–45 cm × 25 cm deep	131.00	0.13	8.05	5.38	150.65	m	**164.08**
Instant Hedge planting; Readyhedge Ltd; individual pre-clipped hedge plants to form mature hedge; planting and backfilling of hedging plants to trenches excavated separately; including minor trimming and shaping after planting							
Feathered hedges; Beech or Hornbeam							
1.75 m high × 300 mm wide; 400 mm centres	45.00	0.20	19.32	1.96	51.75	m	**73.03**
1.75 m high × 300 mm wide; 500 mm centres	36.00	0.16	15.46	1.96	41.40	m	**58.82**
1.75 m high × 300 mm wide; 750 mm centres	25.56	0.11	11.04	1.96	29.39	m	**42.39**
1.75 m high × 300 mm wide; 900 mm centres	19.98	0.09	8.59	1.96	22.98	m	**33.53**
1.75/2.00 m high × 300 mm wide; 600 mm centres	30.00	0.33	32.20	18.79	34.50	m	**85.49**
1.75/2.00 m high × 300 mm wide; 900 mm centres	19.98	0.29	27.60	16.48	22.98	m	**67.06**
2.00/2.25 m high × 300 wide; 600 mm centres	53.34	0.40	38.64	22.68	61.34	m	**122.66**
2.00/2.25 m high × 300 mm wide; 900 mm centres	35.20	0.33	32.20	18.79	40.48	m	**91.47**
2.00/2.25 m high × 300 mm wide; 1.20 m centres	26.67	0.25	24.15	14.74	30.67	m	**69.56**

Prices for Measured Works – Major Works

37 SOFT LANDSCAPING

Item – Overhead and Profit Included	PC £	Labour hours	Labour £	Plant £	Material £	Unit	Total rate £
HEDGE PLANTING – cont							
Hedge planting; Griffin Nurseries Ltd; feathered hedge plants to form mature hedge; planting and backfilling of hedging plants to trenches excavated separately; including minor trimming and shaping after planting							
(Yew) Taxus hedging; rootballed							
600/800 mm high × 300 mm wide; 250 mm centres	51.00	0.20	19.32	1.96	58.65	m	**79.93**
600/800 mm high × 300 mm wide; 500 mm centres	25.50	0.11	10.73	1.96	29.33	m	**42.02**
800 mm/1.00 m high × 300 mm wide; 250 mm centres	66.00	0.20	19.32	1.96	75.90	m	**97.18**
800 mm/1.00 m high × 300 mm wide; 500 mm centres	33.00	0.11	10.73	1.96	37.95	m	**50.64**
1.00/1.20 m high × 300/350 mm wide; 3 nr/m	78.25	0.17	16.10	1.96	89.99	m	**108.05**
1.00/1.20 m high × 300/350 mm wide; 2 nr/m	47.00	0.13	12.07	1.96	54.05	m	**68.08**
1.25/1.50 m high × 300/350 mm wide; 3 nr/m	131.53	0.17	16.10	1.96	151.26	m	**169.32**
1.25/1.50 m high × 300/350 mm wide; 2 nr/m	79.00	0.13	12.88	1.96	90.85	m	**105.69**
1.50/1.75 m high × 400 mm wide; 450 mm centres	128.76	0.33	32.20	14.09	148.07	m	**194.36**
1.50/1.75 m high × 400 mm wide; 600 mm centres	96.67	0.22	21.46	12.74	111.17	m	**145.37**
1.50/1.75 m high × 400 mm wide; 900 mm centres	64.38	0.20	19.32	11.66	74.04	m	**105.02**
1.75/2.00 m high × 500 mm wide; 600 mm centres	135.00	0.33	32.20	18.79	155.25	m	**206.24**
1.75/2.00 m high × 500 mm wide; 900 mm centres	89.91	0.29	27.60	16.48	103.40	m	**147.48**
1.75/2.00 m high × 500 mm wide; 1.20 m centres	67.50	0.25	24.15	14.74	77.63	m	**116.52**
2.00/2.25 m high × 600 mm wide; 600 mm centres	−166668.33	0.40	–	–	–	m	–
2.00/2.25 m high × 600 mm wide; 900 mm centres	−110998.89	0.33	–	–	–	m	–
2.00/2.25 m high × 600 mm wide; 1.20 m centres	–	0.31	29.73	17.55	–	m	**47.28**
(Yew) Taxus hedging; wire rootballed; extra over for wire rootballing	5.00	–	–	–	5.75	ea	**5.75**

37 SOFT LANDSCAPING

Item – Overhead and Profit Included	PC £	Labour hours	Labour £	Plant £	Material £	Unit	Total rate £
Buxus (Box) hedges; rootballed							
400/500 mm high; 200 mm							
rootball; 200 mm centres	44.75	0.09	2.86	–	51.46	m	**54.32**
400/500 mm high; 200 mm							
rootball; 300 mm centres	29.80	0.07	2.15	–	34.27	m	**36.42**
400/500 mm high; 200 mm							
rootball; 500 mm centres	17.90	0.05	1.61	–	20.58	m	**22.19**
500/600 mm high; 200 mm							
rootball; 200 mm centres	62.50	0.09	2.86	–	71.88	m	**74.74**
500/600 mm high; 200 mm							
rootball; 300 mm centres	41.63	0.07	2.15	–	47.87	m	**50.02**
500/600 mm high; 200 mm							
rootball; 500 mm centres	25.00	0.08	2.47	–	28.75	m	**31.22**
600/800 mm high; 250 mm							
rootball; 250 mm centres	70.00	0.09	2.93	–	80.50	m	**83.43**
600/800 mm high; 250 mm							
rootball; 400 mm centres	43.75	0.08	2.68	–	50.31	m	**52.99**
600/800 mm high; 250 mm							
rootball; 600 mm centres	20.83	0.08	2.47	–	23.95	m	**26.42**
800 mm/1.00 m high; 300 mm							
rootball; 300 mm centres	83.25	0.10	3.22	–	95.74	m	**98.96**
800 mm/1.00 m high; 300 mm							
rootball; 500 mm centres	50.00	0.11	3.58	–	57.50	m	**61.08**
800 mm/1.00 m high; 300 mm							
rootball; 750 mm centres	33.25	0.13	4.03	–	38.24	m	**42.27**
1.00 m/1.25 m high; 350 mm							
rootball; 400 mm centres	97.50	0.13	4.03	–	112.13	m	**116.16**
1.00 m/1.25 m high; 350 mm							
rootball; 600 mm centres	65.01	0.11	3.58	–	74.76	m	**78.34**
1.00 m/1.25 m high; 350 mm							
rootball; 900 mm centres	43.29	0.10	3.22	–	49.78	m	**53.00**
Native species and bare root hedge planting							
Hedge planting operations only (excavation of trenches priced separately) including backfill with excavated topsoil; PC £0.55/nr; 40–60 cm							
single row; 200 mm centres	2.75	0.06	2.01	–	3.16	m	**5.17**
single row; 300 mm centres	1.83	0.06	1.79	–	2.10	m	**3.89**
single row; 400 mm centres	1.38	0.04	1.35	–	1.59	m	**2.94**
single row; 500 mm centres	1.10	0.03	1.07	–	1.27	m	**2.34**
double row; 200 mm centres	5.50	0.17	5.37	–	6.32	m	**11.69**
double row; 300 mm centres	3.67	0.13	4.29	–	4.22	m	**8.51**
double row; 400 mm centres	2.75	0.08	2.68	–	3.16	m	**5.84**
double row; 500 mm centres	2.20	0.07	2.15	–	2.53	m	**4.68**
Extra over hedges for incorporating manure; at 1 m³ per 30 m	0.67	0.03	0.81	–	0.77	m	**1.58**

37 SOFT LANDSCAPING

Item – Overhead and Profit Included	PC £	Labour hours	Labour £	Plant £	Material £	Unit	Total rate £
TOPIARY							
Clipped topiary; Lorenz von Ehren; German field grown clipped and transplanted as detailed; planted to plant pit; including backfilling with excavated material and TPMC							
Buxus sempervirens (box); balls							
300 mm dia.; 3 × transplanted; container grown or rootballed	30.00	0.25	8.05	–	45.31	nr	**53.36**
500 mm dia.; 4 × transplanted; wire rootballed	80.00	1.50	48.30	5.39	175.49	nr	**229.18**
900 mm dia.; 5 × transplanted; wire rootballed	320.00	2.75	88.55	5.39	506.72	nr	**600.66**
1300 mm dia.; 6 × transplanted; wire rootballed	960.00	3.10	99.82	6.47	1241.22	nr	**1347.51**
Buxus sempervirens (box); pyramids							
500 mm high; 3 × transplanted; container grown or rootballed	50.00	1.50	48.30	5.39	140.99	nr	**194.68**
900 mm high; 4 × transplanted; wire rootballed	160.00	2.75	88.55	5.39	322.72	nr	**416.66**
1300 mm high; 5 × transplanted; wire rootballed	460.00	3.10	99.82	6.47	666.23	nr	**772.52**
Buxus sempervirens (box); truncated pyramids							
500 mm high; 3 × transplanted; container grown or rootballed	110.00	1.50	48.30	5.39	209.99	nr	**263.68**
900 mm high; 4 × transplanted; wire rootballed	380.00	2.75	88.55	5.39	575.72	nr	**669.66**
Buxus sempervirens (box); truncated cone							
1300 mm high; 5 × transplanted; wire rootballed	1130.00	3.10	99.82	6.47	1436.73	nr	**1543.02**
Buxus sempervirens (box); cubes							
500 mm square; 4 × transplanted; rootballed	140.00	1.50	48.30	5.39	244.49	nr	**298.18**
900 mm square; 5 × transplanted; wire rootballed	470.00	2.75	88.55	5.39	679.22	nr	**773.16**
Taxus baccata (yew); balls							
500 mm dia.; 4 × transplanted; rootballed	80.00	1.50	48.30	5.39	175.49	nr	**229.18**
900 mm dia.; 5 × transplanted; wire rootballed	230.00	2.75	88.55	5.39	403.22	nr	**497.16**
1300 mm dia.; 6 × transplanted; wire rootballed	690.00	3.10	99.82	6.47	930.73	nr	**1037.02**

37 SOFT LANDSCAPING

Item – Overhead and Profit Included	PC £	Labour hours	Labour £	Plant £	Material £	Unit	Total rate £
Taxus baccata (yew); cones							
800 mm high; 4 × transplanted; wire rootballed	80.00	1.50	48.30	5.39	175.49	nr	**229.18**
1500 mm high; 5 × transplanted; wire rootballed	180.00	2.00	64.40	5.39	310.62	nr	**380.41**
2500 mm high; 6 × transplanted; wire rootballed	580.00	3.10	99.82	6.47	804.23	nr	**910.52**
Taxus baccata (yew); cubes							
500 mm square; 4 × transplanted; rootballed	90.00	1.50	48.30	5.39	186.99	nr	**240.68**
900 mm square; 5 × transplanted; wire rootballed	300.00	2.75	88.55	5.39	483.72	nr	**577.66**
Taxus baccata (yew); pyramids							
900 mm high; 4 × transplanted; wire rootballed	150.00	1.50	48.30	5.39	255.99	nr	**309.68**
1500 mm high; 5 × transplanted; wire rootballed	300.00	3.10	99.82	6.47	482.23	nr	**588.52**
2500 mm high; 6 × transplanted; wire rootballed	840.00	4.00	128.80	10.79	1139.63	nr	**1279.22**
Carpinus betulus (common hornbeam); columns; round base							
800 mm wide × 2000 mm high; 4 × transplanted; wire rootballed	190.00	3.10	99.82	6.47	355.73	nr	**462.02**
800 mm wide × 2750 mm high; 4 × transplanted; wire rootballed	370.00	4.00	128.80	10.79	599.13	nr	**738.72**
Carpinus betulus (common hornbeam); cones							
3 m high; 5 × transplanted; wire rootballed	330.00	4.00	128.80	10.79	553.13	nr	**692.72**
4 m high; 5 × transplanted; wire rootballed	690.00	5.00	161.00	10.79	1000.73	nr	**1172.52**
Carpinus betulus 'Fastigiata'; pyramids							
4 m high; 6 × transplanted; wire rootballed	1170.00	5.00	161.00	10.79	1552.73	nr	**1724.52**
7 m high; 7 × transplanted; wire rootballed	3180.00	7.50	241.50	16.17	3914.54	nr	**4172.21**
LIVING WALLS							
Design to living walls; Scotscape Living Walls Ltd							
Fytotextile Living Wall Design							
10–30 m²	–	–	–	–	–	nr	**550.00**
30–70 m²	–	–	–	–	–	nr	**715.00**
70 –100 m²	–	–	–	–	–	nr	**935.00**
100–150 m²	–	–	–	–	–	nr	**1100.00**
over 150 m²	–	–	–	–	–	nr	**1650.00**

37 SOFT LANDSCAPING

Item – Overhead and Profit Included	PC £	Labour hours	Labour £	Plant £	Material £	Unit	Total rate £
LIVING WALLS – cont							
Living Wall; fytotextile and soil-based system; Scotscape Living Walls; self contained, irrigated vertical planting system; Fabric modules, fully established upon installation; saturated weight 40 kg per m²; access costs excluded							
System units 250 × 500 × 100 mm overall; fixed to horizontal hanging rails on softwood batten and DPM; planting depth 150 mm at 30° angle to vertical surface; inclusive of automatic dripline irrigation system and guttering with running outlets to drain (not included); Area of installation							
5–10 m²	–	–	–	–	–	m²	495.00
10–20 m²	–	–	–	–	–	m²	451.00
20–30 m²	–	–	–	–	–	m²	440.00
30–70 m²	–	–	–	–	–	m²	440.00
70–100 m²	–	–	–	–	–	m²	434.50
100–150 m²	–	–	–	–	–	m²	429.00
over 150 m²	–	–	–	–	–	m²	429.00
Fully automatic irrigation systems to the above; comprising breaktank, station controller, solenoid valves, plant feeders and dripline emitters							
10–30 m²	–	–	–	–	–	nr	3025.00
30–70 m²	–	–	–	–	–	nr	4262.50
70–100 m²	–	–	–	–	–	nr	5445.00
100–150 m²	–	–	–	–	–	nr	6875.00
over 150 m²	–	–	–	–	–	nr	8525.00
Rainwater Harvesting System	–	–	–	–	–	nr	2475.00
Preliminary costs for all living walls							
General preliminaries	–	–	–	–	–	nr	385.00
Design costs; site visits, Consultations and presentations	–	–	–	–	–	nr	379.50
Maintenance to living walls							
Annual maintenance to living walls; regular visits to maintain planting and systems; inclusive of feeding; pest control and calibration of irrigation systems							
10–30 m²	–	–	–	–	–	annum	2750.00
30–70 m²	–	–	–	–	–	annum	5225.00
70–100 m²	–	–	–	–	–	annum	8250.00

37 SOFT LANDSCAPING

Item – Overhead and Profit Included	PC £	Labour hours	Labour £	Plant £	Material £	Unit	Total rate £
100–150 m²	–	–	–	–	–	annum	**9900.00**
over 150 m²	–	–	–	–	–	annum	**12100.00**
Accessories to living walls							
planters to base of living wall in GRP; black; inclusive of drainage and planting	–	–	–	–	–	m	**339.63**
Green Screen; Mobilane Ltd							
Fully installed vertical green screen of Hedera hibernica; 65 plants per linear m; all on 5 mm galvanized steel weldmesh; attached to existing wall fence or hoarding							
screen 1.80 m high	–	–	–	–	115.00	m	**115.00**
screen 2.20 m high	–	–	–	–	201.25	m	**201.25**
screen 4.00 m high	–	–	–	–	414.00	m	**414.00**
HERBACEOUS/GROUNDCOVER/ BULB/BEDDING PLANTING							
Groundcover plants							
Groundcover plants in massed areas; PC £1.50/nr; including forming planting holes in cultivated ground (cultivating not included); backfilling with excavated material; 1 litre containers							
average 4 plants per m²; 500 mm centres	6.00	0.09	3.01	–	6.90	m²	**9.91**
average 6 plants per m²; 408 mm centres	9.00	0.14	4.51	–	10.35	m²	**14.86**
average 8 plants per m²; 354 mm centres	12.00	0.19	6.01	–	13.80	m²	**19.81**
Groundcover plants in massed areas; PC £2.20/nr; including forming planting holes in cultivated ground (cultivating not included); backfilling with excavated material; 1 litre containers							
average 4 plants per m²; 500 mm centres	8.80	0.09	3.01	–	10.12	m²	**13.13**
average 6 plants per m²; 408 mm centres	13.20	0.14	4.51	–	15.18	m²	**19.69**
average 8 plants per m²; 354 mm centres	17.60	0.19	6.01	–	20.24	m²	**26.25**

37 SOFT LANDSCAPING

Item – Overhead and Profit Included	PC £	Labour hours	Labour £	Plant £	Material £	Unit	Total rate £
HERBACEOUS/GROUNDCOVER/ BULB/BEDDING PLANTING – cont							
Herbaceous planting							
Herbaceous plants; PC £1.50/nr; including forming planting holes in cultivated ground (cultivating not included); backfilling with excavated material; 1 litre containers							
average 4 plants per m²; 500 mm centres	6.00	0.09	3.01	–	6.90	m²	**9.91**
average 6 plants per m²; 408 mm centres	9.00	0.14	4.51	–	10.35	m²	**14.86**
average 8 plants per m²; 354 mm centres	12.00	0.19	6.01	–	13.80	m²	**19.81**
Note: For machine incorporation of fertilizers and soil conditioners see 'Cultivation'.							
Bulb planting							
Bulbs; including forming planting holes in cultivated area (cultivating not included); backfilling with excavated material							
small	13.00	0.83	26.83	–	14.95	100 nr	**41.78**
medium	22.00	0.83	26.83	–	25.30	100 nr	**52.13**
large	25.00	0.91	29.27	–	28.75	100 nr	**58.02**
Bulbs; in grassed area; using bulb planter; including backfilling with screened topsoil or peat and cut turf plug							
small	13.00	1.67	53.67	–	14.95	100 nr	**68.62**
medium	22.00	1.67	53.67	–	25.30	100 nr	**78.97**
large	25.00	2.00	64.40	–	28.75	100 nr	**93.15**
Aquatic planting							
Aquatic plants; Planting only in prepared growing medium in pool; plant size 2–3 litre containerized (plants not included)	–	0.04	1.29	–	–	nr	**1.29**
MARKET PRICES OF NATIVE SPECIES							
The following prices are typically some of the more popular species used in native plantations; prices vary between species; readers should check the catalogue of the supplier.							

37 SOFT LANDSCAPING

Item – Overhead and Profit Included	PC £	Labour hours	Labour £	Plant £	Material £	Unit	Total rate £
1+1; 1 year seedling transplanted and grown for a year							
30–40 cm	–	–	–	–	0.40	nr	**0.40**
40–60 cm	–	–	–	–	0.47	nr	**0.47**
1+2; 1 year seedling transplanted and grown for 2 years							
60–80 cm	–	–	–	–	0.58	nr	**0.58**
Whips							
80–100 cm	–	–	–	–	0.90	nr	**0.90**
100–125 cm	–	–	–	–	1.35	nr	**1.35**
Feathered trees							
125–150 cm	–	–	–	–	2.10	nr	**2.10**
150–180 cm	–	–	–	–	4.00	nr	**4.00**
180–200 cm	–	–	–	–	5.80	nr	**5.80**
200–250 cm	–	–	–	–	10.00	nr	**10.00**
250–300 cm	–	–	–	–	11.00	nr	**11.00**
Native species planting							
Forming planting holes; in cultivated ground (cultivating not included); by mechanical auger; trimming holes by hand; depositing excavated material alongside holes							
250 mm dia.	–	0.03	0.93	0.07	–	nr	**1.00**
250 × 250 mm	–	0.04	1.12	0.13	–	nr	**1.25**
300 × 300 mm	–	0.08	2.10	0.16	–	nr	**2.26**
Hand excavation; forming planting holes; in cultivated ground (cultivating not included); depositing excavated material alongside holes							
100 × 100 × 100 mm deep; with mattock or hoe	–	0.01	0.19	–	–	nr	**0.19**
250 × 250 × 300 mm deep	–	0.04	1.12	–	–	nr	**1.12**
300 × 300 × 300 mm deep	–	0.06	1.56	–	–	nr	**1.56**

37 SOFT LANDSCAPING

Item – Overhead and Profit Included	PC £	Labour hours	Labour £	Plant £	Material £	Unit	Total rate £
TREE/SHRUB PROTECTION							
Tree planting; tree protection – General							
Preamble: Care must be taken to ensure that tree grids and guards are removed when trees grow beyond the specified dia. of guard.							
Tree planting; tree protection							
Expandable plastic tree guards; including 25 mm softwood stakes							
500 mm high	1.08	0.17	5.36	–	1.24	nr	**6.60**
1.00 m high	1.10	0.17	5.36	–	1.27	nr	**6.63**
English Woodlands; Weldmesh tree guards; nailing to tree stakes (tree stakes not included)							
1800 mm high × 300 mm dia.	26.00	0.25	8.05	–	29.90	nr	**37.95**
J. Toms Ltd; spiral rabbit guards; clear or brown							
610 × 38 mm	0.25	0.03	1.08	–	0.29	nr	**1.37**
English Woodlands; Plastic Mesh Tree Guards; black; supplied in 50 m rolls							
13 × 13 mm small mesh; roll width 60 cm	1.54	0.04	1.15	–	2.35	nr	**3.50**
13 × 13 mm small mesh; roll width 120 cm	3.21	0.06	1.90	–	4.27	nr	**6.17**
Tree guards of 3 nr 2.40 m × 100 mm stakes; driving 600 mm into firm ground; bracing with timber braces at top and bottom; including							
3 strands barbed wire	0.54	1.00	32.20	–	0.62	nr	**32.82**
English Woodlands; strimmer guard in heavy duty black plastic; 225 mm high	4.38	0.07	2.15	–	5.04	nr	**7.19**
Standard Treeshelter inclusive of 25 mm stake; prices shown for quantities of 500 nr							
0.75 m high	1.42	0.05	1.61	–	2.27	nr	**3.88**
Tubex Ltd; Shrubshelter inclusive of 25 mm stake; prices shown for quantities of 500 nr							
0.6 m high	1.71	0.07	2.15	–	1.97	nr	**4.12**
Extra over trees for spraying with anti-desiccant spray; Wiltpruf selected standards; standards;							
light standards	5.72	0.20	6.44	–	6.58	nr	**13.02**
standards; heavy standards	9.54	0.25	8.05	–	10.97	nr	**19.02**

37 SOFT LANDSCAPING

Item – Overhead and Profit Included	PC £	Labour hours	Labour £	Plant £	Material £	Unit	Total rate £
FORESTRY PLANTING							
Forestry planting							
Deep ploughing rough ground to form planting ridges at							
2.00 m centres	–	–	–	58.77	–	100 m²	**58.77**
3.00 m centres	–	–	–	55.30	–	100 m²	**55.30**
4.00 m centres	–	–	–	37.61	–	100 m²	**37.61**
Notching plant forestry seedlings; T or L notch	48.00	0.75	24.15	–	55.20	100 nr	**79.35**
Turf planting forestry seedlings	48.00	2.00	64.40	–	55.20	100 nr	**119.60**
Cleaning and weeding around seedlings; once	–	0.50	16.10	–	–	100 nr	**16.10**
Treading in and firming ground around seedlings planted; at 2500 per ha after frost or other ground disturbance; once	–	0.33	10.73	–	–	100 nr	**10.73**
Beating up initial planting; once (including supply of replacement seedlings at 10% of original planting)	4.80	0.25	8.05	–	5.52	100 nr	**13.57**
OPERATIONS AFTER PLANTING							
Operations after planting							
Initial cutting back to shrubs and hedge plants; including disposal of all cuttings	–	1.00	32.20	–	–	100 m²	**32.20**
Mulch; Melcourt Industries Ltd; Bark Nuggets®; to plant beds; delivered in 80 m³ loads; maximum distance 25 m							
50 mm thick	2.43	0.04	1.43	–	2.79	m²	**4.22**
75 mm thick	3.64	0.07	2.15	–	4.19	m²	**6.34**
100 mm thick	4.85	0.09	2.86	–	5.58	m²	**8.44**
Mulch; Melcourt Industries Ltd; Bark Nuggets®; to plant beds; delivered in 25 m³ loads; maximum distance 25 m							
50 mm thick	2.76	0.03	0.94	–	3.17	m²	**4.11**
75 mm thick	4.67	0.07	2.15	–	5.37	m²	**7.52**
100 mm thick	6.22	0.09	2.86	–	7.15	m²	**10.01**
Mulch; Melcourt Industries Ltd; Amenity Bark Mulch FSC; to plant beds; delivered in 80 m³ loads; maximum distance 25 m							
50 mm thick	1.60	0.04	1.43	–	1.84	m²	**3.27**
75 mm thick	2.40	0.07	2.15	–	2.76	m²	**4.91**
100 mm thick	3.19	0.09	2.86	–	3.67	m²	**6.53**

37 SOFT LANDSCAPING

Item – Overhead and Profit Included	PC £	Labour hours	Labour £	Plant £	Material £	Unit	Total rate £
OPERATIONS AFTER PLANTING – cont							
Operations after planting – cont Mulch; Melcourt Industries Ltd; Amenity Bark Mulch FSC; to plant beds; delivered in 25 m³ loads; maximum distance 25 m							
50 mm thick	2.28	0.04	1.43	–	2.62	m²	**4.05**
75 mm thick	3.42	0.07	2.15	–	3.93	m²	**6.08**
100 mm thick	4.56	0.09	2.86	–	5.24	m²	**8.10**
Mulch mats; English Woodlands; lay mulch mat to planted area or plant station; mat secured with metal J pins 240 × 3 mm Mats to individual plants and plant stations							
Woven Polypropylene; 50 × 50 cm; square mat	0.65	0.05	1.61	–	1.43	nr	**3.04**
Woven Polypropylene; 1 × 1 m; square mat	1.55	0.07	2.15	–	2.46	nr	**4.61**
Hedging mats							
Woven Polypropylene; 1 × 100 m roll; hedge planting	0.51	0.01	0.40	–	0.92	m²	**1.32**
General planting areas							
Plantex membrane; 1 × 14 m roll; weed control	1.43	0.01	0.32	–	1.98	m²	**2.30**
Fertilizers; application during aftercare period Fertilizers; in top 150 mm of topsoil at 35 g/m²							
Mascot Microfine; turf fertilizer; 8+0+6 + 2% Mg + 4% Fe	5.04	0.12	3.94	–	5.80	100 m²	**9.74**
Enmag; controlled release fertilizer (8–9 months); 11+22+09	8.81	0.12	3.94	–	10.13	100 m²	**14.07**
Mascot Outfield; turf fertilizer; 8+12+8	3.91	0.12	3.94	–	4.50	100 m²	**8.44**
Mascot Outfield; turf fertilizer; 9+5+5	3.27	0.12	3.94	–	3.76	100 m²	**7.70**
Fertilizers; in top 150 mm of topsoil at 70 g/m²							
Mascot Microfine; turf fertilizer; 8+0+6 + 2% Mg + 4% Fe	10.07	0.12	3.94	–	11.58	100 m²	**15.52**
Enmag; controlled release fertilizer (8–9 months); 11+22+09	17.63	0.12	3.94	–	20.27	100 m²	**24.21**
Mascot Outfield; turf fertilizer; 8+12+8	7.81	0.12	3.94	–	8.98	100 m²	**12.92**
Mascot Outfield; turf fertilizer; 9+5+5	6.53	0.12	3.94	–	7.51	100 m²	**11.45**

37 SOFT LANDSCAPING

Item – Overhead and Profit Included	PC £	Labour hours	Labour £	Plant £	Material £	Unit	Total rate £
AFTERCARE AS PART OF A LANDSCAPE CONTRACT							
Maintenance operations (Note: the following rates apply to aftercare maintenance executed as part of a landscaping contract only.							
Weeding and hand forking planted areas; including disposing weeds and debris on site; areas maintained weekly	–	–	0.13	–	–	m²	**0.13**
Weeding and hand forking planted areas; including disposing weeds and debris on site; areas maintained monthly	–	0.01	0.32	–	–	m²	**0.32**
Extra over weeding and hand forking planted areas for disposing excavated material off site; loaded to maintenance vehicle and transported to the contractor's yard							
green waste for composting	–	–	–	–	–	m³	**2.86**
green waste to contractor's yard deposited in skips	–	–	–	48.75	2.86	m³	**51.61**
Edging to plant beds; maintaing edge against turfed surround or boundary; edging by half moon tool							
Planting beds							
maintaining edges	–	0.04	1.29	–	–	m	**1.29**
reforming edges; removing arisings	–	0.07	2.15	–	–	m	**2.15**
Tree pits edges							
maintaining edges	–	0.10	3.22	–	–	nr	**3.22**
reforming edges; removing arisings	–	0.20	6.44	–	–	nr	**6.44**
Sweeping and cleaning pathways and surfaces							
Sweep paths or surfaces to high footfall public or commercial areas							
weekly visits	–	0.0067	0.22	–	–	m²	**0.22**
sweeping snow	–	0.03	0.81	–	–	m²	**0.81**
gritting and clearing ice	–	0.03	0.81	–	0.08	m²	**0.89**

37 SOFT LANDSCAPING

Item – Overhead and Profit Included	PC £	Labour hours	Labour £	Plant £	Material £	Unit	Total rate £
AFTERCARE AS PART OF A LANDSCAPE CONTRACT – cont							
Top up mulch at the end of the 12 month maintenance period to the full thickness as shown below							
Mulch; Melcourt Industries Ltd; Bark Nuggets®; to plant beds; delivered in 80 m³ loads; maximum distance 25 m							
50 mm thick	0.49	0.03	0.92	–	0.56	m²	**1.48**
75 mm thick	0.73	0.01	0.16	–	0.84	m²	**1.00**
Mulch; Melcourt Industries Ltd; Bark Nuggets®; to plant beds; delivered in 25 m³ loads; maximum distance 25 m							
50 mm thick	0.62	0.03	0.92	–	0.71	m²	**1.63**
75 mm thick	0.94	0.03	1.07	–	1.08	m²	**2.15**
Mulch; Melcourt Industries Ltd; Amenity Bark Mulch FSC; to plant beds; delivered in 80 m³ loads; maximum distance 25 m							
50 mm thick	0.80	0.03	0.92	–	0.92	m²	**1.84**
75 mm thick	1.12	0.03	1.07	–	1.29	m²	**2.36**
Mulch; Melcourt Industries Ltd; Amenity Bark Mulch FSC; to plant beds; delivered in 25 m³ loads; maximum distance 25 m							
50 mm thick	1.14	0.03	0.81	–	1.31	m²	**2.12**
75 mm thick	1.48	0.03	0.92	–	1.70	m²	**2.62**
Fertilizers; at 35 g/m²							
Mascot Microfine; turf fertilizer; 8+0+6 + 2% Mg + 4% Fe	5.04	0.12	3.94	–	5.80	100 m²	**9.74**
Enmag; controlled release fertilizer (8–9 months); 11+22+09	8.81	0.12	3.94	–	10.13	100 m²	**14.07**
Mascot Outfield; turf fertilizer; 8+12+8	3.91	0.12	3.94	–	4.50	100 m²	**8.44**
Mascot Outfield; turf fertilizer; 9+5+5	3.27	0.12	3.94	–	3.76	100 m²	**7.70**
Fertilizers; at 70 g/m²							
Mascot Microfine; turf fertilizer; 8+0+6 + 2% Mg + 4% Fe	10.07	0.12	3.94	–	11.58	100 m²	**15.52**
Enmag; controlled release fertilizer (8–9 months); 11+22+09	17.63	0.12	3.94	–	20.27	100 m²	**24.21**
Mascot Outfield; turf fertilizer; 8+12+8	7.81	0.12	3.94	–	8.98	100 m²	**12.92**
Mascot Outfield; turf fertilizer; 9+5+5	6.53	0.12	3.94	–	7.51	100 m²	**11.45**

37 SOFT LANDSCAPING

Item – Overhead and Profit Included	PC £	Labour hours	Labour £	Plant £	Material £	Unit	Total rate £
Watering planting; evenly; at a rate of 5 l/m^2							
using hand-held watering equipment	–	0.25	8.05	–	–	100 m^2	**8.05**
using sprinkler equipment and with sufficient water pressure to run one 15 m radius sprinkler	–	0.14	4.47	–	–	100 m^2	**4.47**
using movable spray lines powering three sprinkler heads with a radius of 15 m and allowing for 60% overlap (irrigation machinery costs not included)	–	0.02	0.51	–	–	100 m^2	**0.51**

LONG-TERM MAINTENANCE

Preamble: Long-term landscape maintenance

Maintenance on long-term contracts differs in cost from that of maintenance as part of a landscape contract. In this section the contract period is generally 3–5 years. Staff are generally allocated to a single project only and therefore productivity is higher whilst overhead costs are lower. Labour costs in this section are lower than the costs used in other parts of the book. Machinery is assumed to be leased over a five year period and written off over the same period. The costs of maintenance and consumables for the various machinery types have been included in the information that follows. Finance costs for the machinery have not been allowed for.

The rates shown below are for machines working in unconfined contiguous areas. Users should adjust the times and rates if working in smaller spaces or spaces with obstructions.

37 SOFT LANDSCAPING

Item – Overhead and Profit Included	PC £	Labour hours	Labour £	Plant £	Material £	Unit	Total rate £
MARKET PRICES OF LANDSCAPE CHEMICALS							
The following table provides the areas of spraying in each planting scenario for plants which require 1.00 m dia. weed free circles							
Area of weed free circles required in each planting density							
plants at 500 mm centres; 400 nr/ 100 m²	–	–	–	–	400.00	m²	**400.00**
plants at 600 m centres; 278 nr/ 100 m²	–	–	–	–	278.00	m²	**278.00**
plants at 750 mm centres; 178 nr/ 100 m²	–	–	–	–	178.00	m²	**178.00**
plants at 1.00 m centres; 100 nr/ 100 m²	–	–	–	–	100.00	m²	**100.00**
plants at 1.50 m centres; 4444 nr/ ha	–	–	–	–	3490.00	m²	**3490.00**
plants at 1.75 m centres; 3265 nr/ ha	–	–	–	–	2564.00	m²	**2564.00**
plants at 2.00 m centres; 2500 nr/ ha	–	–	–	–	1963.00	m²	**1963.00**
MARKET PRICES OF LANDSCAPE CHEMICALS AT SUGGESTED APPLICATION RATES							
Note: All chemicals are standard knapsack or backpack applied unless specifically stated as CDA or TDA.							
TOTAL HERBICIDES							
Herbicide applications; CDA; chemical application via low pressure specialized wands to landscape planting; application to maintain 1.00 m dia. clear circles (0.79 m²) around new planting							
Vanquish Biactive; ALS Ltd; Bayer; enhanced movement glyphosate; application rate 15 litre/ha							
plants at 1.50 m centres; 4444 nr/ ha	–	–	–	–	61.24	ha	**61.24**
plants at 1.75 m centres; 3265 nr/ ha	–	–	–	–	45.20	ha	**45.20**
plants at 2.00 m centres; 2500 nr/ ha	–	–	–	–	34.66	ha	**34.66**
mass spraying	–	–	–	–	175.64	ha	**175.64**
spot spraying	–	–	–	–	1.76	ha	**1.76**

37 SOFT LANDSCAPING

Item – Overhead and Profit Included	PC £	Labour hours	Labour £	Plant £	Material £	Unit	Total rate £
Total herbicides; CDA; low pressure sustainable applications Hilite; Nomix Enviro Ltd; glyphosate oil-emulsion; Total Droplet Control (TDC formerly CDA); low volume sustainable chemical applications; formulation controls annual and perennial grasses and non-woody broad-leaved weeds in amenity and industrial areas, parks, highways, paved areas and factory sites; also effective for clearing vegetation prior to planting of ornamental species, around the bases of trees, shrubs and roses, as well as in forestry areas							
10 l/ha	–	–	–	–	2.76	100 m²	**2.76**
7.5 l/ha	–	–	–	–	2.07	100 m²	**2.07**
Gallup Biograde Amenity; Rigby Taylor Ltd; glyphosate 360 g/l formulation							
general use; woody weeds; ash, beech, bracken, bramble; 3 l/ha	–	–	–	–	0.33	100 m²	**0.33**
annual and perennial grasses; heather (peat soils); 4 l/ha	–	–	–	–	0.44	100 m²	**0.44**
pre-planting; general clearance; 5 l/ha	–	–	–	–	0.55	100 m²	**0.55**
heather; mineral soils; 6 l/ha	–	–	–	–	0.66	100 m²	**0.66**
rhododendron; 10 l/ha	–	–	–	–	1.10	100 m²	**1.10**
Gallup Hi-aktiv Amenity; Rigby Taylor Ltd; 490 g/l formulation							
general use; woody weeds; ash, beech, bracken, bramble; 2.2 l/ha	–	–	–	–	0.26	100 m²	**0.26**
annual and perennial grasses; heather (peat soils); 2.9 l/ha	–	–	–	–	0.35	100 m²	**0.35**
pre-planting; general clearance; 3.7 l/ha	–	–	–	–	0.44	100 m²	**0.44**
heather; mineral soils; 4.4 l/ha	–	–	–	–	0.53	100 m²	**0.53**
rhododendron; 7.3 l/ha	–	–	–	–	0.88	100 m²	**0.88**
Discman CDA Biograde; Rigby Taylor Ltd							
application rate; 7 l/ha; light vegetation and annual weeds	–	–	–	–	1.22	100 m²	**1.22**
application rate; 8.5 l/ha; forestry/ woodlands; before planting, to control broad-leaved and grass weeds	–	–	–	–	1.48	100 m²	**1.48**
application rate; 10 l/ha; established deep-rooted annuals, perennial grasses and broad-leaved weeds	–	–	–	–	1.74	100 m²	**1.74**

37 SOFT LANDSCAPING

Item – Overhead and Profit Included	PC £	Labour hours	Labour £	Plant £	Material £	Unit	Total rate £
MARKET PRICES OF LANDSCAPE CHEMICALS AT SUGGESTED APPLICATION RATES – cont							
Roundup Pro Biactive 360; Everris							
application rate; 5 l/ha; land not intended for cropping	–	–	–	–	0.33	100 m²	**0.33**
application rate; 10 l/ha; forestry; weed control	–	–	–	–	0.66	100 m²	**0.66**
Roundup Pro Vantage 480 ; Everris; 480 g/litre glyphosate							
application rate 4 l/ha; natural surfaces not intended to bear vegetation; amenity vegetation control; permeable surfaces overlaying soil; hard surfaces	–	–	–	–	0.29	100 m²	**0.29**
application rate 8 l/ha; forest; forest nursery weed control	–	–	–	–	0.58	100 m²	**0.58**
application rate 16 mm/l; stump application	–	–	–	–	0.12	litre	**0.12**
Roundup Pro Biactive 450; Everris; enhanced movement glyphosate; application rate 4 l/ha							
plants at 1.50 m centres; 4444 nr/ha	–	–	–	–	41.84	ha	**41.84**
plants at 1.75 m centres; 3265 nr/ha	–	–	–	–	30.78	ha	**30.78**
plants at 2.00 m centres; 2500 nr/ha	–	–	–	–	23.55	ha	**23.55**
mass spraying	–	–	–	–	31.82	ha	**31.82**
spot spraying	–	–	–	–	1.19	100 m²	**1.19**
Dual; Nomix Enviro Ltd; glyphosate/sulfosulfuron; TDC; low volume sustainable chemical applications; residual effect lasting up to 6 months; shrub beds, gravelled areas, tree bases, parks, industrial areas, hard areas and around obstacles							
application rate 9 l/ha	–	–	–	–	2.97	100 m²	**2.97**

37 SOFT LANDSCAPING

Item – Overhead and Profit Included	PC £	Labour hours	Labour £	Plant £	Material £	Unit	Total rate £
Total herbicides with residual pre-emergent action							
Pistol; ALS; Bayer; barrier prevention for emergence control and glyphosate for control of emerged weeds at a single application per season; application rate 4.5 litre/ha; 40 g/l diflufenican and 250 g/l glyphosate							
plants at 1.50 m centres; 4444 nr/ha	–	–	–	–	47.57	ha	**47.57**
plants at 1.75 m centres; 3265 nr/ha	–	–	–	–	34.85	ha	**34.85**
plants at 2.00 m centres; 2500 nr/ha	–	–	–	–	34.09	ha	**34.09**
Residual herbicides							
Chikara; Rigby Taylor Ltd; Belchim Crop Protection; flazasulfuron; systemic pre-emergent and early post-emergent herbicide for the control of annual and perennial weeds on natural surfaces not intended to bear vegetation and permeable surfaces overlying soil							
150 g/ha	–	–	–	–	2.25	100 m²	**2.25**
Contact herbicides							
Jewel; Everris; carfentrazone and mecoprop-p; selective weed and mosskiller							
application rate; 1.5 kg/ha	–	–	–	–	1.14	100 m²	**1.14**
Natural Weed and Moss Spray; Headland Amenity Ltd; acetic acid							
application rate; 100 l/ha	–	–	–	–	5.42	100 m²	**5.42**

37 SOFT LANDSCAPING

Item – Overhead and Profit Included	PC £	Labour hours	Labour £	Plant £	Material £	Unit	Total rate £
MARKET PRICES OF LANDSCAPE CHEMICALS AT SUGGESTED APPLICATION RATES – cont							
Selective herbicides CDA Hilite; Nomix Enviro; glyphosate oil-emulsion; TD; low volume sustainable chemical applications; formulation controls annual and perennial grasses and non-woody broad-leaved weeds in amenity and industrial areas, parks, highways, paved areas and factory sites; also effective for clearing vegetation prior to planting of ornamental species, around the bases of trees, shrubs and roses, as well as in forestry areas							
10 l/ha	–	–	–	–	2.76	100 m^2	**2.76**
7.5 l/ha	–	–	–	–	2.07	100 m^2	**2.07**
Contact and systemic combined herbicides Jewel; Everris; carfentrazone and mecoprop-p; selective weed and mosskiller							
application rate; 1.5 kg/ha	–	–	–	–	1.14	100 m^2	**1.14**
Finale; ALS; Bayer; glufosinate-ammonium; controls grasses and broad-leaved weeds; it can also be used for weed control around non-edible ornamental crops including trees and shrubs and in tree nurseries; may be used to 'burn off' weeds in shrub beds prior to treatment with a residual herbicide; is also ideal for preparing sports turf for line marking							
application rate; 5 l/ha; seedlings of all species; established annual weeds and grasses as specified	–	–	–	–	1.97	100 m^2	**1.97**
application rate; 8 l/ha; established annual and perennial weeds and grasses as specified in the product guide	–	–	–	–	3.15	100 m^2	**3.15**

37 SOFT LANDSCAPING

Item – Overhead and Profit Included	PC £	Labour hours	Labour £	Plant £	Material £	Unit	Total rate £
Selective herbicides; power or knapsack sprayer							
Junction; Rigby Taylor Ltd; selective post-emergence turf herbicide containing 6.25 g/litre florasulam plus 452 g/litre 2,4-D; specially designed for use by contractors for maintenance treatments of a wide range of broad-leaved weeds whilst giving excellent safety to turfgrass							
application rate; 1.2 l/ha	–	–	–	–	0.54	100 m²	**0.54**
Greenor; Rigby Taylor Ltd; fluroxypyr, clopyralid and MCPA; systemic selective treatment of weeds in turfgrass							
application rate; 4 l/ha	–	–	–	–	1.09	100 m²	**1.09**
Crossbar Herbicide; Rigby Taylor Ltd; fluroxypyr, 2,4-D and dicamba for control of weeds in amenity turf including fine turf							
application rate; 2 l/ha	–	–	–	–	0.85	100 m²	**0.85**
Praxys; Everris; fluroxypyr-meptyl, clopyralid and florasulam; selective, systemic post-emergence herbicide for managed amenity turf, lawns and amenity grassland with high selectivity to established and young turf							
application rate; 9090–14,285 m²/5 l	–	–	–	–	0.79	100 m²	**0.79**
Cabadex; Headland Amenity Ltd; fluroxypyr and florasulam; newly sown and established turf (particularly speedwell, yarrow and yellow suckling clover)							
application rate 2 l/ha	–	–	–	–	0.69	100 m²	**0.69**
Blaster; Headland Amenity Ltd; triclopyr clopyralid; nettles, docks, thistles, brambles, woody weeds in amenity grassland							
application rate; 4 l/ha	–	–	–	–	2.77	100 m²	**2.77**
Aquatic herbicides; backpack or power sprayers							
Roundup Pro Biactive 360; Everris							
application rate; 6 l/ha; aquatic weed control	–	–	–	–	0.40	100 m²	**0.40**
Roundup Pro Biactive 450; Everris; 450 g/litre glyphosate							
application rate 4.8 l/ha; enclosed waters, open waters, land immediately adjacent to aquatic areas	–	–	–	–	0.35	100 m²	**0.35**

37 SOFT LANDSCAPING

Item – Overhead and Profit Included	PC £	Labour hours	Labour £	Plant £	Material £	Unit	Total rate £
MARKET PRICES OF LANDSCAPE CHEMICALS AT SUGGESTED APPLICATION RATES – cont							
Aquatic herbicides; TDC Conqueror; Nomix Enviro; low volume sustainable chemical applications glyphosate with aquatic approval							
10 l/ha; established annuals, deep-rooted perennials and broad-leaved weeds	–	–	–	–	0.29	100 m²	**0.29**
7.5 l/ha; light vegetation and small annual weeds, aquatic situations	–	–	–	–	2.15	100 m²	**2.15**
12.5 l/ha; aquatic situations, emergent weeds e.g. reed, grasses and watercress	–	–	–	–	0.36	100 m²	**0.36**
15 l/ha; aquatic situations, floating weeds such as water lilies	–	–	–	–	0.43	100 m²	**0.43**
Fungicides Exterris; Rigby Taylor Ltd; broad spectrum fungicide displaying protective and curative activity against fusarium patch and red thread							
application rate; 700 g/ha	–	–	–	–	3.80	100 m²	**3.80**
Fusion; Rigby Taylor Ltd; contact and systemic fungicide with curative and preventative activity against major turf diseases such as fusarium, red thread, dollar spot, anthracnose and rust							
application rate; 700 g/ha	–	–	–	–	7.19	100 m²	**7.19**
Moss control Jewel; Everris; carfentrazone and mecoprop-p; selective weed and moss killer							
application rate; 1.5 kg/ha	–	–	–	–	1.14	100 m²	**1.14**

37 SOFT LANDSCAPING

Item – Overhead and Profit Included	PC £	Labour hours	Labour £	Plant £	Material £	Unit	Total rate £
LANDSCAPE CHEMICAL APPLICATION							
Chemical applications labours only							
Controlled droplet application (CDA)							
plants at 1.50 m centres; 4444 nr/ ha	–	9.26	543.10	–	–	ha	**543.10**
plants at 1.50 m centres; 44.44 nr/ 100 m^2	–	0.93	54.31	–	–	100 m^2	**54.31**
plants at 1.75 m centres; 3265 nr/ ha	–	6.80	398.82	–	–	ha	**398.82**
plants at 1.75 m centres; 3265 nr/ ha	–	0.68	39.88	–	–	100 m^2	**39.88**
plants at 2.00 m centres; 2500 nr/ ha	–	5.21	305.57	–	–	ha	**305.57**
plants at 2.00 m centres; 2500 nr/ ha	–	0.52	30.56	–	–	100 m^2	**30.56**
spot spraying to shrub beds	–	0.20	6.44	–	–	100 m^2	**6.44**
spot spraying to hard surfaces	–	0.13	4.03	–	–	100 m^2	**4.03**
mass spraying	–	2.50	146.63	–	–	ha	**146.63**
mass spraying	–	0.25	14.66	–	–	100 m^2	**14.66**
Power spraying							
mass spraying to areas containing vegetation	–	2.00	64.40	202.40	–	ha	**266.80**
mass spraying to hard surfaces	–	–	0.14	0.41	–	100 m^2	**0.55**
spraying to kerbs and channels along roadways 300 mm wide	–	1.25	40.25	126.50	–	1 km	**166.75**
General herbicides; in accordance with manufacturer's instructions; Knapsack spray application; see 'Market rates of landscape chemicals' for costs of specific applications							
knapsack sprayer; selective spraying around bases of plants	–	0.18	5.65	–	–	100 m^2	**5.65**
knapsack sprayer; spot spraying	–	0.25	8.05	–	–	100 m^2	**8.05**
knapsack sprayer; mass spraying	–	0.40	12.88	–	–	100 m^2	**12.88**
granular distribution by hand or hand applicator	–	0.20	6.44	–	–	100 m^2	**6.44**
Backpack spraying; keeping weed free at bases of plants							
plants at 1.50 m centres; 4444 nr/ ha	–	13.89	814.65	–	–	ha	**814.65**
plants at 1.75 m centres; 3265 nr/ ha	–	10.20	598.23	–	–	ha	**598.23**
plants at 2.00 m centres; 2500 nr/ ha	–	7.82	458.34	–	–	ha	**458.34**
mass spraying	–	7.50	439.88	–	–	ha	**439.88**

37 SOFT LANDSCAPING

Item – Overhead and Profit Included	PC £	Labour hours	Labour £	Plant £	Material £	Unit	Total rate £
LANDSCAPE CHEMICAL APPLICATION – cont							
Herbicide applications; standard backpack spray applicators; application to maintain 1.00 m dia. clear circles (0.79 m²) around new planting							
Everris; Roundup Pro Biactive 360; glyphosate; enhanced movement glyphosate; application rate 5 litre/ha							
plants at 1.50 m centres; 4444 nr/ ha	12.78	12.35	724.04	–	14.70	ha	**738.74**
plants at 1.75 m centres; 3265 nr/ ha	9.35	9.07	531.96	–	10.75	ha	**542.71**
plants at 2.00 m centres; 2500 nr/ ha	7.21	6.95	407.32	–	8.29	ha	**415.61**
mass spraying	–	–	0.05	–	–	m²	**0.05**
GRASS CUTTING							
Grass cutting only; ride-on or tractor drawn equipment; works carried out on one site only such as large fields or amenity areas where labour and machinery is present for a full day; grass cutting only; strimming not included							
Using multiple-gang mower with cylindrical cutters; contiguous areas such as playing fields and the like larger than 3000 m²							
3 gang; 2.13 m cutting width	–	0.01	0.31	0.14	–	100 m²	**0.45**
5 gang; 3.40 m cutting width	–	–	0.28	0.18	–	100 m²	**0.46**
7 gang; 4.65 m cutting width	–	–	0.17	0.15	–	100 m²	**0.32**
Using multiple-gang mower with cylindrical cutters; non-contiguous areas such as verges and general turf areas							
3 gang	–	0.01	0.61	0.14	–	100 m²	**0.75**
5 gang	–	0.01	0.39	0.18	–	100 m²	**0.57**
Using multiple-rotary mower with vertical drive shaft and horizontally rotating bar or disc cutters; contiguous areas larger than 3000 m²							
cutting grass, overgrowth or the like using flail mower or reaper	–	0.01	0.59	0.35	–	100 m²	**0.94**

37 SOFT LANDSCAPING

Item – Overhead and Profit Included	PC £	Labour hours	Labour £	Plant £	Material £	Unit	Total rate £
Grass cutting with strimming; ride-on or tractor drawn equipment; works carried out on one site only such as large fields or amenity areas where labour and machinery is present for a full day; includes accompanying strimmer operative							
Using multiple-gang mower with cylindrical cutters; contiguous areas such as playing fields and the like larger than 3000 m²							
3 gang; 2.13 m cutting width	–	0.01	0.61	0.16	–	100 m²	**0.77**
5 gang; 3.40 m cutting width	–	0.01	0.55	0.21	–	100 m²	**0.76**
7 gang; 4.65 m cutting width	–	0.01	0.35	0.20	–	100 m²	**0.55**
Using multiple-gang mower with cylindrical cutters; non-contiguous areas such as verges and general turf areas							
3 gang	–	0.02	1.23	0.23	–	100 m²	**1.46**
5 gang	–	0.01	0.78	0.22	–	100 m²	**1.00**
Using multiple-rotary mower with vertical drive shaft and horizontally rotating bar or disc cutters; contiguous areas larger than 3000 m²							
cutting grass, overgrowth or the like using flail mower or reaper	–	0.02	1.17	0.44	–	100 m²	**1.61**
Grass cutting only; ride-on or tractor drawn equipment; works carried out on multiple sites; machinery moved by trailer between sites; strimming not included							
Using multiple-gang mower with cylindrical cutters; contiguous areas such as playing fields and the like larger than 3000 m²							
3 gang; 2.13 m cutting width	–	0.01	0.49	0.14	–	100 m²	**0.63**
5 gang; 3.40 m cutting width	–	0.01	0.45	0.18	–	100 m²	**0.63**
7 gang; 4.65 m cutting width	–	–	0.28	0.15	–	100 m²	**0.43**
Using multiple-gang mower with cylindrical cutters; non-contiguous areas such as verges and general turf areas							
3 gang	–	0.02	0.99	0.14	–	100 m²	**1.13**
5 gang	–	0.01	0.63	0.18	–	100 m²	**0.81**

37 SOFT LANDSCAPING

Item – Overhead and Profit Included	PC £	Labour hours	Labour £	Plant £	Material £	Unit	Total rate £
GRASS CUTTING – cont							
Grass cutting only – cont							
Using multiple-rotary mower with vertical drive shaft and horizontally rotating bar or disc cutters; contiguous areas larger than 3000 m²							
cutting grass, overgrowth or the like using flail mower or reaper	–	0.02	0.95	0.35	–	100 m²	**1.30**
Cutting grass, overgrowth or the like; using tractor-mounted side-arm flail mower; in areas inaccessible to alternative machine; on surface							
not exceeding 30° from horizontal	–	0.01	0.69	0.41	–	100 m²	**1.10**
30–50° from horizontal	–	0.02	1.37	0.41	–	100 m²	**1.78**
Grass cutting and strimming; ride-on or tractor drawn equipment; works carried out on multiple sites; machinery moved by trailer between sites; includes for accompanying strimmer operative							
Using multiple-gang mower with cylindrical cutters; contiguous areas such as playing fields and the like larger than 3000 m²							
3 gang; 2.13 m cutting width	–	0.02	1.00	0.17	–	100 m²	**1.17**
5 gang; 3.40 m cutting width	–	0.02	0.91	0.22	–	100 m²	**1.13**
7 gang; 4.65 m cutting width	–	0.01	0.55	0.17	–	100 m²	**0.72**
Using multiple-gang mower with cylindrical cutters; non-contiguous areas such as verges and general turf areas							
3 gang	–	0.03	1.98	0.22	–	100 m²	**2.20**
5 gang	–	0.02	1.28	0.29	–	100 m²	**1.57**
Using multiple-rotary mower with vertical drive shaft and horizontally rotating bar or disc cutters; contiguous areas larger than 3000 m²							
cutting grass, overgrowth or the like using flail mower or reaper	–	0.03	1.91	0.43	–	100 m²	**2.34**
Cutting grass, overgrowth or the like; using tractor-mounted side-arm flail mower; in areas inaccessible to alternative machine; on surface							
not exceeding 30° from horizontal	–	0.02	1.39	0.47	–	100 m²	**1.86**
30° to 50° from horizontal	–	0.05	2.75	0.52	–	100 m²	**3.27**

37 SOFT LANDSCAPING

Item – Overhead and Profit Included	PC £	Labour hours	Labour £	Plant £	Material £	Unit	Total rate £
Grass cutting; pedestrian operated equipment; the following rates are for grass cutting only; there is no allowance for accompanying strimming operations							
Using cylinder lawn mower fitted with not less than five cutting blades, front and rear rollers; on surface not exceeding 30° from horizontal; arisings let fly; width of cut							
51 cm	–	0.03	1.79	0.20	–	100 m	**1.99**
61 cm	–	0.03	1.50	0.22	–	100 m	**1.72**
71 cm	–	0.02	1.29	0.23	–	100 m	**1.52**
91 cm	–	0.02	1.00	0.46	–	100 m	**1.46**
Using rotary self-propelled mower; width of cut							
45 cm	–	0.02	1.16	0.13	–	100 m²	**1.29**
81 cm	–	0.01	0.64	0.21	–	100 m²	**0.85**
120 cm	–	0.01	0.44	0.13	–	100 m²	**0.57**
Add for using grass box for collecting and depositing arisings							
removing and depositing arisings	–	0.03	1.46	–	–	100 m²	**1.46**
Grass cutting; pedestrian operated equipment; the following rates are for grass cutting with accompanying strimming operations							
Using cylinder lawn mower fitted with not less than five cutting blades, front and rear rollers; on surface not exceeding 30° from horizontal; arisings let fly; width of cut							
51 cm	–	0.06	3.58	0.33	–	100 m	**3.91**
61 cm	–	0.05	2.99	0.33	–	100 m	**3.32**
71 cm	–	0.04	2.56	0.33	–	100 m	**2.89**
91 cm	–	0.03	2.00	0.54	–	100 m	**2.54**
Using rotary self-propelled mower; width of cut							
45 cm	–	0.04	2.32	0.22	–	100 m²	**2.54**
81 cm	–	0.02	1.29	0.25	–	100 m²	**1.54**
91 cm	–	0.02	1.14	0.16	–	100 m²	**1.30**
120 cm	–	0.01	0.86	0.16	–	100 m²	**1.02**
Add for using grass box for collecting and depositing arisings							
removing and depositing arisings	–	0.03	1.46	–	–	100 m²	**1.46**
Add for 30–50° from horizontal	–	–	–	–	–	33%	**–**
Add for slopes exceeding 50°	–	–	–	–	–	100%	**–**

37 SOFT LANDSCAPING

Item – Overhead and Profit Included	PC £	Labour hours	Labour £	Plant £	Material £	Unit	Total rate £
GRASS CUTTING – cont							
Grass cutting – cont							
Cutting grass or light woody undergrowth; using trimmer with nylon cord or metal disc cutter; on surface							
not exceeding 30° from horizontal	–	0.10	5.86	0.46	–	100 m²	**6.32**
30–50° from horizontal	–	0.20	11.73	0.92	–	100 m²	**12.65**
exceeding 50° from horizontal	–	0.25	14.66	1.15	–	100 m²	**15.81**
Grass cutting; collecting arisings							
Extra over for tractor drawn and self-propelled machinery using attached grass boxes; depositing arisings							
22 cuts per year	–	0.03	1.46	–	–	100 m²	**1.46**
18 cuts per year	–	0.04	2.20	–	–	100 m²	**2.20**
12 cuts per year	–	0.05	2.93	–	–	100 m²	**2.93**
4 cuts per year	–	0.13	7.34	–	–	100 m²	**7.34**
Arisings collected by trailed sweepers							
22 cuts per year	–	0.01	0.49	0.30	–	100 m²	**0.79**
18 cuts per year	–	0.01	0.66	0.30	–	100 m²	**0.96**
12 cuts per year	–	0.02	0.98	0.30	–	100 m²	**1.28**
4 cuts per year	–	0.03	1.96	0.30	–	100 m²	**2.26**
Disposing arisings on site; 100 m distance maximum							
22 cuts per year	–	–	0.23	0.03	–	100 m²	**0.26**
18 cuts per year	–	–	0.29	0.04	–	100 m²	**0.33**
12 cuts per year	–	0.01	0.43	0.06	–	100 m²	**0.49**
4 cuts per year	–	0.02	1.28	0.17	–	100 m²	**1.45**
Disposal of arisings off site							
22 cuts per year	–	–	0.24	0.06	0.08	100 m²	**0.38**
18 cuts per year	–	0.01	0.30	0.07	0.10	100 m²	**0.47**
12 cuts per year	–	–	0.20	0.18	0.18	100 m²	**0.56**
4 cuts per year	–	0.01	0.59	1.08	0.29	100 m²	**1.96**
GROUNDCARE OPERATIONS							
Harrowing							
Harrowing grassed area with							
drag harrow	–	0.01	0.37	0.17	–	100 m²	**0.54**
chain or light flexible spiked harrow	–	0.01	0.49	0.17	–	100 m²	**0.66**

37 SOFT LANDSCAPING

Item – Overhead and Profit Included	PC £	Labour hours	Labour £	Plant £	Material £	Unit	Total rate £
Scarifying							
Mechanical							
Sisis ARP4; including grass collection box; towed by tractor; area scarified annually	–	0.01	0.69	0.72	–	100 m²	**1.41**
Sisis ARP4; including grass collection box; towed by tractor; area scarified two years previously	–	0.01	0.82	0.87	–	100 m²	**1.69**
pedestrian operated self-powered equipment	–	0.04	2.05	1.15	–	100 m²	**3.20**
add for disposal of arisings	–	0.01	0.74	1.36	1.43	100 m²	**3.53**
By hand							
hand implement	–	0.25	14.66	–	–	100 m²	**14.66**
add for disposal of arisings	–	0.01	0.74	1.36	1.43	100 m²	**3.53**
Rolling							
Rolling grassed area; equipment towed by tractor; once over; using smooth roller	–	0.01	0.43	0.18	–	100 m²	**0.61**
Turf aeration							
By machine							
Vertidrain turf aeration equipment towed by tractor to effect a minimum penetration of 100–250 mm at 100 mm centres	–	0.02	1.14	2.81	–	100 m²	**3.95**
Ryan GA 30; self-propelled turf aerating equipment; to effect a minimum penetration of 100 mm at varying centres	–	0.02	1.14	1.20	–	100 m²	**2.34**
Groundsman; pedestrian operated; self-powered solid or slitting tine turf aerating equipment to effect a minimum penetration of 100 mm	–	0.07	4.12	1.09	–	100 m²	**5.21**
Cushman core harvester; self-propelled; for collection of arisings	–	0.01	0.56	0.87	–	100 m²	**1.43**
By hand							
hand fork; to effect a minimum penetration of 100 mm and spaced 150 mm apart	–	0.67	39.10	–	–	100 m²	**39.10**
hollow tine hand implement; to effect a minimum penetration of 100 mm and spaced 150 mm apart	–	1.00	58.65	–	–	100 m²	**58.65**
collection of arisings by hand	–	1.50	87.97	–	–	100 m²	**87.97**

37 SOFT LANDSCAPING

Item – Overhead and Profit Included	PC £	Labour hours	Labour £	Plant £	Material £	Unit	Total rate £
GROUNDCARE OPERATIONS – cont							
Turf areas; surface treatments and top dressing; British sugar topsoil							
Apply screened topdressing to grass surfaces; spread using Tru-Lute							
Sand soil mixes 90/10 to 50/50	–	–	0.06	0.10	0.56	m²	**0.72**
Apply screened soil 3 mm Kettering loam to goal mouths and worn areas							
20 mm thick	–	0.01	0.30	–	2.28	m²	**2.58**
10 mm thick	2.97	0.01	0.30	–	3.42	m²	**3.72**
Leaf clearance; clearing grassed area of leaves and other extraneous debris							
Using equipment towed by tractor							
large grassed areas with perimeters of mature trees such as sports fields and amenity areas	–	0.01	0.37	0.08	–	100 m²	**0.45**
large grassed areas containing ornamental trees and shrub beds	–	–	0.13	1.13	–	100 m²	**1.26**
Using pedestrian operated mechanical equipment and blowers							
grassed areas with perimeters of mature trees such as sports fields and amenity areas	–	0.01	0.49	0.91	–	100 m²	**1.40**
grassed areas containing ornamental trees and shrub beds	–	0.05	2.93	0.39	–	100 m²	**3.32**
verges	–	0.03	1.96	0.05	–	100 m²	**2.01**
By hand							
grassed areas with perimeters of mature trees such as sports fields and amenity areas	–	0.05	2.93	0.40	–	100 m²	**3.33**
grassed areas containing ornamental trees and shrub beds	–	0.08	4.89	0.67	–	100 m²	**5.56**
verges	–	0.17	9.77	1.35	–	100 m²	**11.12**
Removal of arisings							
areas with perimeters of mature trees	–	–	0.20	0.12	0.11	100 m²	**0.43**
areas containing ornamental trees and shrub beds	–	0.01	0.59	0.45	0.29	100 m²	**1.33**
Litter clearance							
Collection and disposal of litter from grassed area	–	0.01	0.30	–	0.22	100 m²	**0.52**
Collection and disposal of litter from isolated grassed area not exceeding 1000 m²	–	0.02	1.17	–	0.22	100 m²	**1.39**

37 SOFT LANDSCAPING

Item – Overhead and Profit Included	PC £	Labour hours	Labour £	Plant £	Material £	Unit	Total rate £
Tree guards, stakes and ties							
Adjusting existing tree tie	–	0.02	0.98	–	–	nr	**0.98**
Taking up single or double tree stake and ties; removing and disposing	–	0.03	0.81	–	–	nr	**0.81**
SHRUB BED MAINTENANCE							
Edge maintenance							
Maintain edges where lawn abuts pathway or hard surface using							
strimmer	–	–	0.15	0.02	–	m	**0.17**
shears	–	0.01	0.47	–	–	m	**0.47**
Maintain edges where lawn abuts plant bed using							
mechanical edging tool	–	–	0.18	0.04	–	m	**0.22**
shears	–	–	0.31	–	–	m	**0.31**
half moon edging tool	–	0.01	0.56	–	–	m	**0.56**
Pruning shrubs							
Trimming ground cover planting							
soft groundcover; vinca ivy and the like	–	0.33	28.08	–	–	100 m^2	**28.08**
woody groundcover; cotoneaster and the like	–	0.50	42.12	–	–	100 m^2	**42.12**
Pruning massed shrub border (measure ground area)							
shrub beds pruned annually	–	–	0.28	–	–	m^2	**0.28**
shrub beds pruned hard every 3 years	–	0.01	0.78	–	–	m^2	**0.78**
Cutting off dead heads							
bush or standard rose	–	0.02	1.40	–	–	nr	**1.40**
climbing rose	–	0.03	2.35	–	–	nr	**2.35**
Pruning roses							
bush or standard rose	–	0.02	1.40	–	–	nr	**1.40**
climbing rose or rambling rose; tying in as required	–	0.02	1.87	–	–	nr	**1.87**
Pruning ornamental shrub; height before pruning (increase these rates by 50% if pruning work has not been executed during the previous two years)							
not exceeding 1 m	–	0.01	1.13	–	–	nr	**1.13**
1 to 2 m	–	0.02	1.55	–	–	nr	**1.55**
exceeding 2 m	–	0.04	3.52	–	–	nr	**3.52**
Removing excess growth etc. from face of building etc.; height before pruning							
not exceeding 2 m	–	0.01	0.81	–	–	nr	**0.81**
2 to 4 m	–	0.02	1.40	–	–	nr	**1.40**
4 to 6 m	–	0.03	2.35	–	–	nr	**2.35**
6 to 8 m	–	0.04	3.52	–	–	nr	**3.52**
8 to 10 m	–	0.05	4.01	–	–	nr	**4.01**

37 SOFT LANDSCAPING

Item – Overhead and Profit Included	PC £	Labour hours	Labour £	Plant £	Material £	Unit	Total rate £
SHRUB BED MAINTENANCE – cont							
Pruning shrubs – cont							
Removing epicormic growth from base of shrub or trunk and base of tree (any height, any dia.); number of growths							
not exceeding 10	–	0.02	1.40	–	–	nr	**1.40**
10 to 20	–	0.02	1.87	–	–	nr	**1.87**
Shrub beds, borders and planters							
Lifting							
bulbs	–	0.17	14.04	–	–	100 nr	**14.04**
tubers or corms	–	0.13	11.24	–	–	100 nr	**11.24**
established herbaceous plants; hoeing and depositing for replanting	–	0.66	56.17	–	–	100 nr	**56.17**
Cutting down spent growth of herbaceous plant; clearing arisings							
unstaked	–	0.01	0.56	–	–	nr	**0.56**
staked; not exceeding 4 stakes per plant; removing stakes and putting into store	–	0.01	0.70	–	–	nr	**0.70**
Hand weeding; regular visits; costs per occasion							
newly planted areas; mulched beds;	–	0.03	2.35	–	–	100 m^2	**2.35**
newly planted areas; unmulched	–	0.04	3.75	–	–	100 m^2	**3.75**
established areas	–	0.01	0.93	–	–	100 m^2	**0.93**
Removing grasses from groundcover areas	–	0.02	1.87	–	–	100 m^2	**1.87**
Hand digging with fork between shrubs; not exceeding 150 mm deep; breaking down lumps; leaving surface with a medium tilth	–	0.44	37.44	–	–	100 m^2	**37.44**
Hand digging with fork or spade to an average depth of 230 mm; breaking down lumps; leaving surface with a medium tilth	–	0.66	56.17	–	–	100 m^2	**56.17**
Hand hoeing; not exceeding 50 mm deep; leaving surface with a medium tilth	–	0.13	11.24	–	–	100 m^2	**11.24**
Hand raking to remove stones etc.; breaking down lumps: leaving surface with a fine tilth prior to planting	–	0.22	18.72	–	–	100 m^2	**18.72**
Hand weeding; planter, window box; not exceeding 1.00 m^2							
ground level box	–	0.02	1.40	–	–	nr	**1.40**
box accessed by stepladder	–	0.03	2.35	–	–	nr	**2.35**

37 SOFT LANDSCAPING

Item – Overhead and Profit Included	PC £	Labour hours	Labour £	Plant £	Material £	Unit	Total rate £
Spreading only compost, mulch or processed bark to a depth of 75 mm on shrub bed with existing mature planting	–	0.03	2.55	–	–	m²	2.55
recently planted areas	–	0.02	1.87	–	–	m²	1.87
groundcover and herbaceous areas	–	0.02	2.12	–	–	m²	2.12
Clearing cultivated area of litter and other extraneous debris; using hand implement (exludes winter leaf clearance							
weekly maintenance; private areas	–	0.01	0.47	–	–	100 m²	0.47
weekly maintenance; public areas	–	0.01	0.93	–	–	100 m²	0.93
daily maintenance public areas	–	0.01	0.47	–	–	100 m²	0.47
Clearing cultivated area of winter leaf fall, extraneous debris; using hand implement; winter leaf; removal of cleared material to stockpile on site							
per occasion; occasional trees and deciduous shrubs within 100 m²	–	0.01	1.17	–	–	100 m²	1.17
per occasion; dense tall trees and deciduous shrubs in close proximity	–	0.03	2.35	–	–	100 m²	2.35
BEDDING MAINTENANCE							
Works to established bedding areas							
Lifting							
bedding plants; hoeing and depositing for disposal	–	0.90	76.38	–	–	100 m²	76.38
Hand digging with fork; not exceeding 150 mm deep: breaking down lumps; leaving surface with a medium tilth	–	0.24	20.31	–	–	100 m²	20.31
Hand weeding							
newly planted areas	–	0.64	54.17	–	–	100 m²	54.17
established areas	–	0.16	13.54	–	–	100 m²	13.54
Hand digging with fork or spade to an average depth of 230 mm; breaking down lumps; leaving surface with a medium tilth	–	0.16	13.54	–	–	100 m²	13.54
Hand hoeing: not exceeding 50 mm deep; leaving surface with a medium tilth	–	0.13	10.83	–	–	100 m²	10.83
Hand raking to remove stones etc.; breaking down lumps; leaving surface with a fine tilth prior to planting	–	0.21	18.05	–	–	100 m²	18.05

37 SOFT LANDSCAPING

Item – Overhead and Profit Included	PC £	Labour hours	Labour £	Plant £	Material £	Unit	Total rate £
BEDDING MAINTENANCE – cont							
Works to established bedding areas – cont							
Hand weeding; planter, window box; not exceeding 1.00 m²							
ground level box	–	0.02	1.36	–	–	nr	**1.36**
box accessed by stepladder	–	0.03	2.25	–	–	nr	**2.25**
Spreading only; compost, mulch or processed bark to a depth of 75 mm							
on shrub bed with existing mature planting	–	0.03	2.46	–	–	m²	**2.46**
recently planted areas	–	0.02	1.82	–	–	m²	**1.82**
groundcover and herbaceous areas	–	0.02	2.02	–	–	m²	**2.02**
Collecting bedding from nursery	–	0.95	81.25	18.02	–	100 m²	**99.27**
Setting out							
mass planting single variety	–	0.04	3.39	–	–	m²	**3.39**
pattern	–	0.11	9.02	–	–	m²	**9.02**
Planting only							
massed bedding plants	–	0.06	5.42	–	–	m²	**5.42**
Clearing cultivated area of leaves, litter and other extraneous debris; using hand implement							
weekly maintenance	–	0.04	3.39	–	–	100 m²	**3.39**
daily maintenance	–	–	–	–	–	100 m²	**-**
IRRIGATION AND WATERING							
Irrigation and watering							
Hand held hosepipe; flow rate 25 litres per minute; irrigation requirement							
10 litres/m²	–	0.74	19.49	–	–	100 m²	**19.49**
15 litres/m²	–	1.10	29.09	–	–	100 m²	**29.09**
20 litres/m²	–	1.46	38.70	–	–	100 m²	**38.70**
25 litres/m²	–	1.84	48.59	–	–	100 m²	**48.59**
Hand held hosepipe; flow rate 40 litres per minute; irrigation requirement							
10 litres/m²	–	0.46	12.22	–	–	100 m²	**12.22**
15 litres/m²	–	0.69	18.33	–	–	100 m²	**18.33**
20 litres/m²	–	0.91	24.15	–	–	100 m²	**24.15**
25 litres/m²	–	1.15	30.31	–	–	100 m²	**30.31**

37 SOFT LANDSCAPING

Item – Overhead and Profit Included	PC £	Labour hours	Labour £	Plant £	Material £	Unit	Total rate £
HEDGE MAINTENANCE							
Hedge cutting; field hedges cut once or twice annually							
Trimming sides and top using hand tool or hand held mechanical tools							
not exceeding 2 m high	–	0.05	2.93	0.23	–	10 m²	3.16
2 to 4 m high	–	0.17	9.77	0.77	–	10 m²	10.54
Hedge cutting; ornamental							
Trimming sides and top using hand tool or hand held mechanical tools							
not exceeding 2 m high	–	0.06	3.67	0.29	–	10 m²	3.96
2 to 4 m high	–	0.25	14.66	1.15	–	10 m²	15.81
Hedge cutting; reducing width; hand tool or hand held mechanical tools							
Not exceeding 2 m high							
average depth of cut not exceeding 300 mm	–	0.01	0.74	0.06	–	m²	0.80
average depth of cut 300–600 mm	–	0.03	1.63	0.13	–	m²	1.76
average depth of cut 600–900 mm	–	0.03	1.84	0.15	–	m²	1.99
2–4 m high							
average depth of cut not exceeding 300 mm	–	0.02	1.17	0.09	–	m²	1.26
average depth of cut 300–600 mm	–	0.03	1.84	0.15	–	m²	1.99
average depth of cut 600–900 mm	–	0.05	2.93	0.23	–	m²	3.16
4–6 m high							
average depth of cut not exceeding 300 mm	–	0.05	2.93	0.23	–	m²	3.16
average depth of cut 300–600 mm	–	0.08	4.89	0.38	–	m²	5.27
average depth of cut 600–900 mm	–	0.25	14.66	1.15	–	m²	15.81
Hedge cutting; reducing width; tractor mounted hedge cutting equipment							
Not exceeding 2 m high							
average depth of cut not exceeding 300 mm	–	0.02	1.17	0.87	–	10 m²	2.04
average depth of cut 300–600 mm	–	0.03	1.46	1.09	–	10 m²	2.55
average depth of cut 600–900 mm	–	0.10	5.86	4.37	–	10 m²	10.23
2–4 m high							
average depth of cut not exceeding 300 mm	–	0.01	0.37	0.28	–	10 m²	0.65
average depth of cut 300–600 mm	–	0.01	0.74	0.54	–	10 m²	1.28
average depth of cut 600–900 mm	–	0.01	0.59	0.44	–	10 m²	1.03

37 SOFT LANDSCAPING

Item – Overhead and Profit Included	PC £	Labour hours	Labour £	Plant £	Material £	Unit	Total rate £
HEDGE MAINTENANCE – cont							
Hedge cutting; reducing height; hand tool or hand held mechanical tools							
Not exceeding 2 m high							
average depth of cut not exceeding 300 mm	–	0.03	1.96	0.15	–	10 m²	**2.11**
average depth of cut 300–600 mm	–	0.07	3.91	0.31	–	10 m²	**4.22**
average depth of cut 600–900 mm	–	0.20	11.73	0.92	–	10 m²	**12.65**
2–4 m high							
average depth of cut not exceeding 300 mm	–	0.02	0.98	0.08	–	m²	**1.06**
average depth of cut 300–600 mm	–	0.03	1.96	0.15	–	m²	**2.11**
average depth of cut 600–900 mm	–	0.10	5.86	0.46	–	m²	**6.32**
4–6 m high							
average depth of cut not exceeding 300 mm	–	0.03	1.96	0.15	–	m²	**2.11**
average depth of cut 300–600 mm	–	0.06	3.67	0.29	–	m²	**3.96**
average depth of cut 600–900 mm	–	0.13	7.34	0.58	–	m²	**7.92**
Hedge cutting; removal and disposal of arisings							
Sweeping up and depositing arisings							
300 mm cut	–	0.03	1.46	–	–	10 m²	**1.46**
600 mm cut	–	0.10	5.86	–	–	10 m²	**5.86**
900 mm cut	–	0.20	11.73	–	–	10 m²	**11.73**
Chipping arisings							
300 mm cut	–	0.01	0.59	0.48	–	10 m²	**1.07**
600 mm cut	–	0.04	2.45	2.00	–	10 m²	**4.45**
900 mm cut	–	0.10	5.86	4.80	–	10 m²	**10.66**
Disposal of unchipped arisings							
300 mm cut	–	0.01	0.49	0.68	0.19	10 m²	**1.36**
600 mm cut	–	0.02	0.98	1.36	0.29	10 m²	**2.63**
900 mm cut	–	0.04	2.45	3.38	0.72	10 m²	**6.55**
Disposal of chipped arisings							
300 mm cut	–	–	0.10	0.23	0.29	10 m²	**0.62**
600 mm cut	–	0.01	0.49	0.23	0.57	10 m²	**1.29**
900 mm cut	–	0.02	0.98	0.45	1.43	10 m²	**2.86**

37 SOFT LANDSCAPING

Item – Overhead and Profit Included	PC £	Labour hours	Labour £	Plant £	Material £	Unit	Total rate £
CYCLIC MAINTENANCE OPERATIONS							
Note: These items for landscape maintenance may be featured for maintenance in the section Aftercare As Part Of A Landscape Contract.							
Path or hard surface maintenance							
Cyclic maintenance to surfaces based on annual requirements for long-term maintenance contractors							
Check or clear daily for litter or dog faeces	–	12.00	317.40	–	–	100 m^2	**317.40**
Sweeping monthly	–	1.00	26.45	–	–	100 m^2	**26.45**
Jet washing bi-annually	–	0.33	8.73	–	–	100 m^2	**8.73**
Surface grit application; 6 occasions per annum	–	3.00	79.35	–	74.52	100 m^2	**153.87**
Snow clearance; per occasion	–	1.25	33.06	–	–	100 m^2	**33.06**
Childrens play area maintenance							
Visit daily to inspect for safety (Play area as part of larger maintained staffed area)	–	0.10	2.64	–	–	nr	**2.64**
Visit daily to inspect for safety (Play area as isolated play area)	–	0.50	13.23	–	–	nr	**13.23**
Monthly inspections of play equipment integrity and safety	–	1.00	26.45	–	–	nr	**26.45**
Annual repairs and maintenance to equipment in play areas; high density populations	–	4.00	105.80	–	–	nr	**105.80**
Annual repairs and maintenance to equipment in play areas; low density populations	–	1.00	26.45	–	–	nr	**26.45**
Street furniture maintenance – annual cost							
Clean benches weekly	–	2.60	83.72	–	–	nr	**83.72**
Sand timber bench bi-annually; remove splinters	–	0.50	16.10	–	–	nr	**16.10**
Maintain lighting bollard inclusive of cleaning and bulb replacement	–	0.50	16.10	–	4.88	nr	**20.98**
Maintain lighting standard inclusive of cleaning and bulb replacement	–	0.75	24.15	–	4.88	nr	**29.03**
Maintain clean exterior to litter and dog bins	–	3.00	79.35	–	–	nr	**79.35**

37 SOFT LANDSCAPING

Item – Overhead and Profit Included	PC £	Labour hours	Labour £	Plant £	Material £	Unit	Total rate £
CYCLIC MAINTENANCE OPERATIONS – cont							
Path or hard surface maintenance – cont							
Litter maintenance							
Check and clear litter bins daily (high density) based on 240 days per annum	–	20.00	529.00	341.25	16.56	nr	**886.81**
Check and clear litter bins weekly	–	4.33	114.61	228.15	3.59	nr	**346.35**
Clear dog litter bins daily (in conjunction with bin clearance on site)	–	6.00	158.70	331.20	4.14	nr	**494.04**
Clear dog litter bins weekly (in conjunction with bin clearance on site)	–	1.00	26.45	71.76	16.56	nr	**114.77**
Sports maintenance							
Take football posts from store and erect	–	2.00	128.80	–	–	set	**128.80**
Remove football posts and remove to store	–	0.75	48.30	–	–	set	**48.30**
Mark football/rugby field initial mark	20.09	5.00	132.25	–	23.10	nr	**155.35**
Mark football/rugby field remark per occasion	4.63	1.00	26.45	–	5.32	nr	**31.77**
Graffiti maintenance per occasion based on term graffiti removal contract							
Remove graffiti from street furniture	–	–	–	–	53.30	nr	**53.30**
Remove graffiti from masonry	–	–	–	–	53.30	nr	**53.30**
JAPANESE KNOTWEED							
Japanese Knotweed Management Plan; PBA Solutions Ltd							
Formulate Japanese Knotweed Management Plan (KMP) as recommended by the Environment Agency Code of Practice 2006 (EA CoP)	–	–	–	–	–	nr	**1359.61**

37 SOFT LANDSCAPING

Item – Overhead and Profit Included	PC £	Labour hours	Labour £	Plant £	Material £	Unit	Total rate £
Treatment of Japanese Knotweed; PBA Solutions Ltd; removal and/or treatment in accordance with The Knotweed Code of Practice published by the Environment Agency 2006 (EA CoP); Japanese knotweed must be treated by qualified practitioners							
Commercial chemical treatments based on areas less than 50 m² knotweed area; allows for the provision of contractual arrangements including method statements and spraying records; also allows for the provision of risk, environment and COSHH assessments; rates for 3–5 year programme, incl. management plan and post-treatment monitoring							
rate per visit	–	–	–	–	–	nr	**329.70**
rate for complete treatment programme lasting 3 years	–	–	–	–	–	nr	**2968.49**
Commercial chemical treatments based on areas between 50–300 m² knotweed area; allows for the provision of contractual arrangements including method statements and spraying records; also allows for the provision of risk, environment and COSHH assessments; rates for 3–5 year programme, incl. management plan and post-treatment monitoring							
rate per visit	–	–	–	–	–	nr	**409.01**
rate for complete treatment programme lasting 3 years	–	–	–	–	–	nr	**3682.28**
Commercial chemical treatments based on areas greater than 900 m² knotweed area; allows for the provision of contractual arrangements including method statements and spraying records; also allows for the provision of risk, environment and COSHH assessments; rates for 3–5 year programme, incl. management plan and post-treatment monitoring							
rate per visit	–	–	–	–	–	m²	**0.89**
Rate per year (3–5 year)	–	–	–	–	–	m²	**8.11**

37 SOFT LANDSCAPING

Item – Overhead and Profit Included	PC £	Labour hours	Labour £	Plant £	Material £	Unit	Total rate £
JAPANESE KNOTWEED – cont							
Treatment of Japanese Knotweed – cont							
Extra over minimum rate above; for areas over 900 m^2							
rate per visit	–	–	–	–	–	m^2	**0.89**
rate for complete treatment programme lasting 3 years	–	–	–	–	–	m^2	**8.11**
Note: Taking waste off site will require pretreatment. Landfill operators are required by the Environment Agency to obtain evidence of pretreatment to show that the waste has been treated to either reduce its volume or hazardous nature, or facilitate its handling, or enhance its recovery by the waste produce (site of origin) prior to being taken to landfill. Generally pretreatment of Japanese Knotweed can include chemical treatment, cutting or burning. The cost to implement this operation would be similar to other minimum treatment costs above. To enable waste to go to landfill soil testing would need to be completed and consultation with a landfill operator or invasive weed specialist is recommended in the first instance.							
Japanese Knotweed contaminated waste removal; PBA Solutions Ltd (see note above)							
Removal off site to a licensed landfill site using 20 tonne gross vehicle weight lorries; price is for haulage and tipping only; the mileage stated below is the distance from point of collection to licensed landfill site; price excludes landfill tax and loading of material into lorries; rates will fluctuate between landfill operators and the figures given are for initial guidance only							
less than 20 miles	–	–	–	–	–	load	**906.41**
20–40 miles	–	–	–	–	–	load	**963.06**
40–60 miles	–	–	–	–	–	load	**1302.96**

37 SOFT LANDSCAPING

Item – Overhead and Profit Included	PC £	Labour hours	Labour £	Plant £	Material £	Unit	Total rate £
Root barriers; PBA Solutions Ltd; used in conjunction with on site control methods such as bund treatment, cell burial, vertical boundary protection and capping. The following barriers have been reliably used in connection with Japanese Knotweed control and as such are effective when correctly installed as part of a control programme which would usually include for chemical control							
Linear root deflection barriers; installed to trench measured separately							
Flexiroot non-permeable; reinforced LDPE coated polypropylene geomembrane	–	–	–	–	–	m	**5.09**
Cutex Bioroot-barrier X permeable root barrier; mechanically bonded geocomposite consisting of a copper-foil bonded between 2 layers of geotextile	–	–	–	–	–	m²	**13.60**
Landfill tax rates; based on waste being within landfill operator thresholds for controlled waste and with Japanese Knotweed being the only form of contamination; the rates below are based on visual inspections of the load at the gate of the tip							
Rates from April 2018							
standard rate from April 2018 where knotweed is visible at more than 5% on inspection at landfill)	–	–	–	–	105.05	tonne	**105.05**
lower rate (where knotweed is visible at less than 5% on inspection at landfill)	–	–	–	–	3.34	tonne	**3.34**

37 SOFT LANDSCAPING

Item – Overhead and Profit Included	PC £	Labour hours	Labour £	Plant £	Material £	Unit	Total rate £
ECOLOGICAL MITIGATION WORKS							
Ecological mitigation following Ecologist survey; Blackdown Environmental Ltd; Operations shown below are for carrying out recommended operations on a site to receive treatments following or in accord with an ecological consultants survey and recommedations (not included); Items below all exlude any overnight accommodation or subsistence if the site area is remote; All arisings stacked or left on site as is good ecological practice							
Preliminaries							
Rams logistics and the like	–	–	–	–	–	nr	753.53
Badgers and Otters							
Badger artificial sett creation	–	–	–	–	–	nr	13750.41
Badger sett closure 9 entrances footprint 30 × 10 m	–	–	–	–	–	nr	880.03
Badger or Otter fencing 0–200 m	–	–	–	–	–	nr	51.30
Badger or Otter fencing 500 –1000 m	–	–	–	–	–	nr	46.88
Amphibians							
Permanent Amphibian fencing to 200 m	–	–	–	–	–	m	22.24
Permanent Amphibian fencing 500–1000 m	–	–	–	–	–	m	17.50
Semi Permanent Amphibian fencing to 200 m	–	–	–	–	–	m	15.88
Semi Permanent Amphibian fencing 500–1000 m	–	–	–	–	–	m	10.13
Temporary amphibian fencing to 200 m	–	–	–	–	–	m	22.24
Temporary amphibian fencing 500–1000 m	–	–	–	–	–	m	17.50
Pitfall Trap installation to 200 traps	–	–	–	–	–	nr	13.75
Searches; Destructive searches for animals and reptiles							
Destructive search for mammals and wild life 1 ha	–	–	–	–	–	ha	3740.11
Destructive search for mammals and wild life 2 ha	–	–	–	–	–	ha	3063.60
Destructive search for mammals and wild life 5 ha +	–	–	–	–	–	ha	2618.07

37 SOFT LANDSCAPING

Item – Overhead and Profit Included	PC £	Labour hours	Labour £	Plant £	Material £	Unit	Total rate £
Hibernacula construction							
Brick hibernacula	–	–	–	–	–	nr	389.40
Log pile hibernacula	–	–	–	–	–	nr	171.60
Pond construction							
Unlined pond 2.00 m² deep	–	–	–	–	–	450 m²	6435.07
Pond lined with bentonite liner							
2.00 m deep	–	–	–	–	–	450 m²	9900.10
Watervoles; Watervole displacement 1000 m²							
Watervole displacement including vegetation clearance and dewatering	–	–	–	–	–	nr	10010.00
Site preparation for watervole displacement	–	–	–	–	–	nr	10010.10
Watervole displacement cutting operations	–	–	–	–	–	nr	9460.10
Watervole fencing							
Watervole fencing to 200 m²	–	–	–	–	–	m²	41.50
Watervole fencing to 201–500 m²	–	–	–	–	–	m²	41.50
Watervole fencing over 1500 m²	–	–	–	–	–	m²	35.45
Rabbits							
Destructive searches of rabbit burrows	–	–	–	–	–	ha	5390.05

39 ELECTRICAL SERVICES

Item – Overhead and Profit Included	PC £	Labour hours	Labour £	Plant £	Material £	Unit	Total rate £
CLARIFICATION NOTES ON LABOUR COSTS IN THIS SECTION							
General groundworks team							
Generally a two man team is used in this section; The column Labour hours reports team hours. The column Labour £ reports the total cost of the team for the unit of work shown							
2 man expert team	–	1.00	64.40	–	–	hr	**64.40**
craftman	–	1.00	32.20	–	–	hr	**32.20**
subcontract electrician	–	–	–	–	–	hr	**55.00**
GENERAL							
Street area floodlighting – General							
Preamble: There are an enormous number of luminaires available which are designed for small scale urban and garden projects. The designs are continually changing and the landscape designer is advised to consult the manufacturer's latest catalogue. Most manufacturers supply light fittings suitable for column, bracket, bulkhead, wall or soffit mounting. Highway lamps and columns for trafficked roads are not included in this section as the design of highway lighting is a very specialized subject outside the scope of most landscape contracts. The IP reference number refers to the waterproof properties of the fitting; the higher the number the more waterproof the fitting. Most items can be fitted with time clocks or PIR controls.							
CABLES AND SWITCHING							
General groundworks team							
Generally a three man team is used in this section; The column Labour hours reports team hours. The column Labour £ reports the total cost of the team for the unit of work shown							
3 man team	–	1.00	96.60	–	–	hr	**96.60**

39 ELECTRICAL SERVICES

Item – Overhead and Profit Included	PC £	Labour hours	Labour £	Plant £	Material £	Unit	Total rate £
Trenching for electrical services; 3 tonne excavator (bucket volume 0.13 m³); arisings laid alongside							
Straight clear runs in recently excavated or filled ground							
600 mm deep	–	–	–	2.15	–	m	**2.15**
800 mm deep	–	–	–	2.51	–	m	**2.51**
1.00 m deep	–	–	–	3.01	–	m	**3.01**
In brownfield sites or to multiple non-contiguous areas							
600 mm deep	–	–	–	3.76	–	m	**3.76**
800 mm deep	–	–	–	5.65	–	m	**5.65**
1.00 m deep	–	–	–	7.53	–	m	**7.53**
Extra over any types of excavating irrespective of depth for breaking out existing materials; heavy duty 110 V breaker tool							
hard rock	–	5.00	161.00	51.75	–	m³	**212.75**
concrete	–	3.00	96.60	31.05	–	m³	**127.65**
reinforced concrete	–	4.00	128.80	46.92	–	m³	**175.72**
brickwork, blockwork or stonework	–	1.50	48.30	15.52	–	m³	**63.82**
By hand							
600 mm deep	–	0.24	7.74	–	–	m	**7.74**
800 mm deep	–	0.43	13.77	–	–	m	**13.77**
1.00 m deep	–	0.67	21.47	–	–	m	**21.47**
Backfilling of trenches; including laying of electrical marker tape; compacting lightly as work proceeds							
To trenches containing armoured cable							
by machine	0.15	–	–	3.09	0.17	m	**3.26**
by hand	0.15	0.20	6.44	–	0.17	m	**6.61**
To trenches containing ducted cable; including 150 mm sharp sand over the duct							
by machine	2.67	–	–	3.86	3.07	m	**6.93**
by hand	2.67	0.25	8.05	–	3.07	m	**11.12**
Cable to trenches (trenching operations not included); laying only cable							
Twin core steel wire armoured cable; 50 m drums							
1.5 mm core	0.70	0.02	1.29	–	0.81	m	**2.10**
2.5 mm core	0.97	0.03	2.14	–	1.12	m	**3.26**

39 ELECTRICAL SERVICES

Item – Overhead and Profit Included	PC £	Labour hours	Labour £	Plant £	Material £	Unit	Total rate £
CABLES AND SWITCHING – cont							
Cable to trenches (trenching operations not included) – cont							
Twin core steel wire armoured cable; lengths less than 50 m runs							
1.5 mm core	0.73	0.02	1.29	–	0.84	m	**2.13**
2.5 mm core	1.05	0.03	2.14	–	1.21	m	**3.35**
Three core steel wire armoured cable; 50 m drums							
1.5 mm core	0.85	0.02	1.29	–	0.98	m	**2.27**
2.5 mm core	1.21	0.02	1.29	–	1.39	m	**2.68**
4.0 mm core	1.67	0.03	1.61	–	1.92	m	**3.53**
6.0 mm core	2.25	0.03	2.14	–	2.59	m	**4.73**
10.0 mm core	3.60	0.04	2.58	–	4.14	m	**6.72**
Three core steel wire armoured cable; lengths less than 50 m runs							
1.5 mm core	0.91	0.02	1.29	–	1.05	m	**2.34**
2.5 mm core	1.29	0.02	1.29	–	1.48	m	**2.77**
Ducts to trenches; twin wall flexible cable ducts with drawstrings; laid on 150 mm clean sharp sand							
Twin wall duct; laying to trenches							
63 mm × 50 m coils	0.86	0.02	0.64	–	0.99	m	**1.63**
110 mm × 50 m coils	1.78	0.02	0.64	–	2.05	m	**2.69**
Cable drawing through ducts							
straight runs up to 50 m lengths	–	1.00	32.20	–	–	nr	**32.20**
Terminations to armoured cables; cutting coiling and taping length of cable to receive connection to light fitting, transformer and the like; fixing to temporary post							
Armoured cable	–	0.33	10.72	–	–	m	**10.72**
Plain ducted cable	–	0.13	4.03	–	–	m	**4.03**
Junction boxes; fixing to cables to receive connections to light fittings or transformers; to IP68							
Underground jointing box							
2 way	32.80	0.33	13.79	–	37.72	m	**51.51**
3 way	55.35	0.57	23.66	–	63.65	m	**87.31**
Above-ground jointing box							
2 way	11.15	0.33	13.79	–	12.82	m	**26.61**
3 way	14.95	0.40	16.56	–	17.19	m	**33.75**

39 ELECTRICAL SERVICES

Item – Overhead and Profit Included	PC £	Labour hours	Labour £	Plant £	Material £	Unit	Total rate £
Electrical connections							
Connections to existing distribution boards							
per switching circuit located within 1 m of the distribution board; chasing of cables to walls not included	–	–	–	–	–	nr	**73.35**
Connections to switches; external power cable circuits; connections to internal wall switches; drilling through walls and making good; chasing of cables to walls not included							
price for the first circuit	5.95	1.00	32.20	–	80.19	nr	**112.39**
additional circuits	5.95	–	–	–	24.44	nr	**24.44**
POP-UP AND BOLLARD POWER SUPPLY							
Note: All cables ducts and power supply and connections to the distribution boards are not included.							
Pop-up Power supplies; Retractable power supply unit model PUPS02, for power distribution and other services; totally hidden even in operation. Cover in stainless steel, recessed to fill with matching surface finish, or closed; cover size; 570 × 570 × 10 mm opening facilitated by a pivot partially pulled out from the lid surface, and a dual gas actuator. Cover secured in closed position by safety bayonet lock in bronze, with lever lock in stainless steel AISI 304 operated with key (square or triangular footprint). Operates with cover closed and with plug inserted; (max 1 × 63 A), through rubber door for cable outlet. Removable stainless steel plate for water or electrical equipment support (provision for grounding); IP67; RCBO protection							

39 ELECTRICAL SERVICES

Item – Overhead and Profit Included	PC £	Labour hours	Labour £	Plant £	Material £	Unit	Total rate £
POP-UP AND BOLLARD POWER SUPPLY – cont							
PUPS02, 630 × 630 × 1680 mm deep in concrete surround and on concrete base;							
800 × 800 mm × 2.00 m deep;							
Connected to power supply cable and earth sink; (both not included)							
Shows and events							
1 × CEE socket 5p 63 A with protection or 2 × 32 A socket; 1 × power supply cable or terminal block	2125.02	4.00	224.00	39.26	2561.95	nr	**2825.21**
Market places							
1 × 32 A socket with + 3 × 16 A 3P socket or 6 × 16 A 3P sockets + 1× power supply cable or terminal block	2265.02	4.00	224.00	39.26	2701.95	nr	**2965.21**
6 × 16 A 3P sockets + 1× power supply cable or terminal block	2315.02	4.00	224.00	39.26	2751.95	nr	**3015.21**
PUPS 05; 800 × 800 mm; Total power 63 amps single or 3 phase incoming supply. Can be used to supply electricity, water, gas and telephone. Raised by turning handle. Suitable for use in HGV areas.							
Shows and events							
1 no. 63 A 5P or 2 no. 32 A 5P sockets	2388.02	4.00	224.00	39.26	2824.95	nr	**3088.21**
Market places							
6 no.16 A 2P + 2 no. 32 A 2P sockets or 8 no. 16 A 2P sockets	2405.02	4.00	224.00	39.26	2841.95	nr	**3105.21**
8 no.16 A 2P sockets	2465.02	4.00	224.00	39.26	2901.95	nr	**3165.21**
Power supply Bollards; Marshalls Bega; Permanent lockable connecting pillar with installation accessories. Strong construction for power supply in public or industrial facilities							
3 nr safety socket outlets 16 A; 3 CEE-socket outlets 16 A; 2 CEE-socket outlets 16 A; Circuit breakers and connecting pole;							
M70 238 Connecting pillar	2601.29	2.00	112.00	–	2807.95	nr	**2919.95**

39 ELECTRICAL SERVICES

Item – Overhead and Profit Included	PC £	Labour hours	Labour £	Plant £	Material £	Unit	Total rate £
LOW VOLTAGE							
Low voltage lights; connecting to transformers (transformers shown separately); fixing as described							
Hunza Spike adjustable spotlight; 63.5 mm dia. × 75 mm long; c/w 20/35/50 W lamp							
stainless steel	193.00	–	–	–	277.98	nr	**277.98**
copper	120.00	–	–	–	194.03	nr	**194.03**
black	77.00	–	–	–	144.58	nr	**144.58**
Hunza Spike spotlight; 63.5 mm dia. × 75 mm long; c/w 20/35/50 W lamp							
stainless steel	163.00	–	–	–	243.48	nr	**243.48**
copper	114.00	–	–	–	187.13	nr	**187.13**
black	66.00	–	–	–	131.93	nr	**131.93**
Hunza pole lights; single; fixing to concrete base 150 × 150 × 150 mm							
stainless steel	253.00	0.75	24.15	–	347.52	nr	**371.67**
copper	172.00	0.75	24.15	–	254.36	nr	**278.51**
black	109.00	0.75	24.15	–	181.91	nr	**206.06**
Hunza pole lights; double; fixing to concrete base 150 × 150 × 150 mm							
stainless steel	367.00	0.75	24.15	–	478.62	nr	**502.77**
copper	265.00	0.75	24.15	–	361.32	nr	**385.47**
black	367.00	0.75	24.15	–	478.62	nr	**502.77**
Recessed wall lights; eyelid steplights; fixing to brick or stone walls; inclusive of core drilling							
stainless steel	32.80	0.75	24.15	–	327.43	nr	**351.58**
copper	32.80	0.75	24.15	–	296.38	nr	**320.53**
black	32.80	0.75	24.15	–	296.38	nr	**320.53**
Hunza; recessed deck path or lawn lights; inclusive of core drilling excavation and all making good							
stainless steel light installed to deck	228.00	0.50	16.10	–	318.23	nr	**334.33**
stainless steel light installed to path	260.80	0.75	24.15	–	318.23	nr	**342.38**
stainless steel light installed to lawn	260.80	0.40	12.88	–	318.23	nr	**331.11**
stainless steel driveway light	260.80	0.50	16.10	–	318.23	nr	**334.33**

39 ELECTRICAL SERVICES

Item – Overhead and Profit Included	PC £	Labour hours	Labour £	Plant £	Material £	Unit	Total rate £
LOW VOLTAGE – cont							
EXTERIOR LIGHTING TRANSFORMERS							
Outdoor transformers for 12 V exterior lighting; connecting to junction box (not included); IP67							
Single light transformers;							
100 × 68 × 72 mm; 50 vA							
fused	32.49	–	–	–	92.36	nr	**92.36**
Two light transformer;							
130 × 85 × 85 mm; 100 vA							
fused	51.49	–	–	–	114.21	nr	**114.21**
Three light transformer;							
130 × 85 × 85 mm; 150 vA							
fused	56.81	–	–	–	120.33	nr	**120.33**
Four light transformer;							
130 × 85 × 100 mm; 200 vA							
fused	65.08	–	–	–	129.84	nr	**129.84**
MARKET PRICES OF LAMPS							
Market prices of lamps							
Lamps							
70 W HQIT-S	–	–	–	–	12.83	nr	**12.83**
70 W HQIT	–	–	–	–	17.60	nr	**17.60**
70 W SON	–	–	–	–	10.08	nr	**10.08**
70 W SONT	–	–	–	–	8.47	nr	**8.47**
100 W SONT	–	–	–	–	11.70	nr	**11.70**
150 W SONT	–	–	–	–	13.26	nr	**13.26**
28 W 2D	–	–	–	–	11.99	nr	**11.99**
100 W GLS\E27	–	–	–	–	2.16	nr	**2.16**
RECESSED LIGHTING							
Bega Lighting; Recessed step or wall LED Lighting; AC Lighting installed to preformed recesses or cast in by landscape contractor; connections by electrical specialist; (all supply cables, ducts, junctions not included)							
Recessed luminaires for walls and stairs; 65 mm deep; LED							
M 33 053 5.6 W 170 × 70 mm	185.11	0.50	16.10	–	305.60	nr	**321.70**
M 33 054 7.8 W 260 × 70 mm	204.26	0.50	16.10	–	327.62	nr	**343.72**
M 33 055 11.0 W 320 × 70 mm	224.69	0.50	16.10	–	351.11	nr	**367.21**

39 ELECTRICAL SERVICES

Item – Overhead and Profit Included	PC £	Labour hours	Labour £	Plant £	Material £	Unit	Total rate £
Recessed luminaires for walls and stairs; 90 mm deep; Remote controllable; LED							
M 33 058 14.5 W 330 x125 mm	297.10	0.50	16.10	–	448.13	nr	**464.23**
M 33 059 19.5 W 420 × 125 mm	331.82	0.50	16.10	–	488.06	nr	**504.16**
M 33 060 24.0 W 520 × 125 mm	363.83	0.50	16.10	–	524.87	nr	**540.97**
STREET AND PRECINCT LIGHTING							
Pole lighting Bega; Bega lighting poles installed to precinct or parking area; Including all excavations, concrete footings and bases and connections to the underground cabling circuit; connections, underground cabling (not included); pretreated with high quality powdercoating. Aluminium luminaire poles with anchorage unit							
M 70 906 3.50 m high for 76 mm Ø × 135 mm luminaire fitting; 800 mm anchor root	703.97	1.50	144.90	22.57	933.83	nr	**1101.30**
M 70 743 4.50 m high for 76 mm Ø × 135 mm luminaire fitting; 800 mm anchor root	782.98	1.75	169.05	22.57	1024.68	nr	**1216.30**
M 70 749 6.00 m high for 89 mm Ø × 120 mm luminaire fitting; 1.00 m anchor root	1534.04	2.00	193.20	45.15	1935.01	nr	**2173.36**
Aluminium tube-bow poles with 800 mm anchorage unit; Surface lacquered; for traditional style luminaires							
M 70 990 4.30 m high120 Ø mm for 1 luminaire	993.80	1.75	169.05	22.57	1267.12	nr	**1458.74**
M 70 991 4.30 m high120 Ø mm for 2 luminaires	1233.12	1.75	169.05	22.57	1542.34	nr	**1733.96**
M 70 993 4.30 m high120 Ø mm for 3 luminaires	1612.67	1.75	169.05	22.57	1978.83	nr	**2170.45**

39 ELECTRICAL SERVICES

Item – Overhead and Profit Included	PC £	Labour hours	Labour £	Plant £	Material £	Unit	Total rate £
STREET AND PRECINCT LIGHTING – cont							
Pole lighting asymetrical; Bega; luminaires installed to Bega poles (not included); inclusive of all connections to supply via poles; Attack angle adjustable to 0° or 15°; dimmable							
Asymmetrical lighting; 12.0–16.0 m centres							
M 99 515; LED 26 W; 4.0–6.0 m height	567.52	1.00	32.20	7.91	707.65	nr	**747.76**
M 99 519; LED 36.6 W; 6.0–8.0 m height	455.32	1.00	32.20	7.91	578.62	nr	**618.73**
Asymmetrical flat beam; maximum centres and light range shown							
M 99 446 LED 15.7 W; 8 × 18 m centres; 3.5–5.0 m height	621.64	1.00	32.20	7.91	769.89	nr	**810.00**
M 99 491 LED 18.8 W; 8 × 16 m centres; 4.0–6.0 m height	638.30	1.00	32.20	7.91	789.04	nr	**829.15**
M 99 499 LED 28.6w; 12 × 24 m max. centres; 5.0–7.0 m height	648.47	1.00	32.20	7.91	800.74	nr	**840.85**
Symmetrical; 4.00–6.00 m height							
M 84 401 LED 43.5W; 11 m max. centres	1108.82	1.00	32.20	7.91	1330.14	nr	**1370.25**
M 84 402 LED 58.0 W; 22 m max. centres	1407.21	1.00	32.20	7.91	1673.29	nr	**1713.40**
Asymmetrical flat beam; 4.00–6.00 m high							
M 84 403 LED 38.5W; 10 × 17 m max. centres	1214.05	1.00	32.20	7.91	1451.16	nr	**1491.27**
M 84 404 LED 58 W; 14 × 27 m max. centres	1407.21	1.00	32.20	7.91	1673.29	nr	**1713.40**
Pole luminaires 7.00–9.00 m high							
M 99 522; LED; 78.0 W; Single; 12 × 27 m max. centres	1091.49	1.25	40.25	15.81	1323.96	nr	**1380.02**
M 99 533 2 LED; 212.0 W; Double; 12 × 72 m max. centres	2139.36	2.50	80.50	19.77	2597.76	nr	**2698.03**
Traditional style luminaires; Asymmetrical flat beam; 4.0–6.0 m high; LED							
M 77 910; 19.0 W; 8 × 18 max. centres	920.18	1.00	32.20	7.91	1113.21	nr	**1153.32**
M 77 911; 43.0 W; 12 × 22 m max. centres	1111.70	1.00	32.20	7.91	1333.45	nr	**1373.56**
Traditional style luminaires; Asymmetrical flat beam; 4.0–6.00 m high; Metal halide or Sodium vapour lamp; 50 W or 70 W							
M 77 994; 8 × 18 max. centres	722.34	1.00	32.20	7.91	885.69	nr	**925.80**
M 77 996; 12 × 24 m max. centres	874.47	1.00	32.20	7.91	1060.64	nr	**1100.75**

39 ELECTRICAL SERVICES

Item – Overhead and Profit Included	PC £	Labour hours	Labour £	Plant £	Material £	Unit	Total rate £
Extra over for lamps							
Metal halide; HCI-TT 70 W	40.43	–	–	–	46.49	nr	**46.49**
Metal Halide; CDO-TT Plus 50 W 828	25.53	–	–	–	29.36	nr	**29.36**
Sodium Vapour 50 W; SON-T PIA PLUS 50 W	9.04	–	–	–	10.40	nr	**10.40**
Sodium Vapour 70 W; SON-T PIA PLUS 70 W	17.66	–	–	–	20.31	nr	**20.31**

Professional Practice for Landscape Architects, 3rd Edition

TENNANT, GARMORY & WINSCH

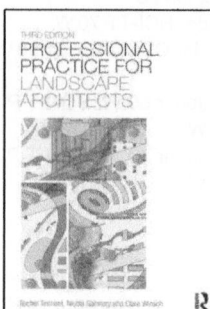

Professional Practice for Landscape Architects third edition deals with the practical issues of being a successful landscape architect professional.

Endorsed by the Landscape Institute, this book is an indispensable guide for licentiate members of the Institute on their Pathway to Chartership. It follows the revised 2013 syllabus covering all aspects of professional judgement, ethics and values, the legal system, organisation and management, legislation and the planning system, environmental policy and control, procurement and implementation. It also serves as a reminder and reference for fully qualified professionals in their everyday practice and for landscape students.

Valuable information is presented in an easy to follow manner with diagrams and schedules, key acts, professional documents and contracts clearly explained and made easy to understand.

A handy list of questions are included to aid with P2C revision, answers of which are found within the text.

September 2015; 234 x 156 mm; 466 pp
Pbk: 978-1-138-78597-7; £32.99

To Order: Tel: +44 (0) 1235 400524 Fax: +44 (0) 1235 400525
or Post: Taylor and Francis Customer Services,
Bookpoint Ltd, Unit T1, 200 Milton Park, Abingdon, Oxon, OX14 4TA UK
Email: book.orders@tandf.co.uk

For a complete listing of all our titles visit:
www.tandf.co.uk

PART 5

Tables and Memoranda

This part contains the following sections:

Building Regulations Pocket Book

Ray Tricker and Samantha Alford

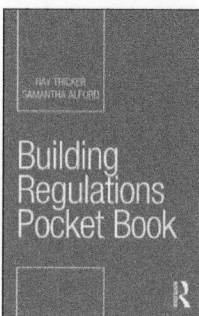

This handy guide provides you with all the information you need to comply with the UK Building Regulations and Approved Documents. On site, in the van, in the office, wherever you are, this is the book you'll refer to time and time again to double check the regulations on your current job.

The *Building Regulations Pocket Book* is the must have reliable and portable guide to compliance with the Building Regulations.

*   Part 1 provides an overview of the Building Act
*   Part 2 offers a handy guide to the dos and don'ts of gaining the Local Council's approval for Planning Permission and Building Regulations Approval
*   Part 3 presents an overview of the requirements of the Approved Documents associated with the Building Regulations
*   Part 4 is an easy to read explanation of the essential requirements of the Building Regulations that any architect, builder or DIYer needs to know to keep their work safe and compliant on both domestic or non-domestic jobs

This book is essential reading for all building contractors and sub-contractors, site engineers, building engineers, building control officers, building surveyors, architects, construction site managers and DIYers. Homeowners will also find it useful to understand what they are responsible for when they have work done on their home (ignorance of the regulations is no defence when it comes to compliance!).

February 2018: 186 x 123 mm: 456 pp

Pb: 978-0-8153-6838-0 : £20.99

To Order: Tel: +44 (0) 1235 400524 Fax: +44 (0) 1235 400525

or Post: Taylor and Francis Customer Services,

Bookpoint Ltd, Unit T1, 200 Milton Park, Abingdon, Oxon, OX14 4TA UK

Email: book.orders@tandf.co.uk

For a complete listing of all our titles visit:
www.tandf.co.uk

CONVERSION TABLES

CONVERSION TABLES

Length	Unit	Conversion factors			
Millimetre	mm	1 in	= 25.4 mm	1 mm	= 0.0394 in
Centimetre	cm	1 in	= 2.54 cm	1 cm	= 0.3937 in
Metre	m	1 ft	= 0.3048 m	1 m	= 3.2808 ft
		1 yd	= 0.9144 m		= 1.0936 yd
Kilometre	km	1 mile	= 1.6093 km	1 km	= 0.6214 mile

Note: 1 cm = 10 mm   1 ft = 12 in
             1 m = 1 000 mm   1 yd = 3 ft
             1 km = 1 000 m   1 mile = 1 760 yd

Area	Unit	Conversion factors			
Square Millimetre	mm^2	1 in^2	= 645.2 mm^2	1 mm^2	= 0.0016 in^2
Square Centimetre	cm^2	1 in^2	= 6.4516 cm^2	1 cm^2	= 1.1550 in^2
Square Metre	m^2	1 ft^2	= 0.0929 m^2	1 m^2	= 10.764 ft^2
		1 yd^2	= 0.8361 m^2	1 m^2	= 1.1960 yd^2
Square Kilometre	km^2	1 $mile^2$	= 2.590 km^2	1 km^2	= 0.3861 $mile^2$

Note: 1 cm^2 = 100 mm^2 1 ft^2 = 144 in^2
             1 m^2 = 10 000 cm^2 1 yd^2 = 9 ft^2
             1 km^2 = 100 hectares   1 acre = 4 840 yd^2
                                          1 $mile^2$ = 640 acres

Volume	Unit	Conversion factors			
Cubic Centimetre	cm^3	1 cm^3	= 0.0610 in^3	1 in^3	= 16.387 cm^3
Cubic Decimetre	dm^3	1 dm^3	= 0.0353 ft^3	1 ft^3	= 28.329 dm^3
Cubic Metre	m^3	1 m^3	= 35.3147 ft^3	1 ft^3	= 0.0283 m^3
		1 m^3	= 1.3080 yd^3	1 yd^3	= 0.7646 m^3
Litre	l	1 l	= 1.76 pint	1 pint	= 0.5683 l
			= 2.113 US pt		= 0.4733 US l

Note: 1 dm^3 = 1 000 cm^3 1 ft^3 = 1 728 in^3 1 pint = 20 fl oz
             1 m^3 = 1 000 dm^3 1 yd^3 = 27 ft^3 1 gal = 8 pints
             1 l = 1 dm^3

Neither the Centimetre nor Decimetre are SI units, and as such their use, particularly that of the Decimetre, is not widespread outside educational circles.

Mass	Unit	Conversion factors			
Milligram	mg	1 mg	= 0.0154 grain	1 grain	= 64.935 mg
Gram	g	1 g	= 0.0353 oz	1 oz	= 28.35 g
Kilogram	kg	1 kg	= 2.2046 lb	1 lb	= 0.4536 kg
Tonne	t	1 t	= 0.9842 ton	1 ton	= 1.016 t

Note: 1 g = 1000 mg   1 oz = 437.5 grains   1 cwt = 112 lb
             1 kg = 1000 g   1 lb = 16 oz   1 ton = 20 cwt
             1 t = 1000 kg   1 stone = 14 lb

Force	Unit	Conversion factors			
Newton	N	1 lbf	= 4.448 N	1 kgf	= 9.807 N
Kilonewton	kN	1 lbf	= 0.004448 kN	1 ton f	= 9.964 kN
Meganewton	MN	100 tonf	= 0.9964 MN		

Tables and Memoranda

CONVERSION TABLES

Pressure and stress	Unit	Conversion factors
Kilonewton per square metre	kN/m^2	1 lbf/in^2 = 6.895 kN/m^2
		1 bar = 100 kN/m^2
Meganewton per square metre	MN/m^2	1 tonf/ft^2 = 107.3 kN/m^2 = 0.1073 MN/m^2
		1 kgf/cm^2 = 98.07 kN/m^2
		1 lbf/ft^2 = 0.04788 kN/m^2

Coefficient of consolidation (Cv) or swelling	Unit	Conversion factors
Square metre per year	m^2/year	1 cm^2/s = 3 154 m^2/year
		1 ft^2/year = 0.0929 m^2/year

Coefficient of permeability	Unit	Conversion factors
Metre per second	m/s	1 cm/s = 0.01 m/s
Metre per year	m/year	1 ft/year = 0.3048 m/year
		= 0.9651 × (10)^8 m/s

Temperature	Unit	Conversion factors	
Degree Celsius	°C	°C = 5/9 × (°F − 32)	°F = (9 × °C)/ 5 + 32

CONVERSION TABLES

SPEED CONVERSION

km/h	m/min	mph	fpm
1	16.7	0.6	54.7
2	33.3	1.2	109.4
3	50.0	1.9	164.0
4	66.7	2.5	218.7
5	83.3	3.1	273.4
6	100.0	3.7	328.1
7	116.7	4.3	382.8
8	133.3	5.0	437.4
9	150.0	5.6	492.1
10	166.7	6.2	546.8
11	183.3	6.8	601.5
12	200.0	7.5	656.2
13	216.7	8.1	710.8
14	233.3	8.7	765.5
15	250.0	9.3	820.2
16	266.7	9.9	874.9
17	283.3	10.6	929.6
18	300.0	11.2	984.3
19	316.7	11.8	1038.9
20	333.3	12.4	1093.6
21	350.0	13.0	1148.3
22	366.7	13.7	1203.0
23	383.3	14.3	1257.7
24	400.0	14.9	1312.3
25	416.7	15.5	1367.0
26	433.3	16.2	1421.7
27	450.0	16.8	1476.4
28	466.7	17.4	1531.1
29	483.3	18.0	1585.7
30	500.0	18.6	1640.4
31	516.7	19.3	1695.1
32	533.3	19.9	1749.8
33	550.0	20.5	1804.5
34	566.7	21.1	1859.1
35	583.3	21.7	1913.8
36	600.0	22.4	1968.5
37	616.7	23.0	2023.2
38	633.3	23.6	2077.9
39	650.0	24.2	2132.5
40	666.7	24.9	2187.2
41	683.3	25.5	2241.9
42	700.0	26.1	2296.6
43	716.7	26.7	2351.3
44	733.3	27.3	2405.9
45	750.0	28.0	2460.6

Tables and Memoranda

CONVERSION TABLES

km/h	m/min	mph	fpm
46	766.7	28.6	2515.3
47	783.3	29.2	2570.0
48	800.0	29.8	2624.7
49	816.7	30.4	2679.4
50	833.3	31.1	2734.0

GEOMETRY

GEOMETRY

Two dimensional figures

Figure	Diagram of figure	Surface area	Perimeter
Square		a^2	$4a$
Rectangle		ab	$2(a+b)$
Triangle		$\frac{1}{2}ch$	$a+b+c$
Circle		πr^2 $\frac{1}{4}\pi d^2$ where $2r = d$	$2\pi r$ πd
Parallelogram		ah	$2(a+b)$
Trapezium		$\frac{1}{2}h(a+b)$	$a+b+c+d$
Ellipse		Approximately πab	$\pi(a+b)$
Hexagon		$2.6 \times a^2$	

GEOMETRY

Figure	Diagram of figure	Surface area	Perimeter
Octagon		$4.83 \times a^2$	6a
Sector of a circle		$\frac{1}{2}rb$ or $\frac{q}{360}\pi r^2$ note b = angle $\frac{q}{360} \times \pi 2r$	
Segment of a circle		S − T where S = area of sector, T = area of triangle	
Bellmouth		$\frac{3}{14} \times r^2$	

GEOMETRY

Three dimensional figures

Figure	Diagram of figure	Surface area	Volume
Cube		$6a^2$	a^3
Cuboid/ rectangular block		$2(ab + ac + bc)$	abc
Prism/ triangular block		$bd + hc + dc + ad$	$\frac{1}{2}hcd$
Cylinder		$2\pi r^2 + 2\pi h$	$\pi r^2 h$
Sphere		$4\pi r^2$	$\frac{4}{3}\pi r^3$
Segment of sphere		$2\pi Rh$	$\frac{1}{6}\pi h(3r^2 + h^2)$ $\frac{1}{3}\pi h^2(3R - H)$
Pyramid		$(a + b)l + ab$	$\frac{1}{3}abh$

Tables and Memoranda

GEOMETRY

Figure	Diagram of figure	Surface area	Volume
Frustum of a pyramid		$l(a + b + c + d) + \sqrt{(ab + cd)}$ [rectangular figure only]	$\frac{h}{3}(ab + cd + \sqrt{abcd})$
Cone		πrl (excluding base) $\pi rl + \pi r^2$ (including base)	$\frac{1}{3}\pi r^2 h$ $\frac{1}{12}\pi d^2 h$
Frustum of a cone		$\pi r^2 + \pi R^2 + \pi l(R + r)$	$\frac{1}{3}\pi(R^2 + Rr + r^2)$

FORMULAE

Formulae

Formula	Description
Pythagoras Theorem	$A^2 = B^2 + C^2$ where A is the hypotenuse of a right-angled triangle and B and C are the two adjacent sides

Simpsons Rule	The Area is divided into an even number of strips of equal width, and therefore has an odd number of ordinates at the division points
	
	$$\text{area} = \frac{S(A + 2B + 4C)}{3}$$
	
	where S = common interval (strip width)
	             A = sum of first and last ordinates
	             B = sum of remaining odd ordinates
	             C = sum of the even ordinates
	
	The Volume can be calculated by the same formula, but by substituting the area of each coordinate rather than its length

Trapezoidal Rule	A given trench is divided into two equal sections, giving three ordinates, the first, the middle and the last
	
	$$\text{volume} = \frac{S \times (A + B + 2C)}{2}$$
	
	where S = width of the strips
	             A = area of the first section
	             B = area of the last section
	             C = area of the rest of the sections

Prismoidal Rule	A given trench is divided into two equal sections, giving three ordinates, the first, the middle and the last
	
	$$\text{volume} = \frac{L \times (A + 4B + C)}{6}$$
	
	where L = total length of trench
	             A = area of the first section
	             B = area of the middle section
	             C = area of the last section

TYPICAL THERMAL CONDUCTIVITY OF BUILDING MATERIALS

(Always check manufacturer's details – variation will occur depending on product and nature of materials)

	Thermal conductivity (W/mK)		Thermal conductivity (W/mK)
Acoustic plasterboard	0.25	Oriented strand board	0.13
Aerated concrete slab (500 kg/m^3)	0.16	Outer leaf brick	0.77
Aluminium	237	Plasterboard	0.22
Asphalt (1700 kg/m^3)	0.5	Plaster dense (1300 kg/m^3)	0.5
Bitumen-impregnated fibreboard	0.05	Plaster lightweight (600 kg/m^3)	0.16
Blocks (standard grade 600 kg/m^3)	0.15	Plywood (950 kg/m^3)	0.16
Blocks (solar grade 460 kg/m^3)	0.11	Prefabricated timber wall panels (check manufacturer)	0.12
Brickwork (outer leaf 1700 kg/m^3)	0.84	Screed (1200 kg/m^3)	0.41
Brickwork (inner leaf 1700 kg/m^3)	0.62	Stone chippings (1800 kg/m^3)	0.96
Dense aggregate concrete block 1800 kg/m^3 (exposed)	1.21	Tile hanging (1900 kg/m^3)	0.84
Dense aggregate concrete block 1800 kg/m^3 (protected)	1.13	Timber (650 kg/m^3)	0.14
Calcium silicate board (600 kg/m^3)	0.17	Timber flooring (650 kg/m^3)	0.14
Concrete general	1.28	Timber rafters	0.13
Concrete (heavyweight 2300 kg/m^3)	1.63	Timber roof or floor joists	0.13
Concrete (dense 2100 kg/m^3 typical floor)	1.4	Roof tile (1900 kg/m^3)	0.84
Concrete (dense 2000 kg/m^3 typical floor)	1.13	Timber blocks (650 kg/m^3)	0.14
Concrete (medium 1400 kg/m^3)	0.51	Cellular glass	0.045
Concrete (lightweight 1200 kg/m^3)	0.38	Expanded polystyrene	0.034
Concrete (lightweight 600 kg/m^3)	0.19	Expanded polystyrene slab (25 kg/m^3)	0.035
Concrete slab (aerated 500 kg/m^3)	0.16	Extruded polystyrene	0.035
Copper	390	Glass mineral wool	0.04
External render sand/cement finish	1	Mineral quilt (12 kg/m^3)	0.04
External render (1300 kg/m^3)	0.5	Mineral wool slab (25 kg/m^3)	0.035
Felt – Bitumen layers (1700 kg/m^3)	0.5	Phenolic foam	0.022
Fibreboard (300 kg/m^3)	0.06	Polyisocyanurate	0.025
Glass	0.93	Polyurethane	0.025
Marble	3	Rigid polyurethane	0.025
Metal tray used in wriggly tin concrete floors (7800 kg/m^3)	50	Rock mineral wool	0.038
Mortar (1750 kg/m^3)	0.8		

EARTHWORK

Weights of Typical Materials Handled by Excavators

The weight of the material is that of the state in its natural bed and includes moisture.
Adjustments should be made to allow for loose or compacted states

Material	Mass (kg/m^3)	Mass (lb/cu yd)
Ashes, dry	610	1028
Ashes, wet	810	1365
Basalt, broken	1954	3293
Basalt, solid	2933	4943
Bauxite, crushed	1281	2159
Borax, fine	849	1431
Caliche	1440	2427
Cement, clinker	1415	2385
Chalk, fine	1221	2058
Chalk, solid	2406	4055
Cinders, coal, ash	641	1080
Cinders, furnace	913	1538
Clay, compacted	1746	2942
Clay, dry	1073	1808
Clay, wet	1602	2700
Coal, anthracite, solid	1506	2538
Coal, bituminous	1351	2277
Coke	610	1028
Dolomite, lumpy	1522	2565
Dolomite, solid	2886	4864
Earth, dense	2002	3374
Earth, dry, loam	1249	2105
Earth, Fullers, raw	673	1134
Earth, moist	1442	2430
Earth, wet	1602	2700
Felsite	2495	4205
Fieldspar, solid	2613	4404
Fluorite	3093	5213
Gabbro	3093	5213
Gneiss	2696	4544
Granite	2690	4534
Gravel, dry ¼ to 2 inch	1682	2835
Gravel, dry, loose	1522	2565
Gravel, wet ¼ to 2 inch	2002	3374
Gypsum, broken	1450	2444
Gypsum, solid	2787	4697
Hardcore (consolidated)	1928	3249
Lignite, dry	801	1350
Limestone, broken	1554	2619
Limestone, solid	2596	4375
Magnesite, magnesium ore	2993	5044
Marble	2679	4515
Marl, wet	2216	3735
Mica, broken	1602	2700
Mica, solid	2883	4859
Peat, dry	400	674
Peat, moist	700	1179

EARTHWORK

Material	Mass (kg/m³)	Mass (lb/cu yd)
Peat, wet	1121	1889
Potash	1281	2159
Pumice, stone	640	1078
Quarry waste	1438	2423
Quartz sand	1201	2024
Quartz, solid	2584	4355
Rhyolite	2400	4045
Sand and gravel, dry	1650	2781
Sand and gravel, wet	2020	3404
Sand, dry	1602	2700
Sand, wet	1831	3086
Sandstone, solid	2412	4065
Shale, solid	2637	4444
Slag, broken	2114	3563
Slag, furnace granulated	961	1619
Slate, broken	1370	2309
Slate, solid	2667	4495
Snow, compacted	481	810
Snow, freshly fallen	160	269
Taconite	2803	4724
Trachyte	2400	4045
Trap rock, solid	2791	4704
Turf	400	674
Water	1000	1685

Transport Capacities

Type of vehicle	Capacity of vehicle	
	Payload	Heaped capacity
Wheelbarrow	150	0.10
1 tonne dumper	1250	1.00
2.5 tonne dumper	4000	2.50
Articulated dump truck (Volvo A20 6 × 4)	18500	11.00
Articulated dump truck (Volvo A35 6 × 6)	32000	19.00
Large capacity rear dumper (Euclid R35)	35000	22.00
Large capacity rear dumper (Euclid R85)	85000	50.00

EARTHWORK

Machine Volumes for Excavating and Filling

Machine type	Cycles per minute	Volume per minute (m³)
1.5 tonne excavator	1	0.04
	2	0.08
	3	0.12
3 tonne excavator	1	0.13
	2	0.26
	3	0.39
5 tonne excavator	1	0.28
	2	0.56
	3	0.84
7 tonne excavator	1	0.28
	2	0.56
	3	0.84
21 tonne excavator	1	1.21
	2	2.42
	3	3.63
Backhoe loader JCB3CX excavator	1	0.28
Rear bucket capacity 0.28 m³	2	0.56
	3	0.84
Backhoe loader JCB3CX loading	1	1.00
Front bucket capacity 1.00 m³	2	2.00

Machine Volumes for Excavating and Filling

Machine type	Loads per hour	Volume per hour (m³)
1 tonne high tip skip loader	5	2.43
Volume 0.485 m³	7	3.40
	10	4.85
3 tonne dumper	4	7.60
Max volume 2.40 m³	5	9.50
Available volume 1.9 m³	7	13.30
	10	19.00
6 tonne dumper	4	15.08
Max volume 3.40 m³	5	18.85
Available volume 3.77 m³	7	26.39
	10	37.70

EARTHWORK

Bulkage of Soils (after excavation)

Type of soil	Approximate bulking of 1 m³ after excavation
Vegetable soil and loam	25–30%
Soft clay	30–40%
Stiff clay	10–20%
Gravel	20–25%
Sand	40–50%
Chalk	40–50%
Rock, weathered	30–40%
Rock, unweathered	50–60%

Shrinkage of Materials (on being deposited)

Type of soil	Approximate bulking of 1 m³ after excavation
Clay	10%
Gravel	8%
Gravel and sand	9%
Loam and light sandy soils	12%
Loose vegetable soils	15%

Voids in Material Used as Subbases or Beddings

Material	m³ of voids/m³
Alluvium	0.37
River grit	0.29
Quarry sand	0.24
Shingle	0.37
Gravel	0.39
Broken stone	0.45
Broken bricks	0.42

Angles of Repose

Type of soil		Degrees
Clay	– dry	30
	– damp, well drained	45
	– wet	15–20
Earth	– dry	30
	– damp	45
Gravel	– moist	48
Sand	– dry or moist	35
	– wet	25
Loam		40

EARTHWORK

Slopes and Angles

Ratio of base to height	Angle in degrees
5:1	11
4:1	14
3:1	18
2:1	27
1½:1	34
1:1	45
1:1½	56
1:2	63
1:3	72
1:4	76
1:5	79

Grades (in Degrees and Percents)

Degrees	Percent	Degrees	Percent
1	1.8	24	44.5
2	3.5	25	46.6
3	5.2	26	48.8
4	7.0	27	51.0
5	8.8	28	53.2
6	10.5	29	55.4
7	12.3	30	57.7
8	14.0	31	60.0
9	15.8	32	62.5
10	17.6	33	64.9
11	19.4	34	67.4
12	21.3	35	70.0
13	23.1	36	72.7
14	24.9	37	75.4
15	26.8	38	78.1
16	28.7	39	81.0
17	30.6	40	83.9
18	32.5	41	86.9
19	34.4	42	90.0
20	36.4	43	93.3
21	38.4	44	96.6
22	40.4	45	100.0
23	42.4		

EARTHWORK

Bearing Powers

Ground conditions		Bearing power		
		kg/m^2	lb/in^2	Metric t/m^2
Rock,	broken	483	70	50
	solid	2415	350	240
Clay,	dry or hard	380	55	40
	medium dry	190	27	20
	soft or wet	100	14	10
Gravel,	cemented	760	110	80
Sand,	compacted	380	55	40
	clean dry	190	27	20
Swamp and alluvial soils		48	7	5

Earthwork Support

Maximum depth of excavation in various soils without the use of earthwork support

Ground conditions	Feet (ft)	Metres (m)
Compact soil	12	3.66
Drained loam	6	1.83
Dry sand	1	0.3
Gravelly earth	2	0.61
Ordinary earth	3	0.91
Stiff clay	10	3.05

It is important to note that the above table should only be used as a guide. Each case must be taken on its merits and, as the limited distances given above are approached, careful watch must be kept for the slightest signs of caving in

CONCRETE WORK

Weights of Concrete and Concrete Elements

Type of material		kg/m³	lb/cu ft
Ordinary concrete (dense aggregates)			
Non-reinforced plain or mass concrete			
Nominal weight		2305	144
Aggregate	– limestone	2162 to 2407	135 to 150
	– gravel	2244 to 2407	140 to 150
	– broken brick	2000 (av)	125 (av)
	– other crushed stone	2326 to 2489	145 to 155
Reinforced concrete			
Nominal weight		2407	150
Reinforcement	– 1%	2305 to 2468	144 to 154
	– 2%	2356 to 2519	147 to 157
	– 4%	2448 to 2703	153 to 163
Special concretes			
Heavy concrete			
Aggregates	– barytes, magnetite	3210 (min)	200 (min)
	– steel shot, punchings	5280	330
Lean mixes			
Dry-lean (gravel aggregate)		2244	140
Soil-cement (normal mix)		1601	100

CONCRETE WORK

Type of material		kg/m² per mm thick	lb/sq ft per inch thick
Ordinary concrete (dense aggregates)			
Solid slabs (floors, walls etc.)			
Thickness:	75 mm or 3 in	184	37.5
	100 mm or 4 in	245	50
	150 mm or 6 in	378	75
	250 mm or 10 in	612	125
	300 mm or 12 in	734	150
Ribbed slabs			
Thickness:	125 mm or 5 in	204	42
	150 mm or 6 in	219	45
	225 mm or 9 in	281	57
	300 mm or 12 in	342	70
Special concretes			
Finishes etc.			
	Rendering, screed etc. Granolithic, terrazzo	1928 to 2401	10 to 12.5
	Glass-block (hollow) concrete	1734 (approx)	9 (approx)
Prestressed concrete		Weights as for reinforced concrete (upper limits)	
Air-entrained concrete		Weights as for plain or reinforced concrete	

CONCRETE WORK

Average Weight of Aggregates

Materials	Voids %	Weight kg/m³
Sand	39	1660
Gravel 10–20 mm	45	1440
Gravel 35–75 mm	42	1555
Crushed stone	50	1330
Crushed granite (over 15 mm)	50	1345
(n.e. 15 mm)	47	1440
'All-in' ballast	32	1800–2000

Material	kg/m³	lb/cu yd
Vermiculite (aggregate)	64–80	108–135
All-in aggregate	1999	125

Applications and Mix Design

Site mixed concrete

Recommended mix	Class of work suitable for	Cement (kg)	Sand (kg)	Coarse aggregate (kg)	Nr 25 kg bags cement per m³ of combined aggregate
1:3:6	Roughest type of mass concrete such as footings, road haunching over 300 mm thick	208	905	1509	8.30
1:2.5:5	Mass concrete of better class than 1:3:6 such as bases for machinery, walls below ground etc.	249	881	1474	10.00
1:2:4	Most ordinary uses of concrete, such as mass walls above ground, road slabs etc. and general reinforced concrete work	304	889	1431	12.20
1:1.5:3	Watertight floors, pavements and walls, tanks, pits, steps, paths, surface of 2 course roads, reinforced concrete where extra strength is required	371	801	1336	14.90
1:1:2	Works of thin section such as fence posts and small precast work	511	720	1206	20.40

CONCRETE WORK

Ready mixed concrete

Application	Designated concrete	Standardized prescribed concrete	Recommended consistence (nominal slump class)
Foundations			
Mass concrete fill or blinding	GEN 1	ST2	S3
Strip footings	GEN 1	ST2	S3
Mass concrete foundations			
Single storey buildings	GEN 1	ST2	S3
Double storey buildings	GEN 3	ST4	S3
Trench fill foundations			
Single storey buildings	GEN 1	ST2	S4
Double storey buildings	GEN 3	ST4	S4
General applications			
Kerb bedding and haunching	GEN 0	ST1	S1
Drainage works – immediate support	GEN 1	ST2	S1
Other drainage works	GEN 1	ST2	S3
Oversite below suspended slabs	GEN 1	ST2	S3
Floors			
Garage and house floors with no embedded steel	GEN 3	ST4	S2
Wearing surface: Light foot and trolley traffic	RC30	ST4	S2
Wearing surface: General industrial	RC40	N/A	S2
Wearing surface: Heavy industrial	RC50	N/A	S2
Paving			
House drives, domestic parking and external parking	PAV 1	N/A	S2
Heavy-duty external paving	PAV 2	N/A	S2

CONCRETE WORK

Prescribed Mixes for Ordinary Structural Concrete

Weights of cement and total dry aggregates in kg to produce approximately one cubic metre of fully compacted concrete together with the percentages by weight of fine aggregate in total dry aggregates

Conc. grade	Nominal max size of aggregate (mm)	40		20		14		10	
	Workability	**Med.**	**High**	**Med.**	**High**	**Med.**	**High**	**Med.**	**High**
	Limits to slump that may be expected (mm)	**50–100**	**100–150**	**25–75**	**75–125**	**10–50**	**50–100**	**10–25**	**25–50**
7	Cement (kg)	180	200	210	230	–	–	–	–
	Total aggregate (kg)	1950	1850	1900	1800	–	–	–	–
	Fine aggregate (%)	30–45	30–45	35–50	35–50	–	–	–	–
10	Cement (kg)	210	230	240	260	–	–	–	–
	Total aggregate (kg)	1900	1850	1850	1800	–	–	–	–
	Fine aggregate (%)	30–45	30–45	35–50	35–50	–	–	–	–
15	Cement (kg)	250	270	280	310	–	–	–	–
	Total aggregate (kg)	1850	1800	1800	1750	–	–	–	–
	Fine aggregate (%)	30–45	30–45	35–50	35–50	–	–	–	–
20	Cement (kg)	300	320	320	350	340	380	360	410
	Total aggregate (kg)	1850	1750	1800	1750	1750	1700	1750	1650
	Sand								
	Zone 1 (%)	35	40	40	45	45	50	50	55
	Zone 2 (%)	30	35	35	40	40	45	45	50
	Zone 3 (%)	30	30	30	35	35	40	40	45
25	Cement (kg)	340	360	360	390	380	420	400	450
	Total aggregate (kg)	1800	1750	1750	1700	1700	1650	1700	1600
	Sand								
	Zone 1 (%)	35	40	40	45	45	50	50	55
	Zone 2 (%)	30	35	35	40	40	45	45	50
	Zone 3 (%)	30	30	30	35	35	40	40	45
30	Cement (kg)	370	390	400	430	430	470	460	510
	Total aggregate (kg)	1750	1700	1700	1650	1700	1600	1650	1550
	Sand								
	Zone 1 (%)	35	40	40	45	45	50	50	55
	Zone 2 (%)	30	35	35	40	40	45	45	50
	Zone 3 (%)	30	30	30	35	35	40	40	45

REINFORCEMENT

Weights of Bar Reinforcement

Nominal sizes (mm)	Cross-sectional area (mm²)	Mass (kg/m)	Length of bar (m/tonne)
6	28.27	0.222	4505
8	50.27	0.395	2534
10	78.54	0.617	1622
12	113.10	0.888	1126
16	201.06	1.578	634
20	314.16	2.466	405
25	490.87	3.853	260
32	804.25	6.313	158
40	1265.64	9.865	101
50	1963.50	15.413	65

Weights of Bars (at specific spacings)

Weights of metric bars in kilogrammes per square metre

Size (mm)	Spacing of bars in millimetres									
	75	100	125	150	175	200	225	250	275	300
6	2.96	2.220	1.776	1.480	1.27	1.110	0.99	0.89	0.81	0.74
8	5.26	3.95	3.16	2.63	2.26	1.97	1.75	1.58	1.44	1.32
10	8.22	6.17	4.93	4.11	3.52	3.08	2.74	2.47	2.24	2.06
12	11.84	8.88	7.10	5.92	5.07	4.44	3.95	3.55	3.23	2.96
16	21.04	15.78	12.63	10.52	9.02	7.89	7.02	6.31	5.74	5.26
20	32.88	24.66	19.73	16.44	14.09	12.33	10.96	9.87	8.97	8.22
25	51.38	38.53	30.83	25.69	22.02	19.27	17.13	15.41	14.01	12.84
32	84.18	63.13	50.51	42.09	36.08	31.57	28.06	25.25	22.96	21.04
40	131.53	98.65	78.92	65.76	56.37	49.32	43.84	39.46	35.87	32.88
50	205.51	154.13	123.31	102.76	88.08	77.07	68.50	61.65	56.05	51.38

Basic weight of steelwork taken as 7850 kg/m³
Basic weight of bar reinforcement per metre run = 0.00785 kg/mm²
The value of π has been taken as 3.141592654

REINFORCEMENT

Fabric Reinforcement

Preferred range of designated fabric types and stock sheet sizes

Fabric reference	Longitudinal wires			Cross wires			
	Nominal wire size (mm)	Pitch (mm)	Area (mm²/m)	Nominal wire size (mm)	Pitch (mm)	Area (mm²/m)	Mass (kg/m²)
Square mesh							
A393	10	200	393	10	200	393	6.16
A252	8	200	252	8	200	252	3.95
A193	7	200	193	7	200	193	3.02
A142	6	200	142	6	200	142	2.22
A98	5	200	98	5	200	98	1.54
Structural mesh							
B1131	12	100	1131	8	200	252	10.90
B785	10	100	785	8	200	252	8.14
B503	8	100	503	8	200	252	5.93
B385	7	100	385	7	200	193	4.53
B283	6	100	283	7	200	193	3.73
B196	5	100	196	7	200	193	3.05
Long mesh							
C785	10	100	785	6	400	70.8	6.72
C636	9	100	636	6	400	70.8	5.55
C503	8	100	503	5	400	49.0	4.34
C385	7	100	385	5	400	49.0	3.41
C283	6	100	283	5	400	49.0	2.61
Wrapping mesh							
D98	5	200	98	5	200	98	1.54
D49	2.5	100	49	2.5	100	49	0.77

Stock sheet size 4.8 m × 2.4 m, Area 11.52 m²

Average weight kg/m³ of steelwork reinforcement in concrete for various building elements

Substructure	kg/m³ concrete	Substructure	kg/m³ concrete
Pile caps	110–150	Plate slab	150–220
Tie beams	130–170	Cant slab	145–210
Ground beams	230–330	Ribbed floors	130–200
Bases	125–180	Topping to block floor	30–40
Footings	100–150	Columns	210–310
Retaining walls	150–210	Beams	250–350
Raft	60–70	Stairs	130–170
Slabs – one way	120–200	Walls – normal	40–100
Slabs – two way	110–220	Walls – wind	70–125

Note: For exposed elements add the following %:
Walls 50%, Beams 100%, Columns 15%

FORMWORK

Formwork Stripping Times – Normal Curing Periods

Conditions under which concrete is maturing	Minimum periods of protection for different types of cement					
	Number of days (where the average surface temperature of the concrete exceeds 10°C during the whole period)			Equivalent maturity (degree hours) calculated as the age of the concrete in hours multiplied by the number of degrees Celsius by which the average surface temperature of the concrete exceeds 10°C		
	Other	SRPC	OPC or RHPC	Other	SRPC	OPC or RHPC
1. Hot weather or drying winds	7	4	3	3500	2000	1500
2. Conditions not covered by 1	4	3	2	2000	1500	1000

KEY
OPC – Ordinary Portland Cement
RHPC – Rapid-hardening Portland Cement
SRPC – Sulphate-resisting Portland Cement

Minimum Period before Striking Formwork

	Minimum period before striking		
	Surface temperature of concrete		
	16°C	17°C	t°C (0–25)
Vertical formwork to columns, walls and large beams	12 hours	18 hours	300 hours t+10
Soffit formwork to slabs	4 days	6 days	100 days t+10
Props to slabs	10 days	15 days	250 days t+10
Soffit formwork to beams	9 days	14 days	230 days t+10
Props to beams	14 days	21 days	360 days t+10

MASONRY

Number of Bricks Required for Various Types of Work per m² of Walling

Description	Brick size	
	215 × 102.5 × 50 mm	215 × 102.5 × 65 mm
Half brick thick		
Stretcher bond	74	59
English bond	108	86
English garden wall bond	90	72
Flemish bond	96	79
Flemish garden wall bond	83	66
One brick thick and cavity wall of two half brick skins		
Stretcher bond	148	119

Quantities of Bricks and Mortar Required per m² of Walling

	Unit	No of bricks required	Mortar required (cubic metres)		
Standard bricks			**No frogs**	**Single frogs**	**Double frogs**
Brick size 215 × 102.5 × 50 mm					
half brick wall (103 mm)	m²	72	0.022	0.027	0.032
2 × half brick cavity wall (270 mm)	m²	144	0.044	0.054	0.064
one brick wall (215 mm)	m²	144	0.052	0.064	0.076
one and a half brick wall (322 mm)	m²	216	0.073	0.091	0.108
Mass brickwork	m³	576	0.347	0.413	0.480
Brick size 215 × 102.5 × 65 mm					
half brick wall (103 mm)	m²	58	0.019	0.022	0.026
2 × half brick cavity wall (270 mm)	m²	116	0.038	0.045	0.055
one brick wall (215 mm)	m²	116	0.046	0.055	0.064
one and a half brick wall (322 mm)	m²	174	0.063	0.074	0.088
Mass brickwork	m³	464	0.307	0.360	0.413
Metric modular bricks			**Perforated**		
Brick size 200 × 100 × 75 mm					
90 mm thick	m²	67	0.016	0.019	
190 mm thick	m²	133	0.042	0.048	
290 mm thick	m²	200	0.068	0.078	
Brick size 200 × 100 × 100 mm					
90 mm thick	m²	50	0.013	0.016	
190 mm thick	m²	100	0.036	0.041	
290 mm thick	m²	150	0.059	0.067	
Brick size 300 × 100 × 75 mm					
90 mm thick	m²	33	–	0.015	
Brick size 300 × 100 × 100 mm					
90 mm thick	m²	44	0.015	0.018	

Note: Assuming 10 mm thick joints

MASONRY

Mortar Required per m² Blockwork (9.88 blocks/m²)

Wall thickness	75	90	100	125	140	190	215
Mortar m³/m²	0.005	0.006	0.007	0.008	0.009	0.013	0.014

Mortar Group	Cement: lime: sand	Masonry cement: sand	Cement: sand with plasticizer
1	1:0–0.25:3		
2	1:0.5:4–4.5	1:2.5-3.5	1:3–4
3	1:1:5–6	1:4–5	1:5–6
4	1:2:8–9	1:5.5–6.5	1:7–8
5	1:3:10–12	1:6.5–7	1:8

Group 1: strong inflexible mortar
Group 5: weak but flexible

All mixes within a group are of approximately similar strength
Frost resistance increases with the use of plasticizers
Cement: lime: sand mixes give the strongest bond and greatest resistance to rain penetration
Masonry cement equals ordinary Portland cement plus a fine neutral mineral filler and an air entraining agent

Calcium Silicate Bricks

Type	Strength	Location
Class 2 crushing strength	14.0 N/mm²	not suitable for walls
Class 3	20.5 N/mm²	walls above dpc
Class 4	27.5 N/mm²	cappings and copings
Class 5	34.5 N/mm²	retaining walls
Class 6	41.5 N/mm²	walls below ground
Class 7	48.5 N/mm²	walls below ground

The Class 7 calcium silicate bricks are therefore equal in strength to Class B bricks
Calcium silicate bricks are not suitable for DPCs

Durability of Bricks	
FL	Frost resistant with low salt content
FN	Frost resistant with normal salt content
ML	Moderately frost resistant with low salt content
MN	Moderately frost resistant with normal salt content

MASONRY

Brickwork Dimensions

No. of horizontal bricks	Dimensions (mm)	No. of vertical courses	Height of vertical courses (mm)
½	112.5	1	75
1	225.0	2	150
1½	337.5	3	225
2	450.0	4	300
2½	562.5	5	375
3	675.0	6	450
3½	787.5	7	525
4	900.0	8	600
4½	1012.5	9	675
5	1125.0	10	750
5½	1237.5	11	825
6	1350.0	12	900
6½	1462.5	13	975
7	1575.0	14	1050
7½	1687.5	15	1125
8	1800.0	16	1200
8½	1912.5	17	1275
9	2025.0	18	1350
9½	2137.5	19	1425
10	2250.0	20	1500
20	4500.0	24	1575
40	9000.0	28	2100
50	11250.0	32	2400
60	13500.0	36	2700
75	16875.0	40	3000

TIMBER

Weights of Timber

Material	kg/m^3	lb/cu ft
General	806 (avg)	50 (avg)
Douglas fir	479	30
Yellow pine, spruce	479	30
Pitch pine	673	42
Larch, elm	561	35
Oak (English)	724 to 959	45 to 60
Teak	643 to 877	40 to 55
Jarrah	959	60
Greenheart	1040 to 1204	65 to 75
Quebracho	1285	80
Material	**kg/m^2 per mm thickness**	**lb/sq ft per inch thickness**
Wooden boarding and blocks		
Softwood	0.48	2.5
Hardwood	0.76	4
Hardboard	1.06	5.5
Chipboard	0.76	4
Plywood	0.62	3.25
Blockboard	0.48	2.5
Fibreboard	0.29	1.5
Wood-wool	0.58	3
Plasterboard	0.96	5
Weather boarding	0.35	1.8

TIMBER

Conversion Tables (for timber only)

Inches	Millimetres	Feet	Metres
1	25	1	0.300
2	50	2	0.600
3	75	3	0.900
4	100	4	1.200
5	125	5	1.500
6	150	6	1.800
7	175	7	2.100
8	200	8	2.400
9	225	9	2.700
10	250	10	3.000
11	275	11	3.300
12	300	12	3.600
13	325	13	3.900
14	350	14	4.200
15	375	15	4.500
16	400	16	4.800
17	425	17	5.100
18	450	18	5.400
19	475	19	5.700
20	500	20	6.000
21	525	21	6.300
22	550	22	6.600
23	575	23	6.900
24	600	24	7.200

Planed Softwood

The finished end section size of planed timber is usually 3/16" less than the original size from which it is produced. This however varies slightly depending upon availability of material and origin of the species used.

Standards (timber) to cubic metres and cubic metres to standards (timber)

Cubic metres	Cubic metres standards	Standards
4.672	1	0.214
9.344	2	0.428
14.017	3	0.642
18.689	4	0.856
23.361	5	1.070
28.033	6	1.284
32.706	7	1.498
37.378	8	1.712
42.050	9	1.926
46.722	10	2.140
93.445	20	4.281
140.167	30	6.421
186.890	40	8.561
233.612	50	10.702
280.335	60	12.842
327.057	70	14.982
373.779	80	17.122

Tables and Memoranda

TIMBER

1 cu metre = 35.3148 cu ft = 0.21403 std

1 cu ft = 0.028317 cu metres

1 std = 4.67227 cu metres

Basic sizes of sawn softwood available (cross-sectional areas)

Thickness (mm)	Width (mm)								
	75	100	125	150	175	200	225	250	300
16	X	X	X	X					
19	X	X	X	X					
22	X	X	X	X					
25	X	X	X	X	X	X	X	X	X
32	X	X	X	X	X	X	X	X	X
36	X	X	X	X					
38	X	X	X	X	X	X	X		
44	X	X	X	X	X	X	X	X	X
47*	X	X	X	X	X	X	X	X	X
50	X	X	X	X	X	X	X	X	X
63	X	X	X	X	X	X	X		
75	X	X	X	X	X	X	X	X	
100		X		X		X		X	X
150				X		X			X
200						X			
250								X	
300									X

* This range of widths for 47 mm thickness will usually be found to be available in construction quality only

Note: The smaller sizes below 100 mm thick and 250 mm width are normally but not exclusively of European origin. Sizes beyond this are usually of North and South American origin

Basic lengths of sawn softwood available (metres)

1.80	2.10	3.00	4.20	5.10	6.00	7.20
	2.40	3.30	4.50	5.40	6.30	
	2.70	3.60	4.80	5.70	6.60	
		3.90			6.90	

Note: Lengths of 6.00 m and over will generally only be available from North American species and may have to be recut from larger sizes

TIMBER

Reductions from basic size to finished size by planning of two opposed faces

Purpose		Reductions from basic sizes for timber			
		15–35 mm	36–100 mm	101–150 mm	over 150 mm
a)	Constructional timber	3 mm	3 mm	5 mm	6 mm
b)	Matching interlocking boards	4 mm	4 mm	6 mm	6 mm
c)	Wood trim not specified in BS 584	5 mm	7 mm	7 mm	9 mm
d)	Joinery and cabinet work	7 mm	9 mm	11 mm	13 mm

Note: The reduction of width or depth is overall the extreme size and is exclusive of any reduction of the face by the machining of a tongue or lap joints

Maximum Spans for Various Roof Trusses

Maximum permissible spans for rafters for Fink trussed rafters

Basic size	Actual size	Pitch (degrees)								
(mm)	(mm)	15 (m)	17.5 (m)	20 (m)	22.5 (m)	25 (m)	27.5 (m)	30 (m)	32.5 (m)	35 (m)
38 × 75	35 × 72	6.03	6.16	6.29	6.41	6.51	6.60	6.70	6.80	6.90
38 × 100	35 × 97	7.48	7.67	7.83	7.97	8.10	8.22	8.34	8.47	8.61
38 × 125	35 × 120	8.80	9.00	9.20	9.37	9.54	9.68	9.82	9.98	10.16
44 × 75	41 × 72	6.45	6.59	6.71	6.83	6.93	7.03	7.14	7.24	7.35
44 × 100	41 × 97	8.05	8.23	8.40	8.55	8.68	8.81	8.93	9.09	9.22
44 × 125	41 × 120	9.38	9.60	9.81	9.99	10.15	10.31	10.45	10.64	10.81
50 × 75	47 × 72	6.87	7.01	7.13	7.25	7.35	7.45	7.53	7.67	7.78
50 × 100	47 × 97	8.62	8.80	8.97	9.12	9.25	9.38	9.50	9.66	9.80
50 × 125	47 × 120	10.01	10.24	10.44	10.62	10.77	10.94	11.00	11.00	11.00

TIMBER

Sizes of Internal and External Doorsets

Description	Internal size (mm)	Permissible deviation	External size (mm)	Permissible deviation
Coordinating dimension: height of door leaf height sets	2100		2100	
Coordinating dimension: height of ceiling height set	2300 2350 2400 2700 3000		2300 2350 2400 2700 3000	
Coordinating dimension: width of all doorsets S = Single leaf set D = Double leaf set	600 S 700 S 800 S&D 900 S&D 1000 S&D 1200 D 1500 D 1800 D 2100 D		900 S 1000 S 1200 D 1800 D 2100 D	
Work size: height of door leaf height set	2090	± 2.0	2095	± 2.0
Work size: height of ceiling height set	2285 2335 2385 2685 2985	± 2.0	2295 2345 2395 2695 2995	± 2.0
Work size: width of all doorsets S = Single leaf set D = Double leaf set	590 S 690 S 790 S&D 890 S&D 990 S&D 1190 D 1490 D 1790 D 2090 D	± 2.0	895 S 995 S 1195 D 1495 D 1795 D 2095 D	± 2.0
Width of door leaf in single leaf sets F = Flush leaf P = Panel leaf	526 F 626 F 726 F&P 826 F&P 926 F&P	± 1.5	806 F&P 906 F&P	± 1.5
Width of door leaf in double leaf sets F = Flush leaf P = Panel leaf	362 F 412 F 426 F 562 F&P 712 F&P 826 F&P 1012 F&P	± 1.5	552 F&P 702 F&P 852 F&P 1002 F&P	± 1.5
Door leaf height for all doorsets	2040	± 1.5	1994	± 1.5

ROOFING

Total Roof Loadings for Various Types of Tiles/Slates

	Roof load (slope) kg/m^2		
	Slate/Tile	Roofing underlay and battens^2	Total dead load kg/m
Asbestos cement slate (600 × 300)	21.50	3.14	24.64
Clay tile interlocking	67.00	5.50	72.50
plain	43.50	2.87	46.37
Concrete tile interlocking	47.20	2.69	49.89
plain	78.20	5.50	83.70
Natural slate (18" × 10")	35.40	3.40	38.80
	Roof load (plan) kg/m^2		
Asbestos cement slate (600 × 300)	28.45	76.50	104.95
Clay tile interlocking	53.54	76.50	130.04
plain	83.71	76.50	60.21
Concrete tile interlocking	57.60	76.50	134.10
plain	96.64	76.50	173.14

ROOFING

Tiling Data

Product		Lap (mm)	Gauge of battens	No. slates per m^2	Battens (m/m^2)	Weight as laid (kg/m^2)
CEMENT SLATES						
Eternit slates	600 × 300 mm	100	250	13.4	4.00	19.50
(Duracem)		90	255	13.1	3.92	19.20
		80	260	12.9	3.85	19.00
		70	265	12.7	3.77	18.60
	600 × 350 mm	100	250	11.5	4.00	19.50
		90	255	11.2	3.92	19.20
	500 × 250 mm	100	200	20.0	5.00	20.00
		90	205	19.5	4.88	19.50
		80	210	19.1	4.76	19.00
		70	215	18.6	4.65	18.60
	400 × 200 mm	90	155	32.3	6.45	20.80
		80	160	31.3	6.25	20.20
		70	165	30.3	6.06	19.60
CONCRETE TILES/SLATES						
Redland Roofing						
Stonewold slate	430 × 380 mm	75	355	8.2	2.82	51.20
Double Roman tile	418 × 330 mm	75	355	8.2	2.91	45.50
Grovebury pantile	418 × 332 mm	75	343	9.7	2.91	47.90
Norfolk pantile	381 × 227 mm	75	306	16.3	3.26	44.01
		100	281	17.8	3.56	48.06
Renown interlocking tile	418 × 330 mm	75	343	9.7	2.91	46.40
'49' tile	381 × 227 mm	75	306	16.3	3.26	44.80
		100	281	17.8	3.56	48.95
Plain, vertical tiling	265 × 165 mm	35	115	52.7	8.70	62.20
Marley Roofing						
Bold roll tile	420 × 330 mm	75	344	9.7	2.90	47.00
		100	–	10.5	3.20	51.00
Modern roof tile	420 × 330 mm	75	338	10.2	3.00	54.00
		100	–	11.0	3.20	58.00
Ludlow major	420 × 330 mm	75	338	10.2	3.00	45.00
		100	–	11.0	3.20	49.00
Ludlow plus	387 × 229 mm	75	305	16.1	3.30	47.00
		100	–	17.5	3.60	51.00
Mendip tile	420 × 330 mm	75	338	10.2	3.00	47.00
		100	–	11.0	3.20	51.00
Wessex	413 × 330 mm	75	338	10.2	3.00	54.00
		100	–	11.0	3.20	58.00
Plain tile	267 × 165 mm	65	100	60.0	10.00	76.00
		75	95	64.0	10.50	81.00
		85	90	68.0	11.30	86.00
Plain vertical tiles (feature)	267 × 165 mm	35	110	53.0	8.70	67.00
		34	115	56.0	9.10	71.00

ROOFING

Slate Nails, Quantity per Kilogram

Length	Type			
	Plain wire	Galvanized wire	Copper nail	Zinc nail
28.5 mm	325	305	325	415
34.4 mm	286	256	254	292
50.8 mm	242	224	194	200

Metal Sheet Coverings

Thicknesses and weights of sheet metal coverings									
Lead to BS 1178									
BS Code No	3	4	5	6	7	8			
Colour code	Green	Blue	Red	Black	White	Orange			
Thickness (mm)	1.25	1.80	2.24	2.50	3.15	3.55			
Density kg/m^2	14.18	20.41	25.40	30.05	35.72	40.26			
Copper to BS 2870									
Thickness (mm)		0.60	0.70						
Bay width									
Roll (mm)		500	650						
Seam (mm)		525	600						
Standard width to form bay	600	750							
Normal length of sheet	1.80	1.80							
Zinc to BS 849									
Zinc Gauge (Nr)	9	10	11	12	13	14	15	16	
Thickness (mm)	0.43	0.48	0.56	0.64	0.71	0.79	0.91	1.04	
Density (kg/m^2)	3.1	3.2	3.8	4.3	4.8	5.3	6.2	7.0	
Aluminium to BS 4868									
Thickness (mm)	0.5	0.6	0.7	0.8	0.9	1.0	1.2		
Density (kg/m^2)	12.8	15.4	17.9	20.5	23.0	25.6	30.7		

Tables and Memoranda

ROOFING

Type of felt	Nominal mass per unit area (kg/10 m)	Nominal mass per unit area of fibre base (g/m^2)	Nominal length of roll (m)
Class 1			
1B fine granule	14	220	10 or 20
surfaced bitumen	18	330	10 or 20
	25	470	10
1E mineral surfaced bitumen	38	470	10
1F reinforced bitumen	15	160 (fibre)	15
		110 (hessian)	
1F reinforced bitumen, aluminium faced	13	160 (fibre)	15
		110 (hessian)	
Class 2			
2B fine granule surfaced bitumen asbestos	18	500	10 or 20
2E mineral surfaced bitumen asbestos	38	600	10
Class 3			
3B fine granule surfaced bitumen glass fibre	18	60	20
3E mineral surfaced bitumen glass fibre	28	60	10
3E venting base layer bitumen glass fibre	32	60*	10
3H venting base layer bitumen glass fibre	17	60*	20

* Excluding effect of perforations

GLAZING

GLAZING

Nominal thickness (mm)	Tolerance on thickness (mm)	Approximate weight (kg/m²)	Normal maximum size (mm)
Float and polished plate glass			
3	+ 0.2	7.50	2140 × 1220
4	+ 0.2	10.00	2760 × 1220
5	+ 0.2	12.50	3180 × 2100
6	+ 0.2	15.00	4600 × 3180
10	+ 0.3	25.00)	6000 × 3300
12	+ 0.3	30.00)	
15	+ 0.5	37.50	3050 × 3000
19	+ 1.0	47.50)	3000 × 2900
25	+ 1.0	63.50)	
Clear sheet glass			
2 *	+ 0.2	5.00	1920 × 1220
3	+ 0.3	7.50	2130 × 1320
4	+ 0.3	10.00	2760 × 1220
5 *	+ 0.3	12.50)	2130 × 2400
6 *	+ 0.3	15.00)	
Cast glass			
3	+ 0.4		
	− 0.2	6.00)	2140 × 1280
4	+ 0.5	7.50)	
5	+ 0.5	9.50	2140 × 1320
6	+ 0.5	11.50)	3700 × 1280
10	+ 0.8	21.50)	
Wired glass			
(Cast wired glass)			
6	+ 0.3	−)	3700 × 1840
	− 0.7	)	
7	+ 0.7	−)	
(Polished wire glass)			
6	+ 1.0	−	330 × 1830

* The 5 mm and 6 mm thickness are known as *thick drawn sheet*. Although 2 mm sheet glass is available it is not recommended for general glazing purposes

METAL

METAL

Weights of Metals

Material	kg/m³	lb/cu ft
Metals, steel construction, etc.		
Iron		
– cast	7207	450
– wrought	7687	480
– ore – general	2407	150
– (crushed) Swedish	3682	230
Steel	7854	490
Copper		
– cast	8731	545
– wrought	8945	558
Brass	8497	530
Bronze	8945	558
Aluminium	2774	173
Lead	11322	707
Zinc (rolled)	7140	446
	g/mm² per metre	**lb/sq ft per foot**
Steel bars	7.85	3.4
Structural steelwork	Net weight of member @ 7854 kg/m³	
riveted	+ 10% for cleats, rivets, bolts, etc.	
welded	+ 1.25% to 2.5% for welds, etc.	
Rolled sections		
beams	+ 2.5%	
stanchions	+ 5% (extra for caps and bases)	
Plate		
web girders	+ 10% for rivets or welds, stiffeners, etc.	
	kg/m	**lb/ft**
Steel stairs: industrial type		
1 m or 3 ft wide	84	56
Steel tubes		
50 mm or 2 in bore	5 to 6	3 to 4
Gas piping		
20 mm or ¾ in	2	1¼

METAL

Universal Beams BS 4: Part 1: 2005

Designation	Mass (kg/m)	Depth of section (mm)	Width of section (mm)	Thickness		Surface area (m²/m)
				Web (mm)	Flange (mm)	
1016 × 305 × 487	487.0	1036.1	308.5	30.0	54.1	3.20
1016 × 305 × 438	438.0	1025.9	305.4	26.9	49.0	3.17
1016 × 305 × 393	393.0	1016.0	303.0	24.4	43.9	3.15
1016 × 305 × 349	349.0	1008.1	302.0	21.1	40.0	3.13
1016 × 305 × 314	314.0	1000.0	300.0	19.1	35.9	3.11
1016 × 305 × 272	272.0	990.1	300.0	16.5	31.0	3.10
1016 × 305 × 249	249.0	980.2	300.0	16.5	26.0	3.08
1016 × 305 × 222	222.0	970.3	300.0	16.0	21.1	3.06
914 × 419 × 388	388.0	921.0	420.5	21.4	36.6	3.44
914 × 419 × 343	343.3	911.8	418.5	19.4	32.0	3.42
914 × 305 × 289	289.1	926.6	307.7	19.5	32.0	3.01
914 × 305 × 253	253.4	918.4	305.5	17.3	27.9	2.99
914 × 305 × 224	224.2	910.4	304.1	15.9	23.9	2.97
914 × 305 × 201	200.9	903.0	303.3	15.1	20.2	2.96
838 × 292 × 226	226.5	850.9	293.8	16.1	26.8	2.81
838 × 292 × 194	193.8	840.7	292.4	14.7	21.7	2.79
838 × 292 × 176	175.9	834.9	291.7	14.0	18.8	2.78
762 × 267 × 197	196.8	769.8	268.0	15.6	25.4	2.55
762 × 267 × 173	173.0	762.2	266.7	14.3	21.6	2.53
762 × 267 × 147	146.9	754.0	265.2	12.8	17.5	2.51
762 × 267 × 134	133.9	750.0	264.4	12.0	15.5	2.51
686 × 254 × 170	170.2	692.9	255.8	14.5	23.7	2.35
686 × 254 × 152	152.4	687.5	254.5	13.2	21.0	2.34
686 × 254 × 140	140.1	383.5	253.7	12.4	19.0	2.33
686 × 254 × 125	125.2	677.9	253.0	11.7	16.2	2.32
610 × 305 × 238	238.1	635.8	311.4	18.4	31.4	2.45
610 × 305 × 179	179.0	620.2	307.1	14.1	23.6	2.41
610 × 305 × 149	149.1	612.4	304.8	11.8	19.7	2.39
610 × 229 × 140	139.9	617.2	230.2	13.1	22.1	2.11
610 × 229 × 125	125.1	612.2	229.0	11.9	19.6	2.09
610 × 229 × 113	113.0	607.6	228.2	11.1	17.3	2.08
610 × 229 × 101	101.2	602.6	227.6	10.5	14.8	2.07
533 × 210 × 122	122.0	544.5	211.9	12.7	21.3	1.89
533 × 210 × 109	109.0	539.5	210.8	11.6	18.8	1.88
533 × 210 × 101	101.0	536.7	210.0	10.8	17.4	1.87
533 × 210 × 92	92.1	533.1	209.3	10.1	15.6	1.86
533 × 210 × 82	82.2	528.3	208.8	9.6	13.2	1.85
457 × 191 × 98	98.3	467.2	192.8	11.4	19.6	1.67
457 × 191 × 89	89.3	463.4	191.9	10.5	17.7	1.66
457 × 191 × 82	82.0	460.0	191.3	9.9	16.0	1.65
457 × 191 × 74	74.3	457.0	190.4	9.0	14.5	1.64
457 × 191 × 67	67.1	453.4	189.9	8.5	12.7	1.63
457 × 152 × 82	82.1	465.8	155.3	10.5	18.9	1.51
457 × 152 × 74	74.2	462.0	154.4	9.6	17.0	1.50
457 × 152 × 67	67.2	458.0	153.8	9.0	15.0	1.50
457 × 152 × 60	59.8	454.6	152.9	8.1	13.3	1.50
457 × 152 × 52	52.3	449.8	152.4	7.6	10.9	1.48
406 × 178 × 74	74.2	412.8	179.5	9.5	16.0	1.51
406 × 178 × 67	67.1	409.4	178.8	8.8	14.3	1.50
406 × 178 × 60	60.1	406.4	177.9	7.9	12.8	1.49

METAL

Designation	Mass (kg/m)	Depth of section (mm)	Width of section (mm)	Thickness		Surface area (m²/m)
				Web (mm)	Flange (mm)	
406 × 178 × 50	54.1	402.6	177.7	7.7	10.9	1.48
406 × 140 × 46	46.0	403.2	142.2	6.8	11.2	1.34
406 × 140 × 39	39.0	398.0	141.8	6.4	8.6	1.33
356 × 171 × 67	67.1	363.4	173.2	9.1	15.7	1.38
356 × 171 × 57	57.0	358.0	172.2	8.1	13.0	1.37
356 × 171 × 51	51.0	355.0	171.5	7.4	11.5	1.36
356 × 171 × 45	45.0	351.4	171.1	7.0	9.7	1.36
356 × 127 × 39	39.1	353.4	126.0	6.6	10.7	1.18
356 × 127 × 33	33.1	349.0	125.4	6.0	8.5	1.17
305 × 165 × 54	54.0	310.4	166.9	7.9	13.7	1.26
305 × 165 × 46	46.1	306.6	165.7	6.7	11.8	1.25
305 × 165 × 40	40.3	303.4	165.0	6.0	10.2	1.24
305 × 127 × 48	48.1	311.0	125.3	9.0	14.0	1.09
305 × 127 × 42	41.9	307.2	124.3	8.0	12.1	1.08
305 × 127 × 37	37.0	304.4	123.3	7.1	10.7	1.07
305 × 102 × 33	32.8	312.7	102.4	6.6	10.8	1.01
305 × 102 × 28	28.2	308.7	101.8	6.0	8.8	1.00
305 × 102 × 25	24.8	305.1	101.6	5.8	7.0	0.992
254 × 146 × 43	43.0	259.6	147.3	7.2	12.7	1.08
254 × 146 × 37	37.0	256.0	146.4	6.3	10.9	1.07
254 × 146 × 31	31.1	251.4	146.1	6.0	8.6	1.06
254 × 102 × 28	28.3	260.4	102.2	6.3	10.0	0.904
254 × 102 × 25	25.2	257.2	101.9	6.0	8.4	0.897
254 × 102 × 22	22.0	254.0	101.6	5.7	6.8	0.890
203 × 133 × 30	30.0	206.8	133.9	6.4	9.6	0.923
203 × 133 × 25	25.1	203.2	133.2	5.7	7.8	0.915
203 × 102 × 23	23.1	203.2	101.8	5.4	9.3	0.790
178 × 102 × 19	19.0	177.8	101.2	4.8	7.9	0.738
152 × 89 × 16	16.0	152.4	88.7	4.5	7.7	0.638
127 × 76 × 13	13.0	127.0	76.0	4.0	7.6	0.537

METAL

Universal Columns BS 4: Part 1: 2005

Designation	Mass (kg/m)	Depth of section (mm)	Width of section (mm)	Thickness		Surface area (m²/m)
				Web (mm)	Flange (mm)	
356 × 406 × 634	633.9	474.7	424.0	47.6	77.0	2.52
356 × 406 × 551	551.0	455.6	418.5	42.1	67.5	2.47
356 × 406 × 467	467.0	436.6	412.2	35.8	58.0	2.42
356 × 406 × 393	393.0	419.0	407.0	30.6	49.2	2.38
356 × 406 × 340	339.9	406.4	403.0	26.6	42.9	2.35
356 × 406 × 287	287.1	393.6	399.0	22.6	36.5	2.31
356 × 406 × 235	235.1	381.0	384.8	18.4	30.2	2.28
356 × 368 × 202	201.9	374.6	374.7	16.5	27.0	2.19
356 × 368 × 177	177.0	368.2	372.6	14.4	23.8	2.17
356 × 368 × 153	152.9	362.0	370.5	12.3	20.7	2.16
356 × 368 × 129	129.0	355.6	368.6	10.4	17.5	2.14
305 × 305 × 283	282.9	365.3	322.2	26.8	44.1	1.94
305 × 305 × 240	240.0	352.5	318.4	23.0	37.7	1.91
305 × 305 × 198	198.1	339.9	314.5	19.1	31.4	1.87
305 × 305 × 158	158.1	327.1	311.2	15.8	25.0	1.84
305 × 305 × 137	136.9	320.5	309.2	13.8	21.7	1.82
305 × 305 × 118	117.9	314.5	307.4	12.0	18.7	1.81
305 × 305 × 97	96.9	307.9	305.3	9.9	15.4	1.79
254 × 254 × 167	167.1	289.1	265.2	19.2	31.7	1.58
254 × 254 × 132	132.0	276.3	261.3	15.3	25.3	1.55
254 × 254 × 107	107.1	266.7	258.8	12.8	20.5	1.52
254 × 254 × 89	88.9	260.3	256.3	10.3	17.3	1.50
254 × 254 × 73	73.1	254.1	254.6	8.6	14.2	1.49
203 × 203 × 86	86.1	222.2	209.1	12.7	20.5	1.24
203 × 203 × 71	71.0	215.8	206.4	10.0	17.3	1.22
203 × 203 × 60	60.0	209.6	205.8	9.4	14.2	1.21
203 × 203 × 52	52.0	206.2	204.3	7.9	12.5	1.20
203 × 203 × 46	46.1	203.2	203.6	7.2	11.0	1.19
152 × 152 × 37	37.0	161.8	154.4	8.0	11.5	0.912
152 × 152 × 30	30.0	157.6	152.9	6.5	9.4	0.901
152 × 152 × 23	23.0	152.4	152.2	5.8	6.8	0.889

METAL

Joists BS 4: Part 1: 2005 (retained for reference, Corus have ceased manufacture in UK)

Designation	Mass (kg/m)	Depth of section (mm)	Width of section (mm)	Thickness		Surface area (m²/m)
				Web (mm)	Flange (mm)	
254 × 203 × 82	82.0	254.0	203.2	10.2	19.9	1.210
203 × 152 × 52	52.3	203.2	152.4	8.9	16.5	0.932
152 × 127 × 37	37.3	152.4	127.0	10.4	13.2	0.737
127 × 114 × 29	29.3	127.0	114.3	10.2	11.5	0.646
127 × 114 × 27	26.9	127.0	114.3	7.4	11.4	0.650
102 × 102 × 23	23.0	101.6	101.6	9.5	10.3	0.549
102 × 44 × 7	7.5	101.6	44.5	4.3	6.1	0.350
89 × 89 × 19	19.5	88.9	88.9	9.5	9.9	0.476
76 × 76 × 13	12.8	76.2	76.2	5.1	8.4	0.411

Parallel Flange Channels

Designation	Mass (kg/m)	Depth of section (mm)	Width of section (mm)	Thickness		Surface area (m²/m)
				Web (mm)	Flange (mm)	
430 × 100 × 64	64.4	430	100	11.0	19.0	1.23
380 × 100 × 54	54.0	380	100	9.5	17.5	1.13
300 × 100 × 46	45.5	300	100	9.0	16.5	0.969
300 × 90 × 41	41.4	300	90	9.0	15.5	0.932
260 × 90 × 35	34.8	260	90	8.0	14.0	0.854
260 × 75 × 28	27.6	260	75	7.0	12.0	0.79
230 × 90 × 32	32.2	230	90	7.5	14.0	0.795
230 × 75 × 26	25.7	230	75	6.5	12.5	0.737
200 × 90 × 30	29.7	200	90	7.0	14.0	0.736
200 × 75 × 23	23.4	200	75	6.0	12.5	0.678
180 × 90 × 26	26.1	180	90	6.5	12.5	0.697
180 × 75 × 20	20.3	180	75	6.0	10.5	0.638
150 × 90 × 24	23.9	150	90	6.5	12.0	0.637
150 × 75 × 18	17.9	150	75	5.5	10.0	0.579
125 × 65 × 15	14.8	125	65	5.5	9.5	0.489
100 × 50 × 10	10.2	100	50	5.0	8.5	0.382

METAL

Equal Angles BS EN 10056-1

Designation	Mass (kg/m)	Surface area (m²/m)
200 × 200 × 24	71.1	0.790
200 × 200 × 20	59.9	0.790
200 × 200 × 18	54.2	0.790
200 × 200 × 16	48.5	0.790
150 × 150 × 18	40.1	0.59
150 × 150 × 15	33.8	0.59
150 × 150 × 12	27.3	0.59
150 × 150 × 10	23.0	0.59
120 × 120 × 15	26.6	0.47
120 × 120 × 12	21.6	0.47
120 × 120 × 10	18.2	0.47
120 × 120 × 8	14.7	0.47
100 × 100 × 15	21.9	0.39
100 × 100 × 12	17.8	0.39
100 × 100 × 10	15.0	0.39
100 × 100 × 8	12.2	0.39
90 × 90 × 12	15.9	0.35
90 × 90 × 10	13.4	0.35
90 × 90 × 8	10.9	0.35
90 × 90 × 7	9.61	0.35
90 × 90 × 6	8.30	0.35

Unequal Angles BS EN 10056-1

Designation	Mass (kg/m)	Surface area (m²/m)
200 × 150 × 18	47.1	0.69
200 × 150 × 15	39.6	0.69
200 × 150 × 12	32.0	0.69
200 × 100 × 15	33.7	0.59
200 × 100 × 12	27.3	0.59
200 × 100 × 10	23.0	0.59
150 × 90 × 15	26.6	0.47
150 × 90 × 12	21.6	0.47
150 × 90 × 10	18.2	0.47
150 × 75 × 15	24.8	0.44
150 × 75 × 12	20.2	0.44
150 × 75 × 10	17.0	0.44
125 × 75 × 12	17.8	0.40
125 × 75 × 10	15.0	0.40
125 × 75 × 8	12.2	0.40
100 × 75 × 12	15.4	0.34
100 × 75 × 10	13.0	0.34
100 × 75 × 8	10.6	0.34
100 × 65 × 10	12.3	0.32
100 × 65 × 8	9.94	0.32
100 × 65 × 7	8.77	0.32

METAL

Structural Tees Split from Universal Beams BS 4: Part 1: 2005

Designation	Mass (kg/m)	Surface area (m²/m)
305 × 305 × 90	89.5	1.22
305 × 305 × 75	74.6	1.22
254 × 343 × 63	62.6	1.19
229 × 305 × 70	69.9	1.07
229 × 305 × 63	62.5	1.07
229 × 305 × 57	56.5	1.07
229 × 305 × 51	50.6	1.07
210 × 267 × 61	61.0	0.95
210 × 267 × 55	54.5	0.95
210 × 267 × 51	50.5	0.95
210 × 267 × 46	46.1	0.95
210 × 267 × 41	41.1	0.95
191 × 229 × 49	49.2	0.84
191 × 229 × 45	44.6	0.84
191 × 229 × 41	41.0	0.84
191 × 229 × 37	37.1	0.84
191 × 229 × 34	33.6	0.84
152 × 229 × 41	41.0	0.76
152 × 229 × 37	37.1	0.76
152 × 229 × 34	33.6	0.76
152 × 229 × 30	29.9	0.76
152 × 229 × 26	26.2	0.76

Universal Bearing Piles BS 4: Part 1: 2005

Designation	Mass (kg/m)	Depth of Section (mm)	Width of section (mm)	Thickness Web (mm)	Flange (mm)
356 × 368 × 174	173.9	361.4	378.5	20.3	20.4
356 × 368 × 152	152.0	356.4	376.0	17.8	17.9
356 × 368 × 133	133.0	352.0	373.8	15.6	15.7
356 × 368 × 109	108.9	346.4	371.0	12.8	12.9
305 × 305 × 223	222.9	337.9	325.7	30.3	30.4
305 × 305 × 186	186.0	328.3	320.9	25.5	25.6
305 × 305 × 149	149.1	318.5	316.0	20.6	20.7
305 × 305 × 126	126.1	312.3	312.9	17.5	17.6
305 × 305 × 110	110.0	307.9	310.7	15.3	15.4
305 × 305 × 95	94.9	303.7	308.7	13.3	13.3
305 × 305 × 88	88.0	301.7	307.8	12.4	12.3
305 × 305 × 79	78.9	299.3	306.4	11.0	11.1
254 × 254 × 85	85.1	254.3	260.4	14.4	14.3
254 × 254 × 71	71.0	249.7	258.0	12.0	12.0
254 × 254 × 63	63.0	247.1	256.6	10.6	10.7
203 × 203 × 54	53.9	204.0	207.7	11.3	11.4
203 × 203 × 45	44.9	200.2	205.9	9.5	9.5

Hot Formed Square Hollow Sections EN 10210 S275J2H & S355J2H

Size (mm)	Wall thickness (mm)	Mass (kg/m)	Superficial area (m²/m)
40 × 40	2.5	2.89	0.154
	3.0	3.41	0.152
	3.2	3.61	0.152
	3.6	4.01	0.151
	4.0	4.39	0.150
	5.0	5.28	0.147
50 × 50	2.5	3.68	0.194
	3.0	4.35	0.192
	3.2	4.62	0.192
	3.6	5.14	0.191
	4.0	5.64	0.190
	5.0	6.85	0.187
	6.0	7.99	0.185
	6.3	8.31	0.184
60 × 60	3.0	5.29	0.232
	3.2	5.62	0.232
	3.6	6.27	0.231
	4.0	6.90	0.230
	5.0	8.42	0.227
	6.0	9.87	0.225
	6.3	10.30	0.224
	8.0	12.50	0.219
70 × 70	3.0	6.24	0.272
	3.2	6.63	0.272
	3.6	7.40	0.271
	4.0	8.15	0.270
	5.0	9.99	0.267
	6.0	11.80	0.265
	6.3	12.30	0.264
	8.0	15.00	0.259
80 × 80	3.2	7.63	0.312
	3.6	8.53	0.311
	4.0	9.41	0.310
	5.0	11.60	0.307
	6.0	13.60	0.305
	6.3	14.20	0.304
	8.0	17.50	0.299
90 × 90	3.6	9.66	0.351
	4.0	10.70	0.350
	5.0	13.10	0.347
	6.0	15.50	0.345
	6.3	16.20	0.344
	8.0	20.10	0.339
100 × 100	3.6	10.80	0.391
	4.0	11.90	0.390
	5.0	14.70	0.387
	6.0	17.40	0.385
	6.3	18.20	0.384
	8.0	22.60	0.379
	10.0	27.40	0.374
120 × 120	4.0	14.40	0.470
	5.0	17.80	0.467
	6.0	21.20	0.465

METAL

Size (mm)	Wall thickness (mm)	Mass (kg/m)	Superficial area (m²/m)
	6.3	22.20	0.464
	8.0	27.60	0.459
	10.0	33.70	0.454
	12.0	39.50	0.449
	12.5	40.90	0.448
140 × 140	5.0	21.00	0.547
	6.0	24.90	0.545
	6.3	26.10	0.544
	8.0	32.60	0.539
	10.0	40.00	0.534
	12.0	47.00	0.529
	12.5	48.70	0.528
150 × 150	5.0	22.60	0.587
	6.0	26.80	0.585
	6.3	28.10	0.584
	8.0	35.10	0.579
	10.0	43.10	0.574
	12.0	50.80	0.569
	12.5	52.70	0.568
Hot formed from seamless hollow	16.0	65.2	0.559
160 × 160	5.0	24.10	0.627
	6.0	28.70	0.625
	6.3	30.10	0.624
	8.0	37.60	0.619
	10.0	46.30	0.614
	12.0	54.60	0.609
	12.5	56.60	0.608
	16.0	70.20	0.599
180 × 180	5.0	27.30	0.707
	6.0	32.50	0.705
	6.3	34.00	0.704
	8.0	42.70	0.699
	10.0	52.50	0.694
	12.0	62.10	0.689
	12.5	64.40	0.688
	16.0	80.20	0.679
200 × 200	5.0	30.40	0.787
	6.0	36.20	0.785
	6.3	38.00	0.784
	8.0	47.70	0.779
	10.0	58.80	0.774
	12.0	69.60	0.769
	12.5	72.30	0.768
	16.0	90.30	0.759
250 × 250	5.0	38.30	0.987
	6.0	45.70	0.985
	6.3	47.90	0.984
	8.0	60.30	0.979
	10.0	74.50	0.974
	12.0	88.50	0.969
	12.5	91.90	0.968
	16.0	115.00	0.959

METAL

Size (mm)	Wall thickness (mm)	Mass (kg/m)	Superficial area (m²/m)
300 × 300	6.0	55.10	1.18
	6.3	57.80	1.18
	8.0	72.80	1.18
	10.0	90.20	1.17
	12.0	107.00	1.17
	12.5	112.00	1.17
	16.0	141.00	1.16
350 × 350	8.0	85.40	1.38
	10.0	106.00	1.37
	12.0	126.00	1.37
	12.5	131.00	1.37
	16.0	166.00	1.36
400 × 400	8.0	97.90	1.58
	10.0	122.00	1.57
	12.0	145.00	1.57
	12.5	151.00	1.57
	16.0	191.00	1.56
(Grade S355J2H only)	20.00*	235.00	1.55

Note: * SAW process

METAL

Hot Formed Square Hollow Sections JUMBO RHS: JIS G3136

Size (mm)	Wall thickness (mm)	Mass (kg/m)	Superficial area (m²/m)
350 × 350	19.0	190.00	1.33
	22.0	217.00	1.32
	25.0	242.00	1.31
400 × 400	22.0	251.00	1.52
	25.0	282.00	1.51
450 × 450	12.0	162.00	1.76
	16.0	213.00	1.75
	19.0	250.00	1.73
	22.0	286.00	1.72
	25.0	321.00	1.71
	28.0 *	355.00	1.70
	32.0 *	399.00	1.69
500 × 500	12.0	181.00	1.96
	16.0	238.00	1.95
	19.0	280.00	1.93
	22.0	320.00	1.92
	25.0	360.00	1.91
	28.0 *	399.00	1.90
	32.0 *	450.00	1.89
	36.0 *	498.00	1.88
550 × 550	16.0	263.00	2.15
	19.0	309.00	2.13
	22.0	355.00	2.12
	25.0	399.00	2.11
	28.0 *	443.00	2.10
	32.0 *	500.00	2.09
	36.0 *	555.00	2.08
	40.0 *	608.00	2.06
600 × 600	25.0 *	439.00	2.31
	28.0 *	487.00	2.30
	32.0 *	550.00	2.29
	36.0 *	611.00	2.28
	40.0 *	671.00	2.26
700 × 700	25.0 *	517.00	2.71
	28.0 *	575.00	2.70
	32.0 *	651.00	2.69
	36.0 *	724.00	2.68
	40.0 *	797.00	2.68

Note: * SAW process

METAL

Hot Formed Rectangular Hollow Sections: EN10210 S275J2h & S355J2H

Size (mm)	Wall thickness (mm)	Mass (kg/m)	Superficial area (m²/m)
50 × 30	2.5	2.89	0.154
	3.0	3.41	0.152
	3.2	3.61	0.152
	3.6	4.01	0.151
	4.0	4.39	0.150
	5.0	5.28	0.147
60 × 40	2.5	3.68	0.194
	3.0	4.35	0.192
	3.2	4.62	0.192
	3.6	5.14	0.191
	4.0	5.64	0.190
	5.0	6.85	0.187
	6.0	7.99	0.185
	6.3	8.31	0.184
80 × 40	3.0	5.29	0.232
	3.2	5.62	0.232
	3.6	6.27	0.231
	4.0	6.90	0.230
	5.0	8.42	0.227
	6.0	9.87	0.225
	6.3	10.30	0.224
	8.0	12.50	0.219
76.2 × 50.8	3.0	5.62	0.246
	3.2	5.97	0.246
	3.6	6.66	0.245
	4.0	7.34	0.244
	5.0	8.97	0.241
	6.0	10.50	0.239
	6.3	11.00	0.238
	8.0	13.40	0.233
90 × 50	3.0	6.24	0.272
	3.2	6.63	0.272
	3.6	7.40	0.271
	4.0	8.15	0.270
	5.0	9.99	0.267
	6.0	11.80	0.265
	6.3	12.30	0.264
	8.0	15.00	0.259
100 × 50	3.0	6.71	0.292
	3.2	7.13	0.292
	3.6	7.96	0.291
	4.0	8.78	0.290
	5.0	10.80	0.287
	6.0	12.70	0.285
	6.3	13.30	0.284
	8.0	16.30	0.279

METAL

Size (mm)	Wall thickness (mm)	Mass (kg/m)	Superficial area (m²/m)
100 × 60	3.0	7.18	0.312
	3.2	7.63	0.312
	3.6	8.53	0.311
	4.0	9.41	0.310
	5.0	11.60	0.307
	6.0	13.60	0.305
	6.3	14.20	0.304
	8.0	17.50	0.299
120 × 60	3.6	9.70	0.351
	4.0	10.70	0.350
	5.0	13.10	0.347
	6.0	15.50	0.345
	6.3	16.20	0.344
	8.0	20.10	0.339
120 × 80	3.6	10.80	0.391
	4.0	11.90	0.390
	5.0	14.70	0.387
	6.0	17.40	0.385
	6.3	18.20	0.384
	8.0	22.60	0.379
	10.0	27.40	0.374
150 × 100	4.0	15.10	0.490
	5.0	18.60	0.487
	6.0	22.10	0.485
	6.3	23.10	0.484
	8.0	28.90	0.479
	10.0	35.30	0.474
	12.0	41.40	0.469
	12.5	42.80	0.468
160 × 80	4.0	14.40	0.470
	5.0	17.80	0.467
	6.0	21.20	0.465
	6.3	22.20	0.464
	8.0	27.60	0.459
	10.0	33.70	0.454
	12.0	39.50	0.449
	12.5	40.90	0.448
200 × 100	5.0	22.60	0.587
	6.0	26.80	0.585
	6.3	28.10	0.584
	8.0	35.10	0.579
	10.0	43.10	0.574
	12.0	50.80	0.569
	12.5	52.70	0.568
	16.0	65.20	0.559
250 × 150	5.0	30.40	0.787
	6.0	36.20	0.785
	6.3	38.00	0.784
	8.0	47.70	0.779
	10.0	58.80	0.774
	12.0	69.60	0.769
	12.5	72.30	0.768
	16.0	90.30	0.759

METAL

Size (mm)	Wall thickness (mm)	Mass (kg/m)	Superficial area (m²/m)
300 × 200	5.0	38.30	0.987
	6.0	45.70	0.985
	6.3	47.90	0.984
	8.0	60.30	0.979
	10.0	74.50	0.974
	12.0	88.50	0.969
	12.5	91.90	0.968
	16.0	115.00	0.959
400 × 200	6.0	55.10	1.18
	6.3	57.80	1.18
	8.0	72.80	1.18
	10.0	90.20	1.17
	12.0	107.00	1.17
	12.5	112.00	1.17
	16.0	141.00	1.16
450 × 250	8.0	85.40	1.38
	10.0	106.00	1.37
	12.0	126.00	1.37
	12.5	131.00	1.37
	16.0	166.00	1.36
500 × 300	8.0	98.00	1.58
	10.0	122.00	1.57
	12.0	145.00	1.57
	12.5	151.00	1.57
	16.0	191.00	1.56
	20.0	235.00	1.55

METAL

Hot Formed Circular Hollow Sections EN 10210 S275J2H & S355J2H

Outside diameter (mm)	Wall thickness (mm)	Mass (kg/m)	Superficial area (m²/m)
21.3	3.2	1.43	0.067
26.9	3.2	1.87	0.085
33.7	3.0	2.27	0.106
	3.2	2.41	0.106
	3.6	2.67	0.106
	4.0	2.93	0.106
42.4	3.0	2.91	0.133
	3.2	3.09	0.133
	3.6	3.44	0.133
	4.0	3.79	0.133
48.3	2.5	2.82	0.152
	3.0	3.35	0.152
	3.2	3.56	0.152
	3.6	3.97	0.152
	4.0	4.37	0.152
	5.0	5.34	0.152
60.3	2.5	3.56	0.189
	3.0	4.24	0.189
	3.2	4.51	0.189
	3.6	5.03	0.189
	4.0	5.55	0.189
	5.0	6.82	0.189
76.1	2.5	4.54	0.239
	3.0	5.41	0.239
	3.2	5.75	0.239
	3.6	6.44	0.239
	4.0	7.11	0.239
	5.0	8.77	0.239
	6.0	10.40	0.239
	6.3	10.80	0.239
88.9	2.5	5.33	0.279
	3.0	6.36	0.279
	3.2	6.76	0.27
	3.6	7.57	0.279
	4.0	8.38	0.279
	5.0	10.30	0.279
	6.0	12.30	0.279
	6.3	12.80	0.279
114.3	3.0	8.23	0.359
	3.2	8.77	0.359
	3.6	9.83	0.359
	4.0	10.09	0.359
	5.0	13.50	0.359
	6.0	16.00	0.359
	6.3	16.80	0.359

METAL

Outside diameter (mm)	Wall thickness (mm)	Mass (kg/m)	Superficial area (m²/m)
139.7	3.2	10.80	0.439
	3.6	12.10	0.439
	4.0	13.40	0.439
	5.0	16.60	0.439
	6.0	19.80	0.439
	6.3	20.70	0.439
	8.0	26.00	0.439
	10.0	32.00	0.439
168.3	3.2	13.00	0.529
	3.6	14.60	0.529
	4.0	16.20	0.529
	5.0	20.10	0.529
	6.0	24.00	0.529
	6.3	25.20	0.529
	8.0	31.60	0.529
	10.0	39.00	0.529
	12.0	46.30	0.529
	12.5	48.00	0.529
193.7	5.0	23.30	0.609
	6.0	27.80	0.609
	6.3	29.10	0.609
	8.0	36.60	0.609
	10.0	45.30	0.609
	12.0	53.80	0.609
	12.5	55.90	0.609
219.1	5.0	26.40	0.688
	6.0	31.50	0.688
	6.3	33.10	0.688
	8.0	41.60	0.688
	10.0	51.60	0.688
	12.0	61.30	0.688
	12.5	63.70	0.688
	16.0	80.10	0.688
244.5	5.0	29.50	0.768
	6.0	35.30	0.768
	6.3	37.00	0.768
	8.0	46.70	0.768
	10.0	57.80	0.768
	12.0	68.80	0.768
	12.5	71.50	0.768
	16.0	90.20	0.768
273.0	5.0	33.00	0.858
	6.0	39.50	0.858
	6.3	41.40	0.858
	8.0	52.30	0.858
	10.0	64.90	0.858
	12.0	77.20	0.858
	12.5	80.30	0.858
	16.0	101.00	0.858
323.9	5.0	39.30	1.02
	6.0	47.00	1.02
	6.3	49.30	1.02
	8.0	62.30	1.02
	10.0	77.40	1.02

Tables and Memoranda

METAL

Outside diameter (mm)	Wall thickness (mm)	Mass (kg/m)	Superficial area (m²/m)
	12.0	92.30	1.02
	12.5	96.00	1.02
	16.0	121.00	1.02
355.6	6.3	54.30	1.12
	8.0	68.60	1.12
	10.0	85.30	1.12
	12.0	102.00	1.12
	12.5	106.00	1.12
	16.0	134.00	1.12
406.4	6.3	62.20	1.28
	8.0	79.60	1.28
	10.0	97.80	1.28
	12.0	117.00	1.28
	12.5	121.00	1.28
	16.0	154.00	1.28
457.0	6.3	70.00	1.44
	8.0	88.60	1.44
	10.0	110.00	1.44
	12.0	132.00	1.44
	12.5	137.00	1.44
	16.0	174.00	1.44
508.0	6.3	77.90	1.60
	8.0	98.60	1.60
	10.0	123.00	1.60
	12.0	147.00	1.60
	12.5	153.00	1.60
	16.0	194.00	1.60

METAL

Spacing of Holes in Angles

Nominal leg length (mm)	Spacing of holes						Maximum diameter of bolt or rivet		
	A	B	C	D	E	F	A	B and C	D, E and F
200		75	75	55	55	55		30	20
150		55	55					20	
125		45	60					20	
120									
100	55						24		
90	50						24		
80	45						20		
75	45						20		
70	40						20		
65	35						20		
60	35						16		
50	28						12		
45	25								
40	23								
30	20								
25	15								

KERBS, PAVING, ETC.

KERBS/EDGINGS/CHANNELS

Precast Concrete Kerbs to BS 7263

Straight kerb units: length from 450 to 915 mm

150 mm high × 125 mm thick		
bullnosed	type BN	
half battered	type HB3	
255 mm high × 125 mm thick		
45° splayed	type SP	
half battered	type HB2	
305 mm high × 150 mm thick		
half battered	type HB1	
Quadrant kerb units		
150 mm high × 305 and 455 mm radius to match	type BN	type QBN
150 mm high × 305 and 455 mm radius to match	type HB2, HB3	type QHB
150 mm high × 305 and 455 mm radius to match	type SP	type QSP
255 mm high × 305 and 455 mm radius to match	type BN	type QBN
255 mm high × 305 and 455 mm radius to match	type HB2, HB3	type QHB
225 mm high × 305 and 455 mm radius to match	type SP	type QSP
Angle kerb units		
305 × 305 × 225 mm high × 125 mm thick		
bullnosed external angle	type XA	
splayed external angle to match type SP	type XA	
bullnosed internal angle	type IA	
splayed internal angle to match type SP	type IA	
Channels		
255 mm wide × 125 mm high flat	type CS1	
150 mm wide × 125 mm high flat type	CS2	
255 mm wide × 125 mm high dished	type CD	

KERBS, PAVING, ETC.

Transition kerb units			
from kerb type SP to HB	left handed	type TL	
	right handed	type TR	
from kerb type BN to HB	left handed	type DL1	
	right handed	type DR1	
from kerb type BN to SP	left handed	type DL2	
	right handed	type DR2	

Number of kerbs required per quarter circle (780 mm kerb lengths)

Radius (m)	Number in quarter circle
12	24
10	20
8	16
6	12
5	10
4	8
3	6
2	4
1	2

Precast Concrete Edgings

Round top type ER	Flat top type EF	Bullnosed top type EBN
150 × 50 mm	150 × 50 mm	150 × 50 mm
200 × 50 mm	200 × 50 mm	200 × 50 mm
250 × 50 mm	250 × 50 mm	250 × 50 mm

KERBS, PAVING, ETC.

BASES

Cement Bound Material for Bases and Subbases

CBM1:	very carefully graded aggregate from 37.5–75 mm, with a 7-day strength of 4.5 N/mm^2
CBM2:	same range of aggregate as CBM1 but with more tolerance in each size of aggregate with a 7-day strength of 7.0 N/mm^2
CBM3:	crushed natural aggregate or blast furnace slag, graded from 37.5–150 mm for 40 mm aggregate, and from 20–75 mm for 20 mm aggregate, with a 7-day strength of 10 N/mm^2
CBM4:	crushed natural aggregate or blast furnace slag, graded from 37.5–150 mm for 40 mm aggregate, and from 20–75 mm for 20 mm aggregate, with a 7-day strength of 15 N/mm^2

INTERLOCKING BRICK/BLOCK ROADS/PAVINGS

Sizes of Precast Concrete Paving Blocks

Type R blocks
200 × 100 × 60 mm
200 × 100 × 65 mm
200 × 100 × 80 mm
200 × 100 × 100 mm

Type S
Any shape within a 295 mm space

Sizes of clay brick pavers
200 × 100 × 50 mm
200 × 100 × 65 mm
210 × 105 × 50 mm
210 × 105 × 65 mm
215 × 102.5 × 50 mm
215 × 102.5 × 65 mm

Type PA: 3 kN
Footpaths and pedestrian areas, private driveways, car parks, light vehicle traffic and over-run

Type PB: 7 kN
Residential roads, lorry parks, factory yards, docks, petrol station forecourts, hardstandings, bus stations

KERBS, PAVING, ETC.

PAVING AND SURFACING

Weights and Sizes of Paving and Surfacing

Description of item	Size	Quantity per tonne
Paving 50 mm thick	900 × 600 mm	15
Paving 50 mm thick	750 × 600 mm	18
Paving 50 mm thick	600 × 600 mm	23
Paving 50 mm thick	450 × 600 mm	30
Paving 38 mm thick	600 × 600 mm	30
Path edging	914 × 50 × 150 mm	60
Kerb (including radius and tapers)	125 × 254 × 914 mm	15
Kerb (including radius and tapers)	125 × 150 × 914 mm	25
Square channel	125 × 254 × 914 mm	15
Dished channel	125 × 254 × 914 mm	15
Quadrants	300 × 300 × 254 mm	19
Quadrants	450 × 450 × 254 mm	12
Quadrants	300 × 300 × 150 mm	30
Internal angles	300 × 300 × 254 mm	30
Fluted pavement channel	255 × 75 × 914 mm	25
Corner stones	300 × 300 mm	80
Corner stones	360 × 360 mm	60
Cable covers	914 × 175 mm	55
Gulley kerbs	220 × 220 × 150 mm	60
Gulley kerbs	220 × 200 × 75 mm	120

KERBS, PAVING, ETC.

Weights and Sizes of Paving and Surfacing

Material	kg/m^3	lb/cu yd
Tarmacadam	2306	3891
Macadam (waterbound)	2563	4325
Vermiculite (aggregate)	64–80	108–135
Terracotta	2114	3568
Cork – compressed	388	24
	kg/m^2	**lb/sq ft**
Clay floor tiles, 12.7 mm	27.3	5.6
Pavement lights	122	25
Damp-proof course	5	1
	kg/m^2 per mm thickness	**lb/sq ft per inch thickness**
Paving slabs (stone)	2.3	12
Granite setts	2.88	15
Asphalt	2.30	12
Rubber flooring	1.68	9
Polyvinyl chloride	1.94 (avg)	10 (avg)

Coverage (m^2) Per Cubic Metre of Materials Used as Subbases or Capping Layers

Consolidated thickness laid in (mm)	Square metre coverage		
	Gravel	Sand	Hardcore
50	15.80	16.50	–
75	10.50	11.00	–
100	7.92	8.20	7.42
125	6.34	6.60	5.90
150	5.28	5.50	4.95
175	–	–	4.23
200	–	–	3.71
225	–	–	3.30
300	–	–	2.47

KERBS, PAVING, ETC.

Approximate Rate of Spreads

Average thickness of course (mm)	Description	Approximate rate of spread			
		Open Textured		Dense, Medium & Fine Textured	
		(kg/m²)	(m²/t)	(kg/m²)	(m²/t)
35	14 mm open textured or dense wearing course	60–75	13–17	70–85	12–14
40	20 mm open textured or dense base course	70–85	12–14	80–100	10–12
45	20 mm open textured or dense base course	80–100	10–12	95–100	9–10
50	20 mm open textured or dense, or 28 mm dense base course	85–110	9–12	110–120	8–9
60	28 mm dense base course, 40 mm open textured of dense base course or 40 mm single course as base course		8–10	130–150	7–8
65	28 mm dense base course, 40 mm open textured or dense base course or 40 mm single course	100–135	7–10	140–160	6–7
75	40 mm single course, 40 mm open textured or dense base course, 40 mm dense roadbase	120–150	7–8	165–185	5–6
100	40 mm dense base course or roadbase	–	–	220–240	4–4.5

KERBS, PAVING, ETC.

Surface Dressing Roads: Coverage (m²) per Tonne of Material

Size in mm	Sand	Granite chips	Gravel	Limestone chips
Sand	168	–	–	–
3	–	148	152	165
6	–	130	133	144
9	–	111	114	123
13	–	85	87	95
19	–	68	71	78

Sizes of Flags

Reference	Nominal size (mm)	Thickness (mm)
A	600 × 450	50 and 63
B	600 × 600	50 and 63
C	600 × 750	50 and 63
D	600 × 900	50 and 63
E	450 × 450	50 and 70 chamfered top surface
F	400 × 400	50 and 65 chamfered top surface
G	300 × 300	50 and 60 chamfered top surface

Sizes of Natural Stone Setts

Width (mm)		Length (mm)		Depth (mm)
100	×	100	×	100
75	×	150 to 250	×	125
75	×	150 to 250	×	150
100	×	150 to 250	×	100
100	×	150 to 250	×	150

SEEDING/TURFING AND PLANTING

Topsoil Quality

Topsoil grade	Properties
Premium	Natural topsoil, high fertility, loamy texture, good soil structure, suitable for intensive cultivation.
General purpose	Natural or manufactured topsoil of lesser quality than Premium, suitable for agriculture or amenity landscape, may need fertilizer or soil structure improvement.
Economy	Selected subsoil, natural mineral deposit such as river silt or greensand. The grade comprises two subgrades; 'Low clay' and 'High clay' which is more liable to compaction in handling. This grade is suitable for low-production agricultural land and amenity woodland or conservation planting areas.

Forms of Trees

Standards:	Shall be clear with substantially straight stems. Grafted and budded trees shall have no more than a slight bend at the union. Standards shall be designated as Half, Extra light, Light, Standard, Selected standard, Heavy, and Extra heavy.
Sizes of Standards	
Heavy standard	12–14 cm girth × 3.50 to 5.00 m high
Extra Heavy standard	14–16 cm girth × 4.25 to 5.00 m high
Extra Heavy standard	16–18 cm girth × 4.25 to 6.00 m high
Extra Heavy standard	18–20 cm girth × 5.00 to 6.00 m high
Semi-mature trees:	Between 6.0 m and 12.0 m tall with a girth of 20 to 75 cm at 1.0 m above ground.
Feathered trees:	Shall have a defined upright central leader, with stem furnished with evenly spread and balanced lateral shoots down to or near the ground.
Whips:	Shall be without significant feather growth as determined by visual inspection.
Multi-stemmed trees:	Shall have two or more main stems at, near, above or below ground.

Seedlings grown from seed and not transplanted shall be specified when ordered for sale as:

1+0	one year old seedling
2+0	two year old seedling
1+1	one year seed bed, one year transplanted = two year old seedling
1+2	one year seed bed, two years transplanted = three year old seedling
2+1	two years seed bed, one year transplanted = three year old seedling
1u1	two years seed bed, undercut after 1 year = two year old seedling
2u2	four years seed bed, undercut after 2 years = four year old seedling

SEEDING/TURFING AND PLANTING

Cuttings

The age of cuttings (plants grown from shoots, stems, or roots of the mother plant) shall be specified when ordered for sale. The height of transplants and undercut seedlings/cuttings (which have been transplanted or undercut at least once) shall be stated in centimetres. The number of growing seasons before and after transplanting or undercutting shall be stated.

0 + 1	one year cutting
0 + 2	two year cutting
0 + 1 + 1	one year cutting bed, one year transplanted = two year old seedling
0 + 1 + 2	one year cutting bed, two years transplanted = three year old seedling

Grass Cutting Capacities in m² per hour

Speed mph	Width of cut in metres												
	0.5	0.7	1.0	1.2	1.5	1.7	2.0	2.0	2.1	2.5	2.8	3.0	3.4
1.0	724	1127	1529	1931	2334	2736	3138	3219	3380	4023	4506	4828	5472
1.5	1086	1690	2293	2897	3500	4104	4707	4828	5069	6035	6759	7242	8208
2.0	1448	2253	3058	3862	4667	5472	6276	6437	6759	8047	9012	9656	10944
2.5	1811	2816	3822	4828	5834	6840	7846	8047	8449	10058	11265	12070	13679
3.0	2173	3380	4587	5794	7001	8208	9415	9656	10139	12070	13518	14484	16415
3.5	2535	3943	5351	6759	8167	9576	10984	11265	11829	14082	15772	16898	19151
4.0	2897	4506	6115	7725	9334	10944	12553	12875	13518	16093	18025	19312	21887
4.5	3259	5069	6880	8690	10501	12311	14122	14484	15208	18105	20278	21726	24623
5.0	3621	5633	7644	9656	11668	13679	15691	16093	16898	20117	22531	24140	27359
5.5	3983	6196	8409	10622	12834	15047	17260	17703	18588	22128	24784	26554	30095
6.0	4345	6759	9173	11587	14001	16415	18829	19312	20278	24140	27037	28968	32831
6.5	4707	7322	9938	12553	15168	17783	20398	20921	21967	26152	29290	31382	35566
7.0	5069	7886	10702	13518	16335	19151	21967	22531	23657	28163	31543	33796	38302

Number of Plants per m² at the following offset spacings

(All plants equidistant horizontally and vertically)

Distance mm	Nr of Plants
100	114.43
200	28.22
250	18.17
300	12.35
400	6.72
500	4.29
600	3.04
700	2.16
750	1.88
900	1.26
1000	1.05
1200	0.68
1500	0.42
2000	0.23

SEEDING/TURFING AND PLANTING

Grass Clippings Wet: Based on 3.5 m³/tonne

Annual kg/100 m²	Average 20 cuts kg/100 m²	m²/tonne	m²/m³
32.0	1.6	61162.1	214067.3

Nr of cuts	22	20	18	16	12	4
kg/cut	1.45	1.60	1.78	2.00	2.67	8.00
Area capacity of 3 tonne vehicle per load						
m²	206250	187500	168750	150000	112500	37500
Load m³	100 m² units/m³ of vehicle space					
1	196.4	178.6	160.7	142.9	107.1	35.7
2	392.9	357.1	321.4	285.7	214.3	71.4
3	589.3	535.7	482.1	428.6	321.4	107.1
4	785.7	714.3	642.9	571.4	428.6	142.9
5	982.1	892.9	803.6	714.3	535.7	178.6

Transportation of Trees

To unload large trees a machine with the necessary lifting strength is required. The weight of the trees must therefore be known in advance. The following table gives a rough overview. The additional columns with root ball dimensions and the number of plants per trailer provide additional information, for example about preparing planting holes and calculating unloading times.

Girth in cm	Rootball diameter in cm	Ball height in cm	Weight in kg	Numbers of trees per trailer
16–18	50–60	40	150	100–120
18–20	60–70	40–50	200	80–100
20–25	60–70	40–50	270	50–70
25–30	80	50–60	350	50
30–35	90–100	60–70	500	12–18
35–40	100–110	60–70	650	10–15
40–45	110–120	60–70	850	8–12
45–50	110–120	60–70	1100	5–7
50–60	130–140	60–70	1600	1–3
60–70	150–160	60–70	2500	1
70–80	180–200	70	4000	1
80–90	200–220	70–80	5500	1
90–100	230–250	80–90	7500	1
100–120	250–270	80–90	9500	1

Data supplied by Lorenz von Ehren GmbH
The information in the table is approximate; deviations depend on soil type, genus and weather

FENCING AND GATES

Types of Preservative

Creosote (tar oil) can be 'factory' applied	by pressure to BS 144: pts 1&2 by immersion to BS 144: pt 1 by hot and cold open tank to BS 144: pts 1&2
Copper/chromium/arsenic (CCA)	by full cell process to BS 4072 pts 1&2
Organic solvent (OS)	by double vacuum (vacvac) to BS 5707 pts 1&3 by immersion to BS 5057 pts 1&3
Pentachlorophenol (PCP)	by heavy oil double vacuum to BS 5705 pts 2&3
Boron diffusion process (treated with disodium octaborate to BWPA Manual 1986)	

Note: Boron is used on green timber at source and the timber is supplied dry

Cleft Chestnut Pale Fences

Pales	Pale spacing	Wire lines	
900 mm	75 mm	2	temporary protection
1050 mm	75 or 100 mm	2	light protective fences
1200 mm	75 mm	3	perimeter fences
1350 mm	75 mm	3	perimeter fences
1500 mm	50 mm	3	narrow perimeter fences
1800 mm	50 mm	3	light security fences

Close-Boarded Fences

Close-boarded fences 1.05 to 1.8 m high
Type BCR (recessed) or BCM (morticed) with concrete posts 140 × 115 mm tapered and Type BW with timber posts

Palisade Fences

Wooden palisade fences
Type WPC with concrete posts 140 × 115 mm tapered and Type WPW with timber posts

For both types of fence:
Height of fence 1050 mm: two rails
Height of fence 1200 mm: two rails
Height of fence 1500 mm: three rails
Height of fence 1650 mm: three rails
Height of fence 1800 mm: three rails

FENCING AND GATES

Post and Rail Fences

Wooden post and rail fences
Type MPR 11/3 morticed rails and Type SPR 11/3 nailed rails
Height to top of rail 1100 mm
Rails: three rails 87 mm, 38 mm

Type MPR 11/4 morticed rails and Type SPR 11/4 nailed rails
Height to top of rail 1100 mm
Rails: four rails 87 mm, 38 mm

Type MPR 13/4 morticed rails and Type SPR 13/4 nailed rails
Height to top of rail 1300 mm
Rail spacing 250 mm, 250 mm, and 225 mm from top
Rails: four rails 87 mm, 38 mm

Steel Posts

Rolled steel angle iron posts for chain link fencing

Posts	Fence height	Strut	Straining post
1500 × 40 × 40 × 5 mm	900 mm	1500 × 40 × 40 × 5 mm	1500 × 50 × 50 × 6 mm
1800 × 40 × 40 × 5 mm	1200 mm	1800 × 40 × 40 × 5 mm	1800 × 50 × 50 × 6 mm
2000 × 45 × 45 × 5 mm	1400 mm	2000 × 45 × 45 × 5 mm	2000 × 60 × 60 × 6 mm
2600 × 45 × 45 × 5 mm	1800 mm	2600 × 45 × 45 × 5 mm	2600 × 60 × 60 × 6 mm
3000 × 50 × 50 × 6 mm	1800 mm	2600 × 45 × 45 × 5 mm	3000 × 60 × 60 × 6 mm
with arms			

Concrete Posts

Concrete posts for chain link fencing

Posts and straining posts	Fence height	Strut
1570 mm 100 × 100 mm	900 mm	1500 mm × 75 × 75 mm
1870 mm 125 × 125 mm	1200 mm	1830 mm × 100 × 75 mm
2070 mm 125 × 125 mm	1400 mm	1980 mm × 100 × 75 mm
2620 mm 125 × 125 mm	1800 mm	2590 mm × 100 × 85 mm
3040 mm 125 × 125 mm	1800 mm	2590 mm × 100 × 85 mm (with arms)

FENCING AND GATES

Rolled Steel Angle Posts

Rolled steel angle posts for rectangular wire mesh (field) fencing

Posts	Fence height	Strut	Straining post
1200 × 40 × 40 × 5 mm	600 mm	1200 × 75 × 75 mm	1350 × 100 × 100 mm
1400 × 40 × 40 × 5 mm	800 mm	1400 × 75 × 75 mm	1550 × 100 × 100 mm
1500 × 40 × 40 × 5 mm	900 mm	1500 × 75 × 75 mm	1650 × 100 × 100 mm
1600 × 40 × 40 × 5 mm	1000 mm	1600 × 75 × 75 mm	1750 × 100 × 100 mm
1750 × 40 × 40 × 5 mm	1150 mm	1750 × 75 × 100 mm	1900 × 125 × 125 mm

Concrete Posts

Concrete posts for rectangular wire mesh (field) fencing

Posts	Fence height	Strut	Straining post
1270 × 100 × 100 mm	600 mm	1200 × 75 × 75 mm	1420 × 100 × 100 mm
1470 × 100 × 100 mm	800 mm	1350 × 75 × 75 mm	1620 × 100 × 100 mm
1570 × 100 × 100 mm	900 mm	1500 × 75 × 75 mm	1720 × 100 × 100 mm
1670 × 100 × 100 mm	600 mm	1650 × 75 × 75 mm	1820 × 100 × 100 mm
1820 × 125 × 125 mm	1150 mm	1830 × 75 × 100 mm	1970 × 125 × 125 mm

Cleft Chestnut Pale Fences

Timber Posts

Timber posts for wire mesh and hexagonal wire netting fences

Round timber for general fences

Posts	Fence height	Strut	Straining post
1300 × 65 mm dia.	600 mm	1200 × 80 mm dia.	1450 × 100 mm dia.
1500 × 65 mm dia.	800 mm	1400 × 80 mm dia.	1650 × 100 mm dia.
1600 × 65 mm dia.	900 mm	1500 × 80 mm dia.	1750 × 100 mm dia.
1700 × 65 mm dia.	1050 mm	1600 × 80 mm dia.	1850 × 100 mm dia.
1800 × 65 mm dia.	1150 mm	1750 × 80 mm dia.	2000 × 120 mm dia.

Squared timber for general fences

Posts	Fence height	Strut	Straining post
1300 × 75 × 75 mm	600 mm	1200 × 75 × 75 mm	1450 × 100 × 100 mm
1500 × 75 × 75 mm	800 mm	1400 × 75 × 75 mm	1650 × 100 × 100 mm
1600 × 75 × 75 mm	900 mm	1500 × 75 × 75 mm	1750 × 100 × 100 mm
1700 × 75 × 75 mm	1050 mm	1600 × 75 × 75 mm	1850 × 100 × 100 mm
1800 × 75 × 75 mm	1150 mm	1750 × 75 × 75 mm	2000 × 125 × 100 mm

FENCING AND GATES

Steel Fences to BS 1722: Part 9: 1992

	Fence height	Top/bottom rails and flat posts	Vertical bars
Light	1000 mm	40 × 10 mm 450 mm in ground	12 mm dia. at 115 mm cs
	1200 mm	40 × 10 mm 550 mm in ground	12 mm dia. at 115 mm cs
	1400 mm	40 × 10 mm 550 mm in ground	12 mm dia. at 115 mm cs
Light	1000 mm	40 × 10 mm 450 mm in ground	16 mm dia. at 120 mm cs
	1200 mm	40 × 10 mm 550 mm in ground	16 mm dia. at 120 mm cs
	1400 mm	40 × 10 mm 550 mm in ground	16 mm dia. at 120 mm cs
Medium	1200 mm	50 × 10 mm 550 mm in ground	20 mm dia. at 125 mm cs
	1400 mm	50 × 10 mm 550 mm in ground	20 mm dia. at 125 mm cs
	1600 mm	50 × 10 mm 600 mm in ground	22 mm dia. at 145 mm cs
	1800 mm	50 × 10 mm 600 mm in ground	22 mm dia. at 145 mm cs
Heavy	1600 mm	50 × 10 mm 600 mm in ground	22 mm dia. at 145 mm cs
	1800 mm	50 × 10 mm 600 mm in ground	22 mm dia. at 145 mm cs
	2000 mm	50 × 10 mm 600 mm in ground	22 mm dia. at 145 mm cs
	2200 mm	50 × 10 mm 600 mm in ground	22 mm dia. at 145 mm cs

Notes: Mild steel fences: round or square verticals; flat standards and horizontals. Tops of vertical bars may be bow-top, blunt, or pointed. Round or square bar railings

Timber Field Gates to BS 3470: 1975

Gates made to this standard are designed to open one way only
All timber gates are 1100 mm high
Width over stiles 2400, 2700, 3000, 3300, 3600, and 4200 mm
Gates over 4200 mm should be made in two leaves

Steel Field Gates to BS 3470: 1975

All steel gates are 1100 mm high
Heavy duty: width over stiles 2400, 3000, 3600 and 4500 mm
Light duty: width over stiles 2400, 3000, and 3600 mm

FENCING AND GATES

Domestic Front Entrance Gates to BS 4092: Part 1: 1966

Metal gates:	Single gates are 900 mm high minimum, 900 mm, 1000 mm and 1100 mm wide

Domestic Front Entrance Gates to BS 4092: Part 2: 1966

Wooden gates:	All rails shall be tenoned into the stiles
	Single gates are 840 mm high minimum, 801 mm and 1020 mm wide
	Double gates are 840 mm high minimum, 2130, 2340 and 2640 mm wide

Timber Bridle Gates to BS 5709:1979 (Horse or Hunting Gates)

Gates open one way only	
Minimum width between posts	1525 mm
Minimum height	1100 mm

Timber Kissing Gates to BS 5709:1979

Minimum width	700 mm
Minimum height	1000 mm
Minimum distance between shutting posts	600 mm
Minimum clearance at mid-point	600 mm

Metal Kissing Gates to BS 5709:1979

Sizes are the same as those for timber kissing gates
Maximum gaps between rails 120 mm

Categories of Pedestrian Guard Rail to BS 3049:1976

Class A for normal use
Class B where vandalism is expected
Class C where crowd pressure is likely

DRAINAGE

Width Required for Trenches for Various Diameters of Pipes

Pipe diameter (mm)	Trench n.e. 1.50 m deep	Trench over 1.50 m deep
n.e. 100 mm	450 mm	600 mm
100–150 mm	500 mm	650 mm
150–225 mm	600 mm	750 mm
225–300 mm	650 mm	800 mm
300–400 mm	750 mm	900 mm
400–450 mm	900 mm	1050 mm
450–600 mm	1100 mm	1300 mm

Weights and Dimensions – Vitrified Clay Pipes

Product	Nominal diameter (mm)	Effective length (mm)	BS 65 limits of tolerance min (mm)	max (mm)	Crushing strength (kN/m)	Weight (kg/pipe)	(kg/m)
Supersleve	100	1600	96	105	35.00	14.71	9.19
	150	1750	146	158	35.00	29.24	16.71
Hepsleve	225	1850	221	236	28.00	84.03	45.42
	300	2500	295	313	34.00	193.05	77.22
	150	1500	146	158	22.00	37.04	24.69
Hepseal	225	1750	221	236	28.00	85.47	48.84
	300	2500	295	313	34.00	204.08	81.63
	400	2500	394	414	44.00	357.14	142.86
	450	2500	444	464	44.00	454.55	181.63
	500	2500	494	514	48.00	555.56	222.22
	600	2500	591	615	57.00	796.23	307.69
	700	3000	689	719	67.00	1111.11	370.45
	800	3000	788	822	72.00	1351.35	450.45
Hepline	100	1600	95	107	22.00	14.71	9.19
	150	1750	145	160	22.00	29.24	16.71
	225	1850	219	239	28.00	84.03	45.42
	300	1850	292	317	34.00	142.86	77.22
Hepduct (conduit)	90	1500	–	–	28.00	12.05	8.03
	100	1600	–	–	28.00	14.71	9.19
	125	1750	–	–	28.00	20.73	11.84
	150	1750	–	–	28.00	29.24	16.71
	225	1850	–	–	28.00	84.03	45.42
	300	1850	–	–	34.00	142.86	77.22

DRAINAGE

Weights and Dimensions – Vitrified Clay Pipes

Nominal internal diameter (mm)	Nominal wall thickness (mm)	Approximate weight (kg/m)
150	25	45
225	29	71
300	32	122
375	35	162
450	38	191
600	48	317
750	54	454
900	60	616
1200	76	912
1500	89	1458
1800	102	1884
2100	127	2619

Wall thickness, weights and pipe lengths vary, depending on type of pipe required

The particulars shown above represent a selection of available diameters and are applicable to strength class 1 pipes with flexible rubber ring joints

Tubes with Ogee joints are also available

DRAINAGE

Weights and Dimensions – PVC-u Pipes

	Nominal size	Mean outside diameter (mm)		Wall thickness	Weight
		min	max	(mm)	(kg/m)
Standard pipes	82.4	82.4	82.7	3.2	1.2
	110.0	110.0	110.4	3.2	1.6
	160.0	160.0	160.6	4.1	3.0
	200.0	200.0	200.6	4.9	4.6
	250.0	250.0	250.7	6.1	7.2
Perforated pipes heavy grade	As above	As above	As above	As above	As above
thin wall	82.4	82.4	82.7	1.7	–
	110.0	110.0	110.4	2.2	–
	160.0	160.0	160.6	3.2	–

Width of Trenches Required for Various Diameters of Pipes

Pipe diameter (mm)	Trench n.e. 1.5 m deep (mm)	Trench over 1.5 m deep (mm)
n.e. 100	450	600
100–150	500	650
150–225	600	750
225–300	650	800
300–400	750	900
400–450	900	1050
450–600	1100	1300

DRAINAGE

DRAINAGE BELOW GROUND AND LAND DRAINAGE

Flow of Water Which Can Be Carried by Various Sizes of Pipe

Clay or concrete pipes

	Gradient of pipeline							
	1:10	1:20	1:30	1:40	1:50	1:60	1:80	1:100
Pipe size	Flow in litres per second							
DN 100 15.0	8.5	6.8	5.8	5.2	4.7	4.0	3.5	
DN 150 28.0	19.0	16.0	14.0	12.0	11.0	9.1	8.0	
DN 225 140.0	95.0	76.0	66.0	58.0	53.0	46.0	40.0	

Plastic pipes

	Gradient of pipeline							
	1:10	1:20	1:30	1:40	1:50	1:60	1:80	1:100
Pipe size	Flow in litres per second							
82.4 mm i/dia.	12.0	8.5	6.8	5.8	5.2	4.7	4.0	3.5
110 mm i/dia.	28.0	19.0	16.0	14.0	12.0	11.0	9.1	8.0
160 mm i/dia.	76.0	53.0	43.0	37.0	33.0	29.0	25.0	22.0
200 mm i/dia.	140.0	95.0	76.0	66.0	58.0	53.0	46.0	40.0

Vitrified (Perforated) Clay Pipes and Fittings to BS En 295-5 1994

Length not specified		
75 mm bore	**250 mm bore**	**600 mm bore**
100	300	700
125	350	800
150	400	1000
200	450	1200
225	500	

Precast Concrete Pipes: Prestressed Non-pressure Pipes and Fittings: Flexible Joints to BS 5911: Pt. 103: 1994

Rationalized metric nominal sizes: 450, 500	
Length:	500–1000 by 100 increments
	1000–2200 by 200 increments
	2200–2800 by 300 increments
Angles: length:	450–600 angles 45, 22.5,11.25°
	600 or more angles 22.5, 11.25°

DRAINAGE

Precast Concrete Pipes: Unreinforced and Circular Manholes and Soakaways to BS 5911: Pt. 200: 1994

Nominal sizes:	
Shafts:	675, 900 mm
Chambers:	900, 1050, 1200, 1350, 1500, 1800, 2100, 2400, 2700, 3000 mm
Large chambers:	To have either tapered reducing rings or a flat reducing slab in order to accept the standard cover
Ring depths:	1. 300–1200 mm by 300 mm increments except for bottom slab and rings below cover slab, these are by 150 mm increments
	2. 250–1000 mm by 250 mm increments except for bottom slab and rings below cover slab, these are by 125 mm increments
Access hole:	750 × 750 mm for DN 1050 chamber
	1200 × 675 mm for DN 1350 chamber

Calculation of Soakaway Depth

The following formula determines the depth of concrete ring soakaway that would be required for draining given amounts of water.

$$h = \frac{4ar}{3\pi D^2}$$

h = depth of the chamber below the invert pipe
a = the area to be drained
r = the hourly rate of rainfall (50 mm per hour)
π = pi
D = internal diameter of the soakaway

This table shows the depth of chambers in each ring size which would be required to contain the volume of water specified. These allow a recommended storage capacity of ⅓ (one third of the hourly rainfall figure).

Table Showing Required Depth of Concrete Ring Chambers in Metres

Area m²	50	100	150	200	300	400	500
Ring size							
0.9	1.31	2.62	3.93	5.24	7.86	10.48	13.10
1.1	0.96	1.92	2.89	3.85	5.77	7.70	9.62
1.2	0.74	1.47	2.21	2.95	4.42	5.89	7.37
1.4	0.58	1.16	1.75	2.33	3.49	4.66	5.82
1.5	0.47	0.94	1.41	1.89	2.83	3.77	4.72
1.8	0.33	0.65	0.98	1.31	1.96	2.62	3.27
2.1	0.24	0.48	0.72	0.96	1.44	1.92	2.41
2.4	0.18	0.37	0.55	0.74	1.11	1.47	1.84
2.7	0.15	0.29	0.44	0.58	0.87	1.16	1.46
3.0	0.12	0.24	0.35	0.47	0.71	0.94	1.18

DRAINAGE

Precast Concrete Inspection Chambers and Gullies to BS 5911: Part 230: 1994

Nominal sizes:	375 diameter, 750, 900 mm deep
	450 diameter, 750, 900, 1050, 1200 mm deep
Depths:	from the top for trapped or untrapped units:
	centre of outlet 300 mm
	invert (bottom) of the outlet pipe 400 mm
Depth of water seal for trapped gullies:	
	85 mm, rodding eye int. dia. 100 mm
Cover slab:	65 mm min

Bedding Flexible Pipes: PVC-u Or Ductile Iron

Type 1 =	100 mm fill below pipe, 300 mm above pipe: single size material
Type 2 =	100 mm fill below pipe, 300 mm above pipe: single size or graded material
Type 3 =	100 mm fill below pipe, 75 mm above pipe with concrete protective slab over
Type 4 =	100 mm fill below pipe, fill laid level with top of pipe
Type 5 =	200 mm fill below pipe, fill laid level with top of pipe
Concrete =	25 mm sand blinding to bottom of trench, pipe supported on chocks, 100 mm concrete under the pipe, 150 mm concrete over the pipe

DRAINAGE

Bedding Rigid Pipes: Clay or Concrete
(for vitrified clay pipes the manufacturer should be consulted)

Class D:	Pipe laid on natural ground with cut-outs for joints, soil screened to remove stones over 40 mm and returned over pipe to 150 m min depth. Suitable for firm ground with trenches trimmed by hand.
Class N:	Pipe laid on 50 mm granular material of graded aggregate to Table 4 of BS 882, or 10 mm aggregate to Table 6 of BS 882, or as dug light soil (not clay) screened to remove stones over 10 mm. Suitable for machine dug trenches.
Class B:	As Class N, but with granular bedding extending half way up the pipe diameter.
Class F:	Pipe laid on 100 mm granular fill to BS 882 below pipe, minimum 150 mm granular fill above pipe: single size material. Suitable for machine dug trenches.
Class A:	Concrete 100 mm thick under the pipe extending half way up the pipe, backfilled with the appropriate class of fill. Used where there is only a very shallow fall to the drain. Class A bedding allows the pipes to be laid to an exact gradient.
Concrete surround:	25 mm sand blinding to bottom of trench, pipe supported on chocks, 100 mm concrete under the pipe, 150 mm concrete over the pipe. It is preferable to bed pipes under slabs or wall in granular material.

PIPED SUPPLY SYSTEMS

Identification of Service Tubes From Utility to Dwellings

Utility	Colour	Size	Depth
British Telecom	grey	54 mm od	450 mm
Electricity	black	38 mm od	450 mm
Gas	yellow	42 mm od rigid 60 mm od convoluted	450 mm
Water	may be blue	(normally untubed)	750 mm

ELECTRICAL SUPPLY/POWER/LIGHTING SYSTEMS

Electrical Insulation Class En 60.598 BS 4533

Class 1:	luminaires comply with class 1 (I) earthed electrical requirements
Class 2:	luminaires comply with class 2 (II) double insulated electrical requirements
Class 3:	luminaires comply with class 3 (III) electrical requirements

Protection to Light Fittings

BS EN 60529:1992 Classification for degrees of protection provided by enclosures.
(IP Code – International or ingress Protection)

1st characteristic: against ingress of solid foreign objects

The figure	2	indicates that fingers cannot enter
	3	that a 2.5 mm diameter probe cannot enter
	4	that a 1.0 mm diameter probe cannot enter
	5	the fitting is dust proof (no dust around live parts)
	6	the fitting is dust tight (no dust entry)

2nd characteristic: ingress of water with harmful effects

The figure	0	indicates unprotected
	1	vertically dripping water cannot enter
	2	water dripping 15° (tilt) cannot enter
	3	spraying water cannot enter
	4	splashing water cannot enter
	5	jetting water cannot enter
	6	powerful jetting water cannot enter
	7	proof against temporary immersion
	8	proof against continuous immersion

Optional additional codes:		A–D protects against access to hazardous parts
	H	high voltage apparatus
	M	fitting was in motion during water test
	S	fitting was static during water test
	W	protects against weather

Marking code arrangement:		(example) IPX5S = IP (International or Ingress Protection)
		X (denotes omission of first characteristic)
		5 = jetting
		S = static during water test

RAIL TRACKS

	kg/m of track	lb/ft of track
Standard gauge Bull-head rails, chairs, transverse timber (softwood) sleepers etc.	245	165
Main lines Flat-bottom rails, transverse prestressed concrete sleepers, etc.	418	280
Add for electric third rail	51	35
Add for crushed stone ballast	2600	1750
	kg/m²	**lb/sq ft**
Overall average weight – rails connections, sleepers, ballast, etc.	733	150
	kg/m of track	**lb/ft of track**
Bridge rails, longitudinal timber sleepers, etc.	112	75

RAIL TRACKS

Heavy Rails

British Standard Section No.	Rail height (mm)	Foot width (mm)	Head width (mm)	Min web thickness (mm)	Section weight (kg/m)
Flat Bottom Rails					
60 A	114.30	109.54	57.15	11.11	30.62
70 A	123.82	111.12	60.32	12.30	34.81
75 A	128.59	114.30	61.91	12.70	37.45
80 A	133.35	117.47	63.50	13.10	39.76
90 A	142.88	127.00	66.67	13.89	45.10
95 A	147.64	130.17	69.85	14.68	47.31
100 A	152.40	133.35	69.85	15.08	50.18
110 A	158.75	139.70	69.85	15.87	54.52
113 A	158.75	139.70	69.85	20.00	56.22
50 'O'	100.01	100.01	52.39	10.32	24.82
80 'O'	127.00	127.00	63.50	13.89	39.74
60R	114.30	109.54	57.15	11.11	29.85
75R	128.59	122.24	61.91	13.10	37.09
80R	133.35	127.00	63.50	13.49	39.72
90R	142.88	136.53	66.67	13.89	44.58
95R	147.64	141.29	68.26	14.29	47.21
100R	152.40	146.05	69.85	14.29	49.60
95N	147.64	139.70	69.85	13.89	47.27
Bull Head Rails					
95R BH	145.26	69.85	69.85	19.05	47.07

Light Rails

British Standard Section No.	Rail height (mm)	Foot width (mm)	Head width (mm)	Min web thickness (mm)	Section weight (kg/m)
Flat Bottom Rails					
20M	65.09	55.56	30.96	6.75	9.88
30M	75.41	69.85	38.10	9.13	14.79
35M	80.96	76.20	42.86	9.13	17.39
35R	85.73	82.55	44.45	8.33	17.40
40	88.11	80.57	45.64	12.3	19.89
Bridge Rails					
13	48.00	92	36.00	18.0	13.31
16	54.00	108	44.50	16.0	16.06
20	55.50	127	50.00	20.5	19.86
28	67.00	152	50.00	31.0	28.62
35	76.00	160	58.00	34.5	35.38
50	76.00	165	58.50	–	50.18
Crane Rails					
A65	75.00	175.00	65.00	38.0	43.10
A75	85.00	200.00	75.00	45.0	56.20
A100	95.00	200.00	100.00	60.0	74.30
A120	105.00	220.00	120.00	72.0	100.00
175CR	152.40	152.40	107.95	38.1	86.92

RAIL TRACKS

Fish Plates

British Standard Section No.	Overall plate length		Hole diameter	Finished weight per pair	
	4 Hole (mm)	6 Hole (mm)	(mm)	4 Hole (kg/pair)	6 Hole (kg/pair)
For British Standard Heavy Rails: Flat Bottom Rails					
60 A	406.40	609.60	20.64	9.87	14.76
70 A	406.40	609.60	22.22	11.15	16.65
75 A	406.40	–	23.81	11.82	17.73
80 A	406.40	609.60	23.81	13.15	19.72
90 A	457.20	685.80	25.40	17.49	26.23
100 A	508.00	–	pear	25.02	–
110 A (shallow)	507.00	–	27.00	30.11	54.64
113 A (heavy)	507.00	–	27.00	30.11	54.64
50 'O' (shallow)	406.40	–	–	6.68	10.14
80 'O' (shallow)	495.30	–	23.81	14.72	22.69
60R (shallow)	406.40	609.60	20.64	8.76	13.13
60R (angled)	406.40	609.60	20.64	11.27	16.90
75R (shallow)	406.40	–	23.81	10.94	16.42
75R (angled)	406.40	–	23.81	13.67	–
80R (shallow)	406.40	609.60	23.81	11.93	17.89
80R (angled)	406.40	609.60	23.81	14.90	22.33
For British Standard Heavy Rails: Bull head rails					
95R BH (shallow)	–	457.20	27.00	14.59	14.61
For British Standard Light Rails: Flat Bottom Rails					
30M	355.6	–	–	–	2.72
35M	355.6	–	–	–	2.83
40	355.6	–	–	3.76	–

Tables and Memoranda

FRACTIONS, DECIMALS AND MILLIMETRE EQUIVALENTS

Fractions	Decimals	(mm)	Fractions	Decimals	(mm)
1/64	0.015625	0.396875	33/64	0.515625	13.096875
1/32	0.03125	0.79375	17/32	0.53125	13.49375
3/64	0.046875	1.190625	35/64	0.546875	13.890625
1/16	0.0625	1.5875	9/16	0.5625	14.2875
5/64	0.078125	1.984375	37/64	0.578125	14.684375
3/32	0.09375	2.38125	19/32	0.59375	15.08125
7/64	0.109375	2.778125	39/64	0.609375	15.478125
1/8	0.125	3.175	5/8	0.625	15.875
9/64	0.140625	3.571875	41/64	0.640625	16.271875
5/32	0.15625	3.96875	21/32	0.65625	16.66875
11/64	0.171875	4.365625	43/64	0.671875	17.065625
3/16	0.1875	4.7625	11/16	0.6875	17.4625
13/64	0.203125	5.159375	45/64	0.703125	17.859375
7/32	0.21875	5.55625	23/32	0.71875	18.25625
15/64	0.234375	5.953125	47/64	0.734375	18.653125
1/4	0.25	6.35	3/4	0.75	19.05
17/64	0.265625	6.746875	49/64	0.765625	19.446875
9/32	0.28125	7.14375	25/32	0.78125	19.84375
19/64	0.296875	7.540625	51/64	0.796875	20.240625
5/16	0.3125	7.9375	13/16	0.8125	20.6375
21/64	0.328125	8.334375	53/64	0.828125	21.034375
11/32	0.34375	8.73125	27/32	0.84375	21.43125
23/64	0.359375	9.128125	55/64	0.859375	21.828125
3/8	0.375	9.525	7/8	0.875	22.225
25/64	0.390625	9.921875	57/64	0.890625	22.621875
13/32	0.40625	10.31875	29/32	0.90625	23.01875
27/64	0.421875	10.71563	59/64	0.921875	23.415625
7/16	0.4375	11.1125	15/16	0.9375	23.8125
29/64	0.453125	11.50938	61/64	0.953125	24.209375
15/32	0.46875	11.90625	31/32	0.96875	24.60625
31/64	0.484375	12.30313	63/64	0.984375	25.003125
1/2	0.5	12.7	1.0	1	25.4

IMPERIAL STANDARD WIRE GAUGE (SWG)

IMPERIAL STANDARD WIRE GAUGE (SWG)

SWG No.	Diameter (inches)	Diameter (mm)	SWG No.	Diameter (inches)	Diameter (mm)
7/0	0.5	12.7	23	0.024	0.61
6/0	0.464	11.79	24	0.022	0.559
5/0	0.432	10.97	25	0.02	0.508
4/0	0.4	10.16	26	0.018	0.457
3/0	0.372	9.45	27	0.0164	0.417
2/0	0.348	8.84	28	0.0148	0.376
1/0	0.324	8.23	29	0.0136	0.345
1	0.3	7.62	30	0.0124	0.315
2	0.276	7.01	31	0.0116	0.295
3	0.252	6.4	32	0.0108	0.274
4	0.232	5.89	33	0.01	0.254
5	0.212	5.38	34	0.009	0.234
6	0.192	4.88	35	0.008	0.213
7	0.176	4.47	36	0.008	0.193
8	0.16	4.06	37	0.007	0.173
9	0.144	3.66	38	0.006	0.152
10	0.128	3.25	39	0.005	0.132
11	0.116	2.95	40	0.005	0.122
12	0.104	2.64	41	0.004	0.112
13	0.092	2.34	42	0.004	0.102
14	0.08	2.03	43	0.004	0.091
15	0.072	1.83	44	0.003	0.081
16	0.064	1.63	45	0.003	0.071
17	0.056	1.42	46	0.002	0.061
18	0.048	1.22	47	0.002	0.051
19	0.04	1.016	48	0.002	0.041
20	0.036	0.914	49	0.001	0.031
21	0.032	0.813	50	0.001	0.025
22	0.028	0.711			

PIPES, WATER, STORAGE, INSULATION

WATER PRESSURE DUE TO HEIGHT

Imperial

Head (Feet)	Pressure (lb/in^2)		Head (Feet)	Pressure (lb/in^2)
1	0.43		70	30.35
5	2.17		75	32.51
10	4.34		80	34.68
15	6.5		85	36.85
20	8.67		90	39.02
25	10.84		95	41.18
30	13.01		100	43.35
35	15.17		105	45.52
40	17.34		110	47.69
45	19.51		120	52.02
50	21.68		130	56.36
55	23.84		140	60.69
60	26.01		150	65.03
65	28.18			

Metric

Head (m)	Pressure (bar)		Head (m)	Pressure (bar)
0.5	0.049		18.0	1.766
1.0	0.098		19.0	1.864
1.5	0.147		20.0	1.962
2.0	0.196		21.0	2.06
3.0	0.294		22.0	2.158
4.0	0.392		23.0	2.256
5.0	0.491		24.0	2.354
6.0	0.589		25.0	2.453
7.0	0.687		26.0	2.551
8.0	0.785		27.0	2.649
9.0	0.883		28.0	2.747
10.0	0.981		29.0	2.845
11.0	1.079		30.0	2.943
12.0	1.177		32.5	3.188
13.0	1.275		35.0	3.434
14.0	1.373		37.5	3.679
15.0	1.472		40.0	3.924
16.0	1.57		42.5	4.169
17.0	1.668		45.0	4.415

1 bar	=	14.5038 lbf/in^2
1 lbf/in^2	=	0.06895 bar
1 metre	=	3.2808 ft or 39.3701 in
1 foot	=	0.3048 metres
1 in wg	=	2.5 mbar (249.1 N/m^2)

PIPES, WATER, STORAGE, INSULATION

Dimensions and Weights of Copper Pipes to BSEN 1057, BSEN 12499, BSEN 14251

Outside diameter (mm)	Internal diameter (mm)	Weight per metre (kg)	Internal diameter (mm)	Weight per metre (kg)	Internal siameter (mm)	Weight per metre (kg)
	Formerly Table X		Formerly Table Y		Formerly Table Z	
6	4.80	0.0911	4.40	0.1170	5.00	0.0774
8	6.80	0.1246	6.40	0.1617	7.00	0.1054
10	8.80	0.1580	8.40	0.2064	9.00	0.1334
12	10.80	0.1914	10.40	0.2511	11.00	0.1612
15	13.60	0.2796	13.00	0.3923	14.00	0.2031
18	16.40	0.3852	16.00	0.4760	16.80	0.2918
22	20.22	0.5308	19.62	0.6974	20.82	0.3589
28	26.22	0.6814	25.62	0.8985	26.82	0.4594
35	32.63	1.1334	32.03	1.4085	33.63	0.6701
42	39.63	1.3675	39.03	1.6996	40.43	0.9216
54	51.63	1.7691	50.03	2.9052	52.23	1.3343
76.1	73.22	3.1287	72.22	4.1437	73.82	2.5131
108	105.12	4.4666	103.12	7.3745	105.72	3.5834
133	130.38	5.5151	–	–	130.38	5.5151
159	155.38	8.7795	–	–	156.38	6.6056

Dimensions of Stainless Steel Pipes to BS 4127

Outside siameter (mm)	Maximum outside siameter (mm)	Minimum outside diameter (mm)	Wall thickness (mm)	Working pressure (bar)
6	6.045	5.940	0.6	330
8	8.045	7.940	0.6	260
10	10.045	9.940	0.6	210
12	12.045	11.940	0.6	170
15	15.045	14.940	0.6	140
18	18.045	17.940	0.7	135
22	22.055	21.950	0.7	110
28	28.055	27.950	0.8	121
35	35.070	34.965	1.0	100
42	42.070	41.965	1.1	91
54	54.090	53.940	1.2	77

PIPES, WATER, STORAGE, INSULATION

Dimensions of Steel Pipes to BS 1387

Nominal Size	Approx. Outside Diameter	Outside diameter				Thickness		
		Light		Medium & Heavy		Light	Medium	Heavy
		Max	Min	Max	Min			
(mm)	(mm)	(mm)	(mm)	(mm)	(mm)	(mm)	(mm)	(mm)
6	10.20	10.10	9.70	10.40	9.80	1.80	2.00	2.65
8	13.50	13.60	13.20	13.90	13.30	1.80	2.35	2.90
10	17.20	17.10	16.70	17.40	16.80	1.80	2.35	2.90
15	21.30	21.40	21.00	21.70	21.10	2.00	2.65	3.25
20	26.90	26.90	26.40	27.20	26.60	2.35	2.65	3.25
25	33.70	33.80	33.20	34.20	33.40	2.65	3.25	4.05
32	42.40	42.50	41.90	42.90	42.10	2.65	3.25	4.05
40	48.30	48.40	47.80	48.80	48.00	2.90	3.25	4.05
50	60.30	60.20	59.60	60.80	59.80	2.90	3.65	4.50
65	76.10	76.00	75.20	76.60	75.40	3.25	3.65	4.50
80	88.90	88.70	87.90	89.50	88.10	3.25	4.05	4.85
100	114.30	113.90	113.00	114.90	113.30	3.65	4.50	5.40
125	139.70	–	–	140.60	138.70	–	4.85	5.40
150	165.1*	–	–	166.10	164.10	–	4.85	5.40

* 165.1 mm (6.5in) outside diameter is not generally recommended except where screwing to BS 21 is necessary
All dimensions are in accordance with ISO R65 except approximate outside diameters which are in accordance with ISO R64
Light quality is equivalent to ISO R65 Light Series II

Approximate Metres Per Tonne of Tubes to BS 1387

Nom. size	BLACK						GALVANIZED					
	Plain/screwed ends			Screwed & socketed			Plain/screwed ends			Screwed & socketed		
	L	M	H	L	M	H	L	M	H	L	M	H
(mm)	(m)	(m)	(m)	(m)	(m)	(m)	(m)	(m)	(m)	(m)	(m)	(m)
6	2765	2461	2030	2743	2443	2018	2604	2333	1948	2584	2317	1937
8	1936	1538	1300	1920	1527	1292	1826	1467	1254	1811	1458	1247
10	1483	1173	979	1471	1165	974	1400	1120	944	1386	1113	939
15	1050	817	688	1040	811	684	996	785	665	987	779	661
20	712	634	529	704	628	525	679	609	512	673	603	508
25	498	410	336	494	407	334	478	396	327	474	394	325
32	388	319	260	384	316	259	373	308	254	369	305	252
40	307	277	226	303	273	223	296	268	220	292	264	217
50	244	196	162	239	194	160	235	191	158	231	188	157
65	172	153	127	169	151	125	167	149	124	163	146	122
80	147	118	99	143	116	98	142	115	97	139	113	96
100	101	82	69	98	81	68	98	81	68	95	79	67
125	–	62	56	–	60	55	–	60	55	–	59	54
150	–	52	47	–	50	46	–	51	46	–	49	45

The figures for 'plain or screwed ends' apply also to tubes to BS 1775 of equivalent size and thickness
Key:
L – Light
M – Medium
H – Heavy

PIPES, WATER, STORAGE, INSULATION

Flange Dimension Chart to BS 4504 & BS 10

Normal Pressure Rating (PN 6) 6 Bar

Nom. size	Flange outside dia.	Table 6/2 Forged Welding Neck	Table 6/3 Plate Slip on	Table 6/4 Forged Bossed Screwed	Table 6/5 Forged Bossed Slip on	Table 6/8 Plate Blank	Raised face Dia.	T'ness	Nr. bolt hole	Size of bolt
15	80	12	12	12	12	12	40	2	4	M10 × 40
20	90	14	14	14	14	14	50	2	4	M10 × 45
25	100	14	14	14	14	14	60	2	4	M10 × 45
32	120	14	16	14	14	14	70	2	4	M12 × 45
40	130	14	16	14	14	14	80	3	4	M12 × 45
50	140	14	16	14	14	14	90	3	4	M12 × 45
65	160	14	16	14	14	14	110	3	4	M12 × 45
80	190	16	18	16	16	16	128	3	4	M16 × 55
100	210	16	18	16	16	16	148	3	4	M16 × 55
125	240	18	20	18	18	18	178	3	8	M16 × 60
150	265	18	20	18	18	18	202	3	8	M16 × 60
200	320	20	22	–	20	20	258	3	8	M16 × 60
250	375	22	24	–	22	22	312	3	12	M16 × 65
300	440	22	24	–	22	22	365	4	12	M20 × 70

Normal Pressure Rating (PN 16) 16 Bar

Nom. size	Flange outside dia.	Table 6/2 Forged Welding Neck	Table 6/3 Plate Slip on	Table 6/4 Forged Bossed Screwed	Table 6/5 Forged Bossed Slip on	Table 6/8 Plate Blank	Raised face Dia.	T'ness	Nr. bolt hole	Size of bolt
15	95	14	14	14	14	14	45	2	4	M12 × 45
20	105	16	16	16	16	16	58	2	4	M12 × 50
25	115	16	16	16	16	16	68	2	4	M12 × 50
32	140	16	16	16	16	16	78	2	4	M16 × 55
40	150	16	16	16	16	16	88	3	4	M16 × 55
50	165	18	18	18	18	18	102	3	4	M16 × 60
65	185	18	18	18	18	18	122	3	4	M16 × 60
80	200	20	20	20	20	20	138	3	8	M16 × 60
100	220	20	20	20	20	20	158	3	8	M16 × 65
125	250	22	22	22	22	22	188	3	8	M16 × 70
150	285	22	22	22	22	22	212	3	8	M20 × 70
200	340	24	24	–	24	24	268	3	12	M20 × 75
250	405	26	26	–	26	26	320	3	12	M24 × 90
300	460	28	28	–	28	28	378	4	12	M24 × 90

PIPES, WATER, STORAGE, INSULATION

Minimum Distances Between Supports/Fixings

Material	BS Nominal pipe size		Pipes – Vertical	Pipes – Horizontal on to low gradients
	(inch)	(mm)	Support distance in metres	Support distance in metres
Copper	0.50	15.00	1.90	1.30
	0.75	22.00	2.50	1.90
	1.00	28.00	2.50	1.90
	1.25	35.00	2.80	2.50
	1.50	42.00	2.80	2.50
	2.00	54.00	3.90	2.50
	2.50	67.00	3.90	2.80
	3.00	76.10	3.90	2.80
	4.00	108.00	3.90	2.80
	5.00	133.00	3.90	2.80
	6.00	159.00	3.90	2.80
muPVC	1.25	32.00	1.20	0.50
	1.50	40.00	1.20	0.50
	2.00	50.00	1.20	0.60
Polypropylene	1.25	32.00	1.20	0.50
	1.50	40.00	1.20	0.50
uPVC	–	82.40	1.20	0.50
	–	110.00	1.80	0.90
	–	160.00	1.80	1.20
Steel	0.50	15.00	2.40	1.80
	0.75	20.00	3.00	2.40
	1.00	25.00	3.00	2.40
	1.25	32.00	3.00	2.40
	1.50	40.00	3.70	2.40
	2.00	50.00	3.70	2.40
	2.50	65.00	4.60	3.00
	3.00	80.40	4.60	3.00
	4.00	100.00	4.60	3.00
	5.00	125.00	5.50	3.70
	6.00	150.00	5.50	4.50
	8.00	200.00	8.50	6.00
	10.00	250.00	9.00	6.50
	12.00	300.00	10.00	7.00
	16.00	400.00	10.00	8.25

PIPES, WATER, STORAGE, INSULATION

Litres of Water Storage Required Per Person Per Building Type

Type of building	Storage (litres)
Houses and flats (up to 4 bedrooms)	120/bedroom
Houses and flats (more than 4 bedrooms)	100/bedroom
Hostels	90/bed
Hotels	200/bed
Nurses homes and medical quarters	120/bed
Offices with canteen	45/person
Offices without canteen	40/person
Restaurants	7/meal
Boarding schools	90/person
Day schools – Primary	15/person
Day schools – Secondary	20/person

Recommended Air Conditioning Design Loads

Building type	Design loading
Computer rooms	500 W/m² of floor area
Restaurants	150 W/m² of floor area
Banks (main area)	100 W/m² of floor area
Supermarkets	25 W/m² of floor area
Large office block (exterior zone)	100 W/m² of floor area
Large office block (interior zone)	80 W/m² of floor area
Small office block (interior zone)	80 W/m² of floor area

PIPES, WATER, STORAGE, INSULATION

Capacity and Dimensions of Galvanized Mild Steel Cisterns – BS 417

Capacity (litres)	BS type (SCM)	Length (mm)	Width (mm)	Depth (mm)
18	45	457	305	305
36	70	610	305	371
54	90	610	406	371
68	110	610	432	432
86	135	610	457	482
114	180	686	508	508
159	230	736	559	559
191	270	762	584	610
227	320	914	610	584
264	360	914	660	610
327	450/1	1220	610	610
336	450/2	965	686	686
423	570	965	762	787
491	680	1090	864	736
709	910	1070	889	889

Capacity of Cold Water Polypropylene Storage Cisterns – BS 4213

Capacity (litres)	BS type (PC)	Maximum height (mm)
18	4	310
36	8	380
68	15	430
91	20	510
114	25	530
182	40	610
227	50	660
273	60	660
318	70	660
455	100	760

PIPES, WATER, STORAGE, INSULATION

Minimum Insulation Thickness to Protect Against Freezing for Domestic Cold Water Systems (8 Hour Evaluation Period)

Pipe size (mm)	Insulation thickness (mm)					
	Condition 1			Condition 2		
	λ = 0.020	λ = 0.030	λ = 0.040	λ = 0.020	λ = 0.030	λ = 0.040
Copper pipes						
15	11	20	34	12	23	41
22	6	9	13	6	10	15
28	4	6	9	4	7	10
35	3	5	7	4	5	7
42	3	4	5	8	4	6
54	2	3	4	2	3	4
76	2	2	3	2	2	3
Steel pipes						
15	9	15	24	10	18	29
20	6	9	13	6	10	15
25	4	7	9	5	7	10
32	3	5	6	3	5	7
40	3	4	5	3	4	6
50	2	3	4	2	3	4
65	2	2	3	2	3	3

Condition 1: water temperature 7°C; ambient temperature –6°C; evaluation period 8 h; permitted ice formation 50%; normal installation, i.e. inside the building and inside the envelope of the structural insulation
Condition 2: water temperature 2°C; ambient temperature –6°C; evaluation period 8 h; permitted ice formation 50%; extreme installation, i.e. inside the building but outside the envelope of the structural insulation
λ = thermal conductivity [W/(mK)]

Insulation Thickness for Chilled And Cold Water Supplies to Prevent Condensation

On a Low Emissivity Outer Surface (0.05, i.e. Bright Reinforced Aluminium Foil) with an Ambient Temperature of +25°C and a Relative Humidity of 80%

Steel pipe size (mm)	t = +10			t = +5			t = 0		
	Insulation thickness (mm)			Insulation thickness (mm)			Insulation thickness (mm)		
	λ = 0.030	λ = 0.040	λ = 0.050	λ = 0.030	λ = 0.040	λ = 0.050	λ = 0.030	λ = 0.040	λ = 0.050
15	16	20	25	22	28	34	28	36	43
25	18	24	29	25	32	39	32	41	50
50	22	28	34	30	39	47	38	49	60
100	26	34	41	36	47	57	46	60	73
150	29	38	46	40	52	64	51	67	82
250	33	43	53	46	60	74	59	77	94
Flat surfaces	39	52	65	56	75	93	73	97	122

t = temperature of contents (°C)
λ = thermal conductivity at mean temperature of insulation [W/(mK)]

PIPES, WATER, STORAGE, INSULATION

Insulation Thickness for Non-domestic Heating Installations to Control Heat Loss

Steel pipe size (mm)	t = 75			t = 100			t = 150		
	Insulation thickness (mm)			Insulation thickness (mm)			Insulation thickness (mm)		
	λ = 0.030	λ = 0.040	λ = 0.050	λ = 0.030	λ = 0.040	λ = 0.050	λ = 0.030	λ = 0.040	λ = 0.050
10	18	32	55	20	36	62	23	44	77
15	19	34	56	21	38	64	26	47	80
20	21	36	57	23	40	65	28	50	83
25	23	38	58	26	43	68	31	53	85
32	24	39	59	28	45	69	33	55	87
40	25	40	60	29	47	70	35	57	88
50	27	42	61	31	49	72	37	59	90
65	29	43	62	33	51	74	40	63	92
80	30	44	62	35	52	75	42	65	94
100	31	46	63	37	54	76	45	68	96
150	33	48	64	40	57	77	50	73	100
200	35	49	65	42	59	79	53	76	103
250	36	50	66	43	61	80	55	78	105

t = hot face temperature (°C)

λ = thermal conductivity at mean temperature of insulation [W/(mK)]

Index

Ebook Single-User Licence Agreement

We welcome you as a user of this Spon Price Book ebook and hope that you find it a useful and valuable tool. Please read this document carefully. **This is a legal agreement** between you (hereinafter referred to as the "Licensee") and Taylor and Francis Books Ltd. (the "Publisher"), which defines the terms under which you may use the Product. **By accessing and retrieving the access code on the label inside the front cover of this book you agree to these terms and conditions outlined herein. If you do not agree to these terms you must return the Product to your supplier intact, with the seal on the label unbroken and with the access code not accessed.**

1. **Definition of the Product**

    The product which is the subject of this Agreement, (the "Product") consists of online and offline access to the VitalSource ebook edition of *Spon's External Works & Landscape Price Book 2020.*

2. **Commencement and licence**

    2.1   This Agreement commences upon the breaking open of the document containing the access code by the Licensee (the "Commencement Date").

    2.2   This is a licence agreement (the "Agreement") for the use of the Product by the Licensee, and not an agreement for sale.

    2.3   The Publisher licenses the Licensee on a non-exclusive and non-transferable basis to use the Product on condition that the Licensee complies with this Agreement. The Licensee acknowledges that it is only permitted to use the Product in accordance with this Agreement.

3. **Multiple use**

    Use of the Product is not provided or allowed for more than one user or for a wide area network or consortium.

4. **Installation and Use**

    4.1   The Licensee may provide access to the Product for individual study in the following manner: The Licensee may install the Product on a secure local area network on a single site for use by one user.

    4.2   The Licensee shall be responsible for installing the Product and for the effectiveness of such installation.

    4.3   Text from the Product may be incorporated in a coursepack. Such use is only permissible with the express permission of the Publisher in writing and requires the payment of the appropriate fee as specified by the Publisher and signature of a separate licence agreement.

    4.4   The Product is a free addition to the book and the Publisher is under no obligation to provide any technical support.

5. **Permitted Activities**

    5.1   The Licensee shall be entitled to use the Product for its own internal purposes;

    5.2   The Licensee acknowledges that its rights to use the Product are strictly set out in this Agreement, and all other uses (whether expressly mentioned in Clause 6 below or not) are prohibited.

6. **Prohibited Activities**

    The following are prohibited without the express permission of the Publisher:

    6.1   The commercial exploitation of any part of the Product.

    6.2   The rental, loan, (free or for money or money's worth) or hire purchase of this product, save with the express consent of the Publisher.

    6.3   Any activity which raises the reasonable prospect of impeding the Publisher's ability or opportunities to market the Product.

    6.4   Any networking, physical or electronic distribution or dissemination of the product save as expressly permitted by this Agreement.

    6.5   Any reverse engineering, decompilation, disassembly or other alteration of the Product save in accordance with applicable national laws.

    6.6   The right to create any derivative product or service from the Product save as expressly provided for in this Agreement.

    6.7   Any alteration, amendment, modification or deletion from the Product, whether for the purposes of error correction or otherwise.

7. **General Responsibilities of the License**
   7.1 The Licensee will take all reasonable steps to ensure that the Product is used in accordance with the terms and conditions of this Agreement.
   7.2 The Licensee acknowledges that damages may not be a sufficient remedy for the Publisher in the event of breach of this Agreement by the Licensee, and that an injunction may be appropriate.
   7.3 The Licensee undertakes to keep the Product safe and to use its best endeavours to ensure that the product does not fall into the hands of third parties, whether as a result of theft or otherwise.
   7.4 Where information of a confidential nature relating to the product of the business affairs of the Publisher comes into the possession of the Licensee pursuant to this Agreement (or otherwise), the Licensee agrees to use such information solely for the purposes of this Agreement, and under no circumstances to disclose any element of the information to any third party save strictly as permitted under this Agreement. For the avoidance of doubt, the Licensee's obligations under this sub-clause 7.4 shall survive the termination of this Agreement.

8. **Warrant and Liability**
   8.1 The Publisher warrants that it has the authority to enter into this agreement and that it has secured all rights and permissions necessary to enable the Licensee to use the Product in accordance with this Agreement.
   8.2 The Publisher warrants that the Product as supplied on the Commencement Date shall be free of defects in materials and workmanship, and undertakes to replace any defective Product within 28 days of notice of such defect being received provided such notice is received within 30 days of such supply. As an alternative to replacement, the Publisher agrees fully to refund the Licensee in such circumstances, if the Licensee so requests, provided that the Licensee returns this copy of *Spon's External Works & Landscape Price Book 2020* to the Publisher. The provisions of this sub-clause 8.2 do not apply where the defect results from an accident or from misuse of the product by the Licensee.
   8.3 Sub-clause 8.2 sets out the sole and exclusive remedy of the Licensee in relation to defects in the Product.
   8.4 The Publisher and the Licensee acknowledge that the Publisher supplies the Product on an "as is" basis. The Publisher gives no warranties:
   8.4.1 that the Product satisfies the individual requirements of the Licensee; or
   8.4.2 that the Product is otherwise fit for the Licensee's purpose; or
   8.4.3 that the Product is compatible with the Licensee's hardware equipment and software operating environment.
   8.5 The Publisher hereby disclaims all warranties and conditions, express or implied, which are not stated above.
   8.6 Nothing in this Clause 8 limits the Publisher's liability to the Licensee in the event of death or personal injury resulting from the Publisher's negligence.
   8.7 The Publisher hereby excludes liability for loss of revenue, reputation, business, profits, or for indirect or consequential losses, irrespective of whether the Publisher was advised by the Licensee of the potential of such losses.
   8.8 The Licensee acknowledges the merit of independently verifying the price book data prior to taking any decisions of material significance (commercial or otherwise) based on such data. It is agreed that the Publisher shall not be liable for any losses which result from the Licensee placing reliance on the data under any circumstances.
   8.9 Subject to sub-clause 8.6 above, the Publisher's liability under this Agreement shall be limited to the purchase price.

9. **Intellectual Property Rights**
   9.1 Nothing in this Agreement affects the ownership of copyright or other intellectual property rights in the Product.
   9.2 The Licensee agrees to display the Publishers' copyright notice in the manner described in the Product.
   9.3 The Licensee hereby agrees to abide by copyright and similar notice requirements required by the Publisher, details of which are as follows:
   "© 2020 Taylor & Francis. All rights reserved. All materials in *Spon's External Works & Landscape Price Book 2020* are copyright protected. All rights reserved. No such materials may be used, displayed, modified, adapted, distributed, transmitted, transferred, published or otherwise reproduced in any form or by any means now or hereafter developed other than strictly in accordance with the terms of the licence agreement enclosed with *Spon's External Works & Landscape Price Book 2020*. However, text and images may be printed and copied for research and private study within the preset program limitations. Please note the copyright notice above, and that any text or images printed or copied must credit the source."

9.4 This Product contains material proprietary to and copyedited by the Publisher and others. Except for the licence granted herein, all rights, title and interest in the Product, in all languages, formats and media throughout the world, including copyrights therein, are and remain the property of the Publisher or other copyright holders identified in the Product.

10. Non-assignment

This Agreement and the licence contained within it may not be assigned to any other person or entity without the written consent of the Publisher.

11. Termination and Consequences of Termination.

11.1 The Publisher shall have the right to terminate this Agreement if:

11.1.1 the Licensee is in material breach of this Agreement and fails to remedy such breach (where capable of remedy) within 14 days of a written notice from the Publisher requiring it to do so; or

11.1.2 the Licensee becomes insolvent, becomes subject to receivership, liquidation or similar external administration; or

11.1.3 the Licensee ceases to operate in business.

11.2 The Licensee shall have the right to terminate this Agreement for any reason upon two month's written notice. The Licensee shall not be entitled to any refund for payments made under this Agreement prior to termination under this sub-clause 11.2.

11.3 Termination by either of the parties is without prejudice to any other rights or remedies under the general law to which they may be entitled, or which survive such termination (including rights of the Publisher under sub-clause 7.4 above).

11.4 Upon termination of this Agreement, or expiry of its terms, the Licensee must destroy all copies and any back up copies of the product or part thereof.

12. General

12.1 *Compliance with export provisions*
The Publisher hereby agrees to comply fully with all relevant export laws and regulations of the United Kingdom to ensure that the Product is not exported, directly or indirectly, in violation of English law.

12.2 *Force majeure*
The parties accept no responsibility for breaches of this Agreement occurring as a result of circumstances beyond their control.

12.3 *No waiver*
Any failure or delay by either party to exercise or enforce any right conferred by this Agreement shall not be deemed to be a waiver of such right.

12.4 *Entire agreement*
This Agreement represents the entire agreement between the Publisher and the Licensee concerning the Product. The terms of this Agreement supersede all prior purchase orders, written terms and conditions, written or verbal representations, advertising or statements relating in any way to the Product.

12.5 *Severability*
If any provision of this Agreement is found to be invalid or unenforceable by a court of law of competent jurisdiction, such a finding shall not affect the other provisions of this Agreement and all provisions of this Agreement unaffected by such a finding shall remain in full force and effect.

12.6 *Variations*
This agreement may only be varied in writing by means of variation signed in writing by both parties.

12.7 *Notices*
All notices to be delivered to: Spon's Price Books, Taylor & Francis Books Ltd., 3 Park Square, Milton Park, Abingdon, Oxfordshire, OX14 4RN, UK.

12.8 *Governing law*
This Agreement is governed by English law and the parties hereby agree that any dispute arising under this Agreement shall be subject to the jurisdiction of the English courts.

If you have any queries about the terms of this licence, please contact:

Spon's Price Books
Taylor & Francis Books Ltd.
3 Park Square, Milton Park, Abingdon, Oxfordshire, OX14 4RN
Tel: +44 (0) 20 7017 6000
www.sponpress.com

Spon Press
an imprint of Taylor & Francis

Ebook options

Your print copy comes with a free ebook on the VitalSource® Bookshelf platform. Further copies of the ebook are available from https://www.crcpress.com/search/results?kw=spon+2020
Pick your price book and select the ebook under the 'Select Format' option.

To buy Spon's ebooks for five or more users of in your organisation please contact:

Spon's Price Books
eBooks & Online Sales
Taylor & Francis Books Ltd.
3 Park Square, Milton Park, Oxfordshire, OX14 4RN
Tel: (020) 337 73480
onlinesales@informa.com

Ebook Set up and Use Instructions

Use of the ebook is subject to the single user licence at the back of this book.

Electronic access to your price book is now provided as an ebook on the VitalSource® Bookshelf platform. You can access it online or offline on your PC/Mac, smartphone or tablet. You can browse and search the content across all the books you've got, make notes and highlights and share these notes with other users.

Setting up

1. Create a VitalSource Bookshelf account at https://online.vitalsource.com/user/new (or log into your existing account if you already have one).

2. Retrieve the code by scratching off the security-protected label inside the front cover of this book. Log in to Bookshelf and click the **Redeem** menu at the top right of the screen. and Enter the code in the **Redeem code** box and press **Redeem**. Once the code has been redeemed your Spon's Price Book will download and appear in your **library**. N.B. the code in the scratch-off panel can only be used once, and has to be redeemed before end December 2020.

When you have created a Bookshelf account and redeemed the code you will be able to access the ebook online or offline on your smartphone, tablet or PC/Mac. Your notes and highlights will automatically stay in sync no matter where you make them.

Use ONLINE

1. Log in to your Bookshelf account at https://online.vitalsource.com).
2. Double-click on the title in your **library** to open the ebook.

Use OFFLINE

Download BookShelf to your PC, Mac, iOS device, Android device or Kindle Fire, and log in to your Bookshelf account to access your ebook, as follows:

On your PC/Mac
Go to https://support.vitalsource.com/hc/en-us/ and follow the instructions to download the free VitalSource Bookshelf app to your PC or Mac. Double-click the VitalSource Bookshelf icon that appears on your desktop and log into your Bookshelf account. Select **All Titles** from the menu on the left – you should see your price book on your Bookshelf. If your Price Book does not appear, select **Update Booklist** from the **Account** menu. Double-click the price book to open it.

On your iPhone/iPod Touch/iPad
Download the free VitalSource Bookshelf App available via the iTunes App Store. Open the Bookshelf app and log into your Bookshelf account. Select **All Titles** - you should see your price book on your Bookshelf. Select the price book to open it. You can find more information at https://support.vitalsource.com/hc/en-us/categories/200134217-Bookshelf-for-iOS

On your Android™ smartphone or tablet
Download the free VitalSource Bookshelf App available via Google Play. Open the Bookshelf app and log into your Bookshelf account. You should see your price book on your Bookshelf. Select the price book to open it. You can find more information at https://support.vitalsource.com/hc/en-us/categories/200139976-Bookshelf-for-Android-and-Kindle-Fire

On your Kindle Fire
Download the free VitalSource Bookshelf App from Amazon. Open the Bookshelf app and log into your Bookshelf account. Select All Titles – you should see your price book on your Bookshelf. Select the price book to open it. You can find more information at https://support.vitalsource.com/hc/en-us/categories/200139976-Bookshelf-for-Android-and-Kindle-Fire

Support

If you have any questions about downloading Bookshelf, creating your account, or accessing and using your ebook edition, please visit http://support.vitalsource.com/

For questions or comments on content, please contact us on sponsonline@tandf.co.uk

Free Updates

with three easy steps…

1. Register today on www.pricebooks.co.uk/updates

2. We'll alert you by email when new updates are posted on our website

3. Then go to www.pricebooks.co.uk/updates
   and download the update.

All four Spon Price Books – *Architects' and Builders'*, *Civil Engineering and Highway Works*, *External Works and Landscape* and *Mechanical and Electrical Services* – are supported by an updating service. Two or three updates are loaded on our website during the year, typically in November, February and May. Each gives details of changes in prices of materials, wage rates and other significant items, with regional price level adjustments for Northern Ireland, Scotland and Wales and regions of England. The updates terminate with the publication of the next annual edition.

As a purchaser of a Spon Price Book you are entitled to this updating service for this 2020 edition – free of charge. Simply register via the website www.pricebooks.co.uk/updates and we will send you an email when each update becomes available.

If you haven't got internet access or if you've some questions about the updates please write to us at Spon Price Book Updates, Spon Press Marketing Department, 3 Park Square, Milton Park, Abingdon, Oxfordshire, OX14 4RN.

Find out more about Spon books
Visit www.pricebooks.co.uk for more details.